Oil Field Production Geology

By Mike Shepherd

AAPG Memoir 91

Published by
The American Association of Petroleum Geologists
Tulsa, Oklahoma

ISBN13: 918-0-89181-372-9

AAPG Editor: Gretchen M. Gillis
AAPG Geoscience Director: James B. Blankenship

On the cover: Top photo—Ekofisk Complex platforms in the Norwegian North Sea (courtesy of ConocoPhillips). Middle left photo—Heterogeneity at the macroscopic scale; tidal channel in the Joulters Cay grainstone shoal, Bahamas. Middle right photo—Heterogeneity at the mesoscopic scale; cross-bedded limestone, Corsica. Bottom photo—A production geologist at work; the dual-screen work station shows a seismic line on the left and a 3-D geological model of an oil field on the right (Petrel 3-D modeling software in use; permission to display courtesy of Schlumberger).

This publication is available from:

The AAPG Bookstore
P.O. Box 979
Tulsa, OK U.S.A. 74101-0979
Phone: 1-918-584-2555 or 1-800-364-AAPG (U.S.A. only)
E-mail: bookstore@aapg.org
www.aapg.org

Canadian Society of Petroleum Geologists
600, 640 – 8th Avenue S.W.
Calgary, Alberta T2P 1G7
Canada
Phone: 1-403-264-5610
Fax: 1-403-264-5898
E-mail: reception@cspg.org
www.cspg.org

Geological Society Publishing House
Unit 7, Brassmill Enterprise Centre
Brassmill Lane, Bath BA13JN
United Kingdom
Phone: +44-1225-445046
Fax: +44-1225-442836
E-mail: sales@geoloc.org.uk
www.geolsoc.org.uk

Affiliated East-West Press Private Ltd.
G-1/16 Ansari Road, Darya Gaaj
New Delhi 110-002
India
Phone: +91-11-23279113
Fax: +91-11-23260538
E-mail: affiliate@vsnl.com

Introduction and Acknowledgments

OIL FIELD PRODUCTION GEOLOGY

Introduction

This book has been written as an introductory text on oil field production geology for use by university students and graduates starting with oil companies. The text aims to give both a general background to the subject and guidance on the practical application of production geology. The idea for the book was suggested by Professor Andrew Hurst of Aberdeen University. Aberdeen University also provided resources to help me with writing the text. This would have been a very difficult book to write without the assistance of my colleagues and associates. The following helped me with writing this book and I am very grateful to them for this:

Janet Almond
Ruben Dominguez
Olivier Dubrule
Svetlana Fisenko
Gareth Freeston-Smith
Stephen Garthley
Caroline Gill
Tim Goodall
Andy Gordon
Kathryn Hardacre
Stuart Harker
Mark Hempton
Andy Hurst
David Jones
Keith King
Samantha Large
Angus MacLellan
Robert Maskell
Dominic McCormick
Tom McKie
John Millington
Rebecca Nash
Joyce Neilson
Keith Rawnsley
Mark Reynolds
Martin Roberts
Peter Schutjens
Dave Soutter
Pete Stark
Edward Telatovich
Graeme Will
Dave Woodland

This book is dedicated to my children, Robert and Laura Shepherd; it's their oil too.

Mike Shepherd
Aberdeen October 2008

About the Author

Born in Aberdeen, Scotland, some time before the North Sea oil boom started, Shepherd always wanted to be a geologist after talking to the workmen in the granite quarry near the family home at the age of ten. They told him about the various minerals in the granite and how the rock had formed in the core of an ancient Scottish mountain, long since disappeared and he was hooked.

As a teenager, Shepherd saw the oil industry come to Aberdeen. The granite quarry shut down and later several oil company offices were built on the site, including those of Shell, Conoco-Phillips, Chevron, and Marathon. BP gave him his first job, and he found himself planning wells for the giant Forties Field, the biggest field in the North Sea. This was the beginning of a lifetime career as a production geologist working on oil and gas fields for variously Shell, Conoco-Phillips, Elf, Occidental, Amerada Hess, and Encana. In between, he managed to find time to take a year out of the 'toil for oil.' This was spent at Aberdeen University, and it was there that he started to write this book.

The theme of Shepherd's working life has been a particular interest in getting more oil out of oil fields and in maximizing a key resource for society; the job is great fun. The task of understanding a reservoir with the aim of getting more oil out is very much like detective work, and he believes that it is also becoming an increasingly more important role as oil resources start to get scarcer.

Table of Contents

Shepherd, M., 2009, The production geologist and the reservoir, *in* M. Shepherd, Oil field production geology: AAPG Memoir 91, p. 1–3.

Section 1

The Production Geologist and the Reservoir

THE LARGE RESOURCE REMAINING IN THE WORLD'S OIL FIELDS

There are widespread concerns about the future of oil resources. The volume added by new oil discoveries has declined since the 1960s while the global demand for hydrocarbons is rising as the world population increases.

Although less oil is being discovered as a result of exploration for new fields, a significant volume of reserves is being added because of improved oil recovery from existing fields. Reserves are the volumes of petroleum that a company expects to produce from a field by the end of its life. Recent data suggest that the volume of exploration finds and reserves growth in producing fields are now roughly similar (Table 1).

So why is so much oil being added as reserves from existing fields? Where is this oil coming from?

To start answering these questions, the basic observation is that the amount of oil recovered from the world's oil fields has historically been poor. Typically, more oil has been left behind in oil fields than they have ever produced. Today, out of the total amount of oil that has been found in the world, it is anticipated that only about 30–35% of this volume is likely to be recovered under current estimates (e.g., Conn, 2006). The remaining 65–70% of the oil is expected to be abandoned in the world's oil fields once these have become unprofitable to produce from.

A combination of factors resulting from geological heterogeneity, physical forces, and economics is responsible for this poor oil recovery. Reservoir complexity creates the situation where it can be: difficult to establish where the oil is to be found within individual reservoirs; hard to estimate what the remaining volumes are; and then a major technical challenge to produce the oil economically. Yet despite these problems, continuing efforts are being made by oil companies to understand their oil fields in more detail with the aim of improving recovery. Every decade since the 1950s has seen new and better techniques for extracting additional reserves from hydrocarbon accumulations in the subsurface (Table 2). Specialist petroleum geology journals are full of examples of how modern reservoir characterization methods have led to a better understanding of the geological configuration of reservoirs. These have resulted in increased production rates in oil fields and substantial reserves additions (Figure 1).

It is interesting to speculate as to how much the recovery factor for conventional oil resources can be improved globally by better reservoir management. The question was addressed by Keith King of Exxon Mobil at the 2006 Hedberg AAPG Oil Resources Conference. The range in the possible increase in recovery was estimated as an additional 4% to close to an additional 13% (based on the maximum resource case). The low case may arise if the current improvements in recovery start to moderate. The high case will depend on future technological advances in enhanced oil recovery techniques and the application of these globally, particularly in the world's giant oil fields.

Worthy of note is that an upside improvement in recovery of 13% would add almost as much oil supply as has been consumed by the world to date. It would take a heroic effort to get this much oil out of our reservoirs, but if we could, then this would go a long way to providing a solution for the world's energy problems. Whether this will happen or not is open to debate. Nevertheless, it is clear from these figures that there is a very large hydrocarbon resource available in our existing fields.

SCOPE OF THE BOOK

This book concerns itself with how the geologist can help to get more hydrocarbons out of existing fields. It is

DOI:10.1306/13161182M913372

Table 1. Recent global reserves additions from exploration discoveries and reserves growth in existing fields.*

	Volume (Billion Barrels of Oil)
2004	
Discovered by exploration wells	10.9
Reserves growth from existing fields	12.1
Total	23
(2004 world production)	(28.5)
2005	
Discovered by exploration wells	13.5
Reserves growth from existing fields	9.5
Total	23
(2005 world production)	(29)

**Data courtesy of Pete Stark and Ken Chew, IHS-Energy, 2007.*

set out in seven sections. The first section deals with the production geologist and the reservoir. This describes the role of the production geologist within an oil company and how they relate to everyone else working in a subsurface team. The first few chapters pull together separate strands so as to understand the job as a whole (chapters 2–5), and includes a chapter on the production geologist and the reservoir life cycle that shows how the nature of the job changes according to the maturity of an asset. The factors responsible for why much of the oil is left behind at the end of field life are discussed. If the geologist can establish where the remaining oil is located relative to the geological framework, then this will enable a greater recovery of oil from the reservoir. Accessing the left-behind oil involves drilling wells, and a preliminary chapter explains how this is done.

The second section of this publication is more thematic and is the start of the portion of the book that describes the production geology workflow, beginning with the production geologist, who is required to establish a conceptual geological scheme for the reservoir. This can be converted into a computer representation to be used by the rest of the subsurface team.

The third section of the book is concerned with what can be termed flow geology. The geologist carries

Table 2. Key oil field techniques that have led to improved field recovery over the last 60 years.* Details can be found in this publication in the chapters indicated.

Decade	*Developments in Subsurface Techniques*
1950s	Use of facies models for modern depositional systems (Chapter 9).
1960s to 1970s	Oil price rise leads to the common use of secondary recovery techniques (Chapter 5).
1960s to the present	Continuing development of new wire-line logging tools (Chapter 6).
1960s onward	An increase in computing power allows numerical reservoir simulation techniques to be used for production forecasts (Chapter 23).
1960s onward	Development of enhanced oil recovery techniques (Chapter 5).
1970s	Seismic stratigraphy coming to the forefront (Chapter 10).
1970s and 1980s	First 3-D seismic survey shot by Exxon in 1967. The use of 3-D for reservoir geophysics starts to become routine in the 1970s and 1980s (Chapter 6).
1980s	High-resolution sequence stratigraphy established. Wire-line logs and outcrop data integrated with seismic data. This extends the sequence stratigraphy concept from the basin to the reservoir scale (Chapter 10).
1980s	The modern phase of horizontal well drilling starts in the United States and Europe in the early 1980s. Kicks off the growing use of a variety of nonconventional well techniques to boost field production (Chapter 28).
1980s	Increasing use of geostatistical methods in reservoir geology (Chapter 19).
1990s	Continuing development of advanced data integration techniques to locate the remaining oil in mature fields (Chapters 15–18, 24–26).
1990s	Common use of 3-D computer models in reservoir geology (Chapter 20).
1990s	Vastly improved resolution of 3-D seismic data. Start of routine shooting of 4-D seismic surveys for evaluating sweep efficiency in reservoirs and for planning infill well campaigns (Chapter 17).
1990s	Development of methods for determining structural controls on fluid compartmentalization, particularly by sealing faults (Chapter 13).
2000 onward	Reservoir management is becoming increasingly more complex as the world's oil fields deplete. To cope with this, subsurface teams have become more integrated and are employing sophisticated technologies such as 4-D seismic, reservoir characterization and computer modeling (Chapter 2).

**After Fisher (1991) and Weimer and Slatt (2004).*

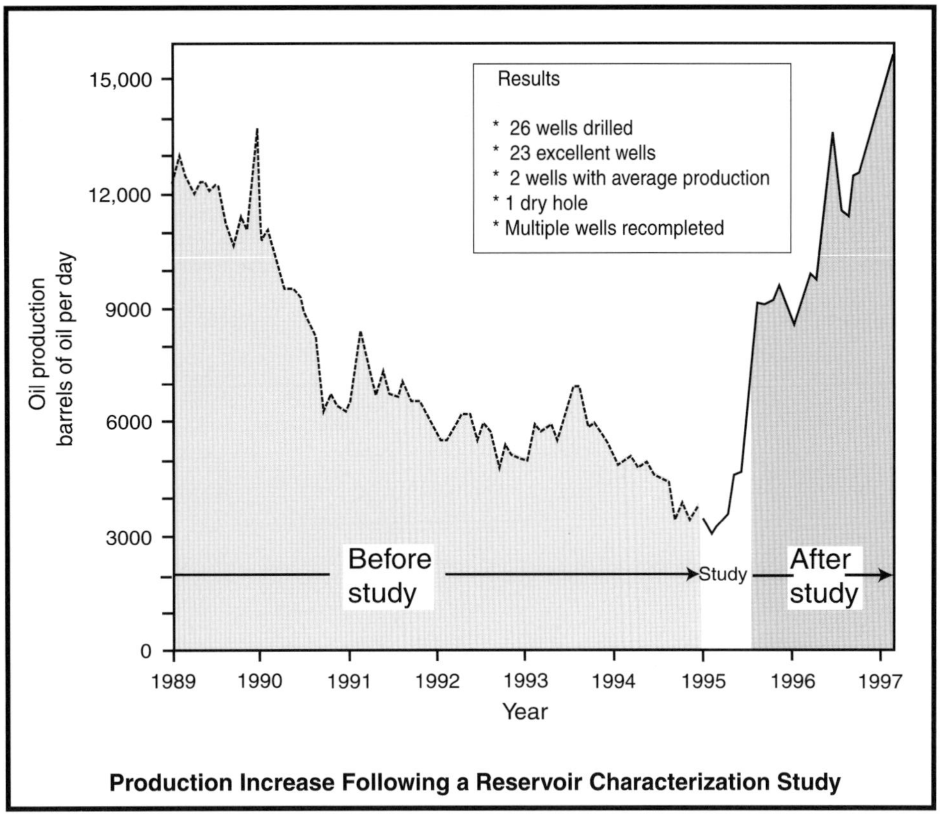

Figure 1. A spectacular example from Venezuela of how a better understanding of the reservoir can lead to increased production rates and better recovery (adapted from Hamilton et al., 2002). Reprinted with permission from the AAPG.

out detailed 'detective' work in trying to understand the controls on production in a field. This helps to modify the geological scheme to one tailored to the fluid flow behavior of the reservoir.

The publication's fourth section describes how three-dimensional (3-D) geological models are built on a computer using geostatistical techniques. The computer model can be used to calculate hydrocarbon volumes, locate the remaining hydrocarbons in a mature field, and help plan new well locations. The geologist can use it to understand reservoir uncertainties, especially those involved in the estimation of the range of the hydrocarbon volumes in place. The computer model may be used by the reservoir engineers as the basis for simulating the dynamic performance of the reservoir and to estimate the field reserves. The geologist will work with the reservoir engineers to set up these models.

Describing the various methods that enable the production geologist to locate the remaining hydrocarbons in a field is the focus of the fifth section of the book. Several patterns by which oil can be stranded within a reservoir are well known. It is possible to frame and then screen the distribution of oil in a producing field so as to locate and quantify the volumes that have been left behind. In this way, the reserves can be maximized for the field.

The sixth section of the publication gives advice on how to plan wells to recover oil and gas. The geologist will be heavily involved in this. Different types of well can be drilled depending on the specific configuration of the remaining hydrocarbon opportunities. Wells are the production geologist's tool kit to help unlock the stranded volumes in a field.

The seventh, and final, section of this book is a summary of the various depositional environments that make up reservoirs, and shows that there are common themes as to how each type will yield oil or gas. This knowledge will help the geologist to look out for similar patterns in the fields they are working on.

Shepherd, M., 2009, The Role of the production geologist, *in* M. Shepherd, Oil field production geology: AAPG Memoir 91, p. 5–6.

1

The Role of the Production Geologist

WHAT DOES A PRODUCTION GEOLOGIST DO?

The production geologist works in a *subsurface team*; a team that manages production for a field and looks for ways of getting more hydrocarbons out of it. He or she has a specific role. The production geologist is responsible for understanding the geological framework of the reservoir and creating a representation of it, typically using computer software (Figure 2). The object of this model is to help understand how the geology both influences fluid flow within a producing reservoir and creates *dead ends* that could potentially trap hydrocarbons. The bigger dead-end pockets may be worth targeting with new wells. If these wells look profitable, the geologist will then take a leading role in planning them with the drilling engineers.

Production geologists assigned to an operated field will find themselves working as part of a multidisciplinary team (Figure 3). In a large company, this will include a subsurface manager, geologists, geophysicists, petrophysicists, reservoir engineers, production engineers, chemists, and technical assistants. Some teams may also include drilling engineers and economists (Table 3).

Teamwork is essential because the staggeringly complex nature of a subsurface operation means that the various disciplines have to integrate their specific areas of expertise for the venture to be successful (Durrani et al., 1994). Some oil companies have separate geology and engineering departments, although this rarely works in practice. Short lines of communication should exist within a subsurface team such that an inclusive atmosphere of shared purpose is created. Any problems that arise can then be quickly recognized and solved by common directed action (Satter et al., 1994; Neate, 1996).

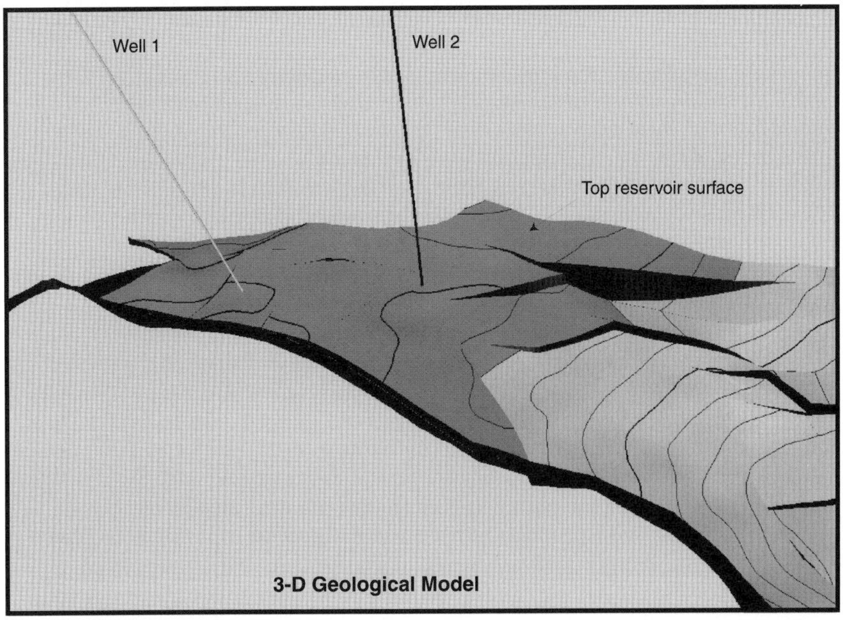

FIGURE 2. Production geologists build three-dimensional (3-D) computer models of the larger fields to represent the geology. The figure shows the relief on the top surface of a reservoir interval. Also shown are the paths of the wells that intersect the top reservoir surface.

DOI:10.1306/13161183M913372

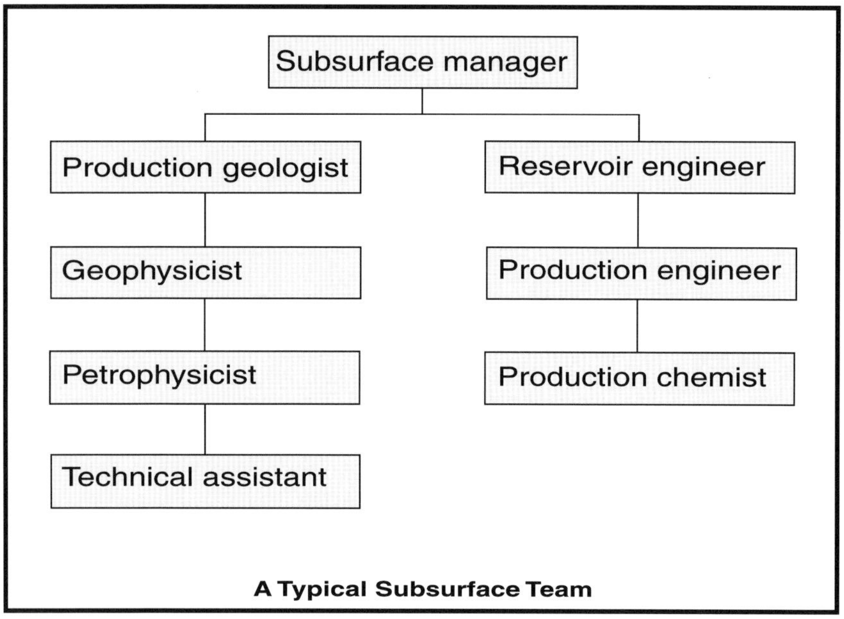

Figure 3. A production geologist works in a team within an oil company. The subsurface operation is so complex that everyone has to integrate their expertise for the project to work efficiently.

Table 3. Professional disciplines within a subsurface team.

Job Title	*Job Description*
Subsurface manager	Manages and coordinates the work of everyone in the subsurface team.
Production geologist	Responsible for understanding and modeling the geological framework of the reservoir. Helps to identify and plan new well locations.
Geophysicist	Spends much of his/her time interpreting seismic data to define the reservoir structure and fault distribution. Where the seismic data allow, depositional environment, rock, and fluid properties can also be characterized.
Petrophysicist	A key task is to analyze wireline logs to quantify the rock and fluid properties of the reservoir at the well scale.
Technical assistant	Provides technical support to the team. This includes data management, data preparation, and computer mapping.
Reservoir engineer	Predicts how much oil and gas a field is likely to produce, and may use a computer simulation of reservoir performance to analyze how the field will behave as well as taking a lead in reservoir management activities.
Production engineer	Responsible for optimizing all the mechanical aspects of hydrocarbon production from the wellbore to the surface facilities.
Production chemist	Analyzes and treats problems related to scale formation, metal corrosion, drilling fluids, wax formation, and solids precipitation between the reservoir and the surface facilities.
Drilling engineer (Well engineer)	Plans the mechanical aspects of any well operations including drilling new wells.
Economist	Costs and evaluates any economic activity relevant to the subsurface.

Shepherd, M., 2009, Drilling a well, *in* M. Shepherd, Oil field production geology: AAPG Memoir 91, p. 7–12.

2

Drilling a Well

INTRODUCTION

Production geologists spend a large part of their career planning wells and monitoring them as they are being drilled. Wells provide the bulk of the geological data for understanding the reservoir. Wells are also the means by which more oil and gas can be produced from an existing field. The production geologist therefore needs to have a reasonably detailed understanding of how wells are drilled and the various operations conducted on wells after they have started producing.

HOW WELLS ARE DRILLED

The most common method of drilling wells uses *rotary drilling* (Rabia, 1985) (Figure 4). A *drilling bit* is attached to the end of a long string of jointed, hollow *drill pipe*, and the whole assembly is rotated by a motorized turntable at the surface, the *rotary table*. Modern rigs use a *top drive system* for rotating the drill pipe, an assembly that is guided up and down rails on the derrick. The rotating bit cuts or crushes the rock. *Drilling mud*, consisting of water or an oil-water mixture, solids, and various additives, is circulated down through the drill pipe and out through nozzles in the drilling bit. The mud returns to the surface up the *annulus*, the space outside of the drill pipe. The mud lubricates the bit, prevents it from getting too hot because of friction, and lifts the drilled *rock cuttings* up the hole. It should be dense enough to *overbalance* any *high-pressure formations* encountered while drilling. If it fails in this last action, the fluid in the formation will displace the mud up the hole. This is called a *kick*. Should this hazardous situation not be dealt with quickly, hydrocarbons will exit at the surface and a *blowout* results. Scenes of oilmen dancing with glee as oil gushes over the drilling rig are for the cinema only. In reality, oil field professionals are acutely aware of the danger involved in the combustion and explosive blast that can result from a hydrocarbon blowout.

The scale and cost of a drilling operation differs between wells onshore and those offshore. An onshore well is drilled with a relatively cheap *land rig* (Figure 5); offshore, the operation is several times more expensive.

In shallow water, typically about 6–45 m (20–150 ft) deep, drilling is conducted by a *jackup rig*. A jackup is a rig that has three or more legs that sit on the sea floor. In moderately deep water (more than 45 m [150 ft] deep), a floating or *semisubmersible rig* is used. The semisubmersible rig is kept in place by several anchors (Reed, 1992).

In deep water, a *drill ship* is the preferred option. *Deep water* is defined as water depths between 500 and 2000 m (1640 and 6562 ft) (Weimer and Slatt, 2004). The drill ship is maintained in place by *dynamic positioning*. Computers constantly calculate the position of the drill ship using global positioning system technology or in response to signals from transducers on the sea bed. Signals are sent to propellers and lateral thrusters on the sides of the vessel. These readjust the location of the ship to keep it stable against the forces of wind and water currents.

THE DRILLING OPERATION

A well starts by being *spudded* as the drill bit encounters the first bit of soil or subsea sediment. A well will not be drilled all the way through in one go; instead, there will be several stages of drilling. Each section will involve drilling the hole to a certain depth and

DOI:10.1306/13161184M913372

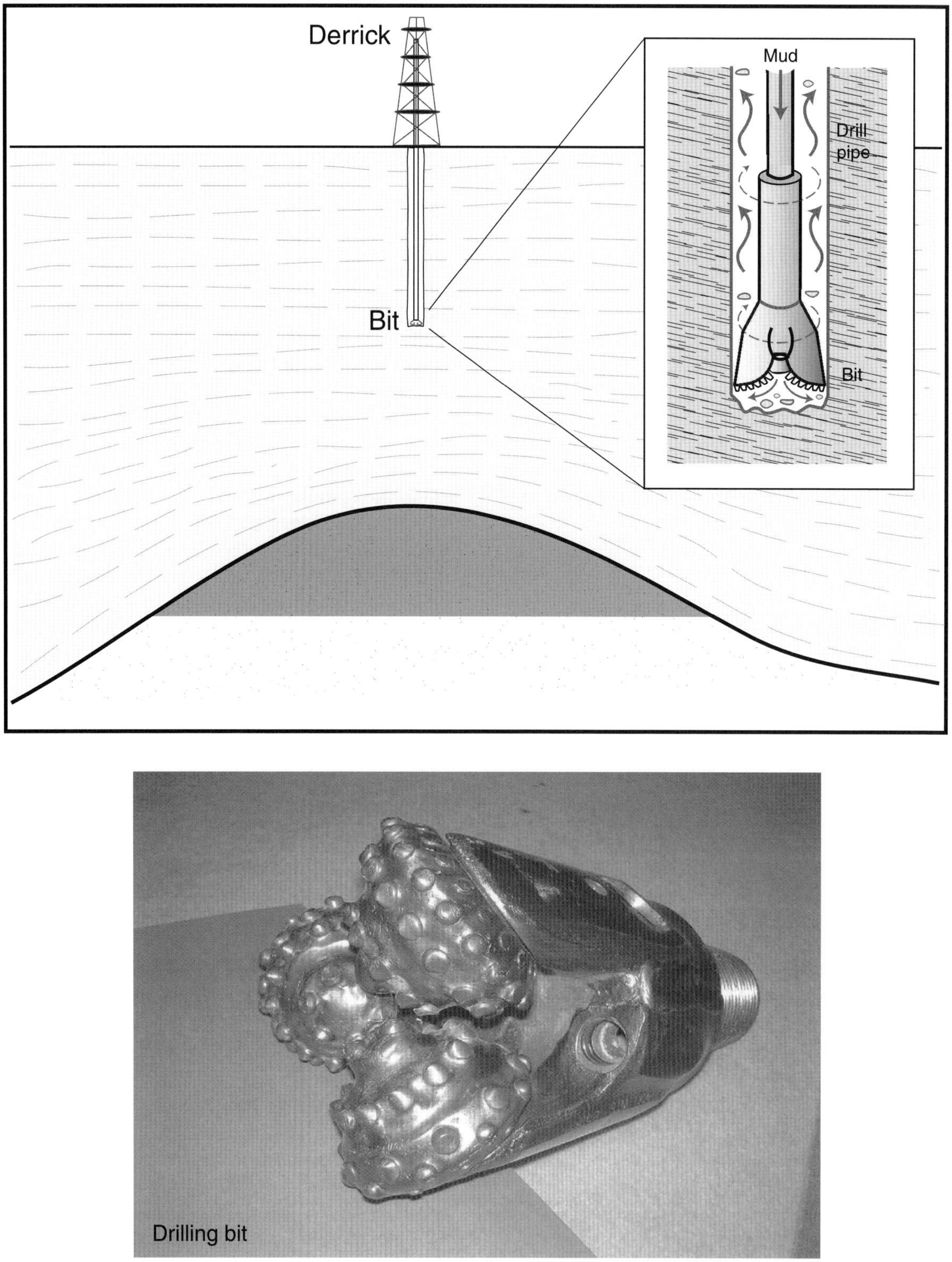

Rotary Drilling

Figure 4. The drilling bit, attached to the end of the drill pipe, rotates, cutting and crushing the rock underneath it. Mud passes out through the nozzles in the drill bit, cools the bit, and acts to lift the rock cuttings up to the surface.

FIGURE 5. Various types of drilling operations, offshore and onshore. Drill ship Jack Ryan courtesy of BP (www.bp.com). Jackup rig courtesy of Maersk Oil and Gas (www.media.maersk.com). Semisubmersible rig and land rig in the Sahara Desert, Libya, courtesy of Woodside Energy Ltd. (www.woodside.com.au).

then running in and cementing metal *casing* onto the rock surface of the borehole wall before going any farther (Figure 6). The simplest reason for doing this is to prevent poorly consolidated sediment from collapsing once the well has been drilled over a long interval, although there may also be good reasons for isolating certain problem formations.

A typical well will have a similar geometry to an inverted telescope, with the hole size and casing diameter decreasing incrementally down the hole. Typical hole sizes and casing diameters are 36-in. (91.44-cm) hole, 24-in. (60.96-cm) hole, cased with 18 5/8-in. (47.29-cm) or 20-in. (50.8-cm) casing; 17 1/2-in. (44.45-cm) hole, cased with 13 3/8-in. (33.95-cm) casing; 12 1/4-in. (31.11-cm) hole, cased with 9 5/8-in. (24.43-cm) casing; and 8 1/2-in. (21.59-cm) hole, cased with a 7-in. (17.78-cm) liner. A *liner* is a type of casing that is not run all the way up the hole; instead, it is *hung off* inside the lower part of a casing string (Figure 6).

It is necessary to change out the bit frequently because it will become worn and inefficient after several days of drilling. When this happens, the entire drill pipe needs to be pulled out of the wellbore and then run in with a new bit attached. This operation is known as *tripping*. A two-way trip, or *round trip*, can take 12 hr or more in the deeper sections of the well.

Specialist service personnel called *mud loggers* monitor the drilling parameters and collect the drill cuttings for analysis. There may also be a *well site geologist* present at the rig site who will draw up a *lithology log* from examination of the cuttings. The objective is to analyze the lithostratigraphy of the interval being drilled in order to help make operational decisions, such as when to run casing. The well site geologist will also examine the cuttings for indications of *hydrocarbon shows*. An ultraviolet light source will be used to check for *hydrocarbon fluorescence* in the samples, a sign that oil is present.

Sometimes the subsurface team will require the reservoir interval to be *cored*. This is carried out with a special *coring barrel* attached to the end of the drilling assembly once the drill bit has been removed. A doughnut shaped *coring head* will cut a cylinder of reservoir rock, and the cut core will slide into the coring barrel, typically about 18–27 m (60–90 ft) long. Once full, the core barrel is pulled back up to the surface for retrieval. Several coring trips may be required to core a reservoir interval of interest. Given the trip time for coring and

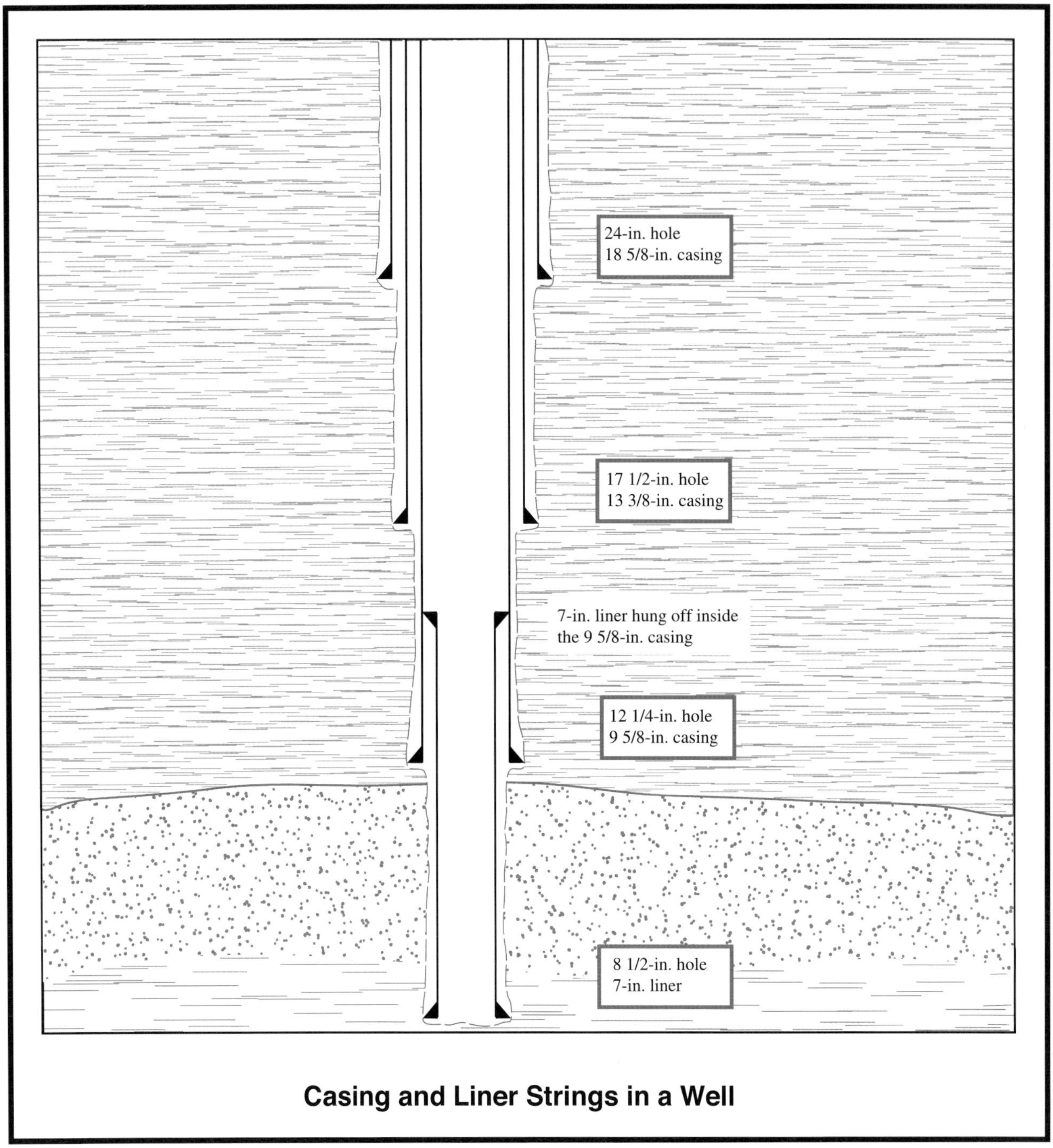

FIGURE 6. A well is not drilled all the way through. Metal casing strings are run to isolate specific sections of the hole before drilling further.

the expensive rig day rates, a coring operation is costly (Whitebay, 1992).

DRILLING PROBLEMS

Occasionally, something will go wrong and a piece of equipment is lost down the well. For example, the drill pipe may *twist off* somewhere along its length and fall to the bottom of the hole. The drilling operation will come to a halt unless the foreign object or *fish*, as it is known, is 'fished' out; that is, physically removed from the hole (Woods, 1992). Specialist tools are available for *fishing operations*. Sometimes the fishing operation can last many days.

Every now and again, the hole will collapse in on itself. This will happen where the earth stresses exceed the rock strength. Salt sections or shale sections at

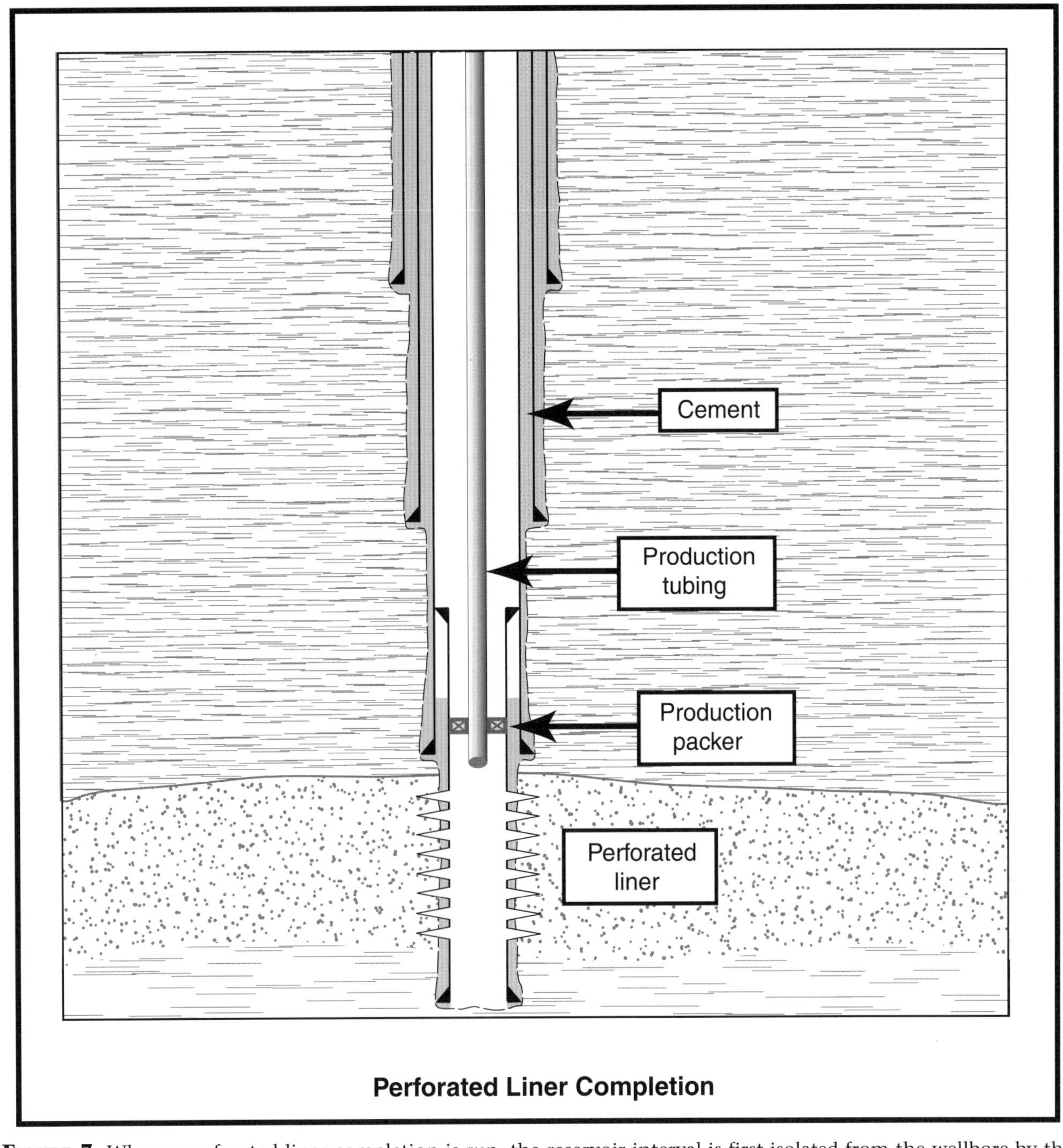

FIGURE 7. When a perforated liner completion is run, the reservoir interval is first isolated from the wellbore by the cemented liner. The liner is then selectively perforated with holes so as to allow fluid to flow into the wellbore from the specific zones required for production.

shallow depths containing *water-sensitive clays* are prone to this. Water-sensitive clays can expand by reacting with drilling fluids, particularly low salinity muds. This can cause the borehole wall to founder and bury the drill bit irretrievably. A decision may then be made to branch off from what hole is left, and this is called *sidetracking*.

Another problem that can occur is *lost circulation*, whereby the drilling mud is lost in large quantities into a fracture or a highly permeable interval. Adding fibrous material to the mud will solve the problem. This clogs up the lost circulation zone and prevents any further losses.

WELL OPERATIONS AFTER DRILLING HAS STOPPED

Once the reservoir has been drilled through, the drilling operation will come to a stop. The end of the well is called the *total depth* or *TD* for short. At this point, the drill pipe is pulled out and *wireline logs* may be run. Wireline logs are measuring tools run on the end of a long cable that record variations in the physical properties of the reservoir rock and fluids behind the borehole. It is possible to build up a detailed picture of the reservoir lithologies and fluids based on an analysis of the various types of *log response*. For instance, some wireline logs will allow the *porosity* of the reservoir interval to be determined. Porosity is the fraction or percentage of the volume of void space in the rock relative to the whole rock volume.

Sometimes the well results are so poor that the well is *plugged and abandoned*. The well is not required for production and has no further use. To avoid any hydrocarbons leaking to the surface, the well is isolated by cement plugs. However, if the wireline logs indicate that the well is likely to produce an economically significant volume of hydrocarbons, then it will be *completed*. In a typical production well, the reservoir will be isolated behind casing or a liner. The annulus between the rock and the liner is filled with cement. This means that it is possible to *perforate*, that is punch holes in, the liner such that a specific interval in the reservoir can be accessed for production or injection (Holditch, 1992) (Figure 7). Alternatively, a preslotted liner can be used, or, in hard competent rocks such as limestones, the hole can be left open. This latter practice is sometimes referred to as a *barefoot completion*.

Production tubing will then be installed in the well. This is a narrow-diameter pipe, which isolates the produced hydrocarbons from the rest of the well on the way up to the surface. The diameter of the tubing can be matched to the optimal flow rate for the producing fluids.

On occasions, the production rate from a new well will be less than expected; techniques are available for improving the production rate when this happens. The reservoir rock can be *hydraulically fractured* by injection of fluids at a high rate into the wellbore. Some reservoirs can be *acidized* by adding acids to dissolve any acid-soluble material in the *near wellbore area*. This can locally improve the *permeability*, the rock's capacity to flow. Acidization can be particularly effective in increasing the near-wellbore permeability of carbonate reservoirs and sandstone reservoir rock with carbonate or acid-soluble cements (Gidley, 1992).

On completion, the well is tied into the *production train* and *brought on stream*. An excellent initial production rate for an offshore well is about 20,000 barrels per day; a barrel is equivalent in volume to 0.159 m^3 (5.6 ft^3) (see chapter 21, this publication). The fluids are produced through a *separator*, a large piece of apparatus for splitting up the produced fluids into oil, gas, water, and unwanted solids (Jennings, 1992).

Later on in the life of a well, there may be reasons for making an *intervention* in the well, for instance to run *production logs*. These can be used to get information on the source of water ingress into a production well. Many wells produce water along with oil. Water mixing in with the lighter oil as it flows to the surface will increase the density of flowing fluids within the tubing, and the production rate will fall. The water-producing perforations can be shut off by running an *isolation plug* in the well.

Wells can also be *worked over*. This involves repairing the pipework or equipment in a well, or carrying out an operation to improve the productivity of a production well. For instance, the engineers may want to install a *gas lift system* in a production well. Gas is injected into the upper part of the wellbore and then through valves in the production tubing. This reduces the density of the producing fluid column and increases the flow rate. The gas will expand as it moves up through the tubing providing additional lift (Smallwood, 1992).

Shepherd, M., 2009, The reservoir life cycle, *in* M. Shepherd, Oil field production geology: AAPG Memoir 91, p. 13–28.

3

The Reservoir Life Cycle

THE RESERVOIR LIFE CYCLE

It is a useful metaphor to refer to the *life cycle of a reservoir* (Figure 8). In the early years, the reservoir will produce vigorously with few management difficulties. As the field matures, however, numerous problems can arise and eventually, as the reservoir energy and production decline, the field is abandoned. This chapter shows how the role of the production geologist varies according to each stage of the reservoir's production life. It also gives a general background as to how reservoir management is conducted.

DISCOVERY AND APPRAISAL

Oil companies are forever exploring for oil and gas to keep the business going, sometimes successfully, sometimes not. Exploration can be a spectacular way of spending millions of dollars with nothing to show for it. Because of this, in countries where drilling concessions are licensed by the government, it is common practice for more than one company to share the risk and expense of exploration (Table 4). A consortium will be formed for the purpose of applying for blocks of land or sea area for exploration from the appropriate government department of the country where the blocks are located. If the companies are successful in securing the exploration acreage, a *joint operating agreement* will be set up between them (Tinkler, 1992).

This specifies the terms and conditions under which costs and any profits resulting from the exploration effort will be shared. Under this agreement, one of the member companies will volunteer or be nominated to look after the block. This company will then become the *operator*, and they will supervise all the exploration work. That is, they will hire a drilling rig, do all the exploration analysis, and then drill any attractive looking *prospects*. The other oil companies will become *partners* in the project. Although they will not normally be directly involved in the day to day operations, the partners will have a voting interest in any major decisions that involve spending money. If the result of the exploration effort is a new find, the operating company will usually continue in the capacity of operator, and they will organize all the subsequent subsurface and engineering work. Regular meetings are held to report on the proceedings to the partners.

After a *discovery well* has been drilled, the next stage is to *appraise* the new find. The aim is to decide whether the new *hydrocarbon pool* will produce enough oil or gas to be profitable for the partnership. The *development* of a new field requires a huge investment, and there is a need to carefully judge whether the investment is worth the risk. This responsibility is assigned to a *development group* within the operating company, whose ultimate objective is to recommend to the management and partners whether or not the field development should proceed.

The first task is to estimate the size of the new find. This is primarily the geologist's responsibility. The basic volume for the geologist to estimate is the *hydrocarbons initially in place*, effectively the total volume of hydrocarbons in the structure. This comprises the *oil initially in place* for an oil field and the *gas initially in place* for a gas field. Once these values have been derived, then the reservoir engineer will calculate the reserves, a determination of approximately how much hydrocarbons the new pool is likely to produce.

To evaluate the hydrocarbons in place, several basic questions need to be answered at this stage. Where are the *fluid contacts*? For example, where is the oil-water contact (Figure 9)?

The *oil-water contact* is the base of the effective producing oil column. Some fields will have a *gas cap* overlying the oil column. The base of the gas cap is the *gas-oil contact*. For a gas field without an oil rim, the base of the gas column is the *gas-water contact*.

DOI:10.1306/13161185M913372

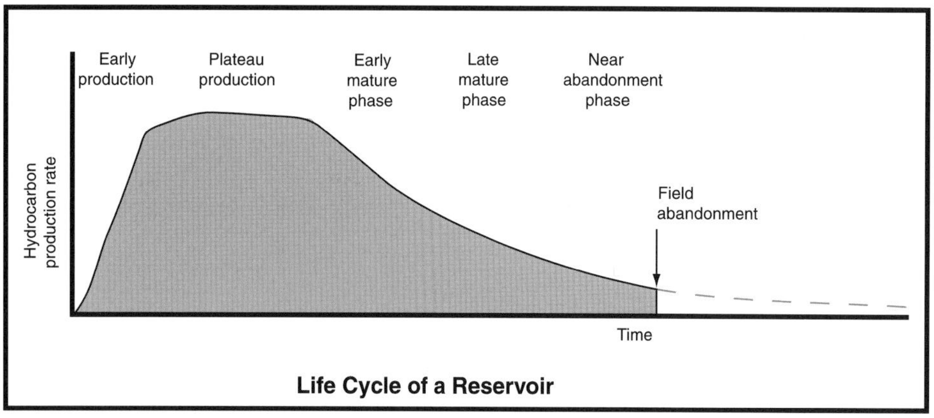

FIGURE 8. The figure shows the life cycle of a reservoir from production start up through to field abandonment. The nature of the production geologist's job will change according to the stage of field development.

The depths of the fluid contacts and the height of the hydrocarbon column are important facts to know when estimating the volume of hydrocarbons in place for a new discovery. However, the fluid contacts will not always be known at this point. The discovery well may have found the reservoir full of hydrocarbons down to the base of the porous rock interval with no evidence of a fluid contact depth (Figure 10). Another well may be needed downdip to determine a fluid contact.

Another question that arises at this time concerns whether the reservoir is in communication throughout or is it split into several isolated *compartments* such that it will take more wells to develop it (Figure 11). If many expensive wells are required for an acceptable hydrocarbon flow rate from the new field, will the project still be economic after the wells have been paid for?

APPRAISAL WELLS

At this stage, the data available to the development team will comprise the results of the discovery well plus a seismic survey. This may not be sufficient to make a sensible decision to develop a new field, even if the initial results do look promising. At least one other well is required to get an adequate understanding of the volumes in place. For very large complex fields, several *appraisal wells* may be required.

Drilling appraisal wells can be expensive, much more so offshore than onshore. For example, the cost of drilling a moderately shallow well onshore in the United States in the first few years of the new millennium was typically less than $1 million, whereas offshore, the costs ranged from about $4 million upward to $30 million plus. The outlay can go as high as $100 million for difficult deep-water, high-pressure wells such as in the Caspian Sea or offshore Gulf of Mexico (Stewart and Holt, 2004). Thus, it is easier to come up with development decisions for onshore fields. The decision to proceed from discovery through appraisal to production can be less of a major hurdle with onshore assets because there will not be as much money at risk (Dake, 1994).

Table 4. Partners in the Mutineer and Exeter fields, Western Australia.*

Company	*Interest in the Field (%)*	*Role*
Santos	33.3977	Operator
Kufpec	33.4023	Partner
Nippon Oil	25.0	Partner
Woodside	8.2	Partner

**As of 2008 (from the Woodside Petroleum Web site: www.woodside .com.au).*

RESERVOIR UNCERTAINTY AT THE APPRAISAL STAGE

Even after a few appraisal wells have been drilled, there will still be much *uncertainty* as to the volume of hydrocarbons in the reservoir and how the field is likely to perform. However, more data will have been acquired, and the overall result will have been to *reduce the risk* of the development losing money, although the chances of a subeconomic performance will not be totally eliminated.

Much of the data gained at this stage are *static data*, data that can be used to understand the shape, storage, and fluid properties of the hydrocarbon pool under investigation. There will also be a need to obtain *dynamic data*, data that give an idea of the likely flow rates for future production wells. *Drill stem tests* may be conducted on the appraisal wells. A drill stem test is a temporary completion of a well that allows the flow rate, pressure, and fluid composition to be determined (Borah, 1992). If a significant uncertainty remains as to the long-term production behavior of the reservoir, a decision may be made to proceed with an *extended well*

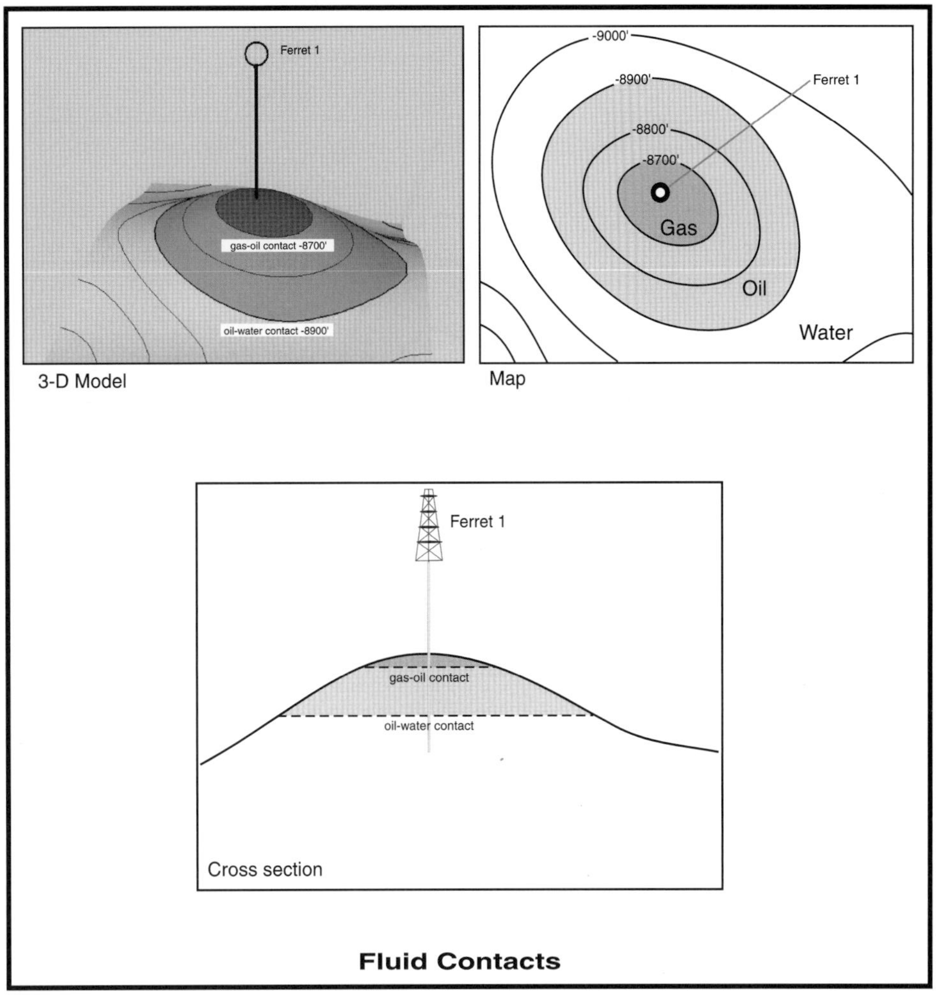

FIGURE 9. Fluid contacts in a newly found hydrocarbon pool with a gas cap and an oil column.

test. A single well is put on production for at least four days and sometimes much longer, several months for instance. This can establish that the hydrocarbon pool has a large enough volume to sustain production for a sufficient period of time without too much *pressure depletion.* If the pressure falls noticeably after only a few days of production, the *contactable volume* is probably small and it is unlikely that the reservoir will be a commercial proposition.

Some types of reservoir require more careful appraisal than others. For many relatively simple reservoirs, including shoreface and sheet-like turbidites, field appraisal can be straightforward. Many of the reservoirs in these systems are of the kind that a SHELL subsurface team would refer to as *high rate high ultimate (HRHU) recovery reservoirs* (Weimer and Slatt, 2004). High initial producing rates will mean that the large sums of money required for investment can be paid back quickly. Expected high ultimate recoveries will give a large degree of comfort that the project will make considerable sums of money with only a low risk of economic failure.

The riskier appraisal targets will include "chopped up" reservoirs, such as those with channel sand bodies or fault-segmented structures. These may require many wells to develop (Figure 12).

It could be said that the results of an appraisal program, particularly offshore, can be no more than an informed guess that the project might make money. In the worst possible case, an enormous amount of cash can be spent as a result of the decision to develop a field only for the project to operate at a loss thereafter.

It is not always that bad, however. A common practice in the industry is to concentrate on the *low side case* when appraising a hydrocarbon pool. The idea is for the geologist to think of the hydrocarbon volumes not as a single value but as a range of values lying between a minimum and maximum. This accounts for the significant uncertainty involved in making volumetric estimates particularly where there is so little information available. Many companies define the low side case as the value that has the probability that 90% of the possible volumes in the range are higher than the value itself (Rose, 1992). This is also referred to as the P_{90} case (Table 5).

The logic here is that if the low side case based on a pessimistic model can be shown to be economic, then the project is robust and will be deemed likely to make

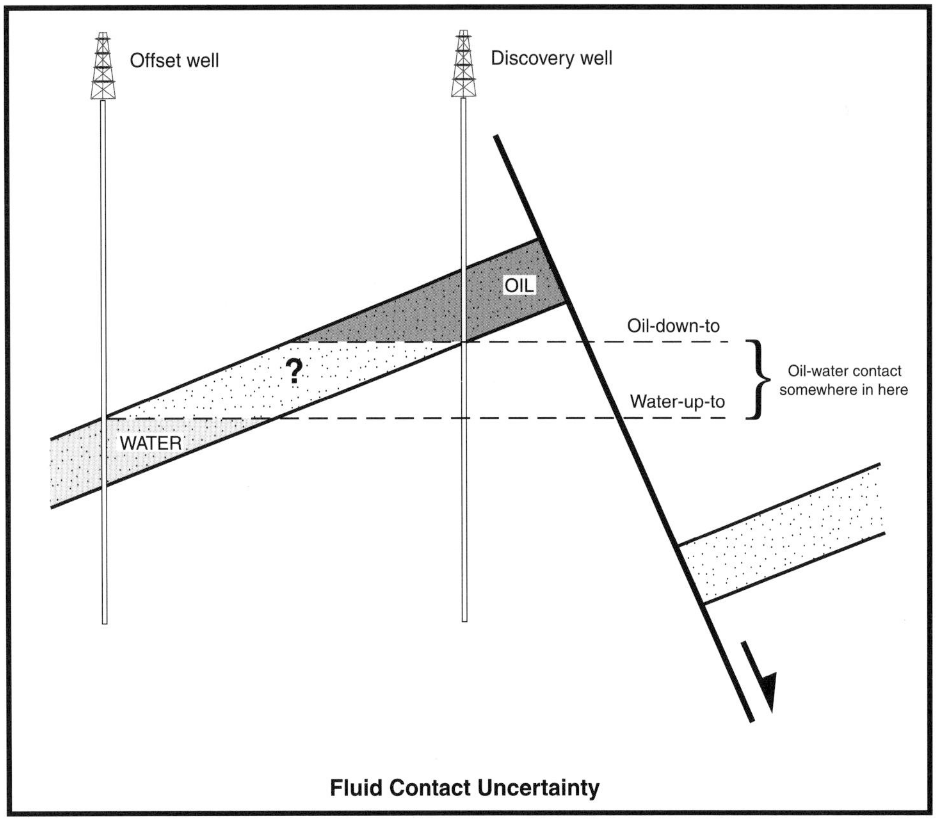

FIGURE 10. Sometimes a well will not find a clear fluid contact. In this example, the discovery well shows an oil-down-to depth at the base of the sand interval, whereas an offset well down flank has found a water-up-to depth at the top of the sand. The oil-water contact lies somewhere between these two depths.

money. If the low side case is not economic, then an effort can be made to find out what it is that will make it profitable. This may involve acquiring more subsurface data or in finding ways of making the development plan less expensive. The various risks affecting the development will be listed and prioritized in terms of their impact on project values. Efforts will be made to try and reduce the risk on those at the top of the list.

A case history for good practice in reservoir appraisal is the Schiehallion (pronounced Shee-hal-leon) field development (Leach et al., 1999). BP and partners discovered the Schiehallion field in 1993 on the United Kingdom continental shelf, within Paleocene channelized turbidite sandstones (Figure 13). It was appraised in the period 1994–1995, and the first oil was produced in 1998.

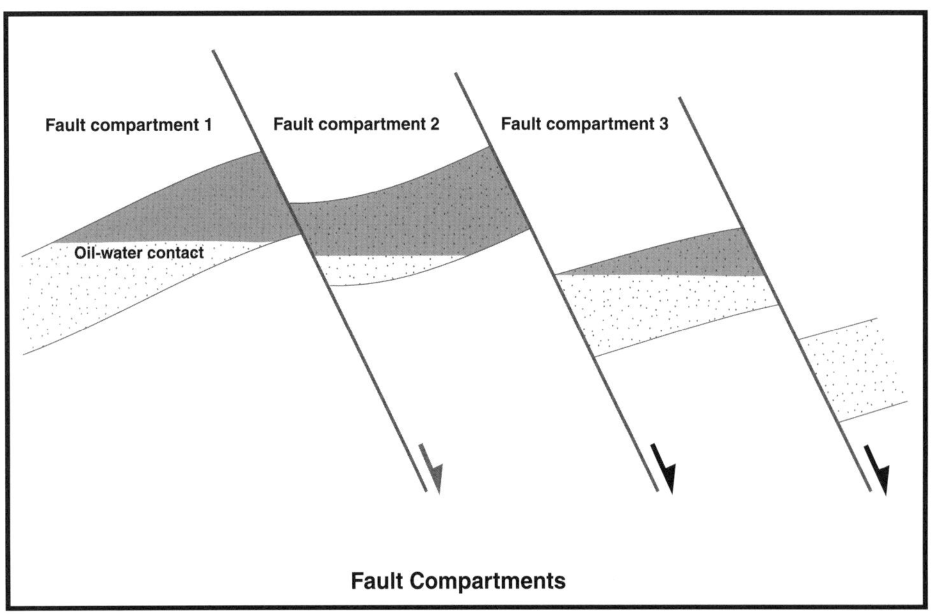

FIGURE 11. Complex reservoirs may be split up into several isolated compartments, each one of which will require a dedicated production well to drain the hydrocarbons within it.

FIGURE 12. A sheet-like turbidite reservoir will need only a few wells to appraise and develop it. Channelized turbidite reservoirs will require several wells for appraisal and development; they may be less profitable because of this.

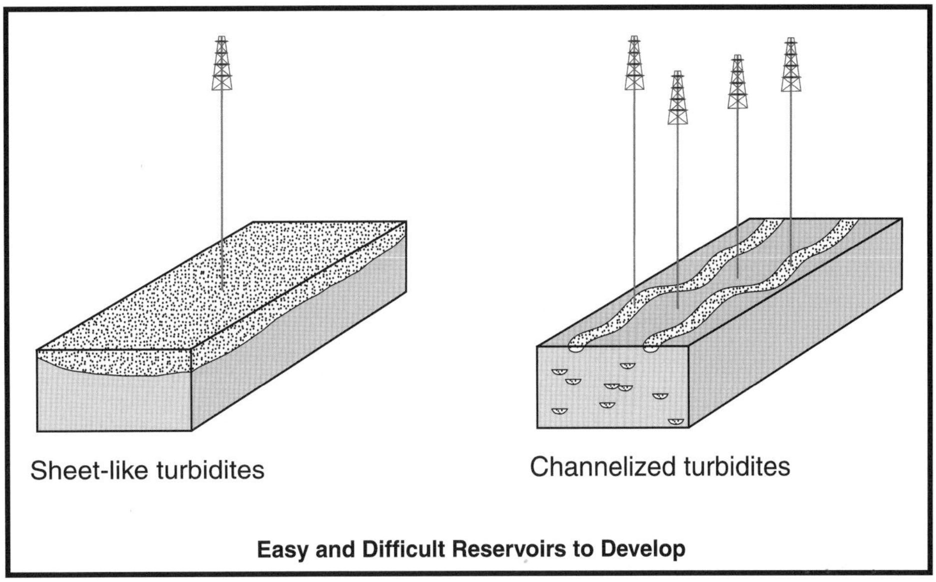

Table 5. Some common terms and acronyms used at the appraisal phase of oil and gas pools.

Term and Acronym	*Definition*
HIIP	Hydrocarbons initially in place
STOIIP	Stock tank oil initially in place (a stock tank is a surface storage vessel for the oil)
GIIP	Gas initially in place
Reserves	The volumes of oil and gas that a company estimates will be produced by a field from a given date forward to the end of field life.
Estimated ultimate recovery	The estimated volume of the total recoverable hydrocarbons from a field. It is equal to the sum of the past production and the reserves volume.
Recovery factor	Estimated ultimate recovery as a percentage of HIIP
bbl	Barrel
BOPD	Barrel of oil per day
MM	Million (10^6)
Billion	1000 million (10^9)
Bcf	Billion cubic feet (gas volume)
Trillion	Million million (10^{12})
Tcf	Trillion cubic feet (gas volume)
P_{10}	10% probability that reserves or HIIP are greater than the quoted number (high side case)
P_{50}	50% probability that reserves or HIIP are greater than the quoted number (medium case)
P_{90}	90% probability that reserves or HIIP are greater than the quoted number (low side case)

The initial appraisal strategy for the Schiehallion field was to define the volume of oil in place. Good quality seismic data helped, and this enabled the oil column thickness to be predicted with confidence using seismic data. Thus, a reliable estimate of oil in place could be made for the Schiehallion field without drilling unnecessary appraisal wells to validate the oil volumes.

Five appraisal wells were drilled, and these were targeted at reducing the key reservoir uncertainties to a level that would be acceptable for project sanction. Given the turbidite channel geometry of the Schiehallion reservoir, a major uncertainty was the degree of communication between the various channel complexes. One appraisal well was drilled with the objective of conducting an extended well test, which lasted 57 days. This was long enough to demonstrate that good connectivity existed. Thus, it looked as if the reservoir could be developed with a minimal number of wells.

Another well was drilled in the crest of the structure to find out whether a gas cap was present or not. Gas handling would have added to the cost of the development, and the need for this had to be determined for evaluating the project economics. No gas cap was found in the well.

The results of the appraisal stage thus indicated that the Schiehallion reservoir would make money. The plan was to develop the field with 12 horizontal and high-angle wells mainly within the channel complexes, and to support production with 10 water injection wells.

The methodology used on Schiehallion is instructional as to the logic involved in planning an appraisal strategy, and this is why it is included here. As a postscript, it is also educational to know what happened

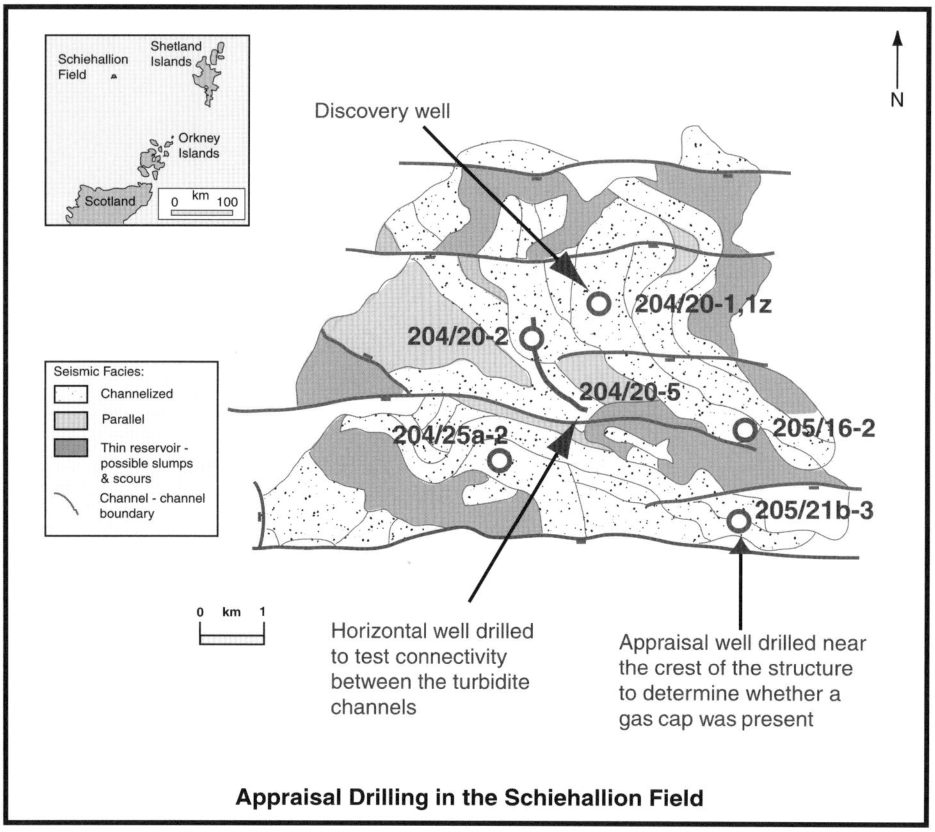

FIGURE 13. Location of appraisal wells in the Schiehallion field, United Kingdom continental shelf, west of the Shetland Islands (from Leach et al., 1999). Reprinted with permission from the Geological Society.

when the field came on production, because this illustrates how much uncertainty can remain despite best practice (Primmer, 2005). Subsequent wells showed that the field is more compartmentalized than was previously indicated. The appraisal well with the extended well test happened to be located in the largest compartment, and the results from this gave the misleading impression that the reservoir was mostly connected. Another production well found a gas cap after all but within a compartment that was structurally lower than the crestal appraisal well. The result of this extra complexity was that BP ended up having to drill more production wells than they expected to.

THE BIG UNCERTAINTY ON FIELD RESERVES AT THE APPRAISAL STAGE

The estimate made at the appraisal stage of a field's ultimate recoverable volumes is a rough approximation with a large error bar. This has been established by numerous studies of oil and gas fields, which have shown significant changes in reserves over the producing lifetime of a field (e.g., North America, Attanasi and Root, 1994; North Sea, Dromgoole and Speers, 1997; the Volga-Ural province of Russia, Verma et al., 2000). A general pattern is for the reserves to grow with time particularly in large, low-complexity fields. Various factors lead to reserves growth:

1) As more wells are drilled, the new data gathered will give a better understanding of how the reservoir is performing. Consequently, reservoir management practices can be optimized to improve recovery.
2) New compartments may have been found. A good example of this occurred during the development drilling of the Ula field in the Norwegian North Sea. A water injection well drilled on the eastern edge of the field unexpectedly found a new reservoir compartment with the base of the oil column approximately 300 m (984 ft) deeper than in the main part of the field (Dromgoole and Speers, 1997).
3) New technology may have been developed during the producing lifetime of the field, which helps to improve recovery. For instance, enhanced oil recovery techniques have led to substantial reserve growth in the United States in recent years (Verma, 2000).
4) Large fields will have sufficient value and long-term production potential to justify further investment in reservoir characterization, new wells, and production technology enhancements (Gluyas and Garrett, 2005).

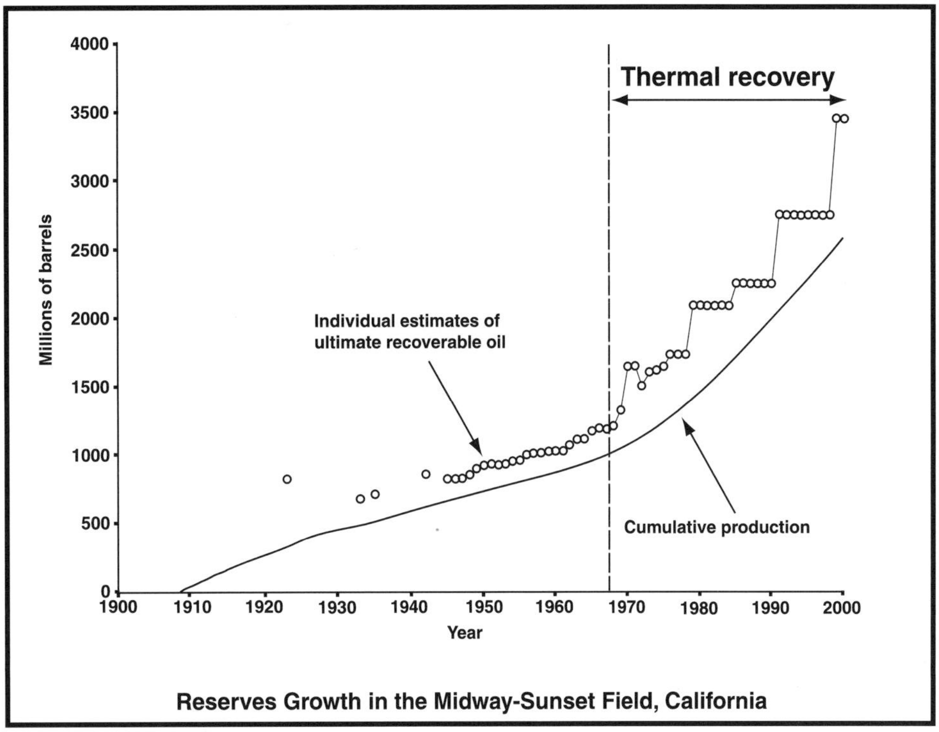

FIGURE 14. The Midway-Sunset field in California is a spectacular example of reserve growth. Each individual circle shows the ultimate recoverable oil expected from the field at the time. However, the field just kept producing more and more oil, and the reserves were required to be revised upward on a regular basis (from Tennyson, 2005). Reprinted with permission from U.S. Geological Survey.

5) Many oil companies are conservative about stating the size of field reserves publicly for reasons concerning the external auditing of the company value. One reason for this is that oil companies trading in the United States are required to conform to a strict definition of proved reserves by the regulatory authorities. They will only define reserves as those that can be rigorously proved from the data, even if they think it likely that the reserves are larger. As new data become available for these fields, the reserves will tend to grow as a result.

A spectacular example of reserve growth is shown by the Midway-Sunset field in California (Figure 14). In 1968, 58 yr after the field was brought on stream, the estimated ultimate recovery was estimated as 1.2 billion barrels. In 2000, the estimated ultimate recovery had grown to close to 3.5 billion barrels (Tennyson, 2005). The main factor behind this reserve growth had been the advent of new technology in the 1960s, which had enabled more of the field's heavy oil to be recovered. Many more wells were drilled after this, and the reserves grew rapidly as a result (Lennon, 1990).

Reserves can also shrink compared to the initial estimate made at field sanction. Examples of this include the Northwest Hutton, Tartan, Thistle, and Dunlin fields in the UK (United Kingdom) North Sea (Demirmen, 2005). Reserves shrinkage generally happens in the more complex fields, particularly where there has been insufficient appraisal wells and data acquisition to fully understand the reservoir heterogeneity (Dromgoole and Speers, 1997). Smaller fields are also prone to reserves shrinkage as they will have less potential to generate surplus cash to justify new technology or to cope with any serious problems that require further investment (Gluyas and Garrett, 2005).

PLANNING THE INFRASTRUCTURE FOR PRODUCTION

Once there is enough data to convince management that the hydrocarbon pool will make money, the task is then to work out what infrastructure is required for development. The infrastructure needs to be engineered to the correct specification to handle the expected hydrocarbon volumes. If the actual volumes produced end up being significantly lower, too much money will have been spent on the facilities. If the field has the potential to flow at higher rates than the facilities were designed for, the company cash flow will be less than it could have been.

For large offshore development projects, one or more *production platforms* may be required (Figure 15). These are commonly fixed structures, with topsides production plant, drilling rig, and living quarters. They can be expensive development options, with the big platforms costing a billion dollars or more. They are also expensive to dismantle and abandon. The advantage of a platform is that existing wells are easy to maintain and

FIGURE 15. Casablanca Oil Production Platform, offshore Spain; courtesy of Repsol, (www.repsolypf.com). The Glas Dowr FPSO, courtesy of Bluewater Ltd., (www.bluewater-offshore.com).

re-enter should there be any problems. If the platform has a drilling rig, the wells will be the cheapest wells offshore because this eliminates the need to hire an expensive floating rig. For the very large offshore fields, a major decision is required at the development stage; how many platforms are required and where are they to be placed within the field area?

For smaller offshore fields, an *FPSO vessel* is an option (Figure 15). This is a floating production vessel that can store the oil it produces. FPSO is an acronym for *floating, production, storage, and offloading*. Once the storage tank fills up with oil, the oil is transferred into a tanker and shipped to an oil refinery. FPSO vessels are suitable for small offshore fields isolated from existing infrastructure or where building a new pipeline back to existing infrastructure is uneconomic. They have also been used for production from deep-water fields, in offshore west Africa for instance. Minimal abandonment costs are involved as the FPSO vessel is simply moved away and used on another field. A disadvantage is that wells cannot be drilled from an FPSO vessel. Expensive semisubmersible drilling rigs need to be brought in to drill new wells and to fix any problems in the existing wells.

A *subsea development* may be an option for small fields in an area of existing infrastructure (Figure 16). Single well tiebacks may be made from small discoveries to an existing offshore platform, or several small fields can be tied back to one central production gathering facility. Oil or gas is collected to a central subsea

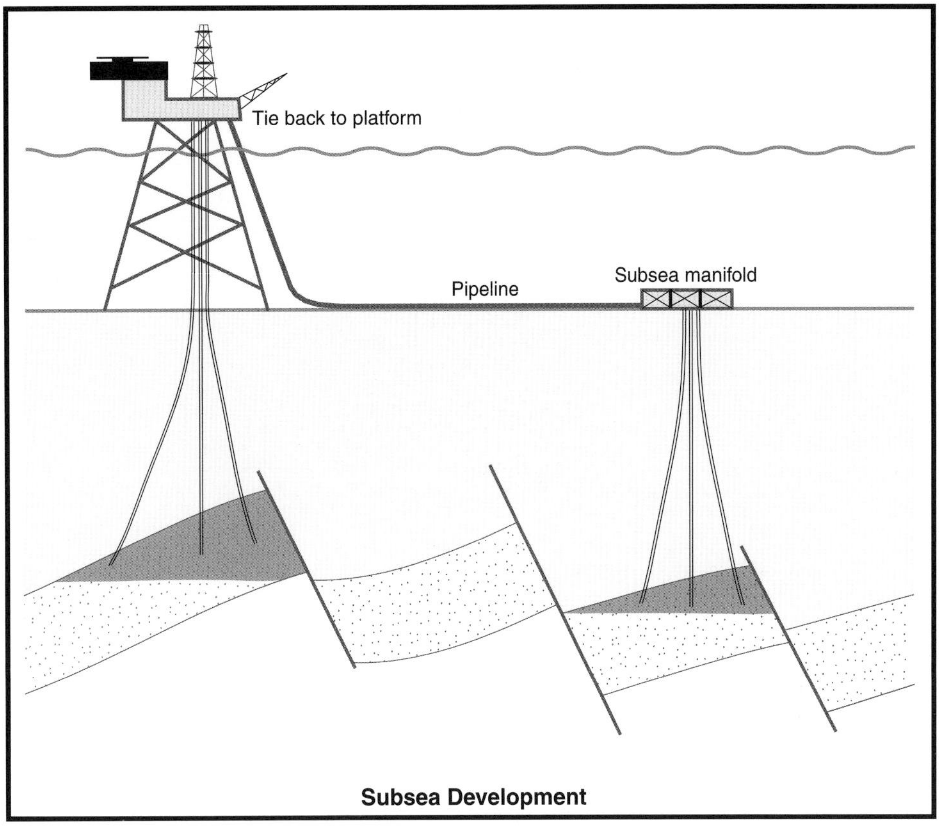

FIGURE 16. A subsea development may be used for small fields in an area of existing infrastructure. Oil or gas is collected to a central subsea manifold system on the seabed and then transported through a pipeline to a production platform.

manifold system on the seabed from where it is transported through a pipeline (generally called a flowline in the context of a subsea development) to a nearby production platform or onshore. This is a relatively inexpensive development option. However, new wells or well interventions require a drilling rig to be brought in.

WELLS

A decision will have to be made as to how many wells are likely to be required for optimal recovery from the field. *Drilling slots* will be allocated on an offshore production platform according to the perceived need for them at this stage. It is a recurring problem to find later on in field life that there are not enough slots for all the infill wells required. Space can be so limited on the topsides of a production platform that it may not be possible to find any more room for additional slots. Hence, it is a good idea to plan for more instead of less drilling slots during the front-end engineering design phase.

An established practice onshore is to drill wells on a regularly spaced grid to cover the field (Figure 17; Table 6). In the United States, onshore fields are generally drilled at a specific spacing relative to the field heterogeneity. A simple oil field or a gas field will have a large well spacing, for example a 160-ac spacing. A more complex or low-permeability reservoir may be drilled on a finer grid, for example, a 40-, 20-, or 10-ac spacing. An acre is equivalent to 43,560 ft^2 or approximately 4047 m^2.

ASSISTED RECOVERY

The need for an assisted recovery of hydrocarbons in the reservoir should be reviewed as part of the field development proposal. Once production starts, there will be some limited pressure support coming from the natural energy inherent in the reservoir and the adjacent aquifer. Fields producing like this undergo *natural depletion*. Where oil fields produce by natural depletion only, this is referred to as the *primary phase of production*. Production under natural depletion is not efficient for oil fields because there is usually only a small amount of natural energy available to drive the oil up to the surface facilities. Recovery factors can be as low as 5–10% for oil from primary production.

When production drops because of the lack of pressure support, then the *secondary phase* of assisted production may be initiated. This involves injecting water or gas into the reservoir to provide additional energy to support the flow of hydrocarbons. The injected fluids

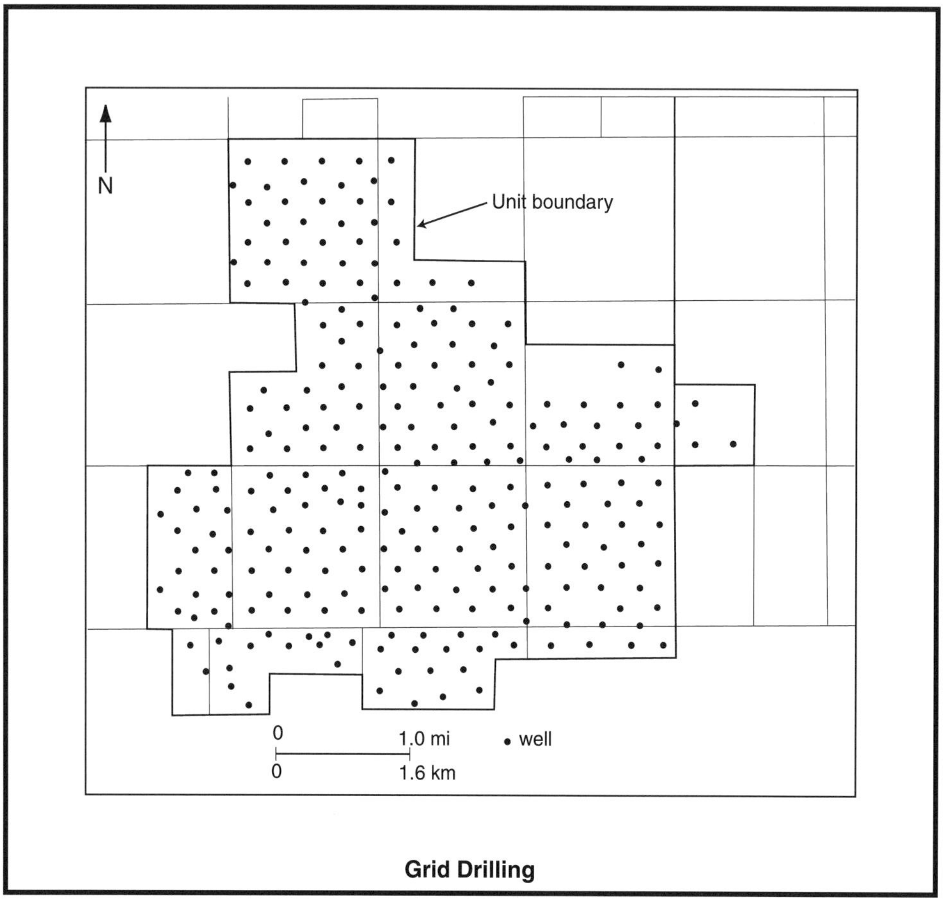

FIGURE 17. Grid drilling in the North Robertson Unit, Texas. The wells have been drilled on a 20-ac spacing to maximize recovery from a highly heterogeneous carbonate reservoir of Permian age (from Montgomery, 1998). Reprinted with permission from the AAPG.

augment production rates by increasing the reservoir pressure, and will boost the recovery by displacing the hydrocarbons toward the wells. *Waterflooding*, where water is injected to support production, is the most common secondary recovery operation. This is because of the easy availability of water. *Injection wells* may be drilled downflank or in a specific pattern relative to the production wells (Figure 18).

Gas reinjection is a typical operation in oil fields where a large volume of gas is produced along with the oil. The gas is recycled into the reservoir to maintain reservoir pressures.

Table 6. Acre spacing and closest interwell distances.

Acre Spacing	*Closest Interwell Distance (ft)*	*Closest Interwell Distance (m)*
160	2640	805
80	1867	569
40	1320	402
20	933	284
10	660	201

The scale of operations offshore generally requires water or gas injection support to the reservoir at the start of production. To make the distinction between primary and secondary phases offshore can be meaningless. Offshore, a ready source of injection water exists; the sea. For onshore fields, the geologist may be asked to find a suitable aquifer as a potential source of water for injection (Figure 19).

UNITIZATION

One situation that can arise as a new field is about to start producing, sometimes later, is the need to define the *equity share* for an oil or gas field (Archer and Wall, 1986). This can happen where a field extends across two or more license areas with a different set of oil company partners in each area (sometimes even across country legislative borders). Thus, a decision is required to determine which percentage of the field reserves applies to each of the license block partnership groups. This process is also known as *unitization*, with the idea that a financial and logistical agreement will be put in place so the whole field is developed as a single unit under one operator. This is rarely a simple task. Revenue and cost

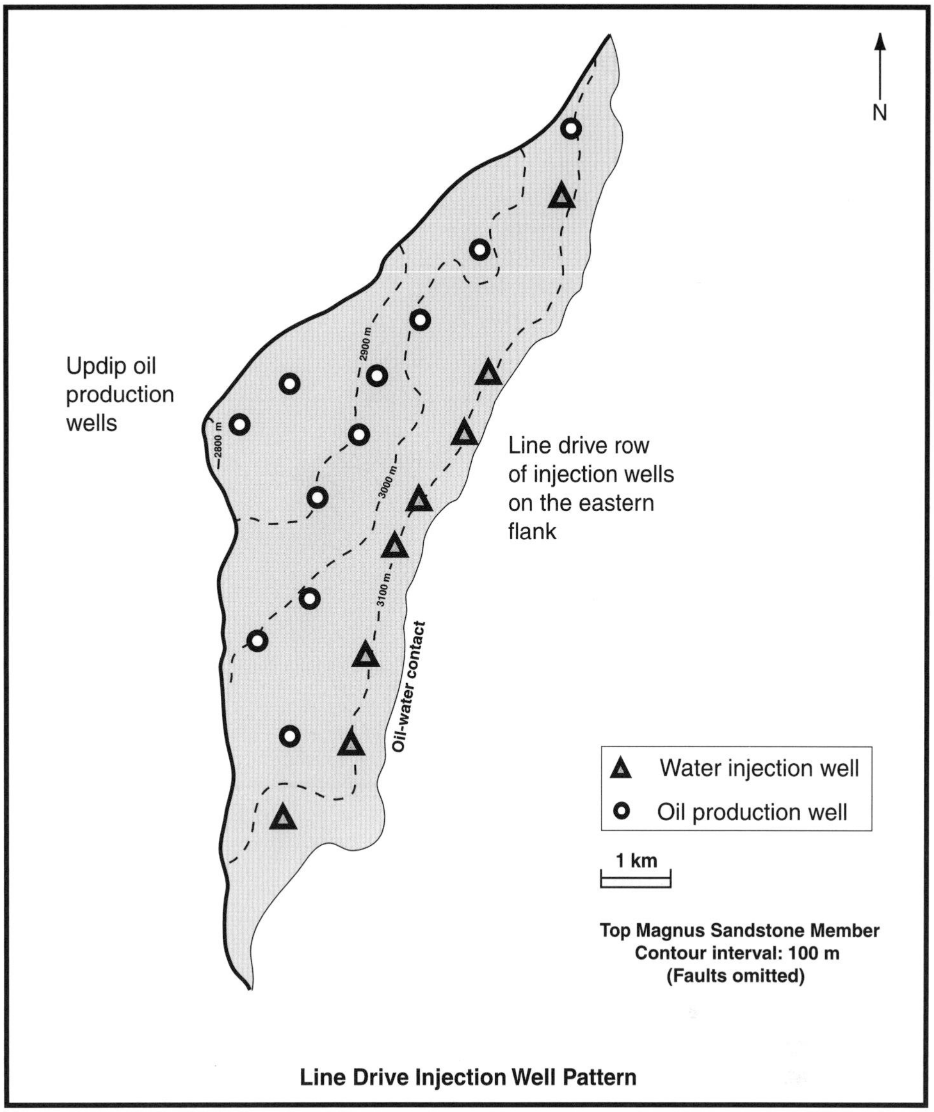

FIGURE 18. An example of a waterflood pattern, a line drive injection well pattern in the Magnus field, United Kingdom North Sea (from Shepherd et al., 1990). The water injectors are located along the flank of the dipping reservoir and provide injection support to production wells updip. Reprinted with permission from the AAPG.

sharing will be split according to the equity share, and very big money is involved. The potential to gain or lose a few percentage points in the share of an oil or gas field will materially affect the value of the oil companies involved. Committee meetings to discuss equity issues are tense affairs, where somewhat biased reservoir schemes will be presented by geologists from the opposing groups. It is not always possible for the companies to come to a meaningful agreement on equity, and it is common practice to call in an expert to decide equity share independently. The expert may be a service company that specializes in these types of disputes. The various partner groups will make their case to the expert as to how they see the equity share being divided. Sensitivity analysis will be conducted on the critical aspects of the subsurface representation that can be biased toward a favorable equity negotiation outcome by either side. For a large field, there may be more than one equity negotiation phase; a later *redetermination* may be made if there is likely to be additional drilling activity. The extra data will allow the equity split to be determined more precisely.

FIELD SANCTION

Once the reservoir has been appraised and evaluated, a *field development plan* is compiled and submitted to the government or appropriate authority for approval (*sanction*). This will include details of the proposed design for the production infrastructure and plans for developing the reservoir.

Oil companies will want to reduce the time between discovery and first production to a minimum by *fast tracking* the development. The idea is to ensure that the enormous sum of money committed by the company to exploration, appraisal, and the building of the infrastructure is paid back as quickly as possible by revenues

Workflow for New Field Development

Appraisal

- Locate the fluid contacts
    - oil-water contact
    - gas-water contact
    - oil-gas contact
- Establish the hydrocarbons initially in place
- What is the potential range of volumes?
- How many compartments?
- Size of the aquifer
- Establish the critical uncertainties that control the success of the development
- Decide on the location and objective of appraisal wells
- Is a long-term well test required?
- Economic evaluation
- Make a decision to develop or not

Engineering Design

- Design the infrastructure for production
- How many wells are needed?
- What are the requirements for assisted recovery?

FIGURE 19. Workflow for new field development.

from the hydrocarbon stream. However, it should be recognized that the desire to ensure a fast development should not be at the cost of a rushed development plan. Any mistakes made at this stage will be expensive to remediate later.

THE EARLY PRODUCTION PHASE

Once approval has been obtained to develop the field, the next stage is to proceed with the drilling of the production and injection wells. For offshore production, some of these wells may be predrilled prior to installing a platform, others may be drilled from the platform once it has been installed. With each new well, production should increase until *plateau production* is achieved. This is a phase when all of the wells are producing at full capacity. Plateau production may last for a year or more for an oil field, and ideally much longer for a gas field. During this stage, more and more wells may be added. For the production geologist, there will be a steady phase of *well planning* as well after well is drilled. The main concern is to establish that there is likely to be a thick, productive section in the planned wells and to avoid any faults that could cut out part or all of the reservoir target intervals. In big, low-complexity fields, production geology gets no easier than this.

THE ENHANCED RESERVOIR MANAGEMENT PHASE

Managing a reservoir will start to get difficult when the field begins to *decline* off plateau production. This happens once production from the wells decreases as the reservoir pressure falls. There may also be some *water or gas breakthrough* to the wells by this time. When water breaks through to an oil producer, the oil column being pushed to the surface by the reservoir pressure gets loaded up with the water. A combination of lower reservoir pressures and heavier fluid columns will cause the well flow rates to drop off noticeably with time.

It is now that the production geologist has to be mindful of *sweep*, the portion of the reservoir that has been contacted by water (Figure 20). Already the *sweep patterns* will be starting to suggest that the reservoir is behaving in a more complex manner than was previously thought.

As the field production rates decline further and further, the geological contribution becomes more and more critical for both understanding and predicting sweep. At this stage, well planning will have to account for not just the expected geology but for the expected sweep patterns also. The reservoir may now show an irregular patchwork of petroleum fluids and water, controlled to a major degree by the heterogeneity of the rock. There is a risk with drilling new wells that

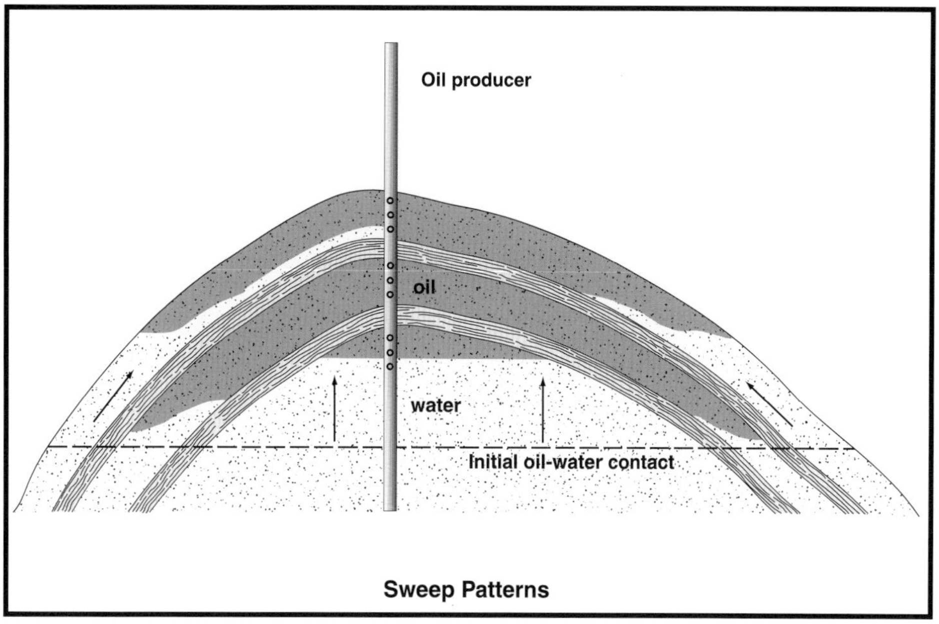

FIGURE 20. The geology influences the fluid flow within a reservoir. The resulting configuration of injected fluids, water, and hydrocarbons forms the sweep patterns in the field.

mostly water swept reservoir intervals will be found if the sweep patterns are not properly understood. Close cooperation between the production geologist and the reservoir engineer is required to analyze the reservoir behavior.

At the *late mature phase* of oil field life, infill production and injection wells may be drilled between the existing wells to access any unswept oil that is present there. Onshore in the United States, this can happen without any geological input. The decision to *reduce the well spacing*, e.g., from a 40-ac to a 20-ac spacing, will be taken by the production engineer. General experience is that this will result in an increase in oil production and recovery (Davis and Shepler, 1969; Driscoll, 1974). Ambrose et al. (1991) criticized this behavior, as the drilling of infill wells on a regular spacing could miss many unproduced or poorly drained reservoir compartments. It can also result in the drilling of numerous wells that are nonproductive or poorly producing with inefficient economic returns. Offshore, wells are too expensive to drill like this. Geological input and more careful well planning are required.

It is also possible to make well interventions to change the perforated intervals in wells. This can involve isolating some or all of the existing perforations and if appropriate, adding perforations to intervals that have not been perforated previously. For instance, a water injection well may be worked over to ensure that only a specific reservoir zone will take water thus enhancing the sweep in that interval.

A sign that a subsurface team is performing at a high level is where a series of successful infill wells and well interventions have kept the field on a subsidiary production plateau, postponing the inevitable decline. A good example of this is the North Cormorant field in the UK North Sea (Figure 21) whereby the field oil production rate was maintained at a level of 30,000 BOPD for 9 yr between 1991 and 2000 (Bater, 2003).

THE NEAR ABANDONMENT PHASE

Eventually, a time comes when the monetary returns from the field production rate decline dangerously close to the *economic cut off* for the project. At this point, serious consideration must be given to abandoning the field and the associated infrastructure. This involves putting cement plugs in the wells and removing the surface facilities. Abandoning a field is not cheap, especially offshore; the costs involved can be substantial. The late near abandonment phase of field life is an anxious time as there will be a certain unwillingness for managers to commit the company to such a large outgoing of expenditure at the same time as losing an asset that has been a major source of cash flow. The subsurface team will be under a lot of pressure to do anything to postpone field abandonment. Many risky infill wells can be drilled during this period.

Oil companies will now be looking for any means of cutting costs to keep the field profitable. A common criterion used at this stage is the *lifting costs*, typically defined as the operating expenditure required to produce one barrel of oil or its equivalent. Some companies will restructure their asset teams at this stage with the specific objective of reducing these costs. The alternative is to sell off the field to another company who may

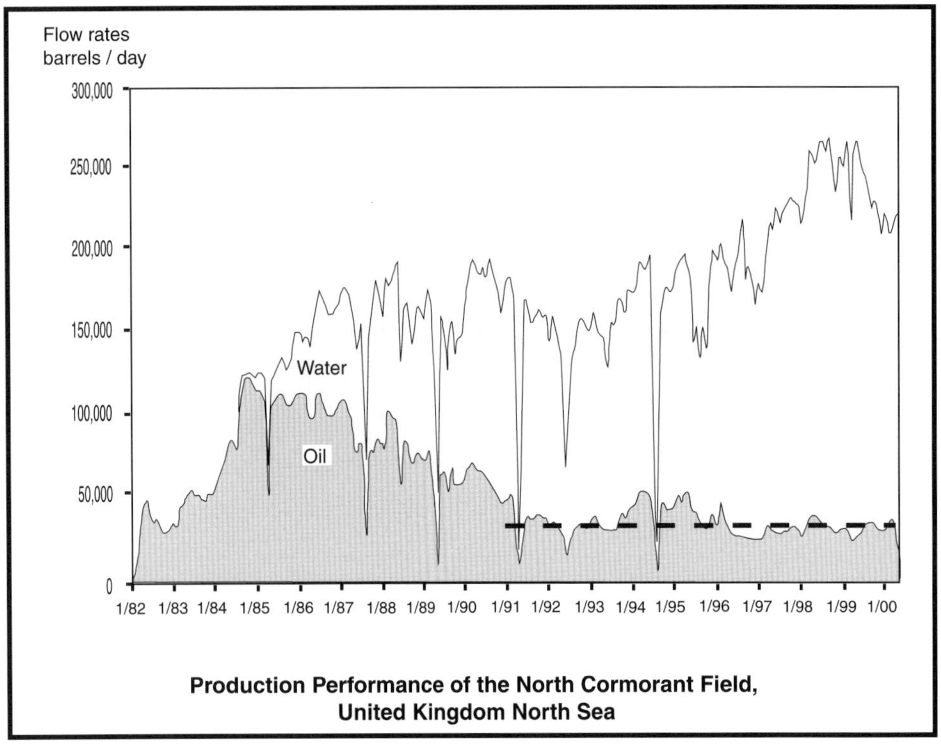

FIGURE 21. A combination of successful new wells and workovers will keep a field on a subsidiary plateau during the mature phase of production (from Bater, 2003). Reprinted with permission from the Geological Society.

have a different view on the economics and the potential to increase the ultimate recovery.

At the late stage of field life, efforts will be made to tie in any potential satellite hydrocarbon pools that have been discovered near the field. Additionally, if another operator has discovered a hydrocarbon pool close to the infrastructure, the oil company may want to discuss *tariffing arrangements* to encourage them to put their hydrocarbons through the facilities. The other company will be paying money to use the plant and pipelines, but it can still be a cheaper option for them to do this than to build new infrastructure themselves.

Satellite production can be especially valuable at the late mature phase. If, for example, an offshore facility producing from an oil field is deemed uneconomic at a threshold oil production of say 7000 BOPD, then it will be shut in. However, if a satellite field can be tied in such that the combined throughput of oil through the facility is kept above 7000 BOPD, then the facility will be kept going longer. If by virtue of bringing on a satellite field, the *hub field* produces for an extra year at say an average of 6000 BOPD, this adds an extra 2 MMbbls of oil on top of any reserves directly attributed to the satellite itself (Figure 22).

It is good practice for oil companies to encourage their subsurface teams to conduct *near-field exploration.* This is where potential oil or gas prospects are defined in the areas around existing fields, and the more promising prospects are drilled. In some major oil companies, the exploration group has the responsibility for near-field exploration. This can cause frustration for the subsurface team as near-field prospects are typically small and will rank at the bottom of a global list of exploration targets; the larger "elephant"-size targets in frontier acreage will get the most attention here. Nevertheless, it has to be recognized that the proximity of any new satellite discovery to existing infrastructure is likely to make it profitable even if the prize looks none too exciting to a dedicated explorationist. Another reason for wanting to ensure some near-field exploration is that the risks on exploration can be low. Experienced professionals consider that "the best place to look for oil is around an oil field." This well-known old adage has been said for a good reason. Exploration geologists will assess the probability of success of an exploration prospect by estimating the chances of it having a reservoir, a trap structure, and an oil migration source route. These are more likely to be present if there is an existing oil field close to the prospect.

Some reservoirs have been discovered by the simple expedience of drilling a well to investigate *deeper reservoir potential* directly underneath an existing field. The trapping structure of an established field can be mirrored by any potential deeper reservoir intervals. A spectacular example of this was the discovery of the Sihil field under the giant Cantarell field, offshore Mexico (Aquino et al., 2003). Seismic indications of a deeper thrust block were tested by a new well in 1998, 22 yr after the Cantarell field was discovered. A reserves estimate of 1136 MMSTB of oil has established the Sihil field as the largest oil field discovered in Mexico in recent years.

A North Sea example is the discovery of the Triassic Alwyn North gas condensate field in 1995 underneath

FIGURE 22. Gannet platform and satellites, United Kingdom North Sea (from Pieters and Por, 1995). The Gannet platform is the hub platform for the three satellite fields shown in this figure. Reprinted with permission from the Society of Petroleum Engineers.

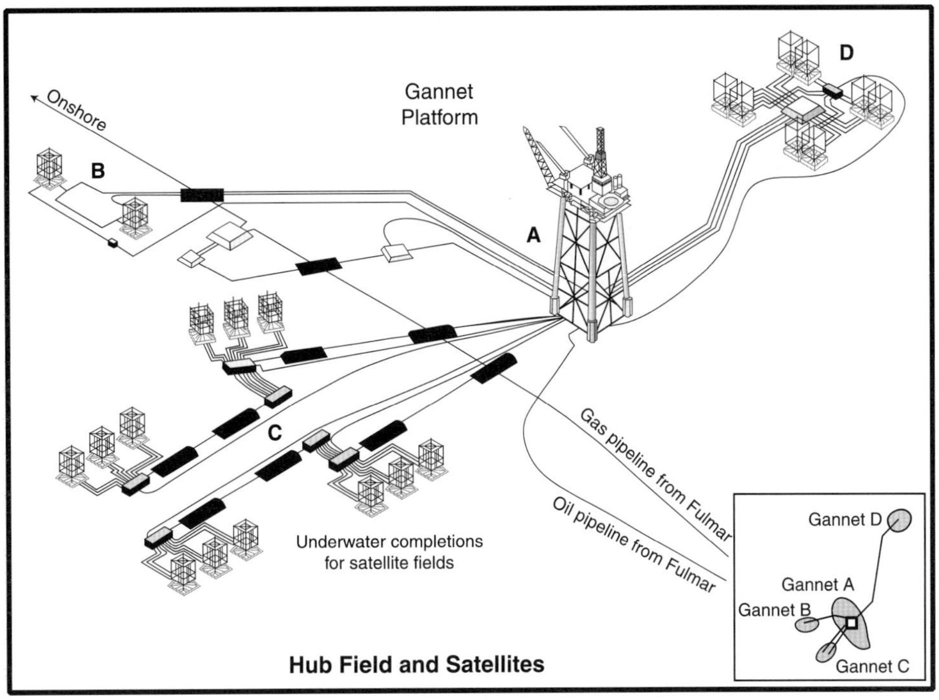

the Alwyn North field, discovered 20 years earlier in 1975 (Figure 23) (Harker et al., 2003).

FIELD ABANDONMENT AND RESURRECTION

Eventually, the decision will be made to abandon the field. However, this may not be the end of the matter. It is becoming more common for abandoned oil fields to be *resurrected*. Perhaps, the original operator did not manage the field as effectively as they could have, leaving behind significant commercial quantities of *bypassed oil*. In addition, many reservoirs with strong aquifers can repressurize with resegregation of oil and water after they have been shut in for several years (Harker, 1998).

An example of field resurrection is the Lakota reservoir of the Lost Soldier field in Wyoming. The reservoir produced about 5 MMbbls between 1922 and 1950s when it was abandoned. In the 1970s, development wells targeting a deeper reservoir found bypassed oil within the Lakota Formation. This spurred a decision to

FIGURE 23. Deeper reservoirs can sometimes be found below existing fields. In the United Kingdom North Sea, a deeper gas condensate field was found under the Alwyn North oil field 20 yr after the first field was discovered (from Harker et al., 2003). Reprinted with permission from the Geological Society.

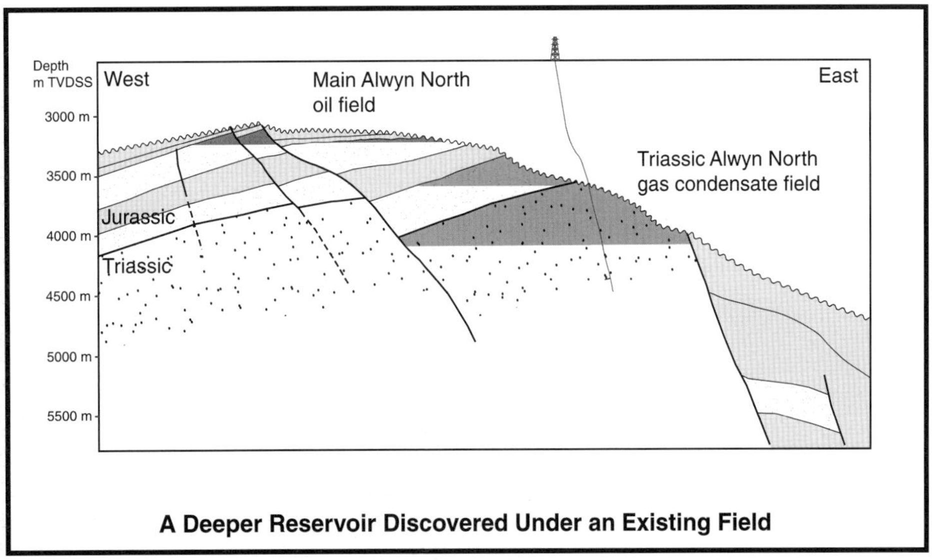

start up production from the reservoir, and first oil was achieved in 1979. A total of 32 new wells and recompleted existing wells are expected to produce a further 1 MMbbl of oil (Schmechel and McGuire, 1986).

Abandoned reservoirs may also have a use for storing unwanted waste products. This includes oil-contaminated drill cuttings, produced water, and toxic wastes. Carbon dioxide gas can be disposed into underground reservoirs (Kheshgi et al., 2006). Emission of carbon dioxide to the atmosphere by industrial processes has given rise to concerns about its effect on global climate change. The process of capturing and storing carbon dioxide in the subsurface is known as *carbon sequestration*. The first commercial operation to store carbon dioxide in the subsurface started in the Sleipner gas field, offshore Norway. Since 1996, about a million tonnes of CO_2/year has been separated from the produced gas and injected into the Utsira aquifer.

Shepherd, M., 2009, Reservoir fluids, *in* M. Shepherd, Oil field production geology: AAPG Memoir 91, p. 29–35.

4

Reservoir Fluids

INTRODUCTION

This chapter describes the physics of how oil, gas, and water interact with each other and the rock. The basic concepts of wettability, capillary pressure, and relative permeability are important. This is knowledge required to understand how reservoirs behave. Physical processes also control the distribution of oil and water in a reservoir, and an understanding of these will help the production geologist to estimate the in-place hydrocarbon volumes.

WETTABILITY

Much of the physical processes that control the fluid distribution in a reservoir occurs at the molecular level. Individual molecules show an attraction for each other resulting from weak intermolecular forces. In a body of liquid, the tendency is for the molecules to be pulled in toward the center of the body; surface tension forces will reduce the surface area of the liquid to a minimum. For an interface between two immiscible liquids, the term *interfacial tension* is used for the force acting to reduce the area of contact between two different fluids.

Where two immiscible liquids, or a liquid and a solid, are in contact with each other, the surface molecules of each substance are also attracted to each other across the interface by weak intermolecular forces.

Therefore, at a solid-liquid boundary interface, the molecules of the liquid are subjected to opposing forces of attraction; in the first instance, the liquid is attracted by its own molecules and secondly by the molecules of the solid across the boundary. The degree to which force is dominant controls what is termed the *wettability* (Vavra et al., 1992). For instance, glass is *water wet*, in that water will spread across the surface of a glass plate as a thin sheet. The adhesive attraction of the water for the glass is greater than the cohesive attraction of the water molecules for each other. A liquid such as mercury will form globules on a glass surface and is *nonwetting*. The cohesive attraction of the mercury molecules for each other is greater than the adhesive attraction of glass and mercury (Figure 24).

Where a reservoir rock is water wet, the water forms a thin film over most of the grain surfaces and will also fill the smaller pores. The oil or gas will occupy the remaining, more central volume of the pore system. Conversely, in a reservoir that is *oil wet*, it is the oil that covers the grain surface and occupies the smaller pores; the water is located centrally within the pore structure (Anderson, 1986).

Most reservoirs were water wet before oil migration started; the major mineral phases in reservoirs such as quartz, carbonate and dolomite are all water wetting prior to coming in contact with oil (Abdallah et al., 2007). Following oil migration, sandstone reservoirs can end up as predominantly water wet, predominantly oil wet, or more frequently in a *mixed-wettability* state, that is, somewhere in between oil wet and water wet. Carbonate reservoirs are commonly described as showing mixed wettability tending to oil wet (Treiber et al., 1972; Chilingar and Yen, 1983). The degree of wettability can vary even within a single reservoir. The rocks in the reservoir will show a variety of mineral types, each mineral with its own wetting characteristics. Other variables affecting wettability include the wetting nature of the numerous compounds comprising crude oil and the degree to which polar compounds from the oil are absorbed onto the rock surface (Anderson, 1986).

Waterfloods produce more efficient sweeps in water-wet reservoirs than in oil-wet systems. Water forced to move through a water-wet pore system will displace

DOI:10.1306/13161186M913372

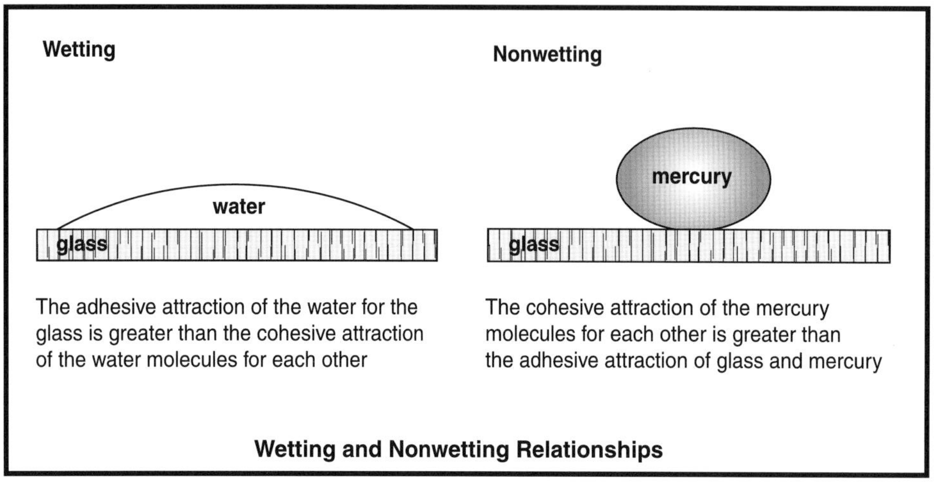

FIGURE 24. Wetting and nonwetting relationships between fluids and rocks have a major effect on the static and dynamic behavior of hydrocarbons in reservoirs.

the oil from the center of the pores relatively efficiently (Figure 25). Water will also be drawn into the smaller pores, displacing oil into the main flow pathways. In an oil-wet sandstone, the oil forms a film around the sand grains and water will move through the center of the pores, particularly the larger connected pores. The pathway for the water here is less tortuous than in water-wet sandstones, and the water will move through the rock more quickly, bypassing a large volume of oil. Rapid water breakthrough to the production wells typically occurs, and oil rates will drop significantly once this happens. Nevertheless, the film of oil around the grains can survive as a continuous path to a production well after water has broken through. Because of this, a continuous flow of oil can still be maintained in oil-wet reservoirs by injecting large volumes of water (Anderson, 1987).

BUOYANCY FORCES IN RESERVOIR FLUIDS

When hydrocarbons migrate into a trap, the *buoyancy force* exerted by the lighter oil (or gas) will push the water that was previously in the pore space sideways and downward. However, not all of the water is displaced; some of it will be held by *capillary forces* within the pores. Narrower *capillaries*, pores with smaller pore throats, hold onto water the strongest.

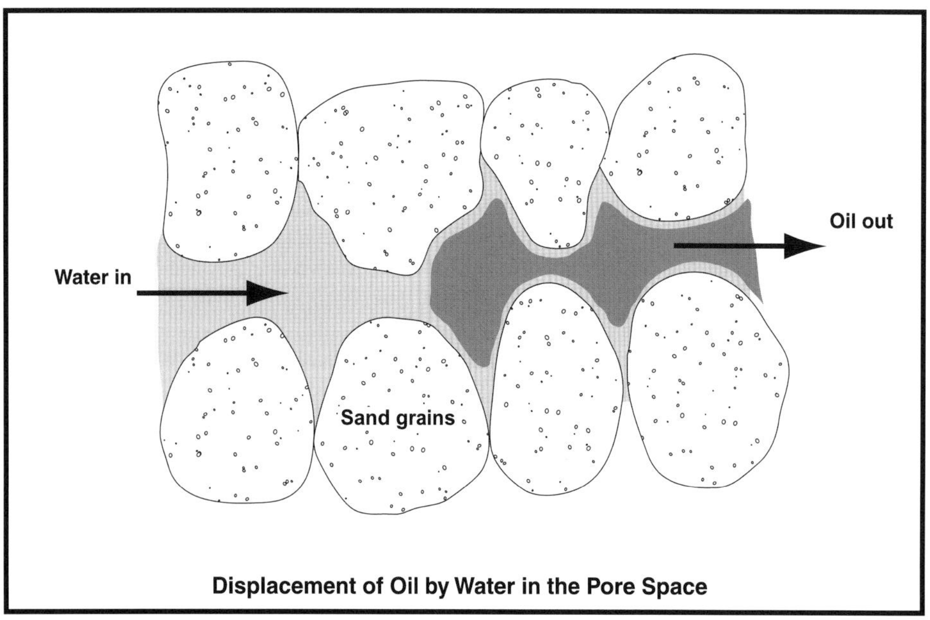

FIGURE 25. In a water-wet reservoir, water wets the surface of the grains, and hydrocarbons occupy the central parts of the pore space. Moving water will displace the oil from the center of the pores (from Clark et al., 1958). Reprinted with permission from the Society of Petroleum Engineers.

FIGURE 26. Water saturation decreases with height in an oil column. The volume of water is a function of the balance of capillary forces pulling the water up from the oil-water interface and the force of gravity acting together with the density contrast between the reservoir fluids, tending to pull the water down.

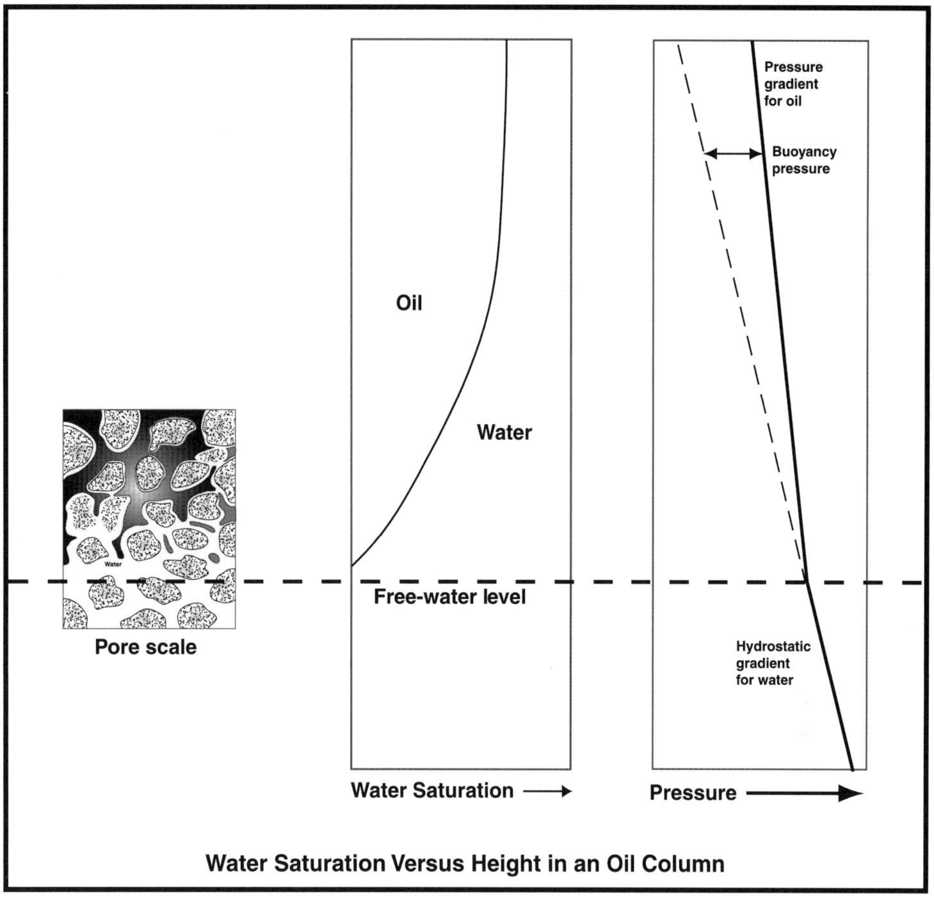

The two forces acting on the fluids in the pore space are controlled by physical laws. The equation for the buoyancy pressure is given by

$$P_{\mathrm{b}} = (\rho_{\mathrm{w}} - \rho_{\mathrm{nw}})gh$$

where P_{b} is the buoyancy pressure; ρ_{w} and ρ_{nw} are the specific gravities of the wetting and nonwetting phases respectively; g is the acceleration of gravity; and h is the height above the free-water level.

The equation for capillary forces is given by

$$P_{\mathrm{c}} = \frac{2\sigma \cos\theta}{r}$$

where P_{c} is the capillary pressure, σ is the interfacial tension, θ is the contact angle between the wetting fluid and the solid surface, and r is the capillary (pore throat) radius (Vavra et al., 1992).

The volume of water remaining at a given height in a reservoir is a function of the balance of capillary forces pulling the water up from the hydrocarbon-water interface and the force of gravity acting together with the density contrast between the reservoir fluids, acting to pull the water down (Arps, 1964). Thus, a given part of the pore space within the hydrocarbon leg can contain both hydrocarbons and water. The fraction (percentage) of water to total fluid volume is termed the *water saturation.*

For an oil field, the capillary-bound water comprises a continuous column of water within the oil leg, which will have a hydrostatic pressure gradient controlled by the water density. The oil is located in the remaining pore space as a continuous phase and will have a pressure gradient controlled by the (lower) oil density (Figure 26). Although oil and water can coexist in the same localized volume of rock, the pressures acting on the two fluids are different. The difference in pressure between the oil and water phases increases with height above the *free-water level.* The free-water level is the level at which the water-hydrocarbon interface would theoretically stand in a large open hole drilled through the oil column (Schowalter, 1979). In this situation, only gravity and buoyancy forces control the fluid distribution in the borehole.

As the buoyancy pressure increases with height above the free-water level, the oil phase will displace more water from increasingly smaller pore volumes. The effect of this is that hydrocarbon saturations increase with height above the hydrocarbon-water contact. The relationship between capillary and buoyancy forces thus controls the static distribution of fluids in oil and gas pools. Knowledge of these relationships is fundamental to the accurate calculation of hydrocarbon volumes within a reservoir.

Capillary pressure is typically measured in the laboratory by injecting mercury under pressure into a core plug. The mercury is a nonwetting phase, which replicates the behavior of hydrocarbons in reservoir rocks. The procedure simulates the entry of hydrocarbons into a water-wet rock and the way in which buoyancy pressure increases with height in the hydrocarbon column.

Mercury will not enter the rock immediately. The pressure required to do this will depend on the radius of the pore throats, the contact angle, and the mercury-air interfacial tension. The pressure at which the mercury effectively enters the pore network is termed the *displacement* or *entry pressure* (Vavra et al., 1992). Lower entry pressures are found in the better quality reservoir rocks, that is, those with larger pore throat diameters. A cap rock with tiny capillaries, shale for instance, has a very high displacement pressure. The displacement pressure for a cap rock can be so high that the tightly bound water in the pore space of the shale will prevent the oil from entering and the oil remains trapped in the underlying reservoir rock (Berg, 1975; Schowalter, 1979).

With increasing injection pressure, more and more mercury is forced into the rock. The shape of the curves on a capillary pressure plot reflects the grain sorting and the connection of pores and pore throats. The longer the plateau shown by the capillary curve, the better the reservoir quality. Poorly sorted, fine-grained sediment with narrow pore throats will retain water to higher pressures than coarser grained, better sorted sediments. A homogenous reservoir rock can be represented by a single capillary pressure curve. By contrast, a heterogenous reservoir will have a family of rock types, each with its own capillary pressure curve (Figure 27).

Petrophysicists will use capillary pressure curves as the basis for deriving a water saturation versus height relationship for a reservoir (Vavra et al., 1992).

RELATIVE PERMEABILITY

Permeability is the measure of the ease of movement of fluid through the pore space in a rock. Where more than one fluid phase is present (e.g., oil and water), the permeability of one phase is reduced by the presence of the other phase within the pore system. In this instance, the permeability to a particular fluid is called the *relative permeability* (Hawkins, 1992).

For water-wet reservoirs, Craig (1971) gave some general *water-oil relative permeability end points*. Water will start flowing along with oil once the water saturation is greater than roughly 20–25%. This value is the *irreducible water saturation*; the volume of water bound and immobilized by adhesive attraction to the surface of the pores. Oil will stop flowing where the water saturation in the rock is about 70–80%. When this happens, there will not be enough oil to provide a continuous volume throughout the rock. Interfacial tension will cause the oil stream to snap off and fragment into immobilized globules and strands of residual oil.

The relative permeability end points may vary significantly between reservoirs; the quoted values can be considered as approximate. The relative permeability of water and oil as a function of water saturation is illustrated by *relative permeability curves* (Figure 28).

THE STATIC DISTRIBUTION OF FLUIDS IN UNPRODUCED RESERVOIRS

The producing behavior in an oil column will vary according to the fluid saturations (Jennings, 1987). Several zones can be defined (Figure 29):

1) *The zone of 100% oil production.* This is located above the height where the water saturation is less than the relative permeability end point to water, e.g., less than 20% water saturation. The water is immobile and only oil will flow.
2) *The oil-water transition zone.* Both water and oil are produced in this interval. The water saturations here lie between the end points at which the relative permeability to water is zero and the relative permeability to oil is zero. In coarse-grained sediments, the transition zone may be less than a meter thick; in very fine-grained sediments, it may be many tens of meters thick or more. In some reservoirs, the entire oil column may be within the transition zone (Fanchi et al., 2002).
3) *The zone of 100% water production.* This is that part of the oil column where there is still a small volume of oil present. However, its relative permeability is zero, so the oil will not flow, whereas the water will.
4) *The 100% water level.* This is the level at which the buoyancy pressure of the oil equals the capillary displacement pressure of the reservoir rock. The 100% water level is the level above which the reservoir rock has a water saturation less than 100% (Schowalter, 1979). It is effectively the base of the oil column, although some authorities state that a small volume of residual oil may be present below this level (e.g., Jennings, 1987).
5) *The free-water level.* At some point below the base of the oil column in a water-wet reservoir is the free-water level. The free-water level is a horizontal surface where theoretically the water would stand in a large open hole unconstrained by the effects of

FIGURE 27. The shape of the curves on a capillary pressure plot reflects the grain sorting and the connection of pores and pore throats within the various rock types. The longer the plateau shown by the capillary curve, the better is the reservoir quality of the rock (from Sneider et al., 1977). Reprinted with permission from the Society of Petroleum Engineers.

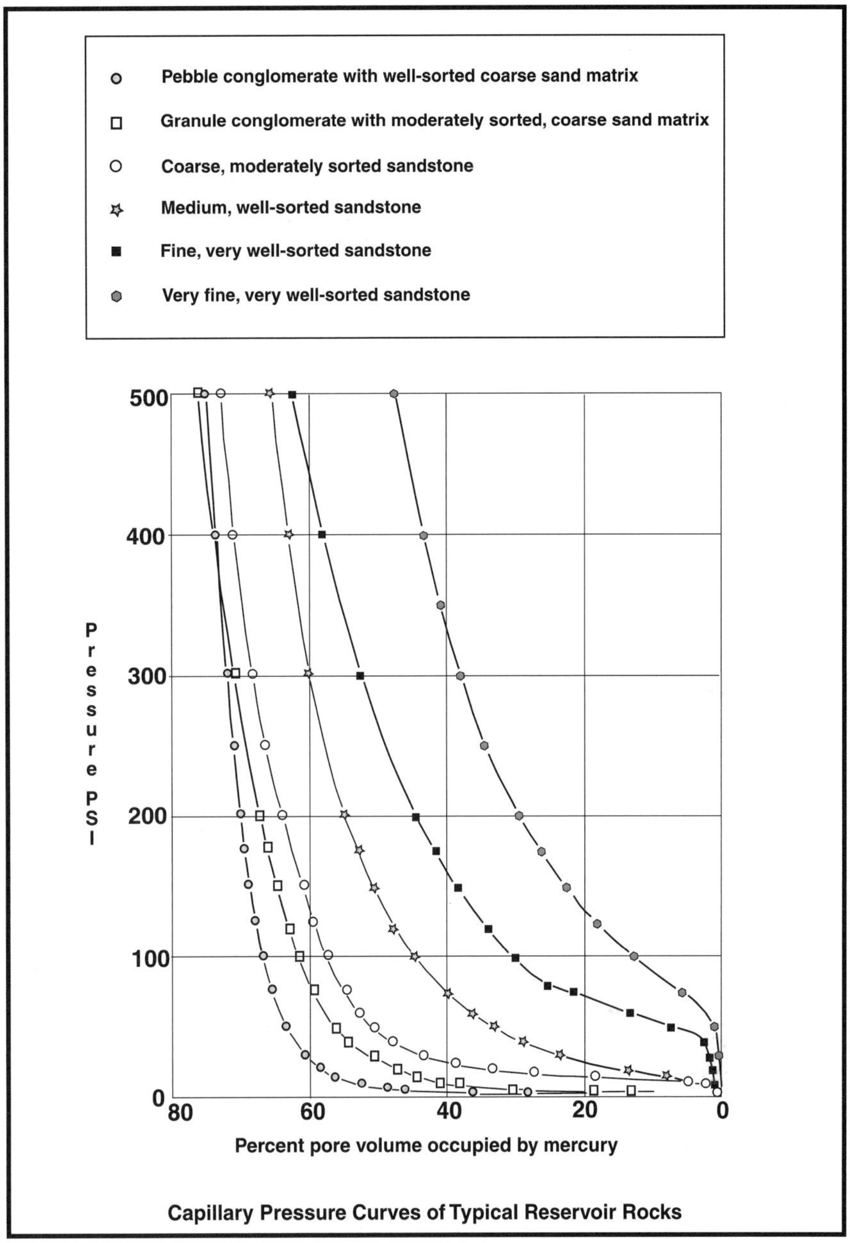

capillary forces. At this point, the buoyancy pressure is zero.

The definition of an *oil-water contact* varies somewhat within the literature (Jennings, 1987). The commonly used definitions are:

1) The base of the oil column corresponding to the 100% water level. This is the most frequently used definition for estimating the oil in place volume.
2) The *producing oil-water contact,* which is located at the base of the transition zone. This is the lowest point at which oil can be produced.
3) The *economic oil-water contact,* the point where enough oil is produced in the total fluid to make the well economically viable. This is nominally the point corresponding to the 50% oil saturation level.
4) The top of the transition zone. Only oil is produced above this point.

TILTED OIL-WATER CONTACTS

Some fields have a *hydrodynamically tilted oil-water contact* (Figure 30). This results from variations in aquifer pressure associated with the movement of water in the subsurface, mainly as a result of a mobile artesian aquifer or basin dewatering (Hubbert, 1953). The tilts are toward the direction of reducing pressure and are generally less than 2° in gradient (Dennis et al., 2000, 2005).

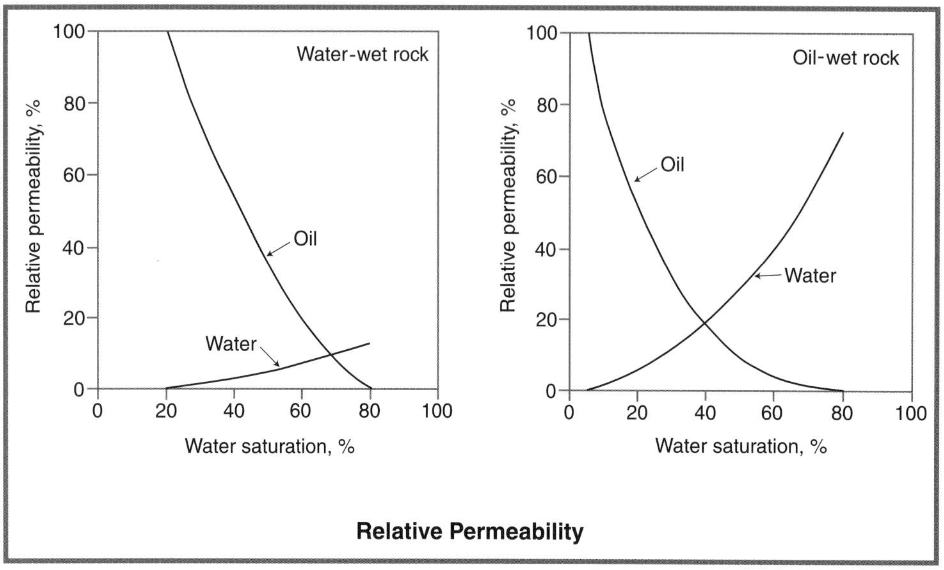

FIGURE 28. When more than one fluid phase is present, the permeability of one phase is reduced by the presence of the other phases within the pore system. Relative permeability curves display these relationships. The plots show a water-displacing-oil relative permeability curve for a water-wet rock and a water-displacing-oil relative permeability curve for an oil-wet rock (modified from Hawkins, 1992). Reprinted with permission from the AAPG.

PERCHED OIL-WATER CONTACTS

Sometimes a new well will find an oil-water contact significantly shallower than the common oil-water contact established in the wells drilled so far in the field, yet there is evidence for pressure communication between the new and old wells (Figure 30). The shallower oil-water contact may be a *perched oil-water contact.* These are common features yet are hardly ever mentioned in the literature. Perched oil-water contacts result from the

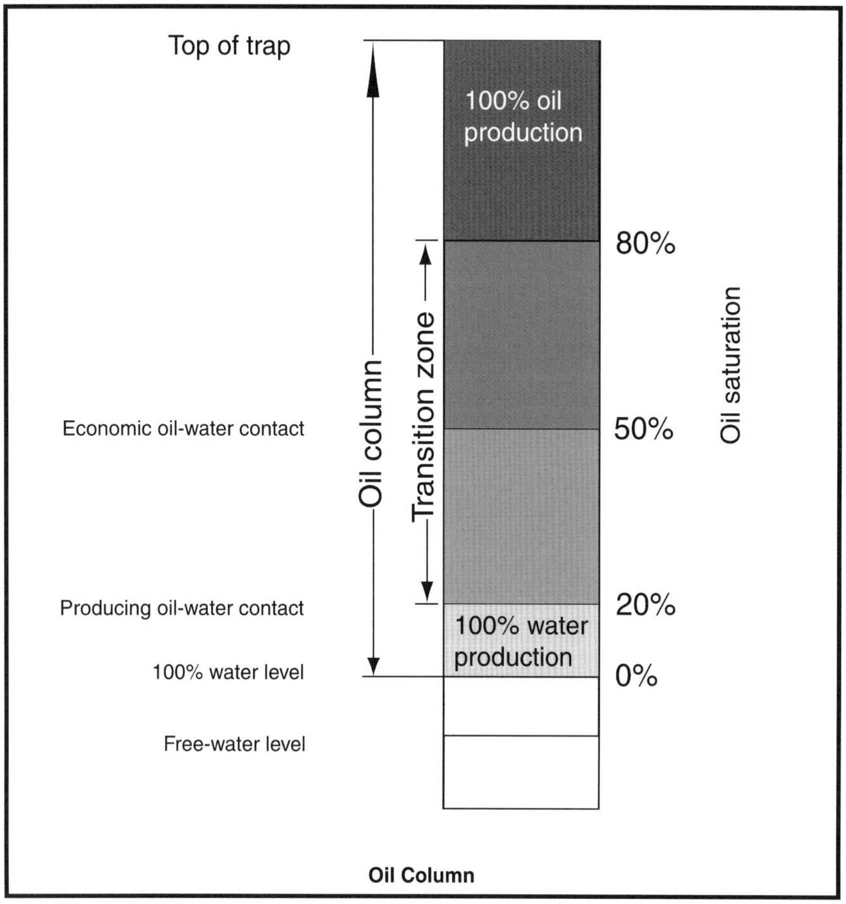

FIGURE 29. The decrease in water saturation with height controls the producing behavior of an oil column. Redrawn from Jennings, 1987, with permission from the AAPG.

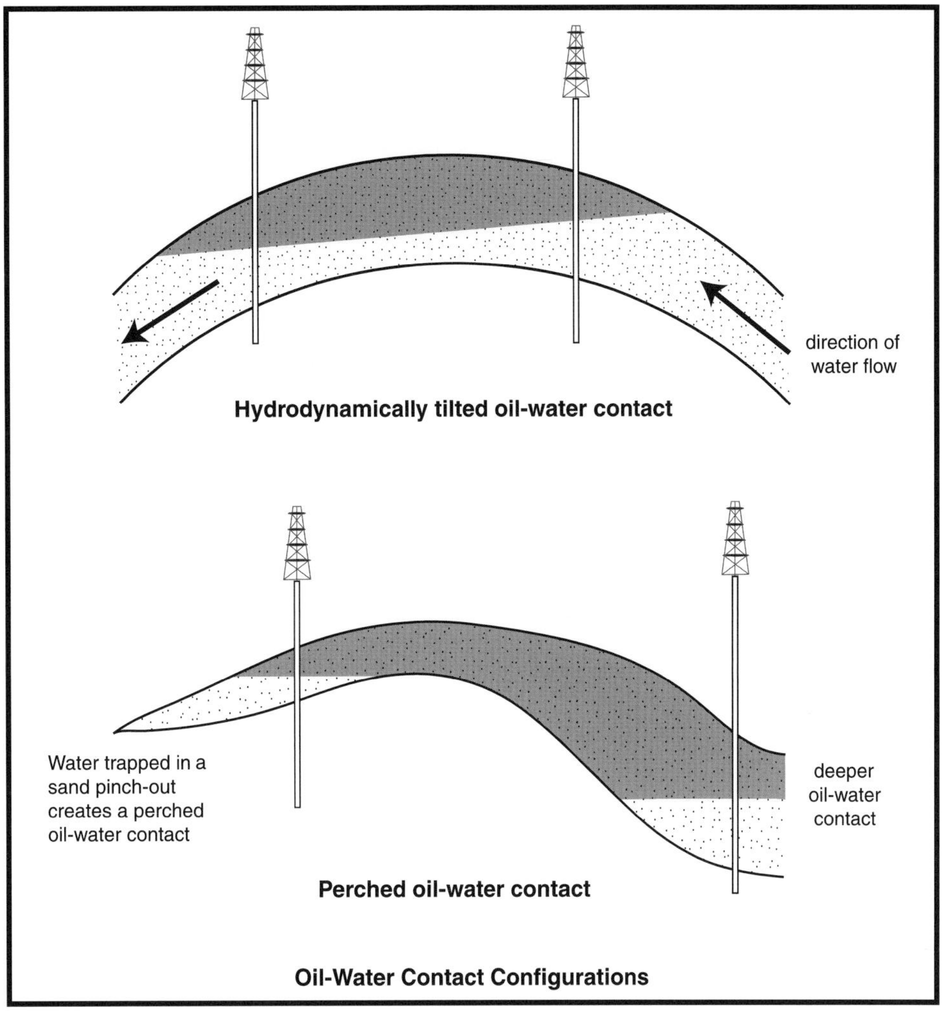

FIGURE 30. Different configurations of oil-water contact are shown. A hydrodynamically tilted oil-water contact results from the movement of water under the oil column. A perched oil-water contact forms where water is locally trapped when the oil migrated into the reservoir.

trapping of small to moderate volumes of water when the oil initially migrated into the reservoir. Normally, the water will be displaced down and sideways as the oil enters. However, if a barrier prevents the water from being moved out of the way, the water will remain where it is.

Sometimes a localized perched oil-water contact can be found more than 50 m (164 ft) higher than the established oil-water contact. For example, in the Fulmar field, UK North Sea, a perched oil-water contact in the north of the field was found at 3228 m (10,590 ft) true vertical depth subsea (TVDSS), 73 m (240 ft) higher than the main oil-water contact at 3301 m (10830 ft) TVDSS. Pressure and production data indicate communication within the oil column between the two areas (Stockbridge and Gray, 1991).

The most common type of perched oil-water contact is a downwarped thin reservoir with the downward flow of water blocked by a sand pinch-out or a sealing fault. Local synclinal areas flanked by sealing faults can also retain water (Weber, 1995).

5

Shepherd, M., 2009, Factors influencing recovery from oil and gas fields, *in* M. Shepherd, Oil field production geology: AAPG Memoir 91, p. 37–46.

Factors Influencing Recovery from Oil and Gas Fields

INTRODUCTION

As mentioned at the start of the book, more oil is left behind in oil fields than will be recovered from them by the end of their field life. Numerous factors influence recovery from an oil field including the geological complexity, fluid physics, and economics. Certain operations can be carried out to enhance oil recovery by changing the physical and chemical nature of the formation fluids. The factors influencing gas recovery are also discussed in this chapter. Gas field recoveries are significantly higher than is the case with oil fields.

RECOVERY FACTORS

Oil companies will want to maximize the value of a field by getting as much of the hydrocarbons out of it as possible. However, it is not feasible to recover all of the hydrocarbons from a reservoir. Only a certain percentage of the total hydrocarbons will be recovered from a field, and this is known as the *recovery factor*.

Recovery factors are higher in gas fields than they are in oil fields. Typical recovery factors for gas are about 50–80% (Jahn et al., 1998). There is more scope to improve oil recovery. Global recovery factors for oil are thought to be in the range of 30–35% (e.g. Conn, 2006). If, for example, you can recover 35% of the oil from an oil field, why can you not produce the other 65%? As mentioned earlier, the answer to this is not simple. The magnitude of the recovery factor for an oil field depends on a complex interplay of geological, physical, and economic elements.

A starting point is to look at the various categories of oil volumes within a typical oil field (i.e., a waterflooded oil field) and represent them on a *maturity pie chart* (Figure 31). These categories are residual oil, cumulative production, remaining recoverable reserves, and unrecovered mobile oil (UMO).

Residual oil saturation is the component of the oil that remains trapped within the pores after an oil-bearing sandstone has been swept by water. Somewhere between about 15 and 35% of the total oil in sandstones can end up as residual oil.

The second category comprises *ultimate recoverable oil*; this is the reservoir engineer's best estimate of what the field will produce by the predicted end of field life. This figure can be split into the volume of oil that has been produced so far (the *cumulative production of hydrocarbons to date)*, and the estimate of what is left to produce (the reserves).

The last category is unrecovered mobile oil (UMO), oil that is movable by primary recovery or water injection, but which will be left behind at the end of field life under current reckoning (Tyler and Finley, 1991). If an oil company wants to improve the recovery factor in a field, then this category is where the oil will normally come from.

The unrecovered mobile oil can be subdivided into three subcategories (Figure 32). Target oil is oil that has a large enough volume to justify the cost of a well to recover it. The phrase locate the remaining oil has been used for the workflow involved in finding these volumes (Wetzelaer et al., 1996). This is discussed in more detail in Section 5 of this publication. *Marginal oil* is the category of trapped oil found in volumes just below the economic threshold to justify an infill well. These volumes will become target oil if the oil price increases or if less expensive ways can be found to access them. The third subcategory is *uneconomic oil*, small volumes of bypassed oil or low oil saturations that cannot be produced economically (Weber, 1999).

DOI:10.1306/13161187M913372

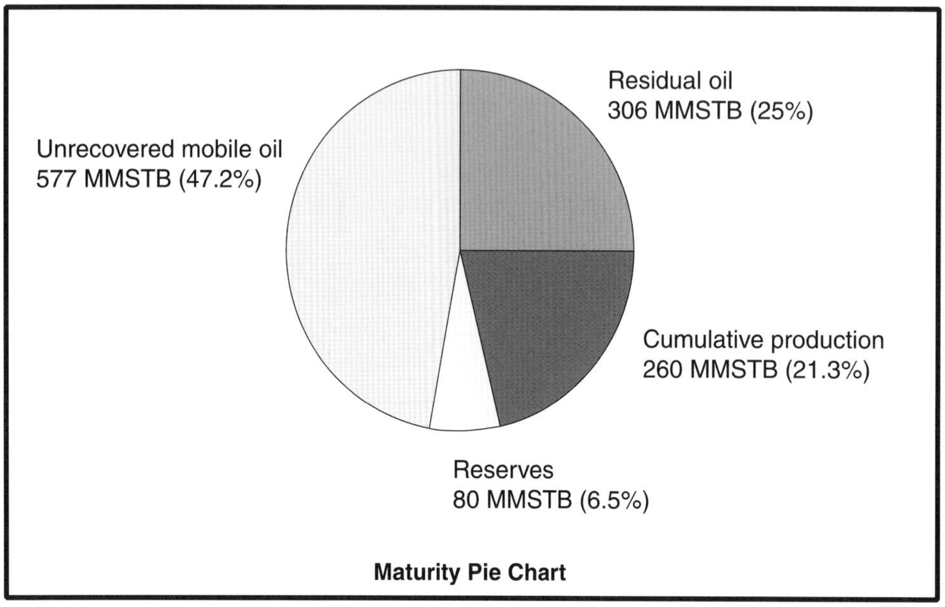

FIGURE 31. An oil volume in a waterflooded reservoir can be divided into categories of residual oil, produced oil, oil reserves, and unrecovered mobile oil. These can be illustrated on a maturity pie chart. Unrecovered mobile oil is the remaining mobile oil left behind at the end of field life if nothing is done to target it. The maturity pie chart illustrates the volumes from the Miocene reservoirs of the Miocene Norte Area, Lake Maracaibo, Venezuela (modified from Ambrose et al., 1997),with permission from the Society of Petroleum Engineers'.

GEOLOGICAL FACTORS CONTROLLING RECOVERY

A key variable controlling the amount of oil recovered from a field is the degree of *geological heterogeneity*. Oil will tend to be stranded within dead ends and low-permeability rock intervals as a consequence of this heterogeneity. An example of a depositional dead end is a back-barrier sandstone thinning and pinching out updip within a lagoonal shale (Figure 33). Patterns of depositional dead ends like this commonly repeat in different fields with similar depositional environments (see Section 7 of this publication for a detailed discussion).

Analysis by the Texas-based Bureau of Economic Geology on Texan oil fields indicates that the type of depositional environment has a major influence on the recovery factor in a reservoir. The less complex and more continuous depositional environments such as barrier-island and wave-dominated deltas commonly show recovery factors of more than 50%. By contrast, the more complex environments, such as fluvial-dominated

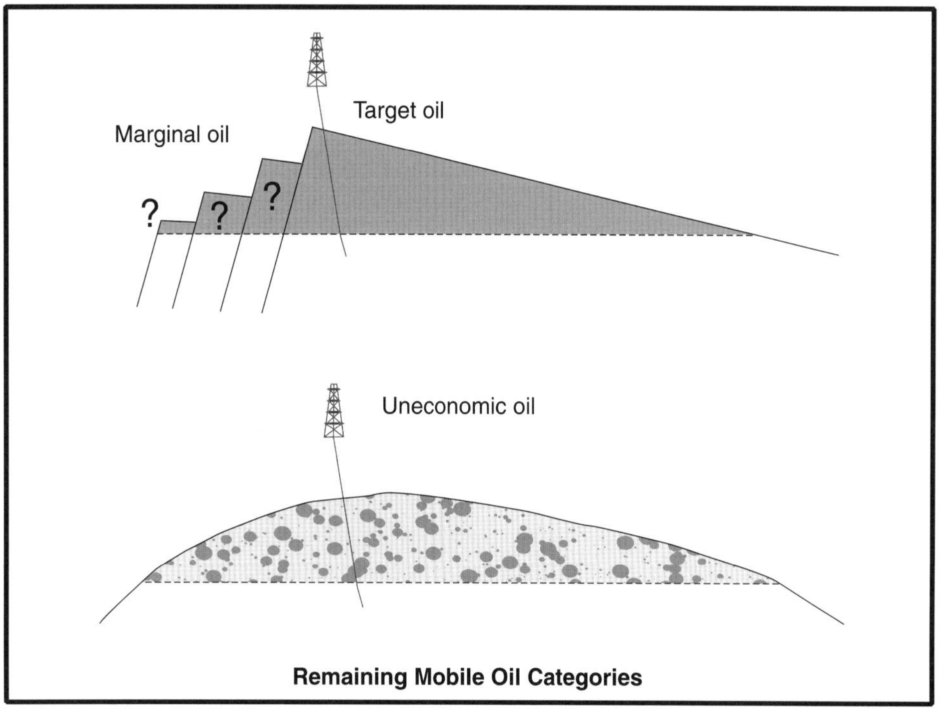

FIGURE 32. The remaining mobile oil in a field can be subdivided into three categories. Target oil columns are large enough to drill with a new well. Marginal oil columns are just below the threshold of profitability to justify an infill well. Uneconomic oil comprises bypassed volumes or patches of low oil saturation that cannot be produced economically.

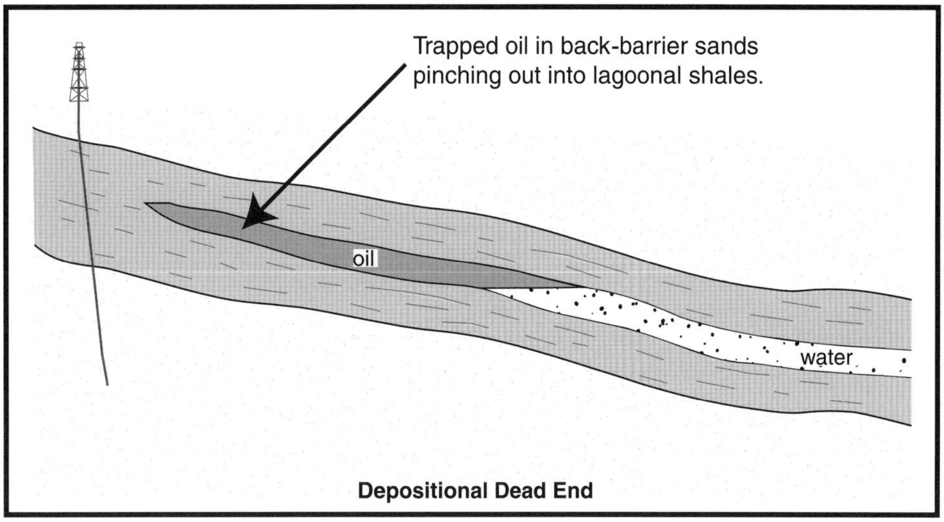

FIGURE 33. A depositional dead end within a barrier-bar depositional environment.

deltas, show recovery factors of between 20 and 40% (e.g., Tyler and Finley, 1991). Carbonates tend to show lower recovery factors than siliclastic reservoir sediments (e.g., Sun and Sloan, 2003).

An extra degree of complexity will result if the reservoir rock has significant volumes of diagenetic cement, particularly pore-filling cement. *Diagenesis*, processes that modify sediments after deposition, can create barriers and baffles within a reservoir in addition to those resulting from primary depositional heterogeneity. Moderate volumes of cement may not cause too many problems with recovery from reservoirs in thick, continuous sandstone intervals. However, in depositional systems where the flow pathways in the reservoir are tortuous and through restricted sand-on-sand apertures, pore-filling cement can destroy large-scale connectivity. The result may be a reservoir with numerous, small, disconnected compartments.

Structural complexity influences the recovery factor from oil fields. Heavily faulted reservoirs will contain numerous structural dead ends, especially if the faults are sealing. If there is a low density of widely spaced sealing faults, the drainage volumes may still end up large enough to remain as oil targets. With an increasing density of faults at a closer spacing, there will be a greater number of marginal and uneconomic volumes, with less target oil volumes.

Where faults are nonsealing and conductive to flow across them, they can increase reservoir connectivity in certain situations. Small nonsealing faults, cutting thick-layered, high net-to-gross reservoir intervals, can create vertical connectivity. However, faults will tend to disconnect reservoirs comprising thin, low net-to-gross channelized systems (Bailey et al., 2002). A network of open fractures can also create widespread connectivity in highly heterogeneous reservoirs such as the more complex carbonate systems.

One type of structural dead end is an *attic oil* accumulation. This is where oil is trapped by a structural culmination above the highest producing interval in a well (Figure 34).

PHYSICAL FACTORS CONTROLLING RECOVERY

Oil Recovery from Primary Depletion

When hydrocarbons are produced from a reservoir, the fluid pressure decreases. As the reservoir pressure is the force pushing the hydrocarbons up to the surface, production rates will start to fall off at the wellheads. Nevertheless, there are mechanisms of natural energy inherent within the reservoir itself, which help to reduce the rate of pressure decline in the wells (Figure 35). The magnitude of this reservoir energy can have a significant influence on primary recovery factors (Levorsen, 1967; Sills, 1992).

A major source of energy is supplied by a large water aquifer in direct contact with an oil zone. This is known as *water drive*. As the oil is produced and the pressure drops, the low-pressure area resulting from production spreads outward into the aquifer. Water has a small compressibility, and the aquifer water will expand as the pressure decreases, flowing into the pore space previously occupied by the oil. Because water compressibility is small, a large aquifer is required for the increase in the volume of the water to be big enough to significantly compress and displace the oil toward the production wells. The volume of aquifer should be at least 10 times the volume of the oil in the oil leg (Jahn et al., 1998). If the water is part of an artesian system with free flowing water, this can also provide a significant source of energy. The primary recovery of oil from water drive reservoirs can be high (35–75%) (Clark, 1969).

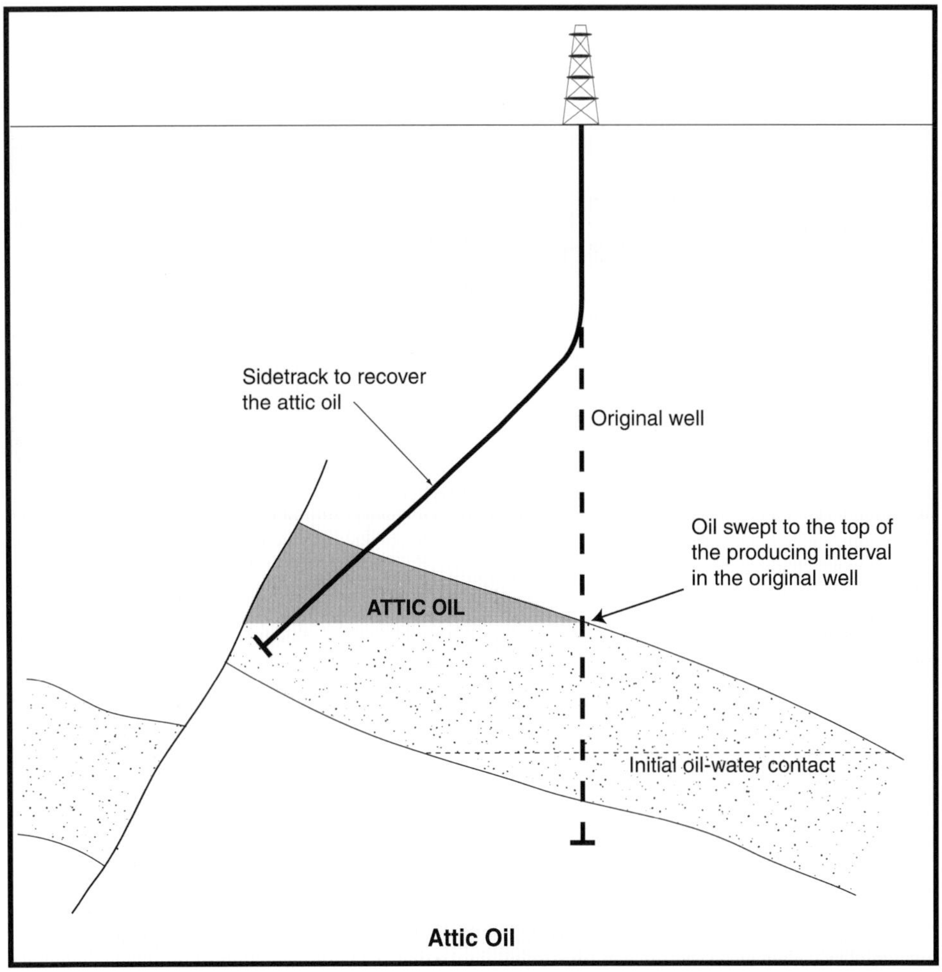

FIGURE 34. Attic oil is oil trapped in a structural dead end above the highest perforated interval in a well. The well can be sidetracked updip to recover this oil.

Water drive is a characteristic of reservoirs with laterally extensive reservoir continuity. A study of fields in Texas found that barrier-island, shoreline, and wave-dominated delta sand bodies, which extend over large areas, show strong water drives with high oil recoveries (Ambrose et al., 1991). A classic example of a water drive reservoir is the giant East Texas field in the United States (Halbouty, 1991). Layer-cake reservoir geometry, high permeabilities, and a large aquifer serve to create an effective water drive. This has resulted in a very high recovery factor with 81.8% of the 7326 MMbbls of oil in place expected to be recovered. Highly heterogeneous reservoirs are less likely to be in good communication with an aquifer and will have weaker drive mechanisms.

Another source of energy in oil reservoirs is provided by gas. Gas will expand as the pressure decreases during depletion. North (1985) commented that although a barrel contains 5.6 ft^3 of oil (0.159 m^3), *solution gas/oil ratios* in the reservoir converted to surface conditions are often expressed as values of about several hundred standard cubic feet per barrel. This shows the enormous degree to which gas can be compressed within oil at reservoir pressures and the large amount of energy stored here.

Solution gas drive is a characteristic of laterally restricted reservoirs, which do not have a gas cap and are not extensive enough to have a significant aquifer. As the pressure drops with production, the oil will have a small compressibility and will expand by a limited amount. Gas in solution in the oil is liberated once the pressure decreases below the *bubble point.* When this happens, gas bubbles emerge as a separate phase from the oil. Gas has high compressibility and will expand on decreasing pressure. This results in the compression and displacement of the oil toward the production wells. Once a critical saturation has built up, the gas starts to move toward the pressure sink in the reservoir, driving some of the oil along with the gas (Dake, 1978).

Solution gas drive is a weaker source of energy than water drive. The reservoir pressure declines rapidly and continuously. Dake (1994) described production at pressures below the bubble point as "messy." Gas viscosity is typically 50 times less than oil viscosities, and gas will flow much faster than oil through the pore space. The gas is nonwetting and will move through the center of the pores, leaving much of the oil undisplaced. According to Clark (1969), primary recoveries are always low

FIGURE 35. Various mechanisms of natural reservoir drive energy can support reservoir pressures to an extent. The magnitude of this energy can have a significant influence on primary recovery factors for oil.

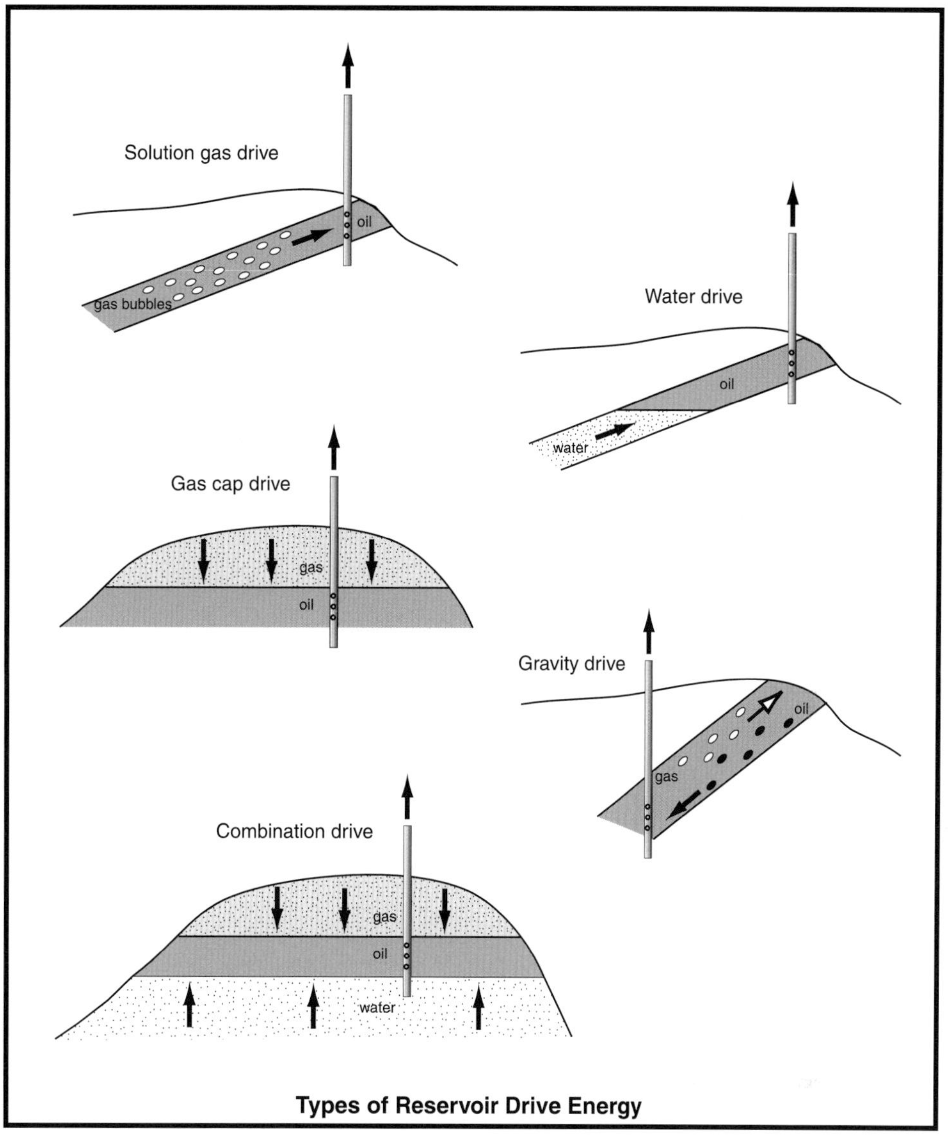

in solution gas drive reservoirs, normally in the range of 5–30% of the oil in place.

Where a gas cap exists above the oil leg, *gas cap drive* provides a source of natural reservoir energy. As the pressure drops in the reservoir, the gas cap expands and acts to slow down the rate of pressure decline. The expansion of the gas also displaces the oil downward toward the producing wells. The efficiency with which this occurs depends on the vertical permeability of the reservoir rock. Where the vertical permeability is high, significant recoveries can result. The producing wells will be perforated at some distance below the gas-oil contact to avoid the gas breaking through too early. If this happens, the wells can "gas out"; that is, they will produce only gas and none of the remaining oil. Pressures are maintained more efficiently with a gas cap drive than in a solution gas drive reservoir. Primary recoveries are in the order of 20–40% (Clark, 1969).

A weak source of energy results from *compaction drive*. Reduction in pore pressure with production results in an increase in the *effective stress* as the weight of the rock lying above the reservoir is incrementally transferred to the grain framework of the reservoir. Although this happens to some extent in most producing fields, the effects are more pronounced in relatively unconsolidated reservoir rock. The pores compact in response to the increased effective stress, compressing the contained fluids and giving some support to the reservoir pressure. An example of compaction drive is the San Diego Norte Pilot Project from the Orinoco Heavy Oil belt of Venezuela (de Rojas, 1987). The reservoir sandstones are friable with high porosity and permeability. An analysis of the *rock compressibility* indicates that the oil recovery resulting from compaction could be 8%. An additional 4% is expected to come from solution gas drive, leading to a total recovery factor from primary production of 12%.

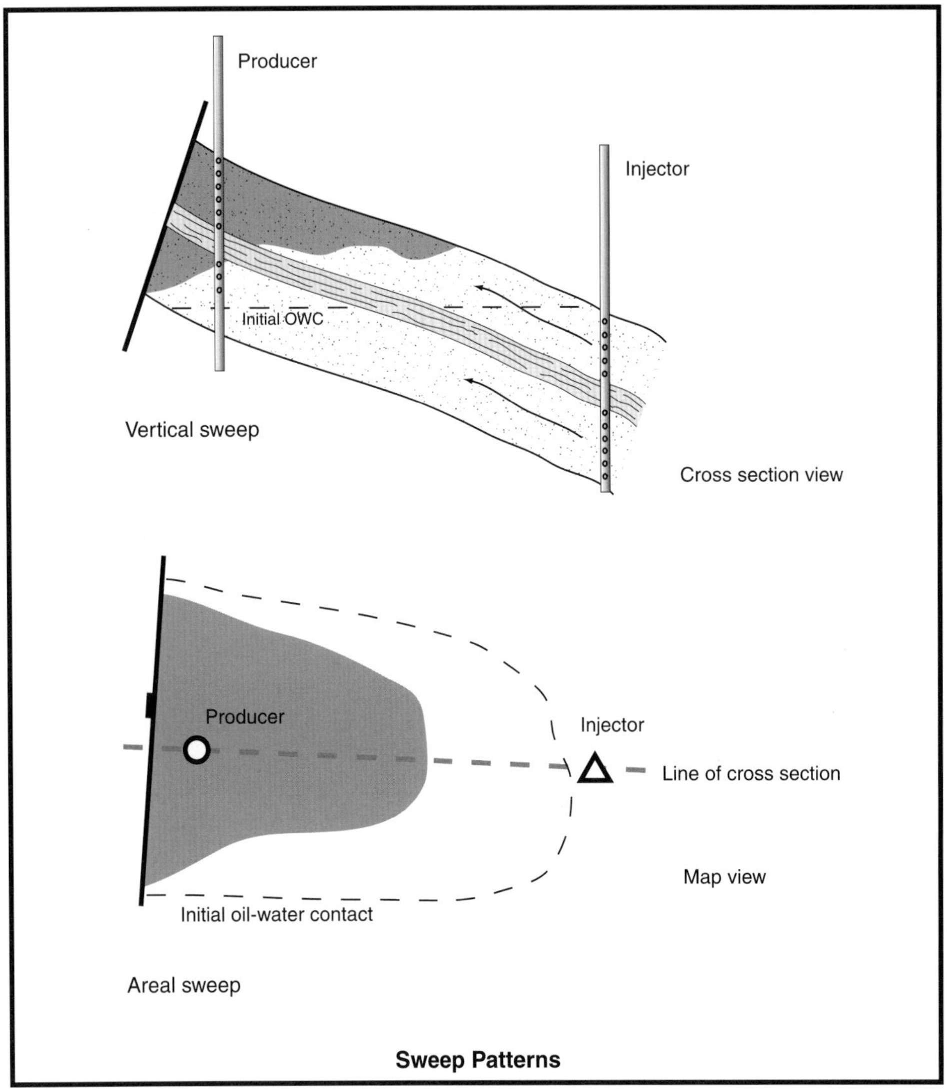

FIGURE 36. Areal sweep is the fraction of the areal extent of the reservoir that has been contacted by injected fluid. Vertical sweep is the fractional part of the reservoir cross section that has been contacted by injected fluid. OWC = oil-water contact.

Gravity drive results from the segregation of oil and gas in a reservoir because of their density differences. It is particularly effective in thick, high-permeability reservoirs or thin reservoirs with steep dips (Sills, 1992).

Many reservoirs show significant primary production resulting from more than one of these processes and this is referred to as a *combination drive.*

Oil Recovery from Waterflooded Reservoirs

Waterfloods increase the recovery from oil fields. Reservoir engineers often refer to the *volumetric sweep efficiency* of a waterflooded reservoir. This is the fraction of the total pore volume in a given part of the reservoir that has been contacted by the injected fluid (Craig, 1971). Common terms used are *areal sweep* and *vertical sweep* (Sarem, 1992). Areal sweep is the fraction of the areal extent of the reservoir that has been contacted by the injected fluid. Similarly, vertical sweep is the fractional part of a reservoir cross section that has been contacted by injected fluid (Figure 36).

The waterflood performance in water-wet reservoirs is largely controlled by the permeability layering at the bed and laminae scale. Water will edge ahead quickly through high-permeability intervals and more slowly through lower permeability rocks. There will then be some readjustment as capillary forces pull water into the smaller pores of the lower permeability intervals; in turn, oil is displaced into the higher permeability rock. The absorption of the wetting phase into a porous rock is called *imbibition.* Once the displaced oil finds a high-permeability pathway through a continuous stream of oil, there is an increased probability of the oil being produced. If the hydrocarbons are produced too quickly, the displacing water volume will advance too fast for efficient recovery by this mechanism. Consequently, many isolated volumes of oil will be left behind in low-permeability rock after the production wells have watered out (Buckley and Leverett, 1942). Thus, in waterflood reservoirs, the ultimate recovery is sensitive to the offtake rate. In reservoirs with significant vertical permeability variations or fractures, very high initial production rates can lead to rapid water breakthrough, poor sweep efficiency, and lower than expected recoveries. Practical reservoir management involves finding a balance between economic production rates and maximizing the recovery.

Cross-bedded sandstones are not efficiently swept because the foresets and the low-permeability bottomsets can act to impede flow (Weber, 1982). In cross-bedded sandstones, alternating finer and coarser grained foreset laminae can result in significant quantities of capillary-trapped oil (Kortekaas, 1985; Corbett et al., 1992). Weber (1982) quoted an early article by Illing (1939) in which it is recognized that the most difficult sediments to sweep with water are those with numerous coarse and fine interfaces. The low-permeability, finer grained laminae rapidly imbibe water and physically trap oil in the coarser grained laminae. The oil is effectively immobilized in the coarser grained laminae as the interfacial tension between the water in the finer grained laminae and the oil prevents the oil from moving through the pores. Oil is produced more readily parallel to the cross-bedding than across it because of this effect. Huang et al. (1995) found that between 30 and 55% of oil was trapped this way in a coreflood experiment on cross-laminated eolian sandstone under conditions of low-rate flooding. Van Lingen and Knight (1997) considered that because of the predominance of cross-bedding in meandering fluvial sediments, capillary-trapped oil could range from 10% to more than 40% of the movable oil volume, depending on the flow direction and the effect of the bottomsets. In braided river systems, the effect is slightly less with an estimated 10–20% capillary-trapped oil. For shoreface sediments, an estimate of 5% capillary-trapped oil is made (Van Lingen and Knight, 1997).

THE EFFECT OF OIL VISCOSITY ON RECOVERY

Oil viscosity has an impact on the recovery factor. Water will readily displace low-viscosity oil to form a *stable flood front*. The oil is pushed ahead of an extensive cushion of water.

Where the oil is heavier and more viscous, the water will tend to finger through the oil column in an irregular manner, breaking through to the production wells rapidly. Large volumes of water will need to be circulated through the reservoir in order to obtain economic oil recovery. This may not be practical offshore given the high production rates required to keep the infrastructure profitable (Dake, 1994). Onshore, it is more efficient to use methods such as steam flooding or in-situ combustion to recover viscous, heavy oils.

ENHANCED OIL RECOVERY

At the mature phase of field life, methods of *enhanced oil recovery* (EOR) may be instigated. This is also called *tertiary recovery*. EOR projects are designed to change the fundamental physics or chemistry of the reservoir conditions in order to improve the recovery. The method used will depend on the fluid type and the reservoir.

The most common EOR operation uses *thermal methods*, involving steam, heat, or combustion to improve oil recovery. These account for 70% of the world's production by EOR techniques (Nind, 1989). Thermal methods are used for recovering heavy (and viscous) oils with gravities between 10 and 25° API units (Nind, 1989). The operation is used in areas with heavy oil such as Venezuela, Canada, the United States, Russia, China, and Indonesia.

Steam can be continuously injected as a flood called a *steam drive*. An alternative method is *steam soaking*, also known as *huff and puff*. This involves a cyclic operation whereby steam is injected into a production well, allowed to soak for a few days to distribute the heat, and is then followed by a period of oil production from the well. Production is increased by several mechanisms. The steam heats the oil and reduces the viscosity allowing it to flow more easily. In addition, the oil expands by swelling, and changes in the surface tension also improves the flow (Briggs, 1987).

In-situ combustion of the oil in the reservoir has also been used, a technique sometimes known as *fire flooding* (Matthews, 1983; Briggs et al., 1987). The oil is ignited in the subsurface with the fire fed by a continuous supply of air via an injector well. The resultant combustion front moves away from the air injection well toward the production wells. The heat of the fire reduces the oil viscosity and vaporizes the water within the reservoir to steam.

For lighter oil, *miscible drive* operations are used for incremental oil recovery. The idea is to inject a fluid such as methane, liquid petroleum gas, CO_2, or nitrogen that is miscible with the oil phase and thus reduce or eliminate the interfacial tension between the injected fluid and the oil. The oil mixes with the injected miscible fluid and flows readily to the producers (Nind, 1989). Carbon dioxide miscible floods have proved so effective in oil fields in North America that oil companies are willing to transport carbon dioxide long distances to enable this. For example, a 205-mi (330-km)-long pipeline brings CO_2 from a coal gasification plant in North Dakota across the United States-Canada border to the Weyburn field in Saskatchewan.

A variation on the theme is a *water alternating gas* (WAG) flood (Christensen et al., 1998), which is a period of water injection alternated with a period of miscible gas injection. A WAG flood is in operation in the

Magnus field in the UK North Sea; water is injected into three wells for a 6-month cycle followed by gas injection for 6 months (MacGregor and Trussell, 2003). The water injection provides a stable floodfront to sweep the oil, whereas the gas displaces residual and bypassed mobile oil at the pore scale. The combined effect can be the recovery of significant volumes of oil. An increase in recovery of between about 5 and 20% has been reported in reservoirs using WAG floods. Examples of where WAG schemes have been implemented include the Dollarhide, Rangeley Weber, and Slaughter Estate fields in the United States.

Polymer flooding is an operation whereby suitable chemicals are added to injection water to increase the viscosity of the waterflood. The use of polymers is intended to create a more stable flood front and thus improve recovery in fields containing moderately viscous oil (Jahn et al., 1998).

Much effort was put into investigating the use of surfectants as an EOR method in the 1980s. Surfectants were added to injection water to reduce the water-oil interfacial tension. Surfactants can be expensive to use in quantity. More recent methods involve combining surfactant with alkali and polymer chemicals. The alkali chemicals react with acids in the oil to form surfectants within the reservoir. The polymer helps to move the mixture along with the water flood.

Bacteria have also been used to produce incremental oil recovery. Bacterial activity in the reservoir can release gases, polymers, acids, surfactants, and other compounds that may mobilize oil (Moses and Springham, 1982).

ECONOMIC FACTORS

An important economic factor controlling the recovery from fields is whether they are onshore or offshore. Wells are much cheaper to drill onshore, and the overall cost of the operation is substantially less.

Recovery factors are higher for onshore fields compared to offshore fields. Onshore fields tend to be drilled with a closer well spacing than is practical offshore. Typical well spacings are 200–500 m (656–1640 ft) onshore and 500–1000 m (1640–3280 ft) offshore (North and Prosser, 1993). The greater density of wells in onshore fields increases the chance that oil in a reservoir dead end will be found when a producer is drilled (Weber, 1999).

The second reason for better recoveries onshore is that the wells are profitable much longer than offshore wells. For instance, it has been estimated by the U.S. Department of Energy that 20% of all the oil produced in the United States comes from wells producing less than 15 BOPD. No offshore well would make any money from rates as low as this. Offshore fields are expensive to run and will be shut in as uneconomic even when the oil production rate is still relatively high. Production tends to decline asymptotically in a predictable manner, and when an offshore field is abandoned at a high rate of production, there is a long tail of potential production beyond this point that would be economic onshore.

UNECONOMIC OIL

The volume deemed to be uneconomic oil is sensitive to the prevailing economic environment. Oil price, equipment costs, taxation, and other factors will determine the nature of oil field economics.

A subsurface team can influence economic factors so as to produce more hydrocarbons. An example of this is the Angus field redevelopment in the UK Central North Sea. Previously produced through an FPSO vessel, the field was shut in and abandoned once the oil rate had dropped below 7000 BOPD. The *operating expenditure (OPEX)*, which is the cost of the vessel, manpower, and associated logistics, was considered too high at these oil rates for the project to make an economic return. The field was reopened 7 yr later. By building a pipeline back to another producing field, the OPEX was minimized as the hub field covered this. The new operation involved the *capital expenditure (CAPEX)* of building a 21-km (13-mi) pipeline and the drilling of a new well. Renewed production from the Angus field paid back the CAPEX within 6 months. Production thereafter continued to provide revenue. The expected reserves were an extra 5.2 MMbbls of oil that had been previously categorized as uneconomic movable oil (Shepherd et al., 2003).

Another way of changing the economic environment is to reduce the drilling costs using cheaper techniques such as coiled tubing or through tubing drilling (see chapter 28 of this publication). The reduced costs can change marginal opportunities into economic targets (Figure 37).

RESERVOIR MANAGEMENT OF GAS FIELDS

Gas fields are managed differently from oil fields. Oil is relatively easy to transport in bulk volumes long distances, whereas this is difficult for gas, unless a very expensive liquid natural gas (LNG) plant is built or there is an extensive regional gas pipeline network. A gas field will typically be developed once a *gas sales contract* has been made to supply the gas to customers living close to the gas field. The contract will involve a commitment to supply a daily volume of gas over a certain period of time. Thus, there is a requirement to be reasonably sure of what a gas field will produce before it starts production. It will be necessary to test every well preproduction to

FIGURE 37. Factors influencing why oil is left behind in oil fields.

Why Oil is Left Behind in Oil Fields

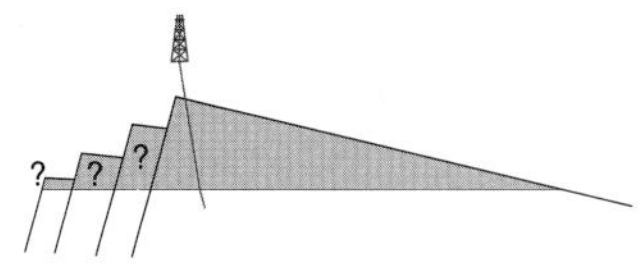

Geological Complexity

- Reservoir dead ends will trap oil uncontactable by existing wells
    - structural dead ends
    - stratigraphic dead ends
- Number of reservoir compartments
- Permeability layering

Fluid Physics

- Residual oil saturation
- Capillary-trapped oil in cross-bedded sandstone
- Oil viscosity
- Type of reservoir drive mechanism
    - water drive
    - solution gas drive
    - gas cap drive
    - compaction drive
    - gravity drive

Economics

- Oil price
- Taxation, etc.
- Onshore vs. offshore
- Operating expenditure for the infrastructure
- Number of wells
- Wells become uneconomic at low production rates (varies onshore vs. offshore)

get a good knowledge of well productivity and flow rates (Ikoku, 1984).

Recovery factors are higher for gas than they are for oil, commonly in the range of 50–80% (Jahn et al., 1998). The recovery factor for gas fields is dependent on factors such as the *abandonment pressure*, the *initial pressure*, and the type of reservoir drive mechanism. Recovery can also be sensitive to the engineering of the surface plant. In big gas fields, the installation of compression equipment can lead to higher recoveries.

A key property of a gas is its *compressibility*; gas compresses readily with increasing pressure. Conversely with decreasing pressure, the gas will expand. The measure of how much a gas will expand between the reservoir and surface conditions is the *gas expansion factor*. A typical value for this is 200 (Jahn et al., 1998). Expansion is the main mechanism by which gas is produced to the surface. Once the pressure drops to reduced levels, then surface flow rates may be too low to be profitable. This is the abandonment pressure, which effectively defines the economic limit of flow from a gas field.

One difference between gas and oil is that, as the pressure decreases, oil with its limited compressibility will stay trapped in a reservoir dead end, whereas gas may not remain trapped for too long. The gas will expand on decreasing pressure and a significant proportion of the trapped volume will eventually escape around the edges of the dead end. This is an important factor contributing to high recovery factors in gas fields.

Another factor is that gas has a lower viscosity than oil and will flow through low-permeability rocks that would not produce oil. Hence, gas can be produced economically from poorer quality reservoir rocks. A wider spectrum of rock types will produce gas by comparison to oil.

A strong water drive is unfavorable to gas recovery as water breakthrough to the production wells will make flow rates sluggish and uneconomic at higher pressures than with closed gas reservoirs (Ikoku, 1984). The intensity of the water drive can be a major factor behind the ultimate recovery of gas; a slower encroachment of water will result in higher recoveries. Permeability is also a critical factor in gas reservoirs. Higher permeability results in a high flow rate for a given pressure drop. Thus, the abandonment pressure can be lower for a high-permeability gas reservoir (Ikoku, 1984).

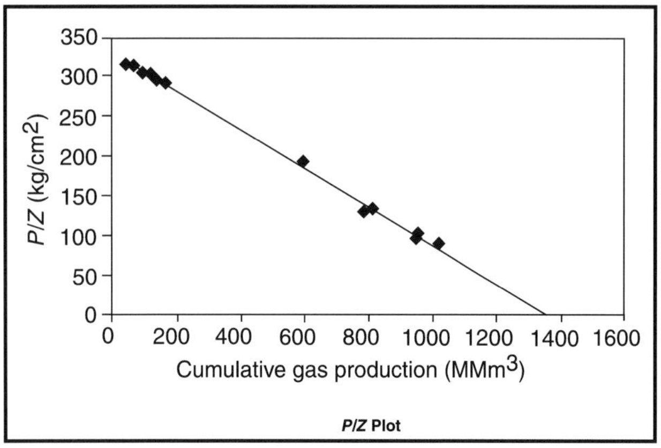

FIGURE 38. *P/Z* plots are used by reservoir engineers to estimate the contacted volumes in a gas field. This example is from the Novillero gas field in the Veracruz basin, Mexico (from Holtz et al., 2002), reprinted with permission from the Gulf Coast Association of Geological Societies. MMm^3 = million cubic meters.

Gas reservoirs without an aquifer have been called *volumetric (or depletion) reservoirs*. This is because the elementary physics of the gas laws allow the volume of gas in the reservoir to be calculated once a certain amount of gas has been produced. Recoveries are higher (~80–90%) as pressure depletion is much more efficient than a water drive regime, because the wells do not load up with water.

A material balance technique used by reservoir engineers to estimate gas volumes is the *P/Z plot* (Dake, 1978). This is based on the gas law relationship whereby if the volume of a gas is reduced within a closed system, then the pressure will drop in a predictable manner. Two parameters are crossplotted on a graph: pressure, divided by the gas deviation (dimensionless compressibility) factor *Z*, used in the equation for the nonideal gas law; and cumulative gas production. In closed system reservoirs, the values will plot on a straight-line trend. The trend is extrapolated to the abandonment pressure to estimate the contactable volume of gas (Figure 38). For example, in the Novillero gas field in the Veracruz basin of Mexico, the *P/Z* plot extrapolates to the base line at a total gas volume of close to 1400 MMm^3 (49.4 bcf) (Holtz et al., 2002). If an aquifer is present, the system cannot be considered closed, and the *P/Z* plot will deviate from a straight-line trend.

GAS CONDENSATE

Gas condensate is a type of petroleum fluid that exists in the reservoir as a gas at initial conditions but once the pressure drops to the dew point, liquids will start to condense (Jahn et al., 1998). Produced to the surface, both gas and condensate liquids are separated out in the production plant.

Pressures can fall below the dew point in the reservoir, with liquid condensate dropping out near the well bore. This *condensate banking* can lead to a reduction in the gas relative permeability, and gas flow rates can drop off significantly as a result (Ayyalasomayajula et al., 2005). Therefore, there is an incentive with a gas condensate reservoir to keep the pressure above the dew point so as to prevent this from happening. A typical strategy is to recycle gas back into the reservoir to maintain pressure.

Shepherd, M., 2009, The geological scheme, *in* M. Shepherd, Oil field production geology: AAPG Memoir 91, p. 47–48

Section 2

The Geological Scheme

GEOLOGICAL SCHEMES AND GEOLOGICAL MODELS

A major role for the production geologist is to establish the *geological scheme* for the field. This is the conceptual scheme, which depicts the sedimentological and structural configuration of the reservoir. Once a geological scheme is in place, it can be integrated with production data to produce a *flow geology scheme*. This is a description of how the various geological elements in the reservoir control the patterns of fluid flow. The conceptual geological scheme for the larger fields will be represented as a *geological model* on a computer to be used by the geologist and the rest of the subsurface team. A computer representation can be built either as a series of 2-D maps or as a 3-D geological model (Figure 39).

The geologist will use these models to locate and target the remaining hydrocarbons within a field. Any large pockets of oil identified in this way will be the basis for infill well drilling. The geological model can also be used to provide the framework for the reservoir engineer's *simulation model*. The reservoir simulation tracks fluid flow within the reservoir and is the basis for predicting future production and reserves.

The workflow given here follows through from establishing a geological scheme to building a geological model and then planning new wells. Details of each stage can be found in this publication in the chapters indicated:

1) All the relevant data are compiled (Chapters 6, 7).
2) Maps and cross sections are drawn to illustrate the internal geometry and rock property variation within the reservoir (Chapter 8).
3) The lithofacies are identified in the cored wells. The equivalent log facies to the lithofacies in cored wells are determined so that these can be recognized on logs in uncored wells (Chapter 9).
4) The gaps between the wells need to be filled in with an inferred 3-D lithofacies scheme, the basis for much of the reservoir description. The initial starting point for this is to establish a sequence stratigraphic framework (Chapter 10).
5) Lithofacies maps are derived for each stratigraphic sequence (Chapter 11).
6) Rock and fluid properties are analyzed in the context of a lithofacies scheme (Chapter 12).
7) The structural framework is then delineated (Chapters 13, 14).
8) The known flow behavior of the reservoir is then integrated with the geological scheme by incorporating production and engineering data (Chapters 15–18).
9) Hydraulic units (Chapter 15) and areal compartments (Chapter 16) are identified, and both are combined to define drainage cells Chapter 18).
10) Rock properties are analyzed (Chapter 12) and can be represented in the reservoir by maps (Chapter 8) or by geostatistical methods (Chapter 19).
11) At this stage, the geological scheme is ready to be represented as a 3-D geocellular computer model (Chapter 20).
12) Once the geological model has been built, the hydrocarbons in place for the reservoir will be calculated (Chapter 21) and the uncertainty can be estimated (Chapter 22).
13) The computer geological model will be used by the reservoir engineers for simulation work, and some iterations may be necessary (Chapter 23).
14) The remaining oil in the reservoir is located using various techniques (Chapters 24–26).
15) This is the basis for compiling an inventory of remaining oil targets that can be screened for drilling new wells or recompleting existing wells (Chapter 27).
16) Wells are planned for the promising target locations (Chapters 28–30).

DOI:10.1306/13161188M913372

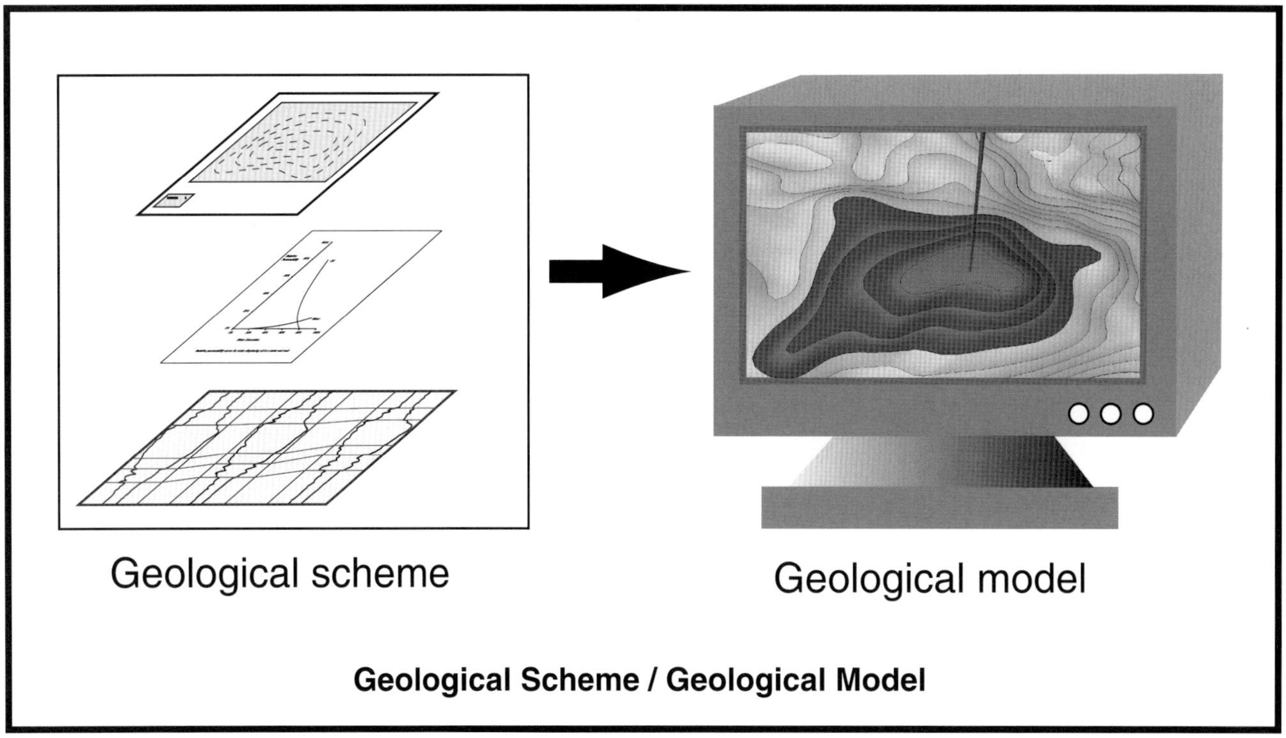

FIGURE 39. The geological scheme is the conceptual scheme for the sedimentology and structural geology of the reservoir. It can then be used as the basis for building a 3-D geological model on a computer.

Shepherd, M., 2009, Sources of data, *in* M. Shepherd, Oil field production geology: AAPG Memoir 91, p. 49–63.

6

Sources of Data

DATA MANAGEMENT

A large amount of data is available to the production geologist for reservoir evaluation. Much of the data will have been expensive to acquire, particularly if obtained from wells offshore. For instance, core taken from a drilling operation on an offshore drilling rig may have cost more than a million dollars to recover. There is an obligation to take good care of the data and to make sure that the information is accessible, either as well-organized paper data files or as data on a computer shared drive. Data files stored on a computer should be labeled with the originator's initials, a date, and some idea of the significance of the data, e.g., "MS August 31, 2008, final top reservoir depth map." Well files should be compiled with all the available data collected on a well-by-well basis. Good data management can make all the difference between a project that is well organized and effective, and one that is disorganized and inefficient.

VALUE OF INFORMATION

Obtaining data in an oil field environment is expensive; therefore, it is necessary to justify the economics of gathering the information. In the early stage of field life, *the value of information* is enormous; the data are essential for reservoir evaluation. Later on in field life, it becomes more important to justify the expense of the data. The new information should be gathered on the basis that it significantly improves the project value and reduces the company's investment risk (Gerhardt and Haldorsen, 1989).

TYPES OF DATA

A production geologist will use data from a variety of sources. These include:

- mud logging data
- core data
- sedimentology and petrography reports
- outcrop analogs/modern depositional environments
- wireline-log and logging-while-drilling (LWD) data
- production-log data
- well-test data
- fluid samples
- production data
- seismic data

MUD LOGGING DATA

The mud loggers on the rig site will monitor the drilling parameters during the well operation, and these are summarized graphically as a *mud log*. The mud log will include a lithology log. This is a depth plot showing in graphical form the percentage of the various lithologies in each cutting sample recovered while drilling the wellbore. A written description will be made for the lithology of the drill cuttings. Accompanying the lithology log is a record of the *rate of penetration* of the drill bit. This is an indication of lithology; sandstone is normally drilled faster than shale for instance. Any drilling problems encountered or changes in the drilling parameters will be reported in the margins of the mud log. The presence of oil shows will be noted. The *gas returns* and *gas chromatography analysis* are monitored and graphed

DOI:10.1306/13161189M913372

against depth. High gas returns are a sign that a hydrocarbon reservoir may have been drilled. Significant concentrations of the higher alkanes on the gas chromatograph can indicate that an oil zone has been penetrated. The mud log is used as a first pass, qualitative indication of reservoir presence and quality. A more detailed and accurate representation will be available once wireline logs have been run and interpreted.

The mud loggers also collect bags of rock cutting samples at regular intervals while the well is being drilled. These may be used later for biostratigraphic and lithological analysis (Whittaker, 1992).

FIGURE 40. Museum core on display.

CORE DATA

The geologist uses core data to provide a sedimentological description and rock property analysis for input to the geological model. Specialist service companies perform the core analysis. Rock properties such as porosity and permeability are measured on *core plugs* cut from the core. These are about 2.5 or 3.8 cm (1 or 1.5 in.) in diameter and about 2.5–5 cm (1–2 in.) long. The plugs are cut horizontally (i.e., bed parallel) at a frequency of three to four samples per meter or every foot for oil companies that use imperial measurements (Monicard, 1980). Vertical core plugs may also be cut every 1.5 m (5 ft) for example. On occasions, large pieces of *full-diameter core* are used for measuring rock properties instead of small core plugs. This can be a more meaningful way of establishing the reservoir characteristics for the more complex lithologies such as carbonates.

Other members of the subsurface team will also use the core data. The petrophysicist uses core data to calibrate the measurement of rock properties from wireline log data. The reservoir engineer obtains data for the various reservoir parameters needed to understand the physics of fluid distribution and flow. Properties such as capillary pressure and relative permeability are measured by *special core analysis*, and this is referred to by the acronym *SCAL*. The geologist will frequently get requests from the production engineer to provide core samples for laboratory tests. The aim is to ensure that the various downhole chemical treatments do not react with the rock or the pore fluid to plug up the pore space and impair productivity.

The core is *slabbed* once all the samples have been taken and the measurements are complete. It will be cut into three vertical sections down the length of the core. The middle slab is kept as a reference core for further study by the geologist. It is placed in a wooden frame and set in resin or glued to a firm base. This part of the core will be kept as a *museum core* (Figure 40). The other two sections of the core, referred to as the *half cut*, are kept for subsequent sampling.

CORING PROGRAMS

Representative cores should be taken from wells throughout the field. The aim should be to have key areas of the field covered. Ideally, all the various reservoir units should be cored. The entire reservoir interval should be cored in at least one well if practical.

Cores are commonly taken at the exploration and appraisal stage, although some of the early production wells may also be cored. It is unusual to take core at the mature stage of field development; however, there may be reasons for doing this if the value of information can be justified.

CORE ANALYSIS REPORTS

The core analysis company will provide two reports once a job has been completed. The first is the *core analysis report*. This can include the following data: *horizontal permeability*, *vertical permeability*, *porosity*, *water saturation*, *oil saturation*, *grain density*, and sometimes a brief description of the core plug lithology (Table 7). A listing will be provided on a foot by foot basis (or its metric equivalent) of the rock properties measured in the lab (Table 8).

The depths at which *preserved core samples* have been picked will also be listed. These are selected pieces of core that are kept to preserve the conditions of the rock as close to those in the reservoir as possible. They may be required for special core analysis such as wettability studies (Bajsarowicz, 1992). One preservation method is to store the samples in sealed jars containing simulated formation brine.

A *core gamma log* will also be included in a core analysis report. The gamma-ray response is measured along the length of the core in the laboratory. It is used to match up the core depths to the depths on the wireline

Table 7. Terms used in core analysis.

Core Data	*What It Means*
Permeability	The measure of the ease of movement of fluid through the pore space in a rock.
Horizontal permeability	Core plugs are cut parallel to the bedding planes in the core, and the horizontal permeability values are measured from these.
Vertical permeability	A lesser amount of core plugs is cut orthogonal to the bedding planes in the core, and the vertical permeability values are measured from these.
Porosity	The decimal fraction or percentage of the void (pore) space volume within the rock to the total rock volume.
Water saturation	The decimal fraction or percentage of the total volume of water relative to the total volume of fluid (hydrocarbon plus water) in the pore space.
Oil saturation	The decimal fraction or percentage of the total volume of oil relative to the total volume of fluid in the pore space.
Grain density	The density of the constituent grains making up the sediment in the core plug.

gamma-ray log run over the cored interval in the reservoir. These can differ from about half a meter to sometimes more than 6 m (18 ft). This is because over a distance of 2000 or 3000 m (6500 or 10,000 ft) within the borehole, the drill string to which the core barrel is attached will stretch under tension a few meters more or less than the wireline to which the log is attached. Also, incomplete recovery of core, particularly unconsolidated core, can lead to discrepancies in the core log. Comparison of the core gamma with the wireline gamma log allows the *core-to-log shift* to be determined. This is important for matching features in the core to the equivalent log response.

The second report received is the *core photography report* (Figure 41). This is a set of color photographs of the slabbed core. The geologist can keep this in the office as a substitute for a trip to the core storage location to see the actual rock. If any oil is present in the core, the core will also be photographed under ultraviolet light. Any oil-saturated intervals will show up as fluorescent patches on the photographs.

THE SEDIMENTOLOGY REPORT

It is good practice to call in an expert sedimentologist to look at the core and to provide a detailed sedimentological report. The report will include a *sedimentological log* with a detailed description of all the sedimentological features seen in the core (Figure 42). Various details will be noted (Blackbourn, 1990).

Table 8. A typical core analysis report.

Core 1 2244.80–2300.10 m (7364.82–7546.25 ft) Drilled Depth							
Sample	*Depth*	*KH**	*KV***	*CPOR*†	*CSO*††	*CSW*‡	*RHOG*‡‡
1	2244.95	838.00		15.7	52.30	3.4	2.65
2	2245.24	2180.00		16.1	46.20	3.3	2.64
3	2245.51	995.00		15.0	50.30	3.6	2.64
4	2245.78	766.00		14.9	48.50	3.5	2.65
5	2246.25	165.00		12.2	41.20	10.2	2.65
6	2246.50	474.00	390.00	15.9	51.50	3.4	2.65
7	2246.95	521.00		15.3	42.60	8.1	2.64
8	2247.75	260.00		12.4	49.70	7.2	2.65
9	2248.05	374.00		15.7	52.40	6.8	2.64
10	2248.25	88.90		14.1	50.10	10.2	2.65
11	2248.45	42.20		10.9	49.70	8.2	2.65
12	2250.42	672.00		15.3	51.30	1.2	2.65
13	2250.75			Preserved sample			
14	2300.00	76.30		9.5	47.90	5.6	2.54

**KH = horizontal permeability to air (md).*
***KV = vertical permeability to air (md).*
†CPOR = core porosity (%).
††CSO = core oil saturation (%).
‡CSW = core water saturation (%).
‡‡RHOG = grain density (g/cm^3).

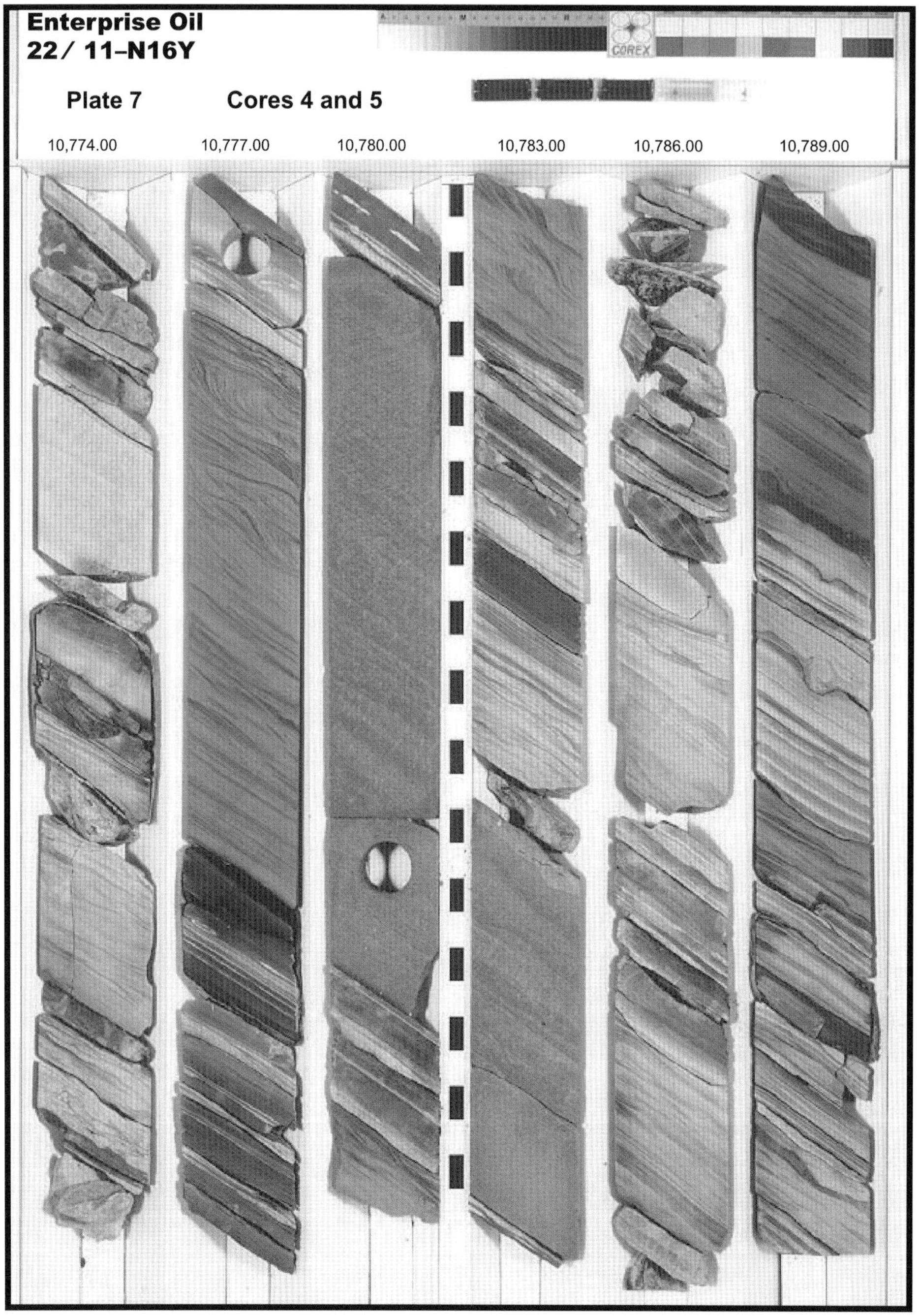

FIGURE 41. Example of a core photograph. The photograph shows the channel margin facies association from deep-water sediments of the Nelson field, UK North Sea (from Kunka et al., 2003). Reprinted with permission from the Geological Society.

These include

- lithology with graphical lithology column
- graphical representation of grain size variation
- accessory minerals and diagenetic cement
- fossils
- diagenetic features
- sedimentary structures
- bioturbation
- nature of bed contacts
- sedimentary texture
- color
- oil staining
- grain sorting
- induration
- lithofacies
- fractures, faults, and other structural features

A written account of the detailed facies description and interpretation will be provided. Interpretations are

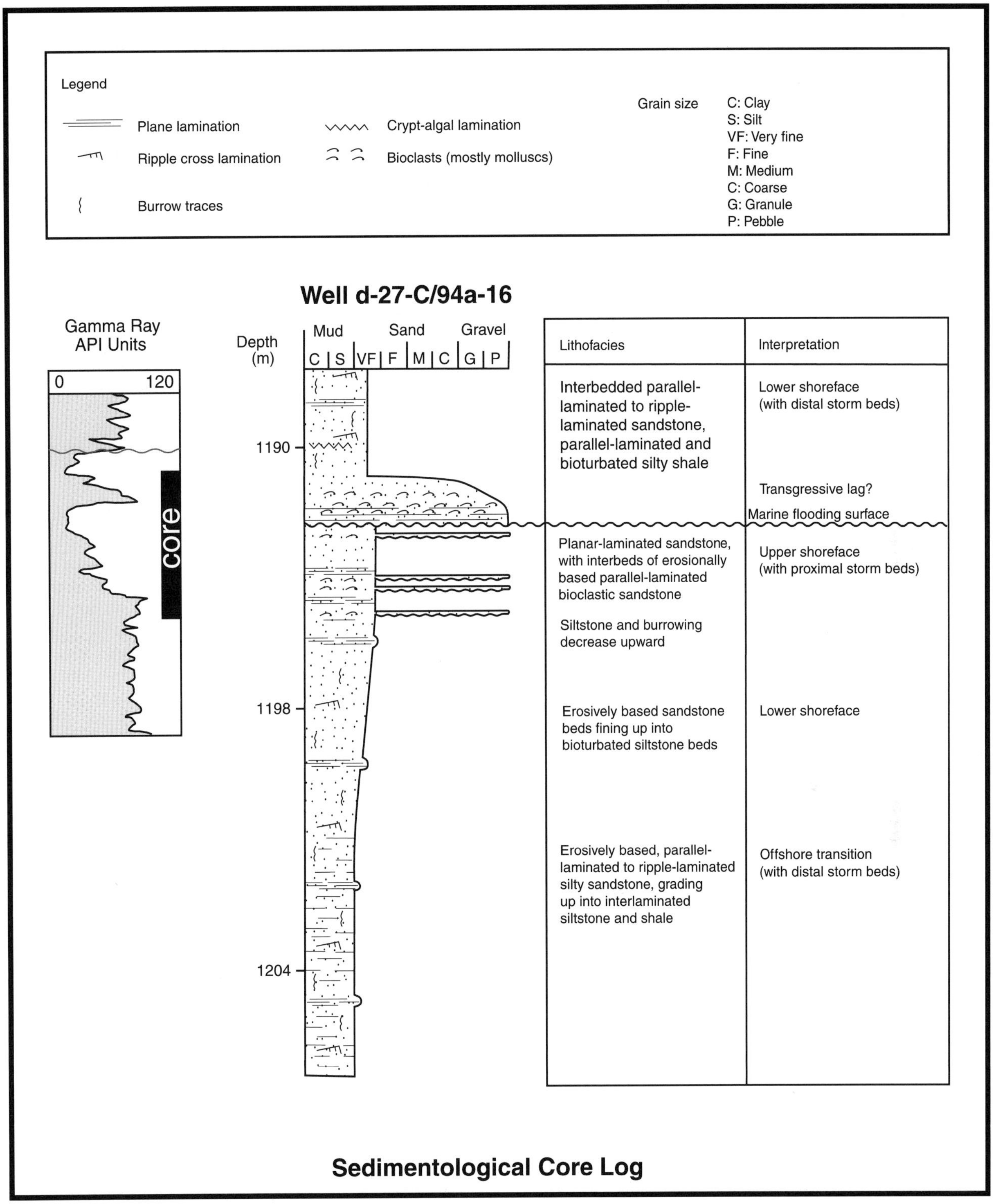

FIGURE 42. Example of a sedimentological core log, Well d-2-C/94a-16, Peejay field, Canada (after Caplan and Moslow, 1999). Reprinted with permission from the AAPG.

made as to the likely sedimentary environment of deposition. A summary of the mineralogy, petrography, porosity types, and diagenetic mineralogy should also be included. The pore scale is also important for the production geologist, especially for an understanding on permeability controls and as to whether there are

Table 9. Main open-hole log types.

Log	*How It Works*	*What It is Used For*
Gamma ray log	Measures the natural gamma-ray response of the rock.	Well-log correlation, lithology identification, ideal for recognizing shales.
Spectral gamma-ray log	As above, but with a more sensitive detector to pick out the individual contribution of potassium, thorium, and uranium to the gamma-ray response.	Knowledge of potassium, thorium, and uranium variation in the rocks can be useful for evaluating mineralogy and depositional environments.
Spontaneous potential log	Measures the potential difference driving the electrical current, that results from salinity differences between the drilling mud and formation water in permeable rocks downhole.	Gives a rough indication of lithology and is used for the evaluation of formation water resistivity.
Electrical logs	Measures the electrical properties of the fluid in the rock.	Can indicate if hydrocarbons are present or not.
Density and neutron logs	Measures the formation density and volume of fluids in the rock respectively.	An estimate of porosity can be made. Also allows the identification of certain lithologies such as limestone, anhydrite, and halite.
Sonic log	Measures how fast an acoustic signal can pass through a rock.	An estimate of porosity can be made. Also used for seismic calibration.
Nuclear magnetic resonance log	Determines the nuclear magnetic response of the fluids in the rock.	Provides data that allows porosity and permeability to be estimated.
Dipmeter logs	Measures the electrical or sonic response of the rocks around the borehole.	Used to calculate formation dip, pick out faults and other structures, and sometimes determine the sedimentary structure for paleocurrent analysis.
Borehole image logs	Measures a detailed profile of the electrical or sonic response of the rocks in the borehole.	Allows micromapping of the rock properties in the borehole wall so that the sedimentary structure, faults, and fractures can be determined.
Caliper log	Measures the diameter and shape of the borehole.	Gives an indication of hole conditions that can affect the reliability of the log responses.
Wireline coring including sidewall coring tool	Takes several short core plugs from the borehole wall.	Lithological determination and rock sampling for biostratigraphy.
Checkshot and vertical seismic profile log	Measures velocity data at specific borehole depths.	Used to calibrate the seismic response.
Formation tester log	Measures pressures at specific points in the reservoir and can allow small volumes of fluid to be sampled.	Establish a pressure profile for the reservoir and define fluid contacts.

significant amounts of clay minerals that could potentially cause formation damage during production operations (see Chapter 31, this publication). The report should include facies photographs, thin section photomicrographs, and, where appropriate, scanning electron microscopy (SEM) photomicrographs.

WIRELINE AND LWD LOGS

Wireline logs are run in wells to determine the physical properties of the rock and fluids in the borehole (Table 9). From this, a detailed interpretation can be made of the geology and fluid saturations in the reservoir interval. A brief summary of these logs is provided here. For more details, the textbooks by Serra (1984), Rider (1996), and Luthi (2001) can be consulted.

GAMMA-RAY LOGS

A gamma-ray log measures the natural radiation in the rocks, much of which is emitted by the elements potassium, uranium, and thorium (Figure 43). The geologist

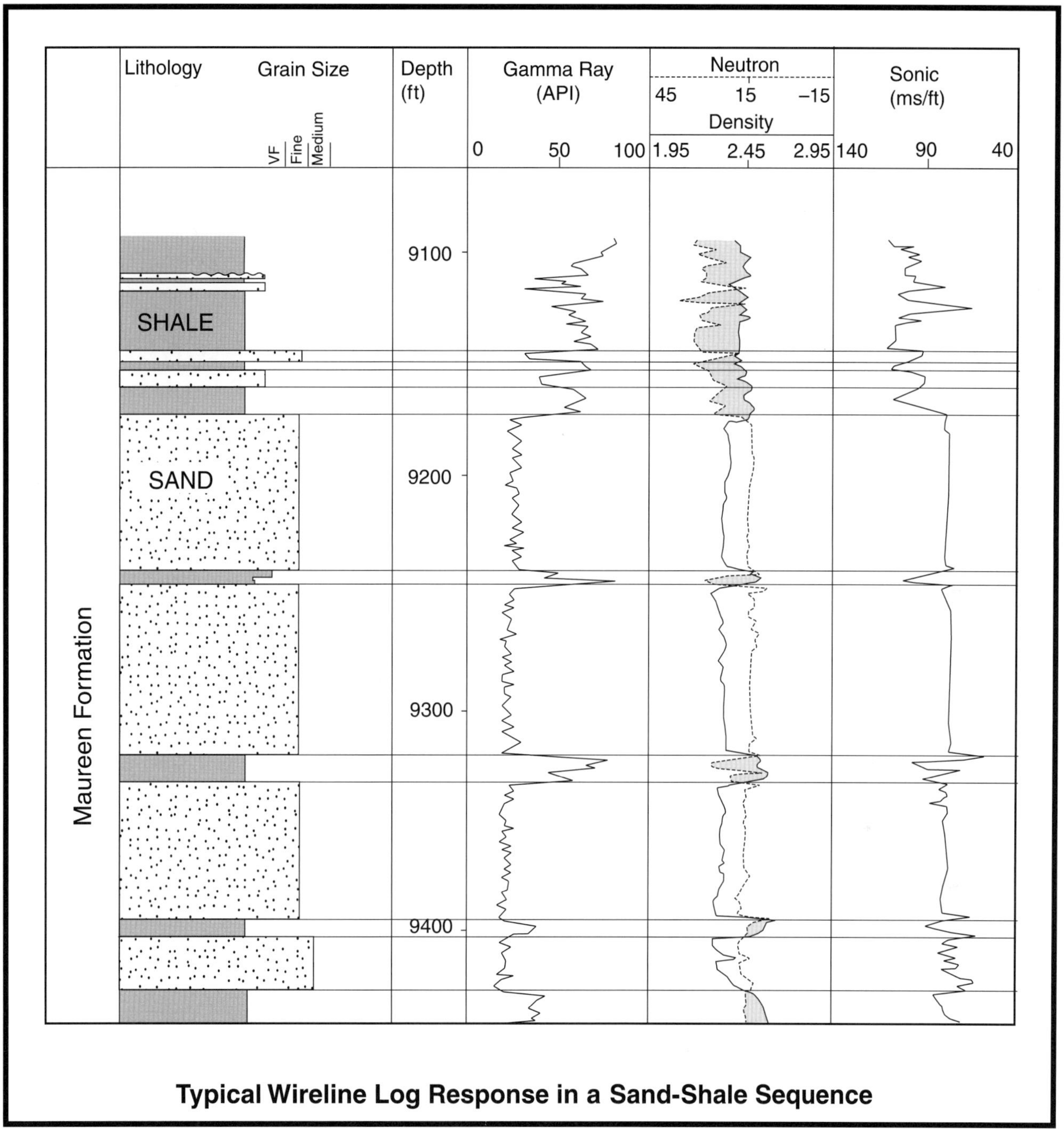

Figure 43. Gamma-ray, density, neutron, and sonic log response of a sandstone and shale sequence. This example is from well 16/29a-9 in the Fleming field, UK North Sea (from Stuart, 2002). Reprinted with permission from the Geological Society.

typically uses the log to differentiate between sandstone and shale for correlation purposes. Sandstones normally show a lower gamma-ray response than shales. The gamma ray is an excellent tool for this, providing it is used in conjunction with other logs to confirm the lithology response. Care should be taken with the interpretation of the gamma-ray log in some sandstones. Sandstones rich in potassium-rich minerals such as potassium feldspar, muscovite mica, illite, or glauconite can give a high gamma response that is easily mistaken for a shale. A gamma spike at the base of a sand-prone upper shoreface profile can be the result of concentrations of heavy, radioactive accessory minerals by wave winnowing.

SPECTRAL GAMMA-RAY LOG

Spectral gamma-ray logs measure the relative contribution of potassium, thorium, and uranium to the overall gamma-ray response. A high potassium content generally indicates the presence of minerals such as potassium feldspar and mica. Thorium is associated with the mineral monazite, a common heavy mineral in sandstones sourced from acid igneous rocks (Hurst and Milodowski, 1996). Uranium is commonly found absorbed onto organic material and clay in marine shales (Serra, 1984).

Spectral gamma-ray logs are used less frequently than the other types of log, although in certain situations they can pick out features that the other logs will not (Hancock, 1992). For example, the spectral gamma-ray log response can be used to identify a zone of potassium feldspar dissolution in leached sandstone below an unconformity.

DENSITY AND NEUTRON LOGS

Density and neutron logs are primarily used for estimating the porosity. Density logs measure the bulk density of a formation, a function of the rock matrix density emitted from the log and the density of the fluids in the pore space, according to the degree by which the energy of gamma rays is progressively absorbed and scattered by electrons in the rock. The principle behind the density log is that, for a rock with a given grain and fluid density, the higher the porosity, the less dense the formation will be. A neutron log bombards the formation with neutrons to detect energy changes as a result of collisions with hydrogen atoms. Hydrogen is found in the water (and oil) molecules filling the pore space. Thus the neutron log gives an indication of the formation porosity (Rider, 1996).

The logs also have specific geological uses. They can be used to pick out cemented intervals in sandstones. Carbonate-cemented intervals will show a distinctive response on these logs.

SONIC LOGS

A sonic log measures the time it takes for a sound pulse to travel from a transmitter to a receiver via the formation (Rider, 1996). Sonic logs can be used for measuring porosity but are more commonly used by the geophysicist as they give velocity information for calibrating seismic data. *Velocity* data allow the geophysicist to convert the time taken for a seismic wave to travel down and back from a specific seismic reflector into an equivalent subsurface depth. The geologist can use sonic logs to pick out coals and poorly consolidated sandstones.

ELECTRICAL LOGS

Electrical logs measure the resistivity of the rock and its contained fluids to the passage of an electrical current (Rider, 1996). A high-resistivity response within a porous rock is an indication of hydrocarbons. The logs can also help to recognize certain lithologies. Tight cemented intervals will have a high-resistivity response and these can be picked out in combination with the density and neutron log response.

NUCLEAR MAGNETIC RESONANCE LOGS

Nuclear magnetic resonance (NMR) logs measure how hydrogen nuclei in a static magnetic field respond to an oscillating radio frequency. The liquid filled porosity, pore size distribution, and volume of movable fluids can be characterized from this. It is also possible to estimate permeability values empirically from NMR log data.

DIPMETER LOGS

Dipmeter logs measure the variation in electrical or sonic response around the circumference of the borehole. From this, formation dip and sometimes the orientation of sedimentary structures can be determined (Bourke, 1992; Cameron, 1992).

BOREHOLE IMAGE LOGS

Borehole image logs give a detailed electrical or sonic map of the borehole wall (Luthi, 1992). This enables geological information such as formation dip, sedimentary structures, faulting, and fracturing to be imaged. The dip and azimuths of these features are measured from the image logs. The logs are especially useful for the structural characterization of heavily faulted and fractured reservoirs. They also show thin beds in reservoir intervals where most conventional logs do not have the resolution to detect them.

FORMATION TESTER LOGS

Wireline pressure test data in infill wells can provide valuable information on the reservoir performance. The formation tester log contains a probe, which is

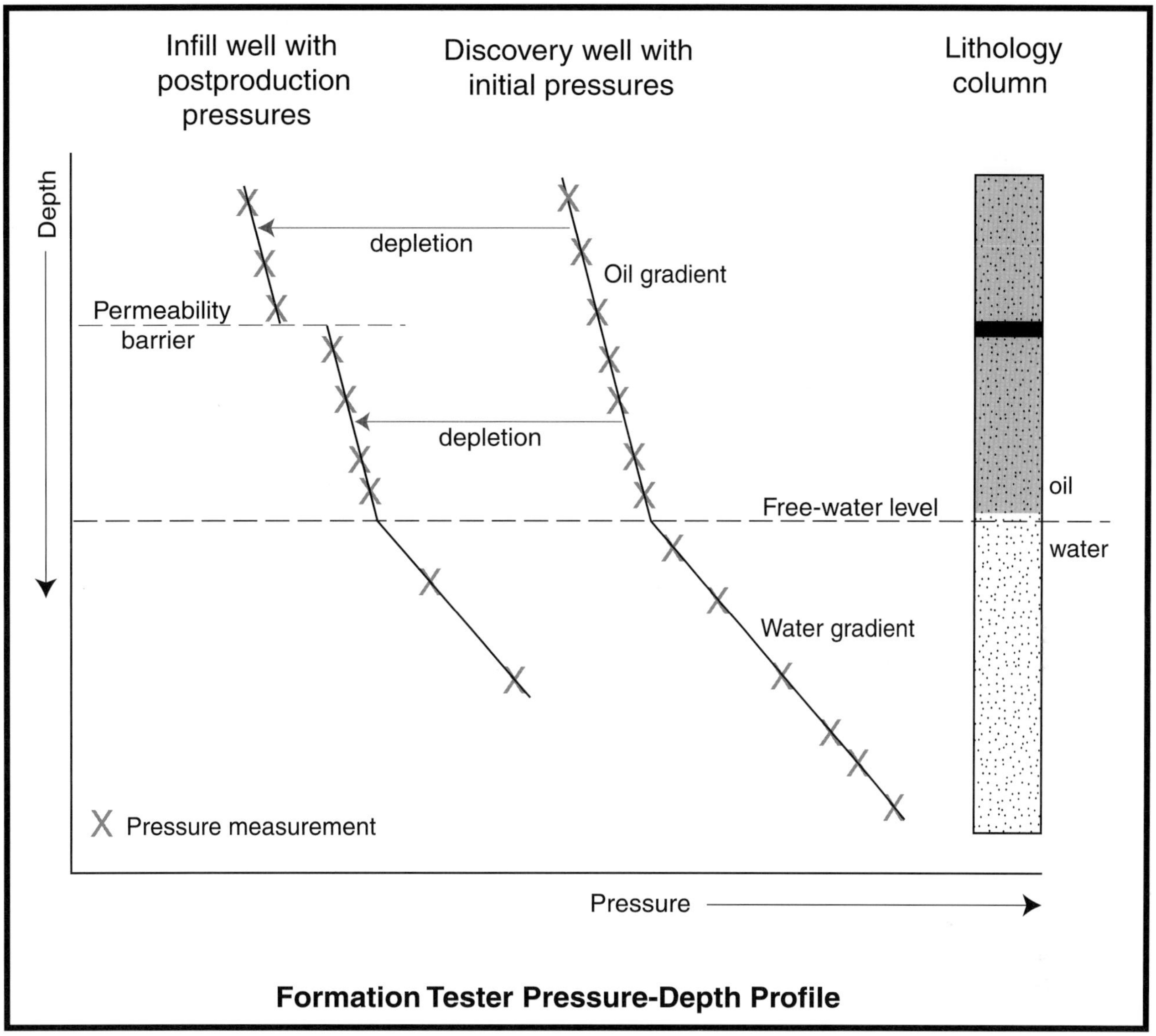

FIGURE 44. Formation pressure measurements are repeated at various depths throughout the reservoir to make a pressure-depth profile. Tests conducted in a virgin reservoir preproduction can allow the free-water level to be defined. Postproduction, formation tester data can give information on where the reservoir may be separating into zones of different pressures as a result of depletion.

pushed horizontally against the formation to take a measurement of the reservoir pressure. A small fluid sample can also be taken if required. The pressure measurements are repeated at various depths throughout the reservoir, enabling a *pressure-depth plot* to be made.

When these tests are conducted in a virgin reservoir preproduction, it may be possible to define the depth of the free-water level. This will correspond to the intersection of the water and oil (gas) gradients. Postproduction, formation tester data can give information on where the reservoir is separating into zones of different pressures as a result of depletion (Figure 44).

The raw log data will show the rate at which the pressure built up for each test, and a crude assessment of the formation permeability can be made from this (Smolen, 1992a).

WIRELINE CORING

Wireline methods such as *sidewall coring* allow the retrieval of several short plug-type cores from the borehole wall. A series of wire-attached, hollow steel bullets are fired horizontally into the borehole wall from the wireline tool (Rider, 1996). Sidewall cores are mainly used for lithology determination and biostratigraphic analysis.

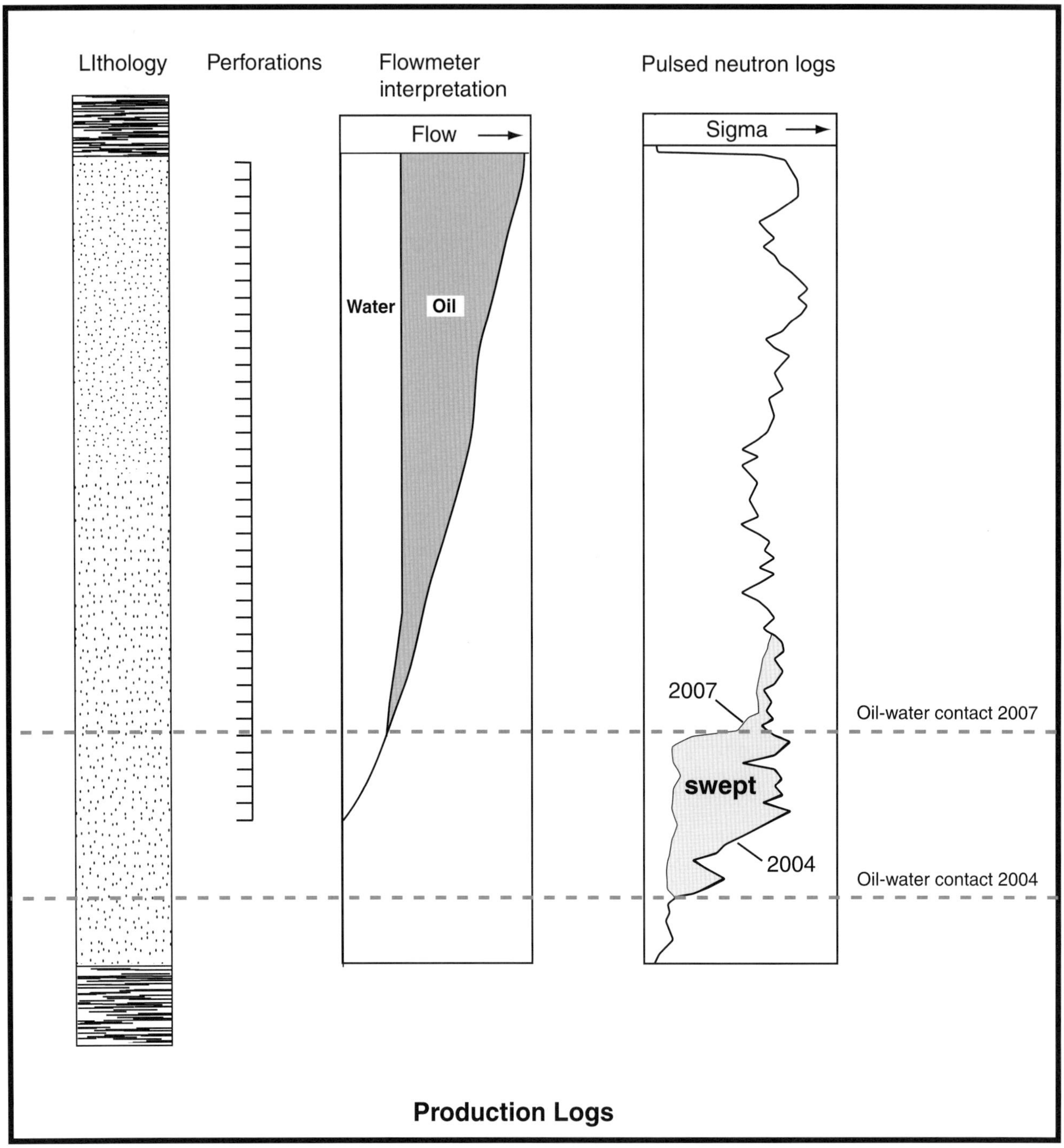

FIGURE 45. Production logs are run in a producing well to determine downhole flow rates and to evaluate reservoir sweep.

CHECKSHOT AND VERTICAL SEISMIC PROFILES

Checkshots and vertical seismic profiles (VSPs) are used by the geophysicist to record velocity information in a well. A *checkshot survey* is taken at different depths down the borehole (Hardage, 1992). A log with a geophone for detecting seismic signals is run in the hole at the same time as a seismic source is activated at the surface. The distance between the source and the log is established, and the time taken for the signal to travel to the log is measured. From this, an accurate velocity can be calculated.

Table 10. The main production log types.

Production Log	*How It Works*	*What It is Used For*
Flowmeter log	Flow from the reservoir turns a spinner; the faster the flow, the faster the spinner turns.	The flow rates and flow profile of a reservoir interval can be derived from the spin rate.
Shut-in flowmeter log	The well is shut in. The tool measures fluid flowing from high- to low-pressured reservoir intervals via the wellbore.	Can be used to pick out vertical permeability barriers.
Pulsed neutron log	The tool bombards the pore fluid with high-energy neutrons that pass through the well casing. The neutrons are captured typically by chlorine atoms, and the gamma radiation emitted in response is measured. The log is sensitive to chlorine abundance and hence the water saturation and the salinity of the pore fluids.	Fluid saturations are determined from the log results and hence the degree of sweep behind the casing can be calculated.
Fluid density log	A variety of tools measure the fluid density by recording either the pressure gradient, the bulk density, or the capacitance of the flowing mixture in the well.	Where more than one flowing phase is present, the fraction of a particular fluid flowing can be calculated, i.e., water, oil, or gas.
Temperature log	Records a temperature profile.	For example, can determine which interval is taking (cold) injection water in an injection well.
Gamma-ray spectrometer log	Measures element concentrations, especially carbon/oxygen ratios.	Indication of hydrocarbon saturations, particularly in low-salinity, clean sandstones.

The major difference between a checkshot survey and a VSP is that the VSP data are recorded at a much closer sampling interval down the well. The data can be processed to produce a seismic image of the near wellbore area (Hardage, 1992). The results will be used to tie reflectors on seismic lines to geological features in the well.

LWD LOGS

Logging-while-drilling (LWD) logs are run as an integral part of the the drill string a short distance behind the drill bit (typically 1.5–24 m [5–80 ft]). The acronyms MWD (monitoring while drilling) or FEWD (formation evaluation while drilling) are also used. These logs enable reservoir measurements to be taken in *real time*, that is, while the well is being drilled (Medeiros, 1992). The log signal is sent up the borehole either by mud pulses or by electromagnetic transmission. The log response can be displayed on monitors at the rig site or transmitted back to the oil company office. Most of the capabilities of wireline logs are available in LWD form.

LWD logs may be used for several reasons:

1) Real time data allow critical decisions to be made before the well has been drilled too far; for example, selection of casing points.
2) The successful run of a suite of LWD logs saves a day or more tying up an expensive rig operation exclusively with wireline logging.
3) They can be run as insurance logs where the need for log data is critical. This can happen in areas where there is a chance that open-hole logs may not be possible because of borehole instability (Meehan, 1994).
4) They are used for steering horizontal wells (see Chapter 28, this publication).

PRODUCTION LOGS

Production logs are run in a producing well to determine downhole flow rates and to evaluate reservoir sweep (Figure 45; Table 10). They give the subsurface team an understanding of how the reservoir is behaving under production. For example, if a well is producing water, the logs can then be analyzed to determine which perforated intervals are sourcing the water. The perforations can then be isolated to restore the well to dry oil production (Smolen, 1992b).

The geologist uses production-log data to determine the flow geology characteristics of the reservoir and to help establish where there may be unswept oil and gas targets.

PRODUCTION WELL-TEST DATA AND INTERFERENCE AND PULSE TESTS

Production well tests are an important for reservoir management because they provide information on flow rates, reservoir architecture, rock properties, and reservoir pressures. A production well test is performed by inducing pressure variations in a well over time. An example of this is where a production well is shut in to conduct a *pressure buildup test.* Fluid will then move into the pressure sink caused by the production, and the pressure will gradually increase in the well. The pressure data are used to assess the properties of the reservoir and the reservoir fluid around the wellbore by a technique known as pressure transient analysis (Lee, 1992). For instance, the higher the permeability, the more rapidly the fluid moves in and the quicker the pressure builds up.

Two types of tests can be run to give an idea of interwell communication. *Interference tests* are set up by assigning one of the wells in a specific sector of the reservoir as an observation well. Then one or a number of wells is produced from or injected into and the pressure response is measured in the monitor well. *Pulse tests* are a variation on the theme of interference tests. The difference is that the active well is shut in, returned to production, shut in, and so on, in a series of pulses. These tests are especially useful in assessing the communication between injection and production wells (Kamal, 1983).

Radioactive or chemical tracers can be put into an injection well and nearby production wells will be monitored to see when and where the tracers are back produced (Bjornstad et al., 1990). For example, radioactive tracers have been used in the Endicott field in Alaska to identify communication pathways between injection and production wells. The data were used to assess the validity of the geological correlation for the reservoir (Shaw et al., 1996).

FLUID SAMPLES

The taking of oil and formation water fluid samples at the appraisal stage of field development can provide valuable data later on in field life. For instance, variation in oil and water geochemistry data can be used to define reservoir compartments within a field (see Chapter 16, this publication).

PRODUCTION DATA

Production data can be used to make inferences about reservoir continuity and connectivity. The geologist should have direct access to the well-by-well production profiles (Figure 46). These show the rate of production against time for each well including the *total fluid flow rate, hydrocarbon flow rate, water flow rate,* and the *water cut* (percentage of water flowing relative to total flow). The idea is to look out for any unexplained changes in production or unexpected anomalies. Sometimes this happens for mechanical reasons, but, typically the anomaly may give an insight into the fluid pathways within the reservoir. For instance, a new injection well may be brought on stream, and this will cause the flow rates to increase in nearby producers. This demonstrates reservoir connectivity between the injector and the producers.

SEISMIC DATA

Seismic data allow subsurface structures to be identified and mapped. It provides structural information for determining suitable places to drill in an oil field. Seismic data will also help to determine the nature of the reservoir between wells, albeit at a relatively low *resolution* both spatially and vertically (Figure 47). Horizontally, a data point is typically acquired every 12.5 m (41 ft) with modern seismic acquisition methods offshore. Vertical resolution will mostly depend on the depth of the reservoir and to some extent on the seismic acquisition and processing parameters. The resolution decreases with increasing depth with the higher frequency component of the signal progressively getting filtered out as the sound wave passes through the subsurface. The shape of the seismic pulse will also change as a function of depth, further distorting the signal. At typical Jurassic reservoir depths in the North Sea for instance, the frequency content of the signal corresponds to a vertical resolution of about 20–40 m (66–132 ft). Features smaller than this will not be seen on seismic sections at these depths. This resolution is sufficient to make an interpretation of the reservoir structure and the position of the larger faults. The geologist will use the seismic interpretation as the basis for the structural framework in their geological scheme. An analysis of seismic data can also occasionally give an indication of the nature of reservoir porosity, fluid type, and an outline of sediment bodies.

In most companies, geophysicists are responsible for interpreting and analyzing seismic data. However, it is becoming more common for geologists in oil companies to make some of the seismic interpretation.

Seismic data are acquired in the broadest sense by sending sound waves into the subsurface and then detecting the echo. Most of the energy will be transmitted deeper into the subsurface, but part of the energy will be reflected at interfaces of different densities and velocities within the rock layers. The reflected energy returns to the surface where it is recorded by *geophones,* electronic receivers that convert ground motion into electronic signals. The offshore equivalent of a geophone is a *hydrophone,* which records the pressure pulses returning

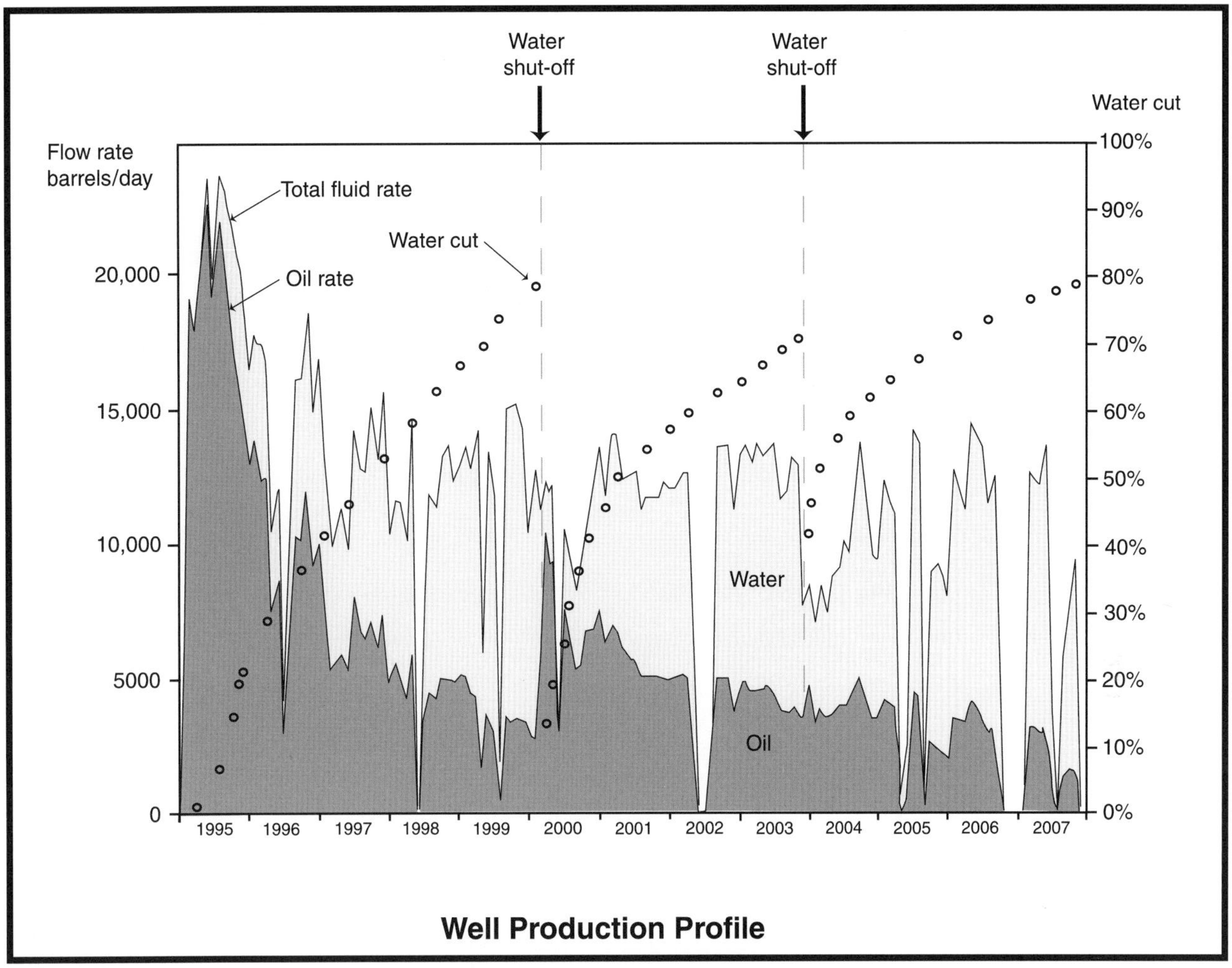

FIGURE 46. Production data for an individual oil producer. The well has a variable history of water production. Water shut-offs in 2000 and 2004 were partially successful in reducing the water cut.

through the sea. The strength of the reflected seismic energy depends on the *acoustic impedance* (AI) contrast at the boundary between two layers of rock. The AI is the product of the rock density and the transmission velocity. The higher the AI contrast, the greater the strength of the reflected signal.

Seismic data can be acquired both on land and at sea. On land, a variety of sound sources have been used, including dynamite, a heavy weight repeatedly dropped on the ground, or a vibrating steel plate on the ground surface. *Airguns* are typically used in the marine environment.

Recording devices on land consist of arrays of connected geophones laid out in long lines. At sea, hydrophones are strung together within a long plastic sheath known as a *streamer*. The streamer can be several kilometers long. At the end of the 20th century, a streamer was typically 3500–4000 m (11,500–13,000 ft) long. The trend today is for increasingly longer cables to allow a greater distance between the source and the furthest hydrophone on the streamer (known as the *far offset*, the distance between the source and the nearest hydrophone being known as the *near offset*). This greater distance allows for better discrimination of the variation in the recorded amplitudes for a given reflector with increasing offset, a technique known as *amplitude versus offset* or AVO. This can be helpful in determining whether hydrocarbons are present at a given location (Russell, 2002). Several sources and several streamers can be towed behind the seismic boat at one time (Figure 48).

Land and marine acquisition techniques differ slightly but in principle are mostly the same. The following describes the marine case. A seismic boat acquires data by sailing as carefully as it can along a predetermined line over the area of interest. When it reaches the end of this line, it turns around and acquires data along a parallel line in the opposite direction. The boat will steam back and forth line after line acquiring the seismic survey for up to months at a time depending on how large an area is to be acquired and on the weather conditions.

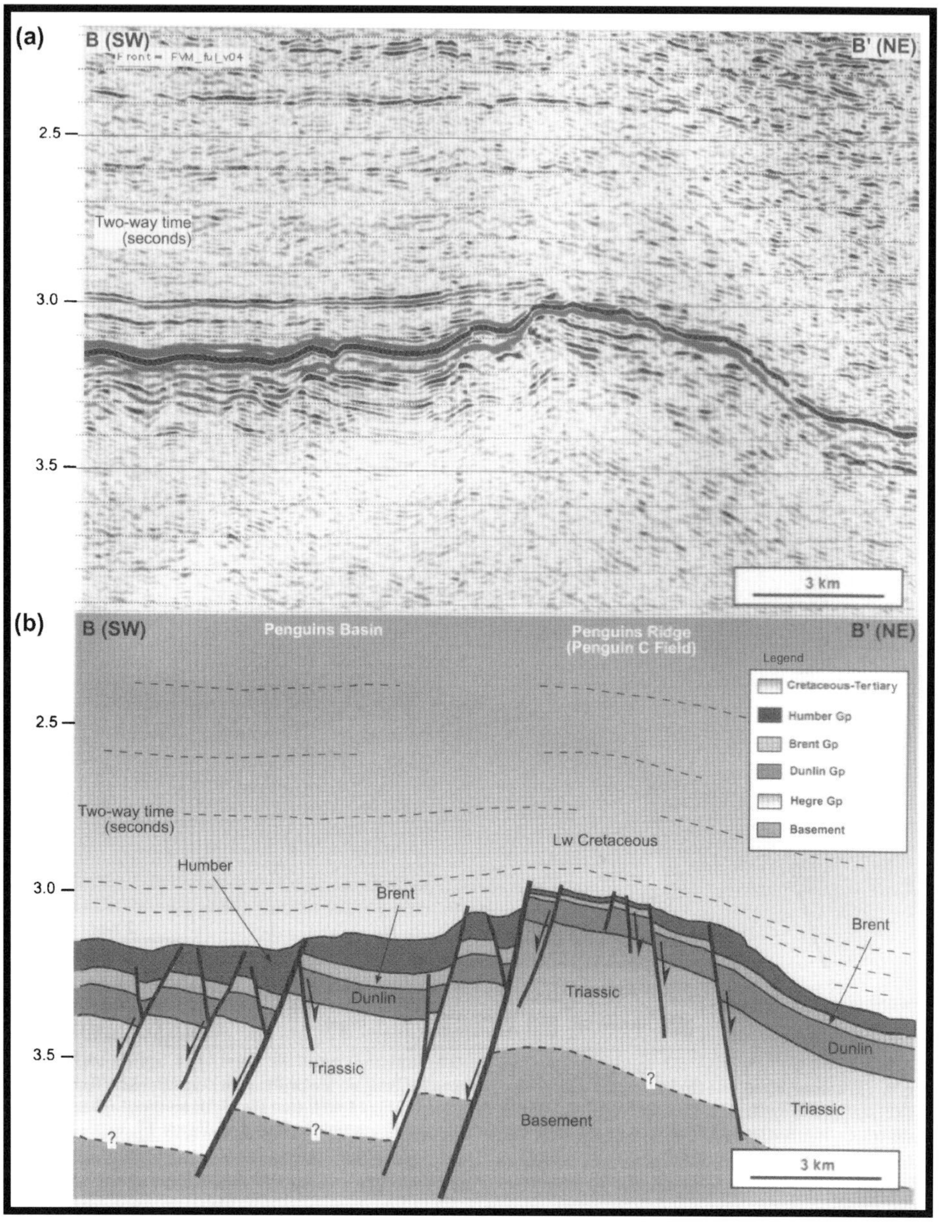

FIGURE 47. Seismic line and equivalent interpretation through the Penguin C South field, UK North Sea (from Domínguez, 2007). Reprinted with permission from the Geological Society.

The boat travels slowly along the predetermined line, and periodically (every 12.5 m [41 ft] or perhaps every 25 m [82 ft]) discharges the airgun. The point on the line where this occurs is known as a *shotpoint*. The hydrophones then record the reflection echoes from the subsurface. Simultaneously, compressors will recharge the airgun ready for the next discharge, and the process repeats over and over again. The result is a record of a large number of shot and receiver pairs for each reflection point in the subsurface. The data are recorded digitally and will include the time it takes for the seismic pulse to return to the surface, the waveform of the seismic signal, and the sound and source location. The time that the seismic energy takes to travel from the source to the reflection and back to the surface again is called the *two-way traveltime* (TWT). This can take 2–3 s or more. Because of the rapid velocity of seismic waves through the subsurface, seismic intervals are measured in milliseconds; 1000 ms equals 1 s.

The seismic data are interpreted with the principal objective of mapping out the structure of the reservoir. If the top of the reservoir gives a usable seismic reflection, a seismic time surface is mapped out. The map will be contoured in two-way time. It can be *depth converted* using velocity information to create a depth map in meters or feet.

3-D SEISMIC SURVEYS

The most common method of acquiring seismic data involves shooting a *3-D survey*. This is where a dense

FIGURE 48. Seismic boat and streamers (courtesy of Woodside Petroleum, Web site: www.woodside.com.au.)

coverage of seismic data has been collected over an area with the objective of determining spatial relations in three dimensions. The data are collected such that it can be processed to get as close to the correct spatial representation of the subsurface as can be practically achieved. This involves migrating the seismic data to correct for oblique reflections from dipping surfaces and faults. After processing, a 3-D data set will consist of a dense box-shaped grid of seismic data covering the field area. The grid comprises a series of inlines and crossline traces at regular intervals, every 12.5 m (41 ft) for instance.

The data are stored on a computer. The interpreter can call up the data set on the screen. It is possible to display any *vertical or horizontal slice* through the data as required. Vertical slices are typically used for picking horizons and faults. Two types of horizontal slices can be derived from a 3-D seismic data set. A *time slice* is a horizontal slice through a volume of 3-D data, which can show areal amplitude variation. Under favorable conditions, this can reveal geometrical patterns related to the depositional environment. A *horizon slice* is a reflection that has been flattened and then redisplayed as a time slice. It shows areal amplitude variation along the reflection.

7

Shepherd, M., 2009, Rock and fluid properties, *in* M. Shepherd, Oil field production geology: AAPG Memoir 91, p. 65–68.

Rock and Fluid Properties

INTRODUCTION

Rock and fluid properties are used to assess both the volume of hydrocarbons in the reservoir and the ability of fluid in the rock to flow. Although it is the responsibility of the petrophysicist to derive these properties from log and core data, it is the geologist who will map out these properties for the whole reservoir. Rock and fluid properties are required for the geologist to make an estimate of the in-place hydrocarbon volumes.

THE PETROPHYSICAL EVALUATION

After the logs have been run in a new well, the petrophysicist will perform a quality control on the data and will then interpret the logs to produce a *petrophysical evaluation*. This illustrates the interpreted rock properties such as gross sand, net sand, net to gross, porosity, Vshale, estimated permeability from logs, and fluid properties, including hydrocarbon and water saturations (Figure 49).

The petrophysicist will also produce sums and averages for the rock properties in each of the reservoir intervals (Table 11).

GROSS AND NET THICKNESS

The *gross thickness* is the total thickness of rock in the interval of interest. The *net thickness* is the thickness of the *net reservoir* rock, i.e., those intervals with useful reservoir properties. *Net pay* is the part of a reservoir unit that has the ability to flow hydrocarbons at economic rates given a specific production method (Gaynor and Sneider, 1992). *Non-net reservoir* is the volume of poor-quality reservoir rock that has no hydrocarbons or cannot produce them at economic rates. As such, it is not worth including non-net rock in any calculated hydrocarbon volumes for a reservoir. Non-net intervals are ignored for the purposes of calculating net thickness.

DEFINITION OF WHAT CONSTITUTES NET THICKNESS

There is a pragmatic way of determining net thickness and a more rigorous method of doing so. Many petrophysicists define an arbitrary permeability of 1 md as the net rock cutoff for oil and 0.01 md for gas. Common practice is to crossplot porosity against permeability and to find the equivalent porosity value to the permeability cutoff. Thus, net reservoir will be defined as any rock with porosity greater than a given value, e.g., greater than a porosity of 10% (Worthington and Cosentino, 2005).

A more rigorous approach to net thickness determination involves a detailed analysis of the rock properties (Gaynor and Sneider, 1992; Worthington and Cosentino, 2005). The flow characteristics of individual lithofacies can be determined from collectively analyzing capillary pressure data, including: entry pressures, pore throat size distributions, core data, the petrophysical interpretation, thin section petrography, mud-log data, and the production log responses.

Net sand thickness calculated from wireline-log data should be cross checked against the net sand estimated from core or image logs. This ensures that there is no systematic overestimate or underestimate of net sand in the wells. Some companies insist that their geologists produce *core-to-log comparison charts* for this purpose.

DOI:10.1306/13161190M913372

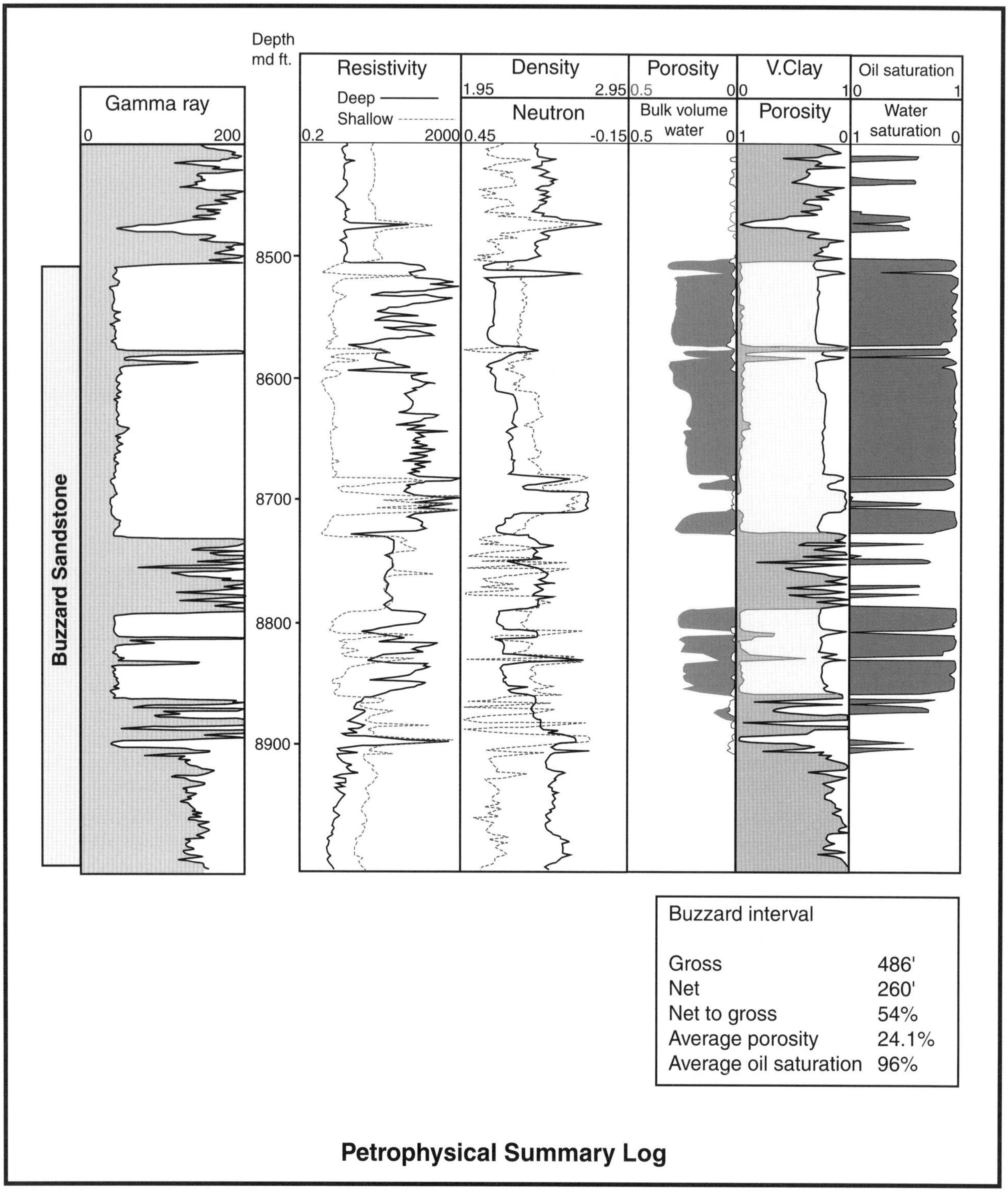

FIGURE 49. Petrophysical summary log for well 20/6-3, Buzzard field, UK North Sea (modified from Doré and Robbins, 2005).

Where there are numerous thin sandstones in a reservoir interval, wireline logs may have too low a vertical resolution to pick them out, and net sand can be significantly underestimated.

NET TO GROSS

The *net to gross* is the decimal fraction or percentage of a specific rock interval calculated from dividing the

Table 11. Example of a petrophysical sums and averages tabulation.

Unit	*Gross (m)*	*Net Sand (m)*	*Net to Gross*	*Average Porosity*	*Average Water Saturation*	*Average Permeability (md)*
F1	62	41	0.66	0.18	0.17	87
F2	100	81	0.81	0.23	0.11	345
F3	56	22	0.39	0.18	0.19	66

net thickness by the gross thickness. A common abbreviation for net to gross is N/G.

VSHALE

Vshale is the petrophysicist's estimate of the volume of shale and clay minerals within the reservoir interval (Alberty, 1992). It is sometimes used to define a net sand cutoff so as to screen out shales and poor quality silty sandstones. For example, it may be deemed that any rock with more than 50% Vshale is non-net. More typically, it is used along with a porosity cutoff for net sand determination. For instance, rock with more than 12% porosity and less than 40% Vshale may be assigned as net sand with these cutoffs applied.

POROSITY

Porosity is defined as the ratio of void space volume, commonly called pore volume, within the rock to the total rock volume. It is quoted as a fraction or a percentage (Cone and Kersey, 1992). Levorsen (1967) gave a rough field guide to porosities in reservoirs:

Negligible: 0–5%
Poor: 5–10%
Fair: 10–15%
Good: 15–20%
Very good: 20–25%
Exceptional: 25–50%

Porosity is given the Greek symbol ϕ (phi). Petrophysicists will make a distinction between *total porosity*, which is the ratio of the volume of the entire void space to the total rock and the *effective porosity*. The effective porosity is the ratio of the volume of the connected void space to the total rock and will not include any isolated pores (Tiab and Donaldson, 2004). The porosity determines how much oil or gas a reservoir rock can store. As such, it is required as input for the calculation of hydrocarbon volumes in a reservoir. Some companies will use total porosity for calculating volumetrics whereas other companies prefer to use effective porosity.

Microporosity is a term used for characterizing the pore volume at the micron scale. With very small pores, capillary effects dominate and much of the pore space will contain bound water. It is possible for rocks with abundant microporosity to have high water saturations, yet they will flow dry oil from the larger pores as the water is tightly held within the micropores. Microporosity is a common characteristic of carbonates and is also associated with clay coatings on sand grains in siliciclastic sediments (Hurst and Nadeau, 1995).

PERMEABILITY

Permeability is the measure of the ease of movement of fluid through the pore space in a rock. It is given the symbol K. The units of measurement of permeability are either expressed in *darcies* (d) or *millidarcies* (md). A thousand *millidarcies* equals one darcy.

The measurement of permeability derives from *Darcy's law*, one of the fundamental reservoir engineering concepts. The equation, the generalized form of which applies to horizontal flow, is given in Figure 50 (Dake, 1994). The horizontal flow of fluid through a rock is a function of the permeability of the rock, the viscosity of the flowing fluid, and the length and cross sectional area of the volume of rock that is taking the flow and the differential pressure.

Levorsen (1967) gave a rough field guide to permeabilities in oil fields:

Fair: 1.0–10 md
Good: 10–100 md
Very good: 100–1000 md
Exceptional: 1000 md plus

The raw measurements taken in the laboratory record the permeability to air at surface conditions. Some modifications may be necessary to correct core permeabilities to the permeability of the rock in the reservoir. *Klinkenberg corrections* may be required. When a gas is used to measure the permeability in the laboratory, this leads to measured permeabilities that are

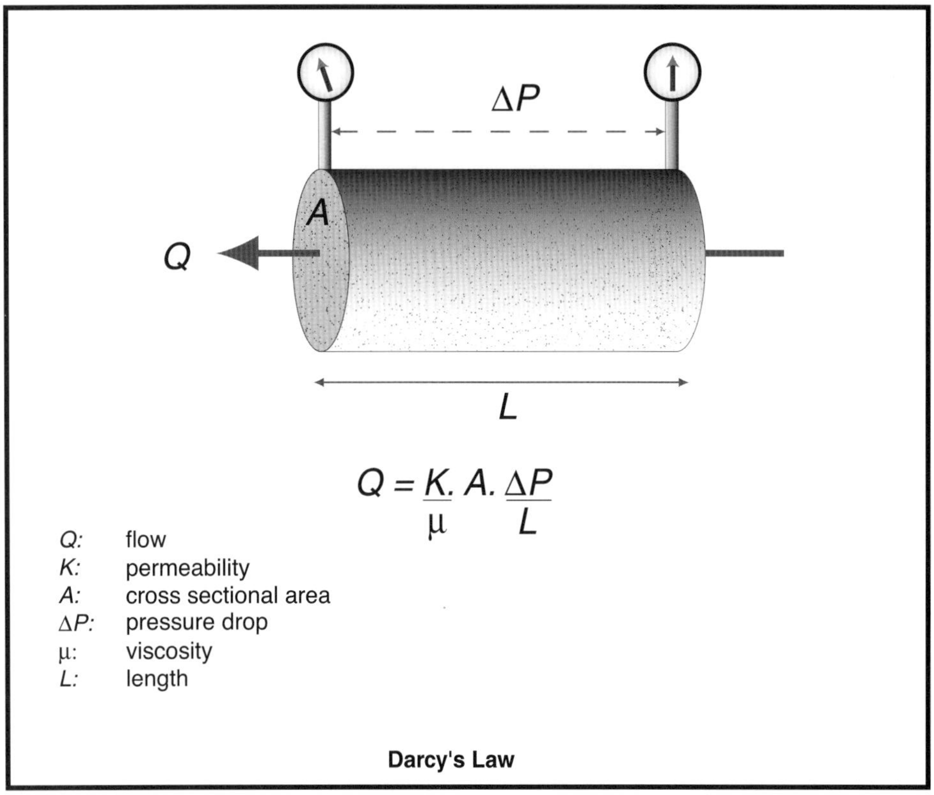

FIGURE 50. Darcy's law relates the rate of fluid flow to the rock permeability, cross sectional area, the pressure drop, the fluid viscosity, and the length of the rock interval contributing to flow. Permeability is the measure of the conductivity of fluid through the pore space in a rock.

too high because of gas slippage. The corrected permeability is known as the *Klinkenberg permeability* or the *equivalent, nonreactive liquid permeability* (Ohen and Kersey, 1992). A correction will also be required for the effect of the *overburden pressure*. This is the pressure exerted by the weight of the rock that lies above the reservoir, which acts to keep the pore space in the reservoir under compression. A correction is needed as the core will expand during retrieval to the surface and the permeabilities will be enhanced as a result (Cosentino, 2001).

HYDROCARBON AND WATER SATURATION

The *hydrocarbon saturation* is the decimal fraction or percentage of the total volume of hydrocarbons relative to the total volume of fluid (hydrocarbon plus water) within the pore space. The *water saturation* is the decimal fraction or percentage of the total volume of water relative to the total volume of fluid (hydrocarbon plus water) in the pore space.

8

Shepherd, M., 2009, Maps and cross sections, *in* M. Shepherd, Oil field production geology: AAPG Memoir 91, p. 69–73.

Maps and Cross Sections

INTRODUCTION

Maps and cross sections are the essential day to day tools by which a geologist can illustrate the spatial relationships within a reservoir. Each field requires a series of maps and cross sections as basic reference material.

MAPPING

Maps show the areal variation of a specific aspect of the reservoir character. For example, maps may be made of the depth to the top of a surface, the thickness of a given reservoir interval, or rock property variation across the field. The most common maps made by a geologist are contour maps. These show at a glance where the important spatial features are located within the reservoir area. Contour maps are easy to make and several software packages are available for the geologist to produce maps on the computer.

STRUCTURE MAPS

Structure maps show the structure of a stratigraphic surface as represented by contours of the subsurface depth. The contours are shown at regular intervals across the map, every 20 m for example. It is important that all the depths are referenced to a common datum, such as sea level; *mean sea level*, the average value of hourly readings of sea level taken over a tidal cycle, is often used as a reference level. A subsurface depth may be measured from different types of reference levels. Common practice on the rig site is to use the drill floor as the reference surface for measuring well depths. Different drilling rigs may have been used to drill the wells in a field, each rig having a different drill floor height (also known as the *rotary table elevation*) from the other. The use of a flat datum, such as sea level, corrects for this difference.

Subsurface depth maps use depths measured vertically from a datum. Where only vertical wells have been drilled in a reservoir, this is not a problem. If the wells are deviated at an angle from vertical, then it will be necessary to convert the depths along the wellbore into their equivalent vertical depths below the datum. These are referred to as *true vertical depths* or TVDs. A TVD below a subsea datum is given the acronym TVDSS. *Deviation surveys* are run in the well to determine the location, orientation, and angle of deviation of the borehole in the subsurface. These are measurement devices using pendulums, accelerometers, or gyroscopes to establish the subsurface location in the well. The TVD is worked out from the survey data using trigonometry.

THICKNESS MAPS

Thickness maps are a common map type used by production geologists (Figure 51). They are made to show the thickness of the reservoir interval and individual reservoir units.

The interval thickness of a reservoir unit can be defined in several ways (Figure 52; Table 12). Many production wells are deviated and will penetrate a reservoir unit at an oblique angle. The distance along the well bore between the top and base of the reservoir unit is the *measured thickness (MT)*. This is not a useful parameter for making maps. The measured thickness will be affected as much by the well angle as it will be by the thickness variation of the unit.

It is more meaningful to make thickness maps by calculating the *true vertical thickness* (TVT) or the *true stratigraphic thickness* (TST) of a reservoir unit. The true vertical thickness is the thickness along a vertical line between the top and base of the unit. The true stratigraphic thickness is the thickness orthogonal to the top and base of the unit. If a reservoir unit is dipping, then the TVT is greater than the TST. Bed dip angle, dip azimuth,

DOI:10.1306/13161191M913372

the well deviation, and well azimuth are required to calculate TVT and TST values. Trigonometrical formulae exist for working out TVTs in these situations (see for instance Boak, 1992a, or Tearpock and Bischke, 2003). Alternatively, the appropriate spreadsheet program can be downloaded from various sites on the Internet.

TVT maps are the most common type of map used in subsurface work as they are the simplest type of

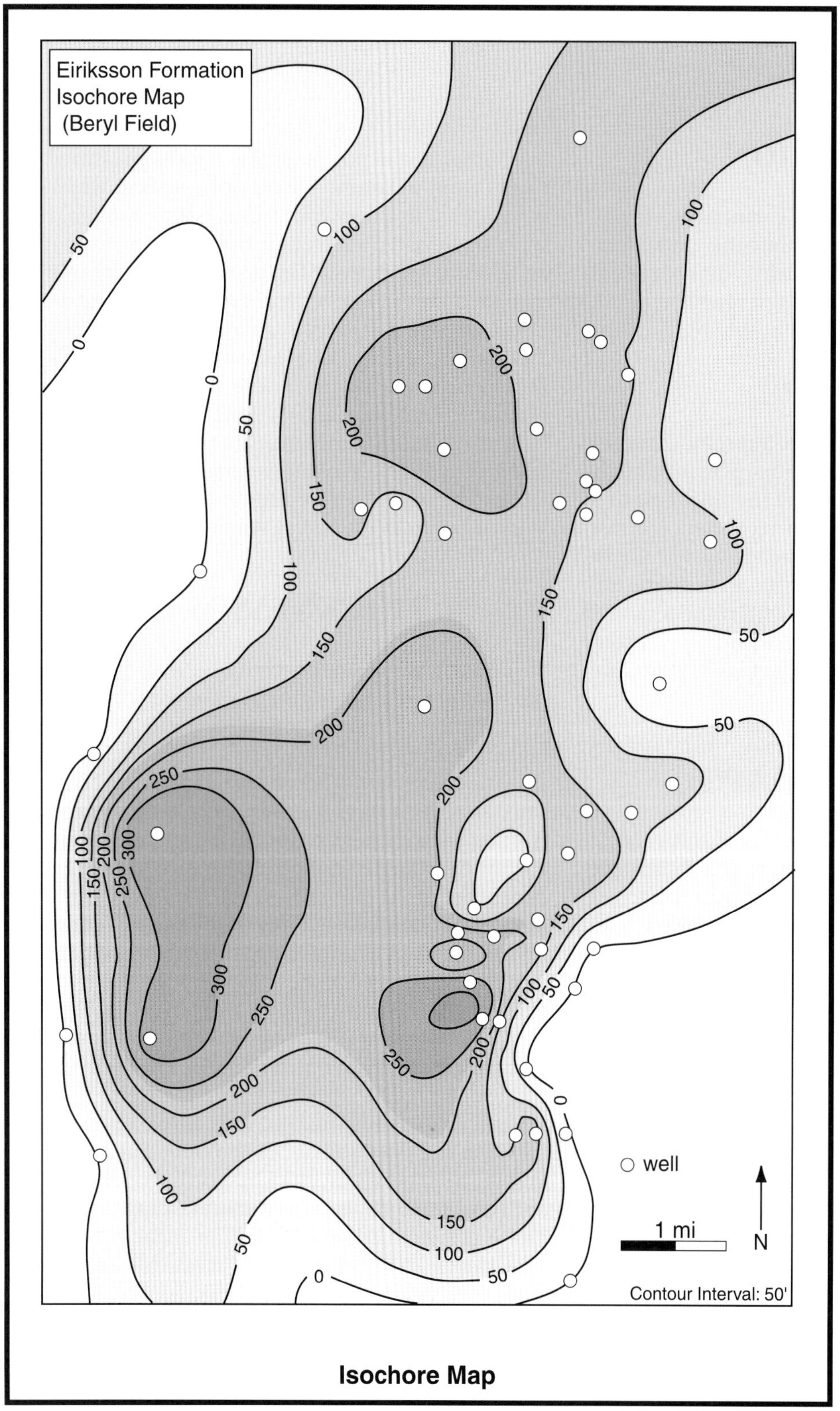

Figure 51. An isochore map shows the true vertical thickness variation of a particular reservoir interval across the mapped area. The example shown is an isochore map of the Eiriksson formation in the Beryl field area, UK North Sea (from Karasek et al., 2003; reprinted with permission from Geological Society).

FIGURE 52. Because many wells are deviated, this results in several different ways of measuring the thickness of a reservoir unit. True vertical thickness (TVT) is the most common way of defining the thickness of a reservoir interval in production geology.

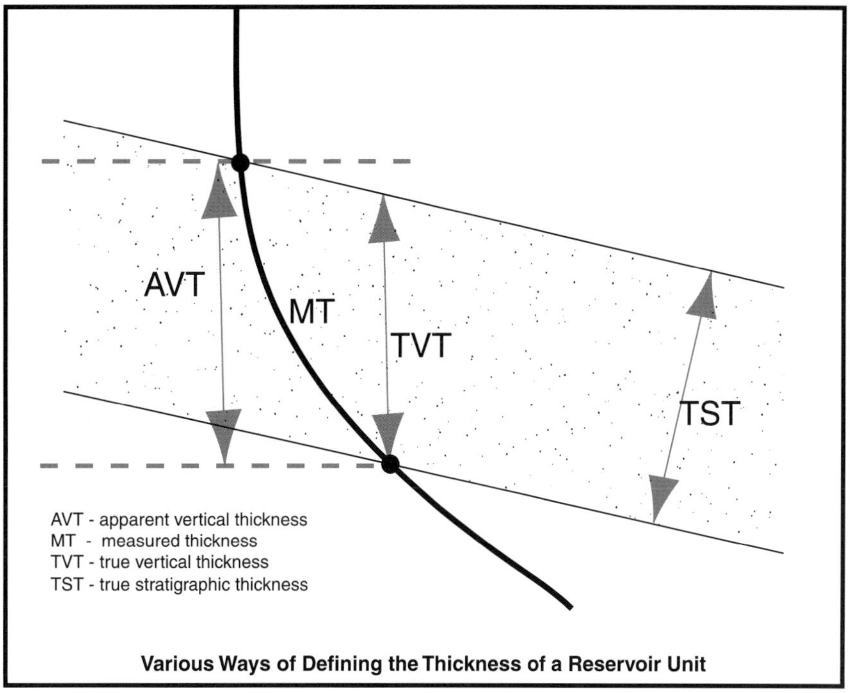

thickness maps to make. *Isochore* maps are interval thickness maps using TVT values, whereas *isopach* maps use TST values (Tearpock and Bischke, 2003).

Where a deviated well is drilled through a dipping unit, subtraction of the TVD depths at the top and base penetrations of the unit will not give the TVT value. The thickness that has been calculated is an *apparent vertical thickness (AVT)*, which is an incorrect value to use for mapping. The AVTs are not comparable from well to well because they will be influenced by the deviation angle of the wells.

OTHER TYPES OF MAPS

Net thickness maps can give a better representation of the areal distribution of the producing reservoir than the gross interval isochores, particularly where the net to gross (N/G) varies significantly across the field.

A variation on the theme is a *net pay map*, a map showing the thickness of the net interval containing hydrocarbons. *Rock property maps* include porosity, N/G, water saturation, and permeability maps.

Table 12. Some acronyms used in mapping.

Acronym	*Term*
MD	Measured depth
MD BRT	Measured depth below rotary table (the drill floor)
TVD	True vertical depth
TVDSS	True vertical depth subsea
TVT	True vertical thickness
TST	True stratigraphic thickness
AVT	Apparent vertical thickness

COMPUTER MAPPING

Various commercial contouring programs are available to make maps on a computer. They work as follows (Davies, 1980):

1) An *area of interest* is defined for the property to be mapped. The corner points, normally a rectangle, are defined with *x*, *y* coordinates. The variable to be mapped over this area is assigned to the *z* axis, e.g., depth.
2) The data values are assigned as *x*, *y*, and *z* coordinates.
3) The computer sets up a detailed 2-D grid covering the area of interest, typically a mesh of rectangles or triangles.
4) Values are assigned to specific *grid nodes* using the well values within a user-defined search radius from the grid node and a weighting function depending on the distance to the well *control points*. The *weighting function* can vary according to the algorithm used, but a common method uses the *inverse weighted distance* (Davis, 1986). Control points closer to a specific grid node will have more influence on the assigned value than those farther away.

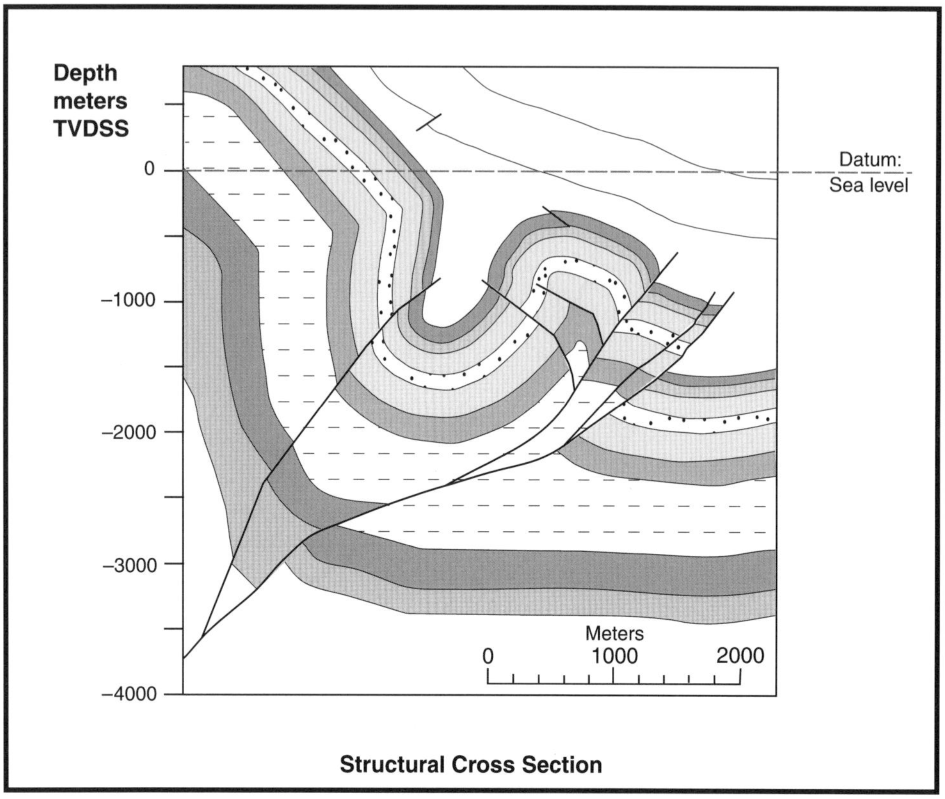

FIGURE 53. True-scale structural cross section through the El Portón field structure, onshore Argentina. The section is hung relative to a sea level datum. The depths are assigned as negative in the subsurface section below sea level and positive in the section above (from Zamora Valcarce et al., 2006; reprinted with permission from AAPG).

5) The program creates the contour map by fitting contours between the grid node values.

When a computer produces a map, there are occasions when the geologist may not like the result. For instance, the map does not look 'geological' or the contours are showing some aesthetically unpleasant *bull's-eyes*. This is where an otherwise regular looking map shows a circular concentration of closely spaced contours around one or more well control points. The contours can

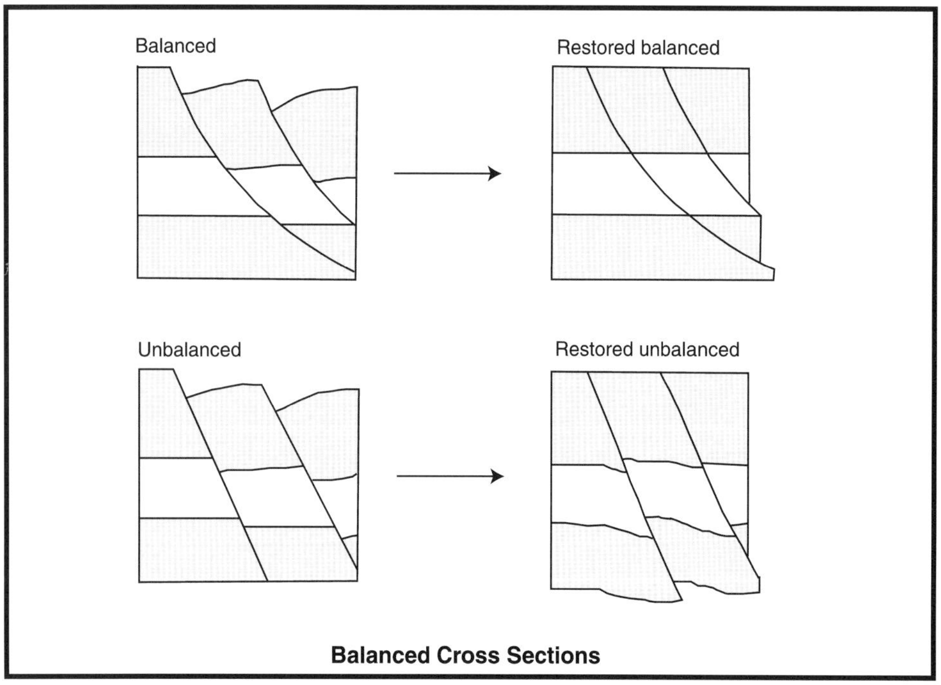

FIGURE 54. A balanced cross section for a faulted, conformable succession can be restored to a simple configuration. Where horizon and fault geometries do not balance, footwall distortion will be seen on the restored section (from Nunns, 1991; reprinted with permission from AAPG).

be edited to a give more reasonable appearance. Most programs allow this to be done interactively on the computer screen (Hamilton, 1992).

STRUCTURAL CROSS SECTIONS

Structural cross sections show a representation of the geology in the vertical plane (Figure 53). They have specific uses in reservoir management. At a glance, the relationships can be seen between the fluid distribution and the wells. They will also give an idea of stratigraphic juxtaposition relationships across faults. Cross sections are useful for illustrating proposed new well locations where structural features near the well are important when planning the well trajectory.

The section should be hung on a level datum, sea level for instance. All vertical depths used to create the cross section should be measured relative to the datum. Structural cross sections should be ideally constructed using no or very little vertical exaggeration such that true dips and the geometry of an interval can be represented with validity. Structural cross sections are drawn with wells on or close to the plane of section as data control. It is preferable for the sections to be linear instead of doglegging as this will give a more reasonable representation of the structure. A recommendation is to draw cross sections orthogonal to fault planes if possible. Following these rules will produce a cross section showing a good representation of the reservoir structure (Boak, 1992b).

VALIDATING CROSS SECTIONS

Methods are available to validate the geometrical integrity of structural cross sections through faulted, conformable stratigraphy. One method involves cutting out the various fault panels on the cross section with a pair of scissors, and then fitting them together to show what the reservoir looked like before it was faulted. This is called *restoring the section* (Figure 54). If the cross section is *balanced*, there should be no distortion, gaps, or overlap of rock on the restored section. Note that the section should be perpendicular to faults for this to work properly. Software packages are available to help the geologist make a fault restoration.

9

Shepherd, M., 2009, Predicting the geology in the gaps between the wells, *in* M. Shepherd, Oil field production geology: AAPG Memoir 91, p. 75–79.

Predicting the Geology in the Gaps Between the Wells

INTRODUCTION

Not that much information is available in reservoirs for the geologist to evaluate and understand them. Nevertheless, it is possible to make a holistic geological scheme for a reservoir using a sparse data set. The geologist makes a prediction of the geology in the gaps between the wells, principally by establishing a sedimentological scheme. Deposition of sediments involves a continuity of process from a small to large scale. This characteristic allows the geologist to extrapolate from a few tens of meters of core in a small number of wells to a depositional scheme covering the entire reservoir.

THE SAMPLING PROBLEM

The basic problem facing the reservoir geologist in establishing a geological scheme is the *sampling or the incomplete data problem* (Budding et al., 1992). Knowledge of the reservoir consists of well data, effectively 1-D well tracks, and the seismic response, which may vary anywhere between a clear to a somewhat indistinct image of the reservoir structure.

A field can extend over an area the size of a small city, yet it may only have 10 to 30 wells in it. The *sampling ratio* of wellbore to total reservoir volume can be very low, possibly one of the lowest sampling ratios of any scientific discipline. A well can be considered to have sampled a cylindrical volume of rock from the borehole cutting the hydrocarbon column. An offshore field can show a sampling ratio of the volume of total wellbore penetration in the hydrocarbon column to the total reservoir gross rock volume of perhaps about 1:3,000,000. This is such a low sampling ratio that if this ratio was applied to the world population of 6.5 billion, the subset would total 2167 people (the population of the small towns of Mammoth in Arizona or El Pont de Suert in Spain, for example). A sampling ratio of this order of magnitude is typical for reservoirs. The gaps between the wells are enormous.

THE RESERVOIR HETEROGENEITY AND THE SCALE PROBLEM

The detail of what is in the gaps between the wells is important because it is the internal architecture of the reservoir that controls the recovery of hydrocarbons. Reservoir architecture determines both the fluid pathways for oil and gas to the wells and the location of dead ends that locally trap them. Hence, working out the details of the reservoir architecture is critical to understanding how the field behaves. The geometry, degree of interconnectedness, communication, volume of dead ends, and compartmentalization of the various reservoir rock bodies all go together to make up *reservoir heterogeneity* (Figure 55). Reservoir heterogeneity exists at all scales from geological features extending over many kilometers down to the smallest pore volume.

Alpay (1972) recognized three levels of heterogeneity that control fluid flow within a reservoir. *Microscopic heterogeneity* involves the pore scale (microns) and is the level at which variation in the fluid saturation and the volume of residual oil occurs. *Macroscopic heterogeneity* is found at the interwell scale (tens to hundreds of meters) and is the result of sedimentary, diagenetic, and structural variation. *Megascopic heterogeneity* occurs at a

DOI:10.1306/13161192M913372

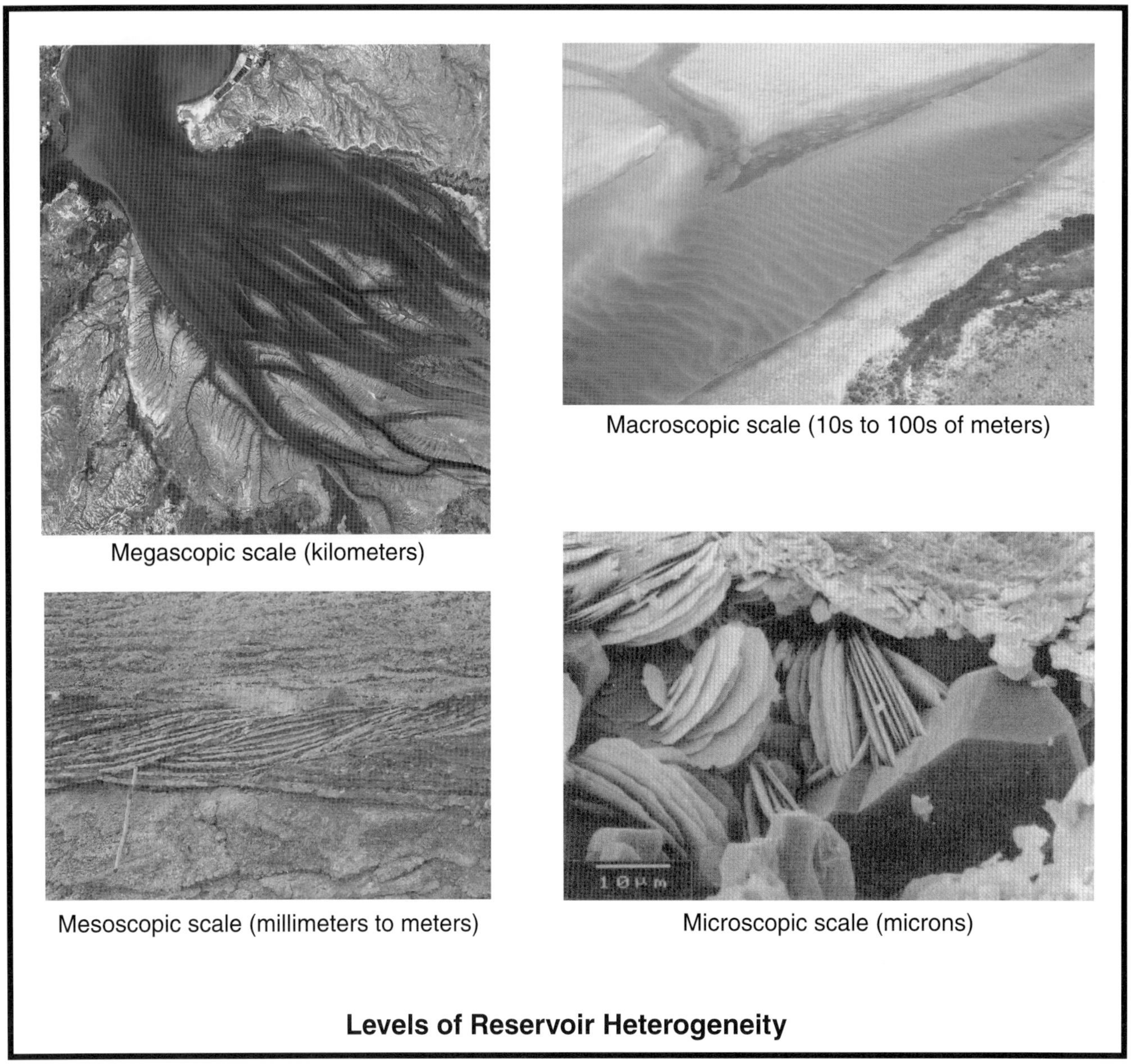

Figure 55. Heterogeneity occurs at all levels in a reservoir from the field scale down to the pore scale (picture of the Bombetoka Bay, Madagascar (top left), courtesy of NASA: www.earthobservatory.nasa.gov. SEM photograph (bottom right) from Salem et al., 2000), reprinted with permission from the AAPG.

field-wide to regional scale (kilometers) and is controlled by the large-scale stratigraphic and structural framework. Some technical papers also refer to *mesoscopic heterogeneity*, which is an intermediate level of heterogeneity at the laminar to bed scale (e.g., Haldorsen, 1986).

Sedimentary features can be categorized on the basis that they are large enough to constitute significant features at a specific level of heterogeneity. These have been given the terms *microforms*, *mesoforms*, and *macroforms* (Jackson, 1975; Miall, 1988). Macroform is the most useful term for the production geologist because macroforms are found on a scale where they are potentially mappable and are generally large enough to contain target oil volumes. Examples of macroforms are fluvial point bars, eolian dunes, or turbidite channels. Related terms are *genetic unit*, *architectural element*, or *geobody*, the latter term is often used by seismic interpreters for sediment bodies recognizable on seismic displays.

If the concept of the sampling problem is combined with the concept of reservoir heterogeneity at all scales of measurement, it is apparent that trying to explain what is going on between wells several hundreds of meters apart will be a somewhat challenging problem. A useful exercise on geology field excursions is to imagine a complex outcrop as lying between two wells 500 m (1640 ft) to 1 km (0.62 mi) apart with the height of the cliff face

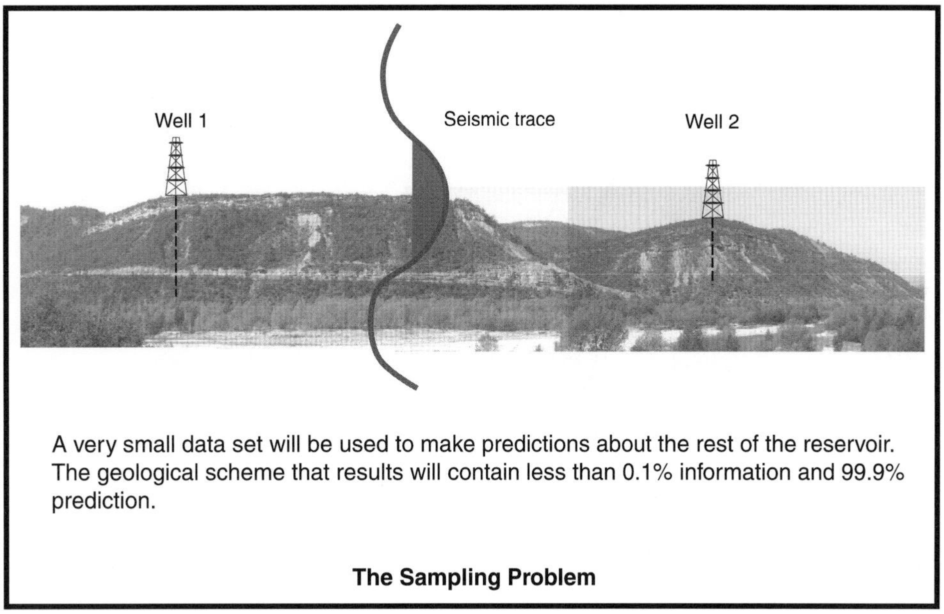

FIGURE 56. Only a very limited data set is available to help the geologist construct a reservoir scheme. A photo of channelized turbidite outcrops, Ainsa, Northern Spain, courtesy of John Millington. The cliff face is about 30–35 m (98–115 ft) high.

corresponding to part of a seismic wiggle (Figure 56). This is an excellent way of visualizing the sampling versus heterogeneity dilemma. The job of the production geologist is to solve the problem in a meaningful manner, daunting as it appears at first glance.

FILLING IN THE GAPS

The challenge given the sparse data distribution is to make a prediction of what happens in the gaps between the wells and to build up a 3-D picture of

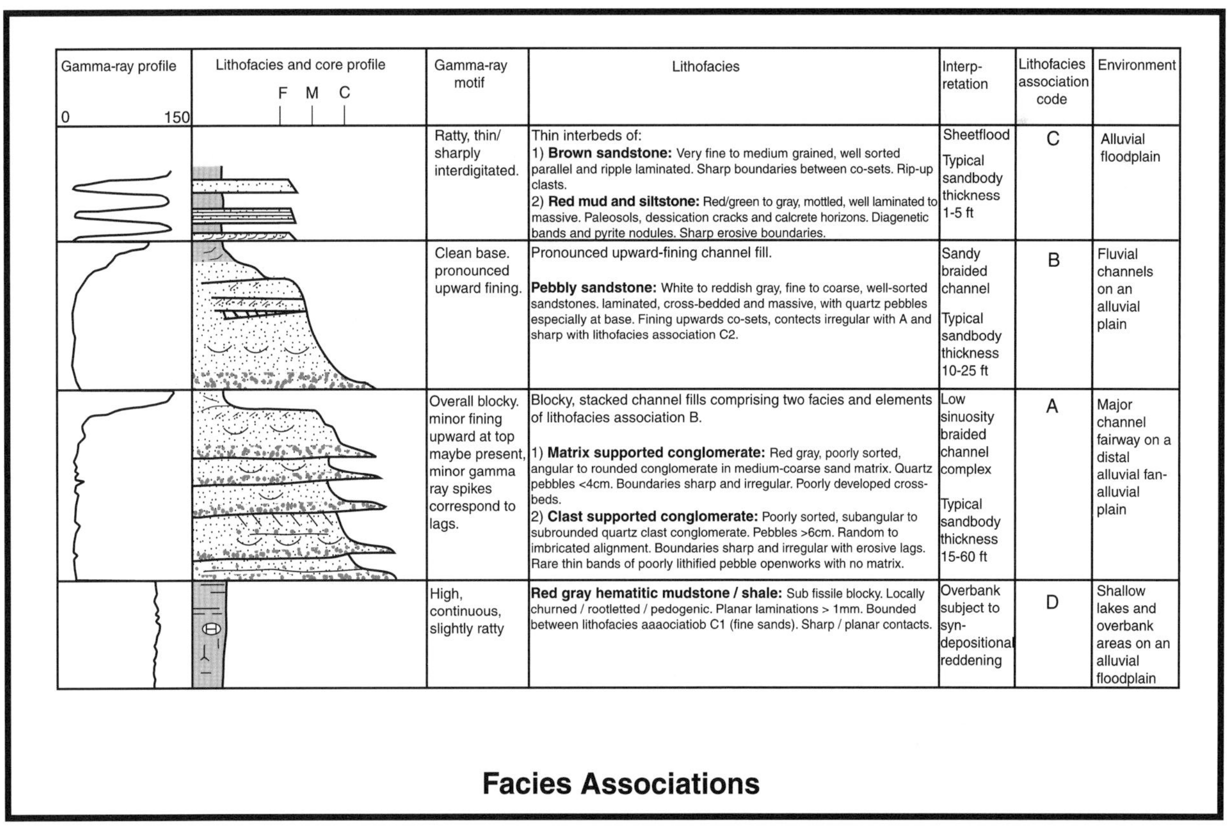

Gamma-ray profile (0–150)	Lithofacies and core profile (F M C)	Gamma-ray motif	Lithofacies	Interp- retation	Lithofacies association code	Environment
		Ratty, thin/ sharply interdigitated.	Thin interbeds of: 1) **Brown sandstone:** Very fine to medium grained, well sorted parallel and ripple laminated. Sharp boundaries between co-sets. Rip-up clasts. 2) **Red mud and siltstone:** Red/green to gray, mottled, well laminated to massive. Paleosols, dessication cracks and calcrete horizons. Diagenetic bands and pyrite nodules. Sharp erosive boundaries.	Sheetflood Typical sandbody thickness 1-5 ft	C	Alluvial floodplain
		Clean base. pronounced upward fining.	Pronounced upward-fining channel fill. **Pebbly sandstone:** White to reddish gray, fine to coarse, well-sorted sandstones. laminated, cross-bedded and massive, with quartz pebbles especially at base. Fining upwards co-sets, contects irregular with A and sharp with lithofacies association C2.	Sandy braided channel Typical sandbody thickness 10-25 ft	B	Fluvial channels on an alluvial plain
		Overall blocky. minor fining upward at top maybe present, minor gamma ray spikes correspond to lags.	Blocky, stacked channel fills comprising two facies and elements of lithofacies association B. 1) **Matrix supported conglomerate:** Red gray, poorly sorted, angular to rounded conglomerate in medium-coarse sand matrix. Quartz pebbles <4cm. Boundaries sharp and irregular. Poorly developed cross-beds. 2) **Clast supported conglomerate:** Poorly sorted, subangular to subrounded quartz clast conglomerate. Pebbles >6cm. Random to imbricated alignment. Boundaries sharp and irregular with erosive lags. Rare thin bands of poorly lithified pebble openworks with no matrix.	Low sinuosity braided channel complex Typical sandbody thickness 15-60 ft	A	Major channel fairway on a distal alluvial fan-alluvial plain
		High, continuous, slightly ratty	**Red gray hematitic mudstone / shale:** Sub fissile blocky. Locally churned / rootletted / pedogenic. Planar laminations > 1mm. Bounded between lithofacies aaaociatiob C1 (fine sands). Sharp / planar contacts.	Overbank subject to syn-depositional reddening	D	Shallow lakes and overbank areas on an alluvial floodplain

Facies Associations

FIGURE 57. Facies associations in the Tyne field, UK southern North Sea (from O'Mara et al., 2003), reprinted with permission from the Geological Society.

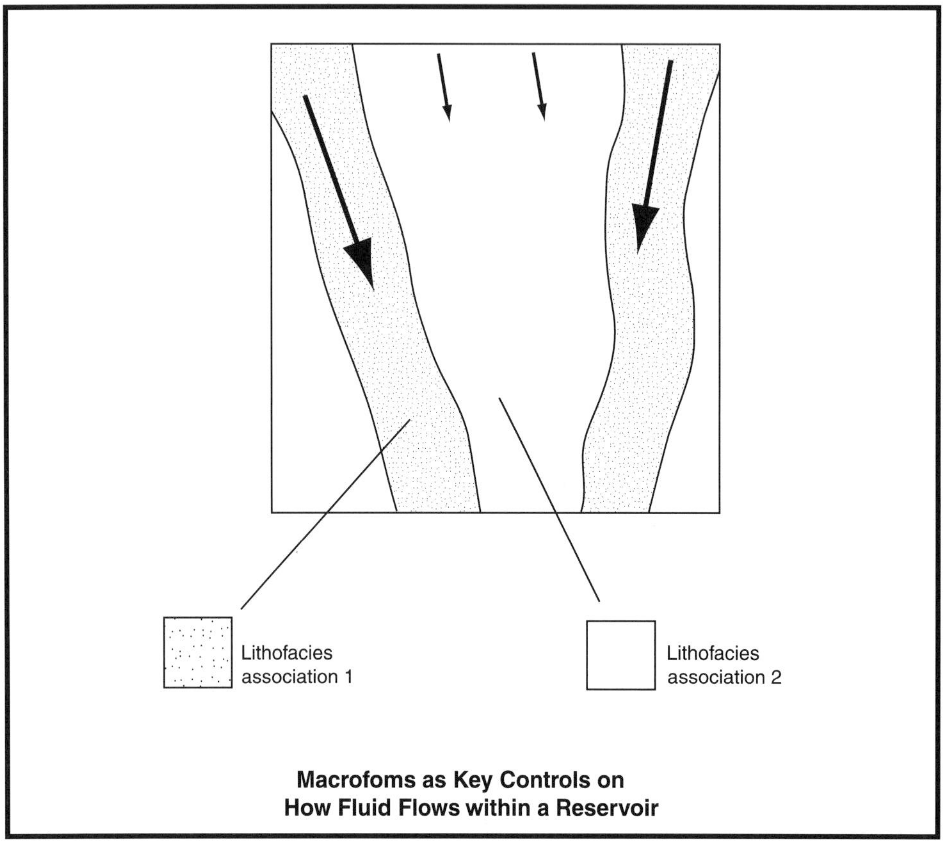

FIGURE 58. A macroform is the geometrical body corresponding to a facies association. As discrete packages showing predictable patterns of grain size, sorting, and permeability variation, macroforms are the major elements controlling how fluid flows within a reservoir.

the geology of the reservoir. A very small data set will be used to make predictions about the rest of the reservoir. The geological scheme that results will contain less than 0.1% information and more than 99.9% prediction (North, 1996).

Although the geologist will have very little information available, they are obliged to make expensive decisions on the basis of predicting the geology using this data. The geologist for instance may be involved in locating a $25 million infill well in an offshore field. The saving grace here is that there is usually enough information to make a usable geological scheme provided that every available data source is used.

Sometimes seismic data can help. Under favorable circumstances, seismic data can be of good enough quality to determine sediment geometry at the reservoir scale. Lithological and rock property attributes can also be modeled where the seismic data allows.

Production and reservoir engineering techniques provide data that will allow an assessment of fluid connectivity and geological continuity between the wells. This information provides feedback to the geologist to validate the interpolated geology in the gaps. For example, if two wells show excellent connectivity yet the interval between the two wells is mostly interpolated as shale in the geological scheme, then further investigation is required.

The main method for filling in the gaps is to establish a *depositional scheme* for the reservoir. The sedimentological description of a small number of cores will be used to predict the 3-D rock framework for the field. This method is practicable because sedimentary environments typically show a continuity of process that allows predictions to be made. Sediments tend to be organized into discrete packages with a specific range of rock properties and with a predictable geometry. These packages can be very large, more often than not, larger than normal well spacing.

The building blocks for a depositional scheme are sedimentary *facies*, which are packages of rock that can be defined on the basis of common lithology, sedimentary structure, and organic features (de Raaf et al., 1965). *Lithofacies* are distinguished from *biofacies*; the former by its physical and chemical characteristics, the latter by its organic make up (Reading, 1996). They can be given an informal designation such as lithofacies 'A' or 'cross-bedded sandstone.'

Lithofacies that occur together may be grouped and interpreted in terms of the *environment of deposition*. Such a grouping is referred to as a *facies association* (Collinson, 1969). For example, O'Mara et al. (2003) defined four facies associations in a core taken from the Carboniferous fluvial braidplain reservoir of the Tyne field in the United Kingdom southern North Sea (Figure 57).

The geometrical body corresponding to a facies association is a macroform. To a sedimentologist, macroforms are the building blocks that make up a depositional

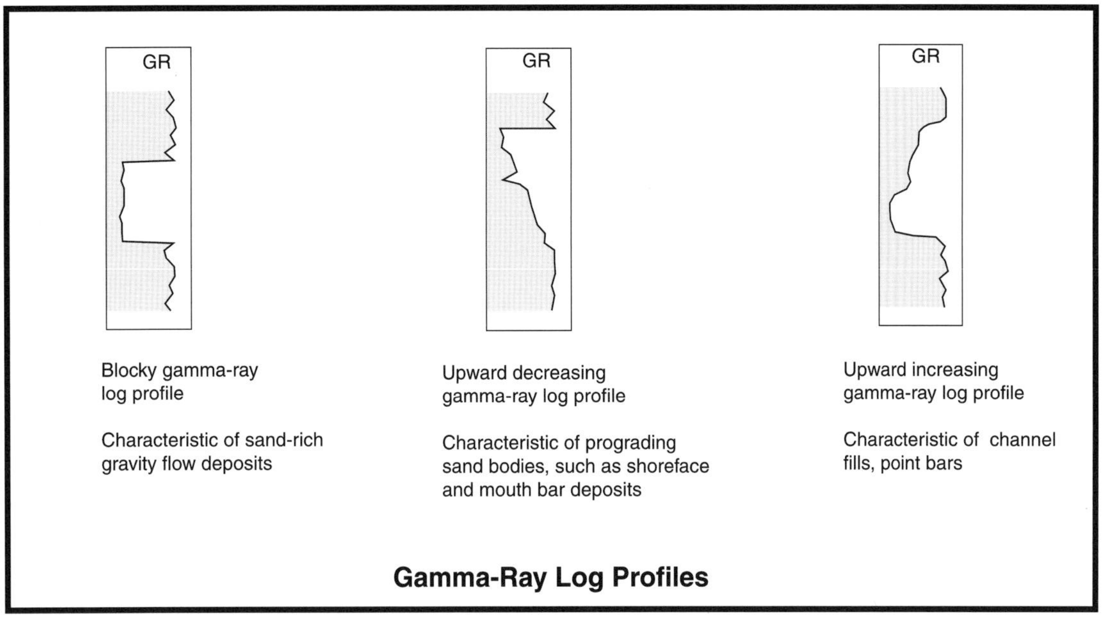

Figure 59. The gamma-ray log commonly shows patterns that give an indication of the depositional process. GR = gamma ray.

environment. A production geologist also sees them as discrete packages showing predictable patterns of grain size, sorting, and permeability variation (Slatt and Galloway, 1992). They are key elements in controlling how fluid flows within a reservoir (Figure 58).

CORE TO LOG CORRELATION OF LITHOFACIES

Lithofacies are defined from the core. However, not all of the wells in a field will be cored because it is too expensive to do this; instead, only wireline logs will normally be available. Ideally, the geologist would like to define lithofacies in the uncored wells and in the uncored intervals of any wells that are only partially cored. To do this, the lithofacies in the cored intervals are compared to the equivalent wireline log response. By correlating the lithofacies with (hopefully) a distinctive wireline log response, the equivalent log facies can be identified. A *log facies* is a distinctive wireline response or pattern.

The geologist will look for a correspondence between the lithofacies and the shape of the logs. The gamma-ray log profile can be used as a first pass indication of grain size variation and hence the environment of deposition (Figure 59). Distinctive patterns are commonly seen, and these allow the depositional environments to be inferred with some degree of reliability (Hancock, 1992).

Petrophysical software is available to make a core-to-log facies comparison automatically; for example, *neural network analysis* can be used for this. This is a computer technique that uses parallel processors to solve complex problems. In this way, a computer can be trained to identify discrete *electrofacies* by classifying distinctive parcels of log responses corresponding to the lithofacies in the cored intervals (Rogers et al., 1992; Bhatt and Helle, 2002). Once this has been done, electrofacies can then be identified from the log response in the uncored intervals.

Gupta and Johnson (2002) used electrofacies analysis on cores of a tidal sandstone reservoir interval in the Gullfaks field in the Norwegian North Sea. Five electrofacies categories were defined, and, for most of these, they consider that there is a greater than 70% probability that they correctly identify the equivalent core lithofacies.

10

Shepherd, M., 2009, The reservoir framework, *in* M. Shepherd, Oil field production geology: AAPG Memoir 91, p. 81–92.

The Reservoir Framework

INTRODUCTION

Once the lithofacies have been established, the next step is to build the depositional scheme. Initially, this involves establishing a sequence-stratigraphic framework for the reservoir interval. The recognition of the various sequences in a reservoir involves making a well correlation. The extent to which a detailed correlation can be made will depend on the complexity of the reservoir geometry, whether layer-cake, jigsaw puzzle, or labyrinthine in nature. Techniques, such as the use of micropaleontology, are available to help the geologist make a reliable correlation.

THE SEQUENCE-STRATIGRAPHIC FRAMEWORK

A major advance in petroleum geoscience took place in the 1960s and 1970s when the concepts of *seismic stratigraphy* were combined into a coherent framework by a team at Exxon led by Peter Vail. Much of this work was published by the AAPG in Memoir 26: Seismic Stratigraphy—Applications to Hydrocarbon Exploration (Payton, 1977). The main observations are that depositional episodes or sequences in a sedimentary basin are related to cyclical changes in relative sea level, and that the effect of these changes can be recognized on basin-scale seismic data. The identification of surfaces that are partly unconformities is the way by which the tops and bases of individual sequences are defined. A *depositional sequence* is therefore defined as a stratigraphic unit composed of a relatively conformable succession of genetically related strata and bounded at its top and base by unconformities, or their correlative conformities (Mitchum, 1977). The surface defined by an unconformity and its correlative conformity is the *sequence boundary*. They form as a result of a relative fall in sea level. Sequence boundaries can be identified on seismic lines, particularly where there is an angular discordance between the underlying truncated reflectors and the overlying reflectors that lap onto the surface. The identification of *seismic sequences* (a depositional sequence identified on seismic data) allows the stratigraphic subdivision of a sedimentary basin to be worked out in detail and this has proved of immense value as an analytical tool for explorationists looking for new prospects to drill.

A depositional sequence of sediments deposited at a continental shelf margin can be further subdivided on the basis of an idealized sequence stratigraphic model. A depositional sequence consists of a vertical succession of sediment packages and bounding surfaces, which from the bottom upwards comprises a sequence boundary, a lowstand systems tract, a transgressive surface, a transgressive systems tract, a maximum flooding surface, a highstand sequence tract, and the subsequent sequence boundary. A *systems tract* is the overall package of sediment deposited during a specific phase of the relative sea level cycle and has been defined as a linkage of contemporaneous depositional systems, where a depositional system is a three-dimensional assemblage of lithofacies, genetically linked by active or inferred processes and environments (Brown and Fisher, 1977).

The *lowstand systems* tract comprises the depositional systems that developed when relative sea level was low following the formation of the sequence boundary. It is characterized by a falling, stable and slowly rising sea level regime. High sediment supply and limited accommodation space results in a sedimentary geometry characterized by low-rate aggradational (building upwards) and progradational (building out seawards) stacking motifs. There will then follow a phase when relative sea level is rising at a rate where more accommodation space is created than can be filled by sediments. The result is an interval of sediments showing a retrogradational motif, that is, a back-stepping pattern. This is the *transgressive systems tract*. The base of the

DOI:10.1306/13161193M913372

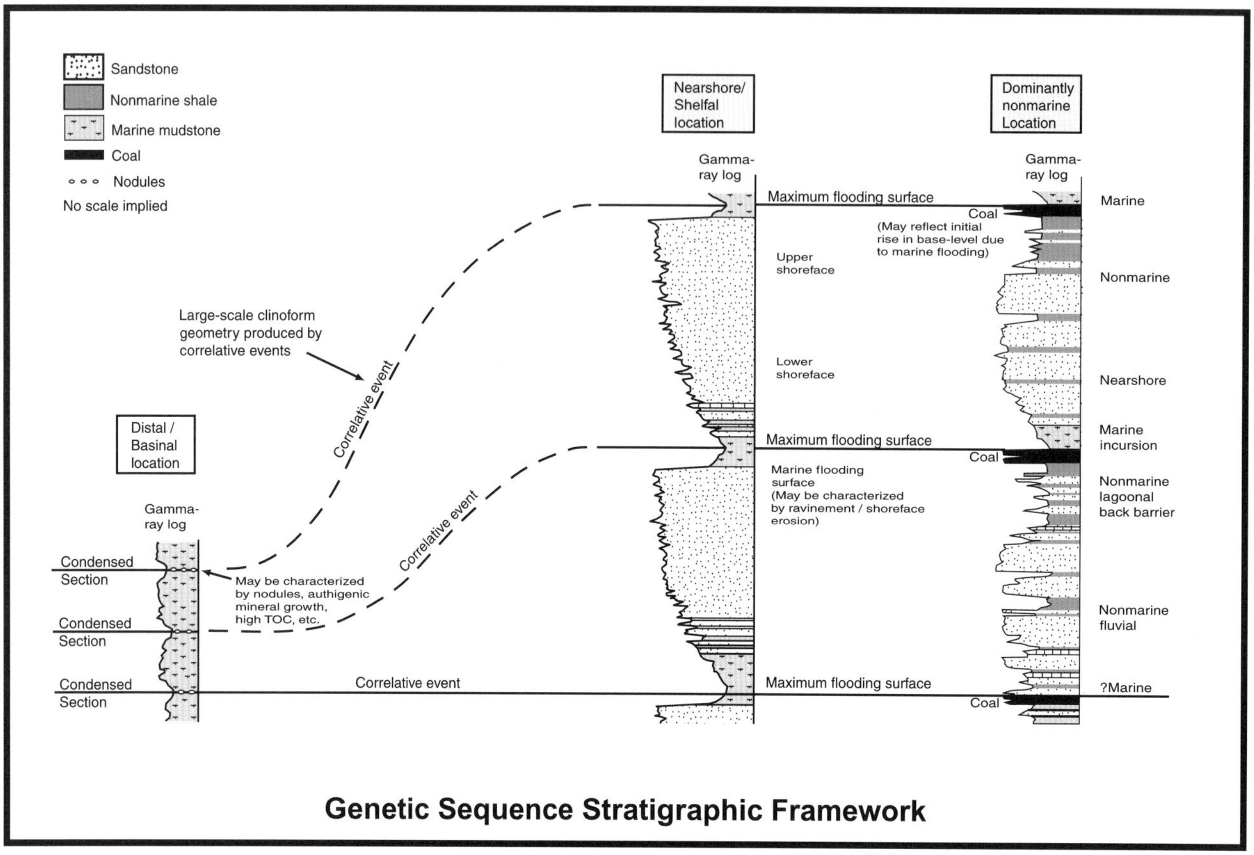

FIGURE 60. Sequence stratigraphy provides a framework for defining envelopes, which contain discrete depositional episodes. Reservoir correlation is ideally conducted within a sequence-stratigraphic scheme. In a genetic sequence stratigraphic scheme, maximum flooding surfaces are used as the tops and bases of genetic sequences (from Underhill and Partington, 1993). Reprinted with permission from the AAPG. TOC = total organic content.

transgressive systems tract is the *transgressive surface*, a prominent flooding surface and commonly represented by condensation.

The transgressive systems tract is succeeded by the *highstand systems tract*. This happens once the relative rise in sea level starts to diminish and accommodation space is either created or destroyed at a relatively slow rate. This causes the sediments to aggrade and then prograde. The base of the highstand systems tract is the *maximum flooding surface*. It is commonly marked by condensation and the farthest landward extend of deep-water sediments.

Many production geologists prefer to use a variation on the theme of sequence stratigraphy, the *genetic sequence stratigraphy* framework as defined by Galloway (1989). This scheme uses maximum flooding surfaces instead of sequence boundaries to define the top and bases of depositional episodes, as they are easier to pick out on logs and in cores. These are commonly identified within shales, typically rich in organic material, or a condensed marine facies with a distinctive biostratigraphic assemblage (Myers and Milton, 1996).

HIGH-RESOLUTION SEQUENCE STRATIGRAPHY

Seismic stratigraphy was originally applied using seismic data at the basin scale. A few years later, efforts were made to understand the effects of relative sea level changes in sediments at a scale closer to that of the hydrocarbon reservoir. This scale of analysis has been called *high-resolution sequence stratigraphy* (Van Wagoner et al., 1990). It involves determining sedimentary sequences from core, log, and outcrop studies (Figure 60).

The laterally extensive bounding surfaces recognized were typically found to correspond to the tops and bases of reservoir and intrareservoir permeability barriers. They also provide the 'envelope' containing discrete depositional episodes. This clearly shows the value of using a sequence-stratigraphic analysis as the framework for building the geological scheme for a reservoir. Stratigraphic sequences are the containers for the assemblages of the various macroforms controlling the rock property variation and fluid flow patterns within the reservoir.

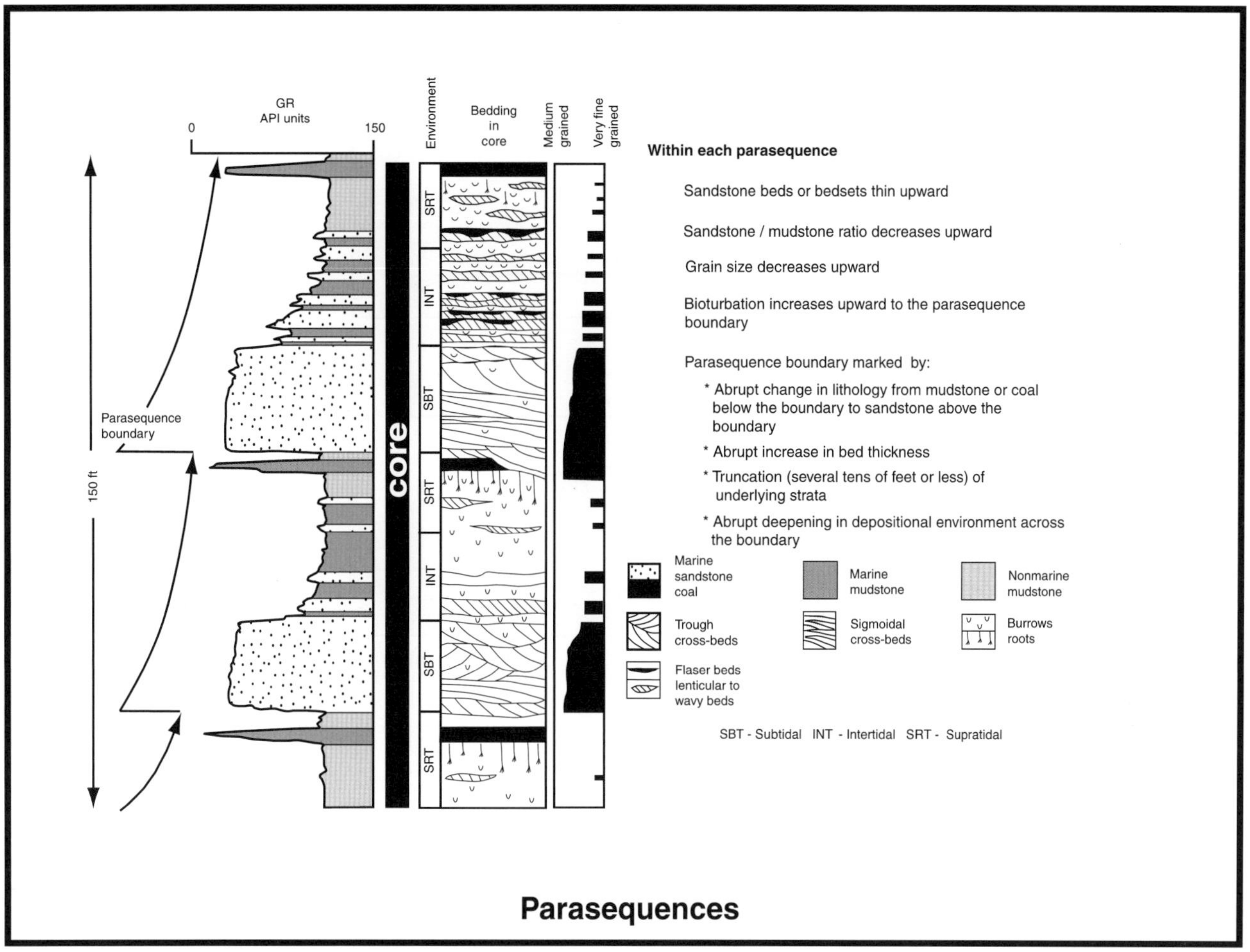

FIGURE 61. Parasequences are relatively conformable successions of genetically related beds or bedsets bounded by marine flooding surfaces and their correlative surfaces (from Van Wagoner et al., 1990). Reprinted with permission from the AAPG. GR = gamma ray.

PARASEQUENCES

Van Wagoner et al. (1988) recognized that a depositional sequence of marine sediments can internally comprise stacked cycles separated by thin flooding surfaces. These cycles have been termed *parasequences*, defined as relatively conformable successions of genetically related beds or bedsets bounded by *marine flooding surfaces* and their correlative surfaces (Figure 61). A marine flooding surface is a surface separating younger from older strata across which there is evidence of an abrupt increase in water depth. The definition of a parasequence does not refer to their thickness variation; nevertheless this can be on a scale of meters to tens of meters, and sometimes thicker.

Parasequence sets are a series of stacked parasequences. Parasequences and parasequence sets are the building blocks for depositional sequences. Individual parasequences show a predictable internal make up, with a series of coastal to offshore facies belts. Parasequences can have distinctive stacking patterns. Where a well cuts through several parasequences, the facies in each parasequence should be compared with those in the parasequences above and below. If a parasequence with a coastal facies association directly overlies a parasequence showing offshore marine sediments, then the facies belts in the upper parasequence are displaced toward the coast compared to the lower parasequence. This is a *stacking pattern*. These can be *progradational, aggradational*, or *retrogradational* depending on the balance between sediment supply and the accommodation space available for the sediment (Figure 62). Once the stacking pattern has been recognized, then this can be used to predict the seaward or landward shift in facies belts for successive parasequences. In practice, this means that it may be possible to predict depositional patterns in marine sediments in those areas beyond the existing well control.

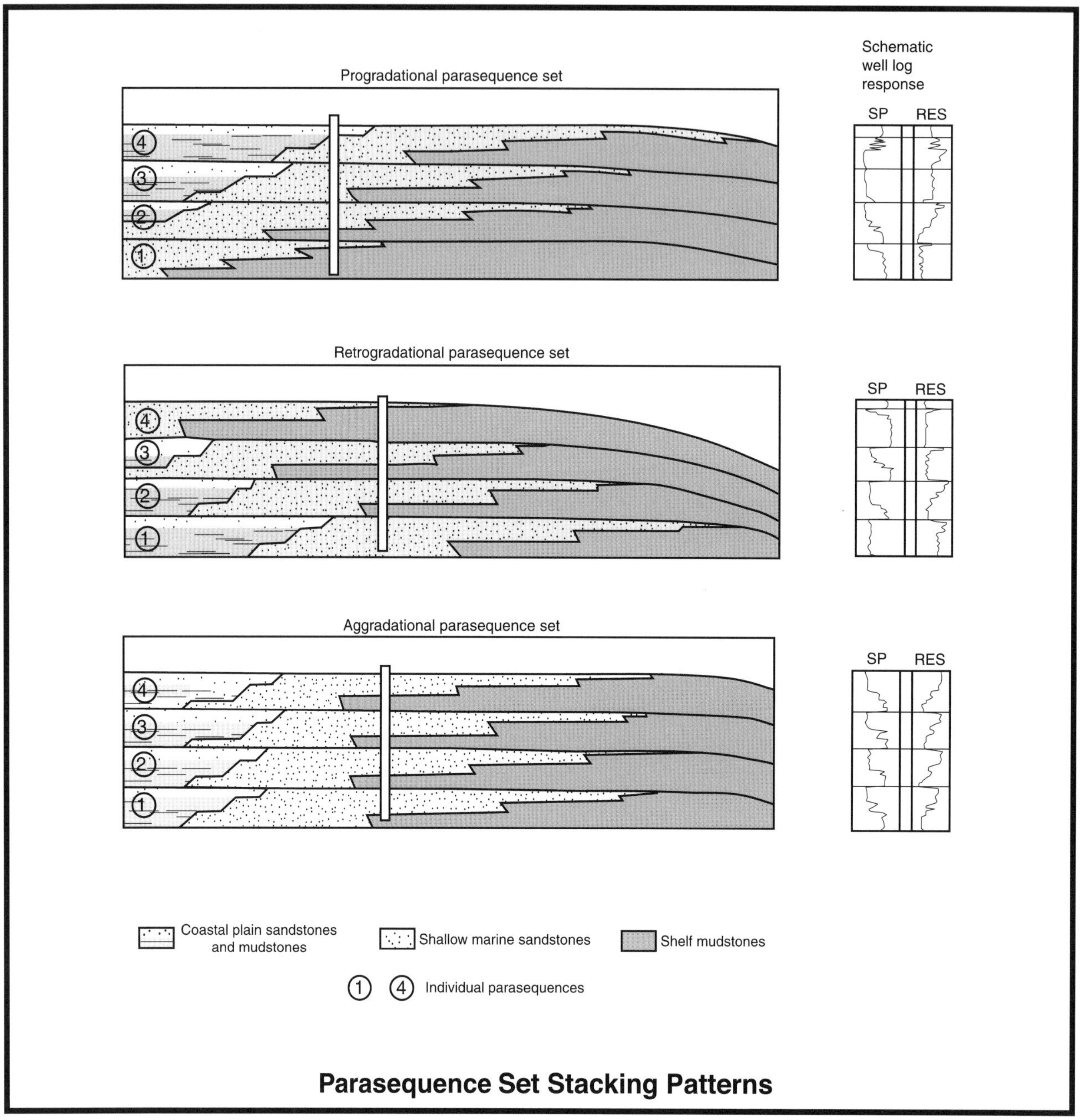

Figure 62. Parasequence stacking patterns can be used to predict the location of landward and seaward facies belts within a reservoir (from Van Wagoner et al., 1990). Reprinted with permission from the AAPG. SP = spontaneous potential log; RES = resistivity log.

ESTABLISHING THE SEQUENCE-STRATIGRAPHIC FRAMEWORK

The work involved in establishing a sequence-stratigraphic framework for a reservoir is iterative in nature. The analysis involves looping through the seismic patterns, the log correlation, biostratigraphy, and the sedimentological scheme, eventually tying the data together into a coherent interpretation. If the basin-scale sequence stratigraphic framework has been established, then this should be used to help understand the stratigraphic framework at the reservoir scale.

The first step is to determine the key stratal surfaces in the wells. Significant flooding surfaces are defined followed by the identification of smaller scale bounding surfaces, including any incision surfaces (Holtz and Hamilton, 1998). The procedure involves determining the vertical facies profile in the individual wells. Flooding events are established by inferring marked shifts in water depths from the more shoreward facies upward to the more basinal facies in cores and logs (Van

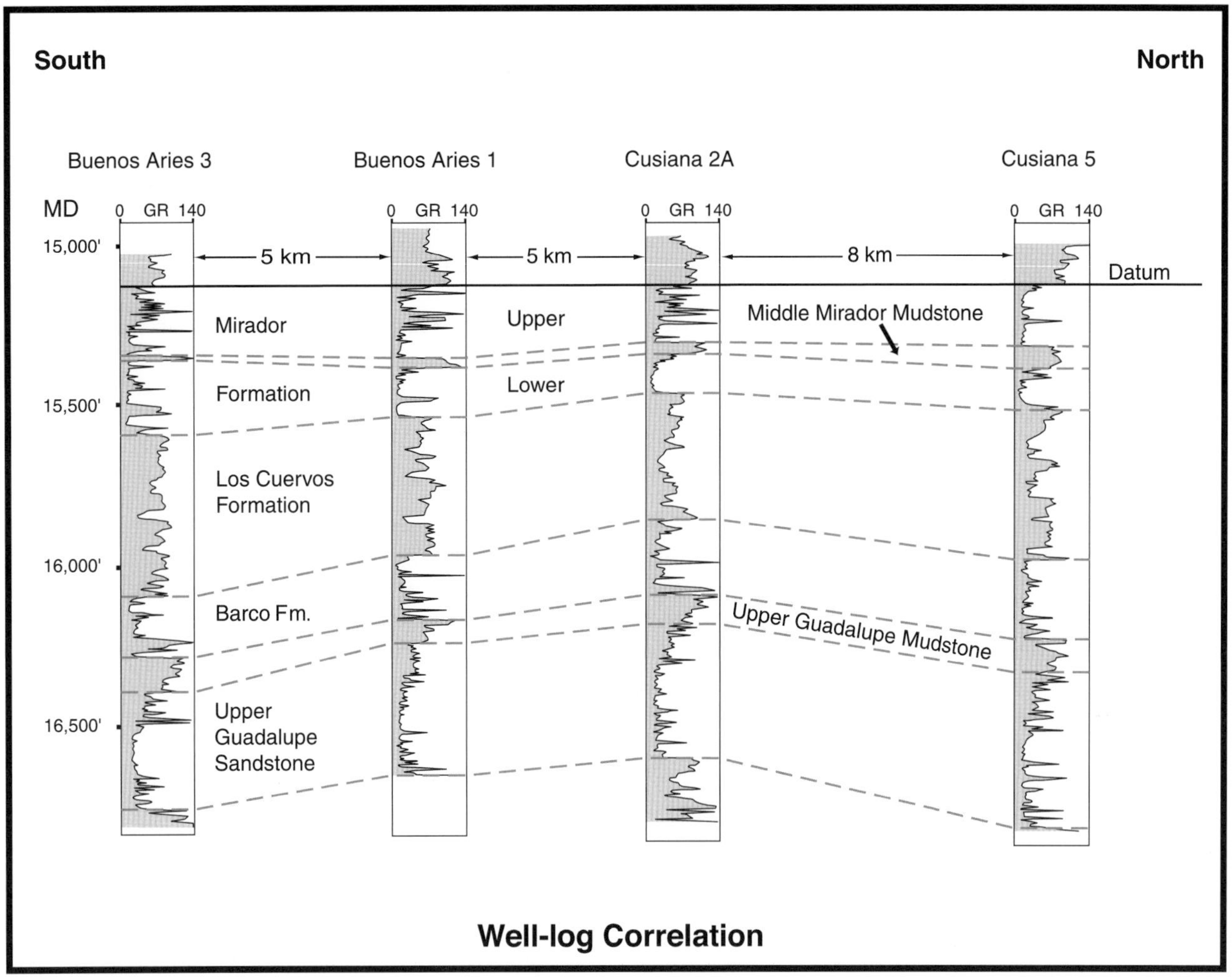

FIGURE 63. Log correlation of wells in the Cusiana field, Colombia. From Cazier et al. (1995). Reprinted with permission from the AAPG. GR = gamma ray.

Wagoner et al., 1990). Upper shoreface sandstones may be overlain by deeper water mudstones and siltstones for instance. An example of the identification of flooding surfaces is given by Shaw et al. (1996) from the lower delta plain sediments of the upper subzones of the Endicott field in Alaska. Sharply defined surfaces are seen in the core where paleosols marking subaerial exposure are overlain by mudstones with a trace fossil assemblage indicating brackish water conditions.

Care has to be taken in establishing that a change upward to muddy facies results from a flooding event and is not the result of a lateral facies progression from shallow-water sand deposition to shallow-water mud deposition; e.g., from a barrier bar sandstone to a lagoonal mudstone (Larue and Legarre, 2004). This is easier to establish in cores than it is on logs. Nevertheless, even where mostly logs are available, well-log correlation can help as marine flooding surfaces are generally regionally extensive.

Establishing sequences or their equivalent in nonmarine sediments is more difficult. Nevertheless, Hamilton et al. (1998) considered that hiatal events in continental sediments can be recognized and that these result from rapid climate change or a shift in base level. Such features include subaerial erosional unconformities, lacustrine shale markers, and regional paleosols.

WELL-LOG CORRELATION

Having established the key stratal surfaces in the wells, the next step is to determine the genetic sequence-stratigraphic framework across the field. The data that will be used to do this will usually be available from vertical and near-vertical wells. Because of this, geologists will have a good representation of the vertical variation in the reservoir geology but they may not have much data to show what is happening laterally (Bryant and Flint, 1993). The geologist is therefore required to infer the lateral reservoir character by various means. The starting point for this is to make a *well-log correlation* (Figure 63).

Table 13. Typical reservoir geometries.*

	Terrestrial	*Coastal*	*Marine*
Layer-cake	Sheet flood deposits, lacustrine sheet sand, eolian dunes	Barrier bars, chenier deposits, transgressive sandstones	Shallow marine sheet sandstones, offshore bars, outer fan turbidites
Jigsaw puzzle	Braided river deposits, point bars, mixed lacustrine and fluvial sediment, mixed eolian and wadi reservoirs	Combined facies complexes, barrier bar plus tidal channel-fill combinations in high net-to-gross intervals	Storm sand lenses, midfan turbidites
Labyrinth	Fluvioglacial deposits with low net-to-gross intervals, low-sinuosity channel fills	Low-sinuosity distributary channel fills	Upper fan turbidites, slumps, storm deposits in low net-to-gross intervals

**From Weber and van Geuns (1990). Reprinted with permission from the Society of Petroleum Engineers.*

As core coverage will normally be sparse, the data source used for well correlation will be wireline logs. The aim in well correlation is to look for *log signatures* that are similar from well to well. These are distinctive patterns or grouping of patterns that can be recognized on the logs. The log signatures help to identify and map out laterally continuous sections within the reservoir. Log correlation is used to subdivide the reservoir vertically into reservoir units based on genetic sequences.

The main logs used in correlating log signatures are the gamma-ray, resistivity, sonic, density, and neutron logs. Other logs may also help, especially if they are harboring a log signature not immediately obvious on the 'main' logs. A correlation plot is made by cutting and pasting well logs onto a sheet of paper or using a computer application designed for this purpose. The logs are hung from a datum, top reservoir for instance, or a marker horizon. Lines are drawn on the log correlation plot connecting the tops of each sequence between the wells. Once the correlation has been defined, a table of reservoir tops is made for each of the units. This is used as the data-set for making depth and thickness maps of the reservoir units (see Chapter 8, this publication).

WELL-LOG CORRELATION AND RESERVOIR GEOMETRY

A correlation between wells is usually possible because many sedimentary environments have a consistent character over a larger distance than the well spacing. On other occasions, the correlation between wells can be much more difficult to make. In certain sedimentary environments, many of the individual sediment bodies can show a length scale that is shorter than the typical interwell spacing. Only the longer ranging features will be correlatable here.

Reservoirs can vary enormously in their character from high continuity and simple at one end of the spectrum to very complex at the other. As a generalization, reservoirs with a strong wave-dominated marine influence such as shoreline systems, wave-dominated deltas, and barrier bars are relatively simple with good lateral continuity. Reservoirs with channel systems and tidal influence show rapid facies variation, exhibit a high degree of complexity, and may be continuous over short ranges only (Tyler and Finley, 1991). Weber and van Geuns (1990) proposed a very useful and simple classification of reservoir geometries. These are the layer-cake, jigsaw-puzzle, and labyrinth reservoir types (Table 13).

Layer-cake reservoirs consist of laterally extensive sandstone units having no major discontinuities or changes in horizontal permeability (Figure 64). Intervals within a layer-cake reservoir do not necessarily need to show a constant thickness to satisfy the definition, but any thickness changes should be gradual. The reservoir units will have good areal connectivity, and waterflooding will result in an efficient areal sweep (Dromgoole and Speers, 1997).

Jigsaw-puzzle reservoirs are made up of a series of sand bodies that fit together without any major gaps between the units (Figure 65). An occasional low or nonpermeable body may be embedded in the reservoir, and nonpermeable baffles may exist between superimposed sand bodies. In these types of reservoirs, several wells per square kilometer are needed before the reservoir architecture can be defined adequately. On waterflooding, there will be some reservoir dead ends that will be bypassed and can trap oil.

Labyrinth reservoirs are complex arrangements of sand pods, lenses, and channels (Figure 66). A detailed correlation is only possible in these reservoir systems with close well spacings, if it is possible at all. Connectivity is commonly anisotropic and strongest in the paleoslope or current flow direction. Reservoir management is difficult in a labyrinth reservoir. The planning of injector-producer well configurations will be problematic because the connectivity between the sand bodies will often be poorly understood. The injection and production wells

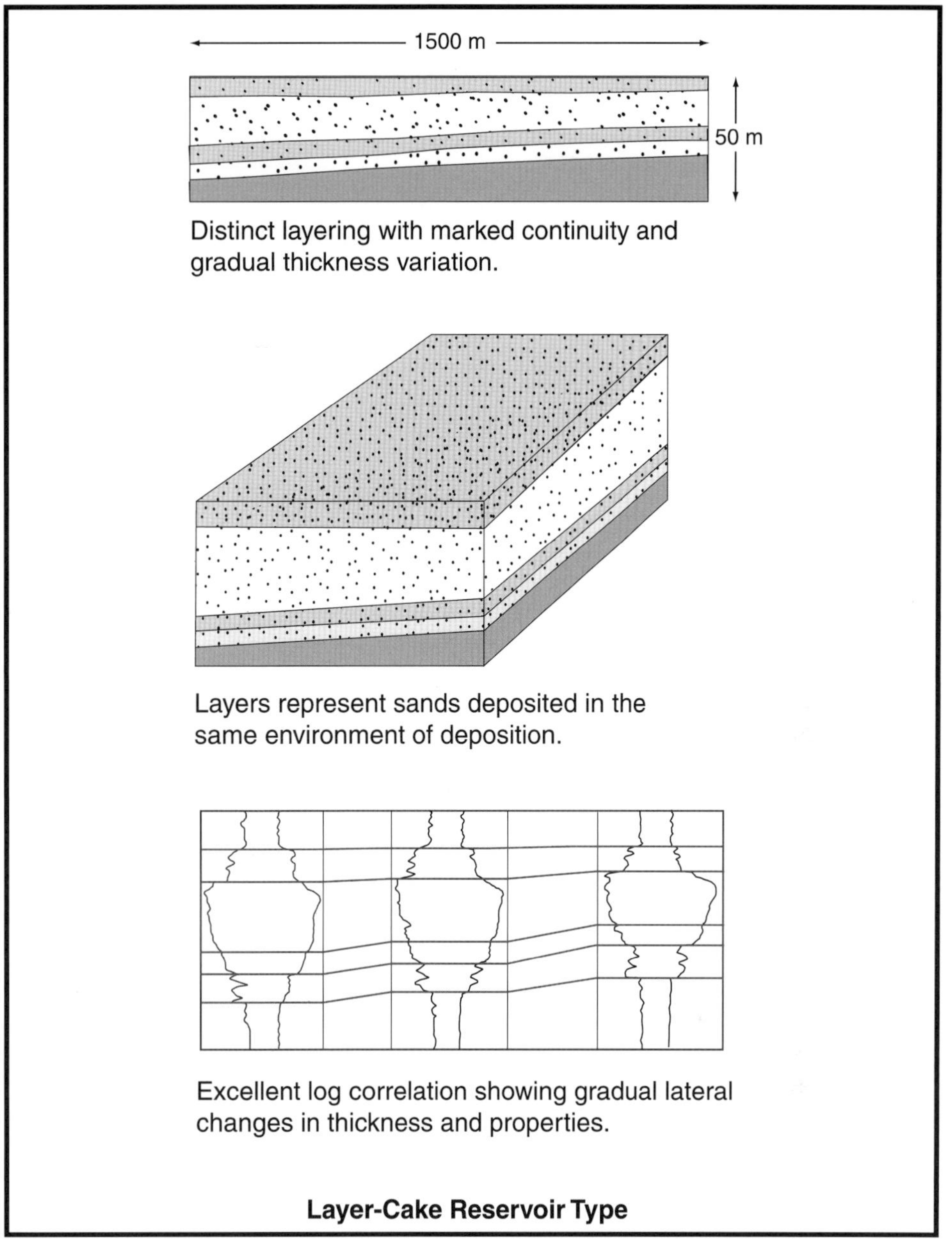

FIGURE 64. Layer-cake reservoirs consist of units of laterally extensive sandstone that show only gradual changes in thickness and rock property characteristics (from Weber and van Geuns, 1990). Reprinted with permission from the Society of Petroleum Engineers.

will be difficult to locate with accuracy, and, in some cases, they can be isolated in different sand bodies (Dromgoole and Speers, 1997).

SHINGLED GEOMETRY

A very common geometric configuration in sediments is the *shingle or clinoform geometry*. This is a geometric pattern where individual inclined sediment bodies downlap onto an underlying flat surface (Figure 67).

It is a common mistake in production geology to force a layer-cake correlation through a sedimentary system that comprises a series of shingled units. If shales or carbonate cements bound each shingle, then potential compartmentalization may go unheeded. Experienced production geologists have made a career out of resurrecting old fields where bypassed oil in shingle geometry reservoirs went unnoticed by the previous operator (Sneider and Sneider, 2001). The Sirikit field of Thailand is an example of a reservoir where the production behavior was better understood once the shingled nature of the fluvial-deltaic sandstones was defined (Ainsworth et al., 1999). Oolite shoals commonly show a shingled geometry in carbonate reservoirs (Sneider and Sneider, 2001). Shingled geometries are also found in some prodelta mass flow deposits.

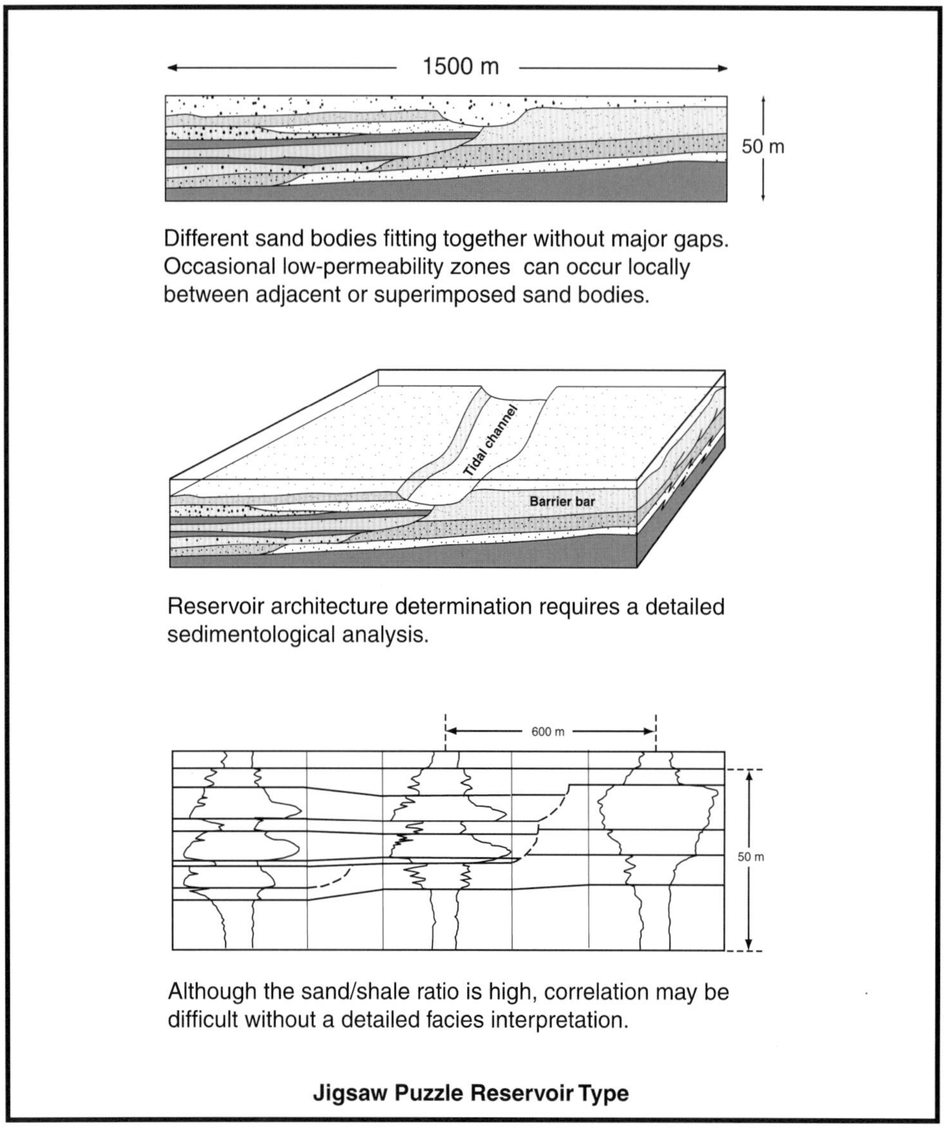

FIGURE 65. Jigsaw-puzzle reservoirs comprise an interlocking complex of sand bodies with the occasional low-permeability interval acting as a baffle (from Weber and van Geuns, 1990). Reprinted with permission from the Society of Petroleum Engineers.

MARKER HORIZONS

Certain intervals can be used as *marker horizons*. These are beds with distinctive features that allow them to be identified and correlated with a high degree of confidence between several wells. Thus, they can be mapped field wide or over large areas of the field. Beds that can be used for marker horizons include coals, paleosols, ash beds, conglomerates, marine limestones, hard grounds, condensed sequences, and organic shales. Many of these are significant in terms of the sequence-stratigraphic framework.

For example, the correlation of fluvial sand bodies is very difficult in meander belt depositional environments. However, if coals are present, they are usually excellent marker beds for defining reservoir zones. Coals form over long periods of time and can be considered as approximate time lines separating individual depositional packages. They are considered to show the essential characteristics of genetic sequence boundaries (Hamilton and Tadros, 1994).

BIOSTRATIGRAPHY

Biostratigraphy involves the identification of fossils so as to age date strata for correlation purposes. In the oil field, microfossils are used because these are small enough to be recovered from drill bit cuttings. *Micropaleontology* is the study and analysis of faunal populations (e.g., foraminifera and ostracods), and *palynology* addresses floral populations (e.g., dinoflagellates, pollen and spores).

Biostratigraphy helps to understand the sequence stratigraphy of a basin. It is used in production geology when it is not possible to correlate the log character field

FIGURE 66. Labyrinth reservoirs are made up of a complex arrangement of sand pods, lenses, and channels. The connectivity between the various sand bodies is tortuous, that is, if it exists at all. This can be the most difficult of the reservoir types to develop (from Weber and van Geuns, 1990). Reprinted with permission from the Society of Petroleum Engineers.

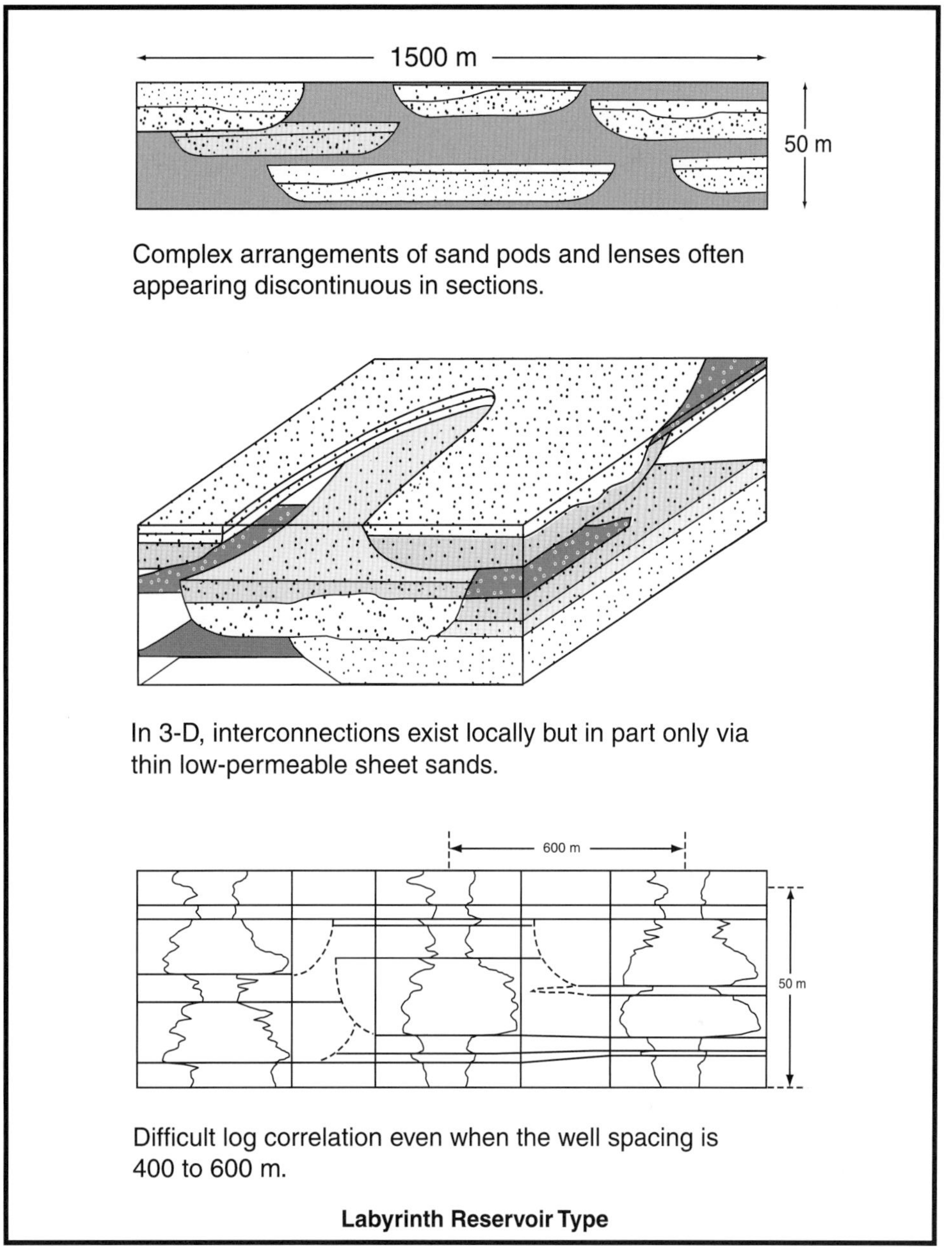

wide with a high degree of confidence. The various *biostratigraphic markers* are posted on well logs and are used to guide the correlation. For example, Morris et al. (1999) used biostratigraphic analysis to pick nine *bioevents* within the reservoir interval of the Magnus field, UK North Sea (Figure 68). These guided the construction of the stratigraphic framework.

Biostratigraphic analysis is made on a well-by-well basis. Biostratigraphy reports will include tables of the *biostratigraphic zonation* for each well and a detailed interval by interval breakdown of all the significant events and markers. They will also contain detailed *biostratigraphic analysis charts* along with a summary wireline log showing the key bioevents and the chronostratigraphic zonation (Figure 68).

The analysis of biostratigraphic data is not an easy task, and the depths assigned to biostratigraphic markers can be prone to a large range of uncertainty. The method depends on picking up changes in assemblages of microfossils, which are either *abundance events*, or the last downhole occurrence (*inception events*), or the first downhole occurrence (*extinction events*). The recognition of these events is frequently based on a sparse data set. The source material can be of variable quality; core is the best, sidewall cores are the next best, and drill cuttings are the worst for sampling. The cuttings will not be precisely on depth. The depth at which the cuttings were drilled by the drill bit is estimated from the amount of time it takes for the cuttings to come to the surface using the drilling mud circulation rates for the well.

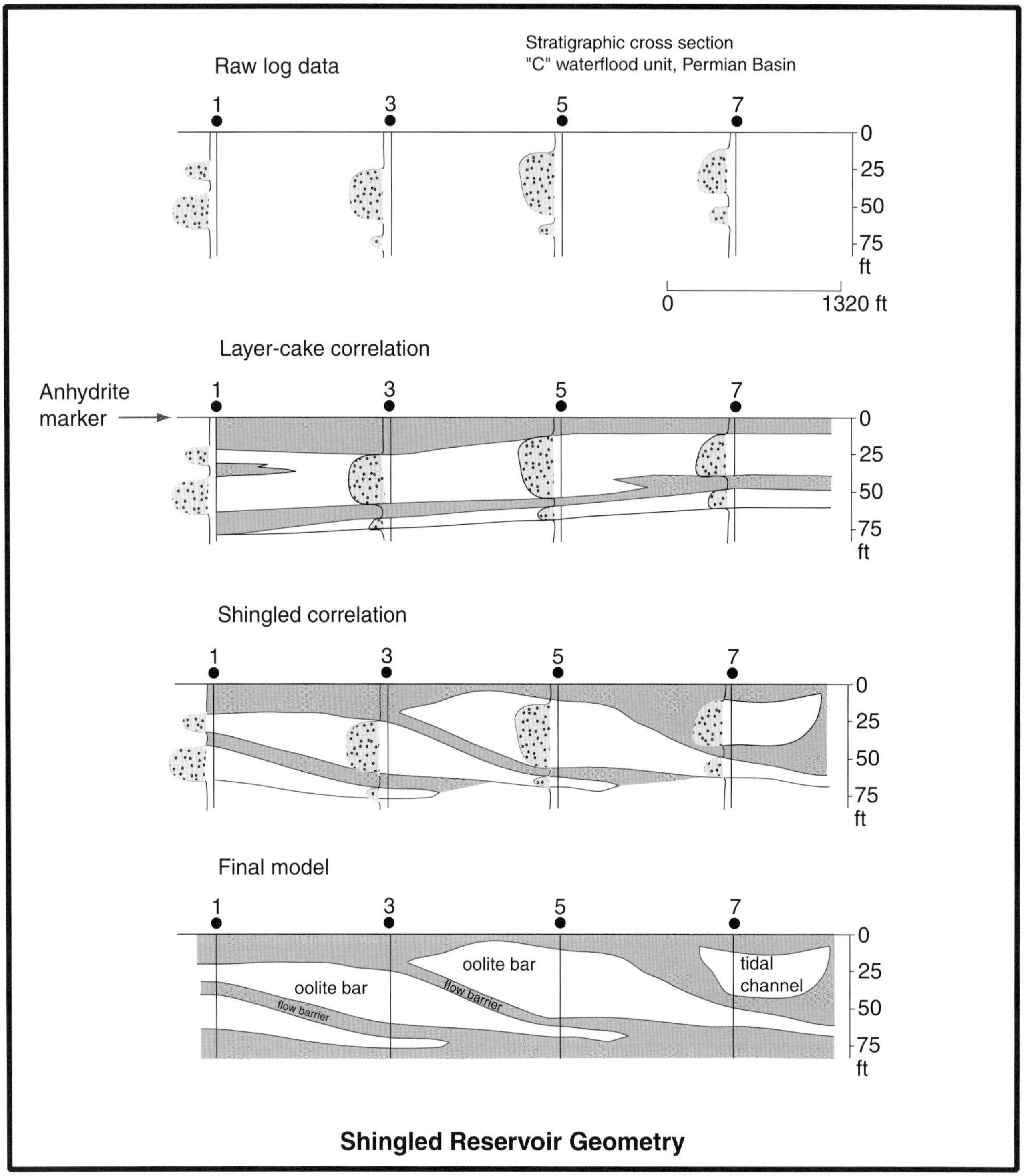

FIGURE 67. Shingled geometries are common in certain depositional environments and can result in a number of isolated reservoir segments. However, this type of geometry is easy to overlook, and a layer-cake geometry is often erroneously imposed (from Sneider and Sneider, 2001). Reprinted with permission from the AAPG.

Slippage can occur down through the drilling mud, and the resultant churning and mixing of the cuttings may lead to a loss of data coherence. Extinction events, as the first downhole occurrence of a species, are thought to be more reliably represented in cuttings than inception events (Sturrock, 1996). Sometimes *cavings*, heavier rock fragments falling down from higher up the borehole, will add to the general data disorder.

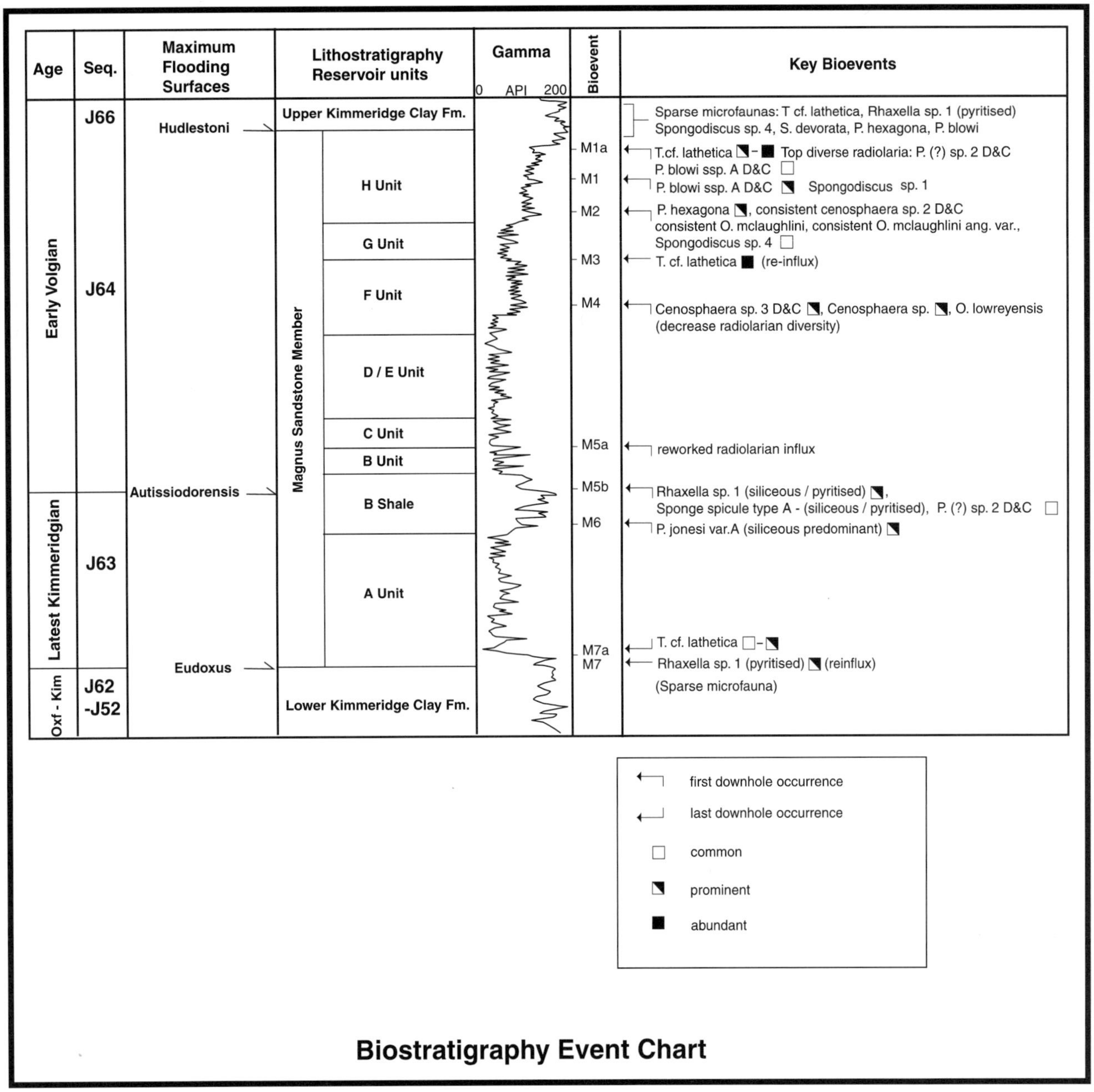

FIGURE 68. Biostratigraphic event chart for the Magnus field, UK North Sea (from Morris et al., 1999). Reprinted with permission from the Geological Society. Seq. = Sequence.

Micropaleontological and palynological assemblages can be used as an aid to the interpretation of depositional environments, since different species may be found associated with different water depths, salinities, and the relative volume of terrestrial sourced sediment in the sample. These can help identify the depositional environment of a reservoir interval particularly where this cannot be confidently inferred from other data. In marginal marine environments, they can be used to differentiate between sediments deposited under fresh water, brackish, and open marine conditions. For example, Parry et al. (1981) used palynology to help establish various environments of deposition within the fluviodeltaic succession in the Brent Group of the UK North Sea.

OTHER AIDS TO RESERVOIR CORRELATION

Several other techniques are available to help make a reservoir correlation, especially in rocks where microfossils are sparse or absent. The vertical variation in the abundance and ratios of *heavy minerals* in sandstone intervals can be used to create a framework for the correlation of reservoir units (Morton et al., 2002). For example, Morton and Hurst (1995) used heavy mineral data to validate the correlation within the fluvial to coastal reservoir sequence of the Statfjord Formation in the Snorre field, Norwegian North Sea.

Whole-rock trace element geochemistry was used to help correlate continental red bed successions in the Triassic of the Beryl field in the North Sea (Preston et al., 1998). These sediments are devoid of fossils and would be otherwise difficult to correlate. Similar techniques have also been used to subdivide the Carboniferous fluvial reservoir interval of the Schooner field in the southern North Sea (Moscariello, 2003) and the Triassic of the Berkine Basin in Algeria (Ratcliffe et al., 2006).

Magnetostratigraphy is based on the measurement of the natural remnant magnetization in sediments. The stratigraphy is tied to known magnetic polarity and reversal events from field outcrops. Magnetostratigraphy was used to help with the reservoir correlation on the Johnston field in the Southern North Sea (Lawton and Roberson, 2003). A framework of 13 magnetic zones was defined in the Permian eolian sandstone reservoir.

Trace fossils are the indications that various organisms have left behind in or on sediments such as burrows or trails. These can be distinctive and may indicate a particular type of sedimentary environment. Where trace fossils are present in sediments, they may define assemblages related to the water salinity, rate of sedimentation, and energy of deposition (Frey and Pemberton, 1984; Bromley, 1996). Because they are sensitive to environmental change, they are of value in high-resolution sequence stratigraphy. Trace fossils may be used to define flooding surfaces and erosion surfaces, each of which can show a distinctive *ichnofacies*. Ichnofacies are characterized by a particular assemblage of trace fossils.

WELL CORRELATION ANOMALIES

The geologist, on making a well-log correlation, may find that some of the wells do not fit easily in to the overall scheme. A well may be missing a field-wide marker bed, present in all the other wells; or another well may have an unusually thin section of an interval that shows a constant layer-cake thickness everywhere else. It is good practice to make hand-drawn thickness maps between the main marker horizons after a first pass correlation. Anomalies will stand out as "bull's-eyes" on the contour maps.

Anomalies can result from four main causes:

1) There may be a mispicked correlation in a particular well and this can become obvious when the anomaly is recognized as a bull's-eye on thickness maps.
2) Anomalous thinning can result from a stratigraphic pinch-out or condensing of the interval over a preexisting basement high.
3) A reservoir interval may be locally absent because of erosion beneath a regional unconformity or an incised valley or channel.
4) If a vertical or high-angle well penetrates a normal fault, the fault will cut out an interval of stratigraphic section. The amount cut out will depend on how big the fault is. Cross checking with dipmeter, image, and seismic data may confirm whether a fault is present (see Section 13, this publication). If a normal fault cutting a well is overlooked, any thickness maps using the well data will show an interval that is too thin in the area around the well.

Shepherd, M., 2009, Lithofacies maps, *in* M. Shepherd, Oil field production geology: AAPG Memoir 91, p. 93–98.

11

Lithofacies Maps

INTRODUCTION

Once the correlation framework has been established, the next stage is to derive a series of lithofacies maps for each genetic sequence (or systems tract if this is practicable). Reservoir analogs are used to help in extrapolating from the lithofacies in the cored wells to a depositional scheme for the reservoir as a whole.

LITHOFACIES MAPS

Lithofacies maps show the areal variation in the depositional patterns that make up each genetic sequence within the reservoir interval. The method for constructing lithofacies maps involves extrapolating the lithofacies from the wells into the gaps between the wells. This is not easy, as there will be mostly vertical or near vertical well data in the field. The vertical facies profile may be determined reasonably confidently; however, the lateral facies progression will have to be inferred by analog and other means. Methods for doing so are described in this chapter.

USE OF ANALOGS FOR LITHOFACIES MAPPING

The practice of making lithofacies maps with the help of *modern analogs* is an effective method for filling in the gaps in the subsurface (Galloway and Hobday, 1996). Modern analogs show the geometrical interrelationship of the various sedimentary bodies; they also permit the width and length of the various macroforms to be readily measured.

However, there are limitations to the use of modern analogs. Certain macroforms lack *preservation potential* and may not be common in the subsurface because of erosion. Conditions today may not be anything like the prevailing conditions when a given interval of reservoir sediments formed. The present day has a specific climate, tectonic variability, relative position of sea level, and rate of sea level change (Grammer et al., 2004). For example, the continents are mountainous and widely dispersed, with a tendency for shorelines to cut predominantly north–south across the world's climate zones. The Earth currently has ice caps but has lacked them for large parts of its geological history.

Modern and ancient analogs can be investigated by referring to technical papers and outcrop studies, or the examination of aerial photographs (Tye, 2004). The latter is a particularly vivid source of information (Figure 69). Geologists working on carbonate reservoirs, for example, should have a look at the AAPG publication on modern day carbonate environments (Harris and Kowalik, 1994). The book includes a series of aerial photographs along with a transparent overlay giving the outlines of typical carbonate fields. This gives a very good impression of the size of depositional environments relative to the field scale. The photographs illustrate the fact that many depositional environments cover a much larger area than oil fields. This is an important observation. Because depositional environments extend over a greater area than the field, the geologist should investigate the sedimentology at the larger (basin) scale to get a realistic idea of the basinal controls influencing the sediments at the field scale.

Satellite photographs are available for consultation on the Internet; for example, Google Earth™ (www. Earth .google.com). The site provides an easily accessible integrated network of satellite photographs spanning the globe. These can be examined at any scale. The program also provides a measuring tool, which allows sediment body dimensions to be readily derived. Table 14 provides a list of some potential reservoir analogs to look at on satellite photographs.

A useful source of data comes from modern-day sedimentary analogs where shallow borehole data, shallow seismic data, or ground penetrating radar surveys are available. An example of a modern day analog with boreholes is the study by Van Heerden and Roberts (1988)

DOI:10.1306/13161194M913372

Figure 69. An alluvial fan in China's Xinjuang Province. The diameter of the fan is about 170 km (106 mi). Courtesy of the NASA Web site: www.earthasart.gsfc.nasa.gov. Shown is the approximate size of a billion-barrel oil field.

of the Atchafalaya delta lobe of the Mississippi delta. Similarly, 3-D seismic data at shallow depths can give a detailed image of depositional systems at a resolution suitable for reservoir analogs (Fowler et al., 2004; Posamentier, 2004; Steffens et al., 2004).

Outcrops allow the geometrical patterns shown by lateral facies variation to be inspected. They may be less reliable as a direct analog for reservoirs by comparison to modern-day environments because their origin requires interpretation, and this introduces an element of subjectivity. Nevertheless, they will provide sedimentological information directly comparable to cores from the subsurface. Perhaps the ideal analog database involves a combination of images from modern-day environments for geometrical data and outcrops for a sedimentological comparison.

THREE-DIMENSIONAL SEISMIC GEOMORPHOLOGY

It may be possible under favorable circumstances to obtain sedimentological information from seismic data (Figure 70). For this to happen, sedimentary bodies should be of sufficient thickness relative to the seismic frequency. They will also need to show an acoustic impedance contrast large enough for them to be seen on seismic sections (Weber, 1993). For a sand body to be picked out seismically, this means in practice that it should be thick enough to be seismically resolvable (tens of meters thick) and surrounded by shale.

Geophysicists can create amplitude maps of a horizon very quickly by *autotracking* a seismic pick on a workstation. Having picked a few representative lines to define the horizon, the computer software proceeds

Table 14. Locations of some potential modern-day reservoir analogs.

Depositional Environment	*Location*	*Latitude*	*Longitude*
Desert dune field	Namibia	24°49′27.07″S	15°22′50.95″E
Barchan dunes	China	39°51′13.70″N	102°35′24.13″E
Sabkha lake	Tunisia	33°41′28.31″N	8°28′21.40″E
Alluvial fans	United States	36°10′42.30″N	116°54′12.46″W
Braided river	Madagascar	21°45′32.20″S	43°53′57.55″E
Braided river	New Zealand	43°42′57.69″S	171°57′56.03″E
Meander belt	Russia	58°49′39.10″N	81°30′57.08″E
Meander belt	Brazil	6°55′21.12″S	64°39′25.27″W
Fluvial-dominated delta	United States	29°08′59.92″N	89°03′50.92″W
Wave-dominated delta	Egypt	31°00′49.72″N	31°11′27.04″E
Wave-dominated delta	Brazil	21°37′51.28″S	41°03′04.50″W
Tidal-dominated delta	Papua and New Guinea	8°34′51.89″S	143°25′20.90″E
Clastic tidal flat	Netherlands	53°11′31.51″N	4°59′52.95″E
Barrier-bar coastline	United States	36°02′30.32″N	75°46′33.57″W
Beach strand plain	Mexico	21°56′28.24″N	105°34′13.71″W
Barrier reef and lagoon	Turks and Caicos	22°45′57.95″N	74°10′43.93″W
Ooid shoal	Bahamas	25°16′01.93″N	78°08′55.80″W
Carbonate tidal flat	Bahamas	25°01′17.43″N	78°10′53.89″W

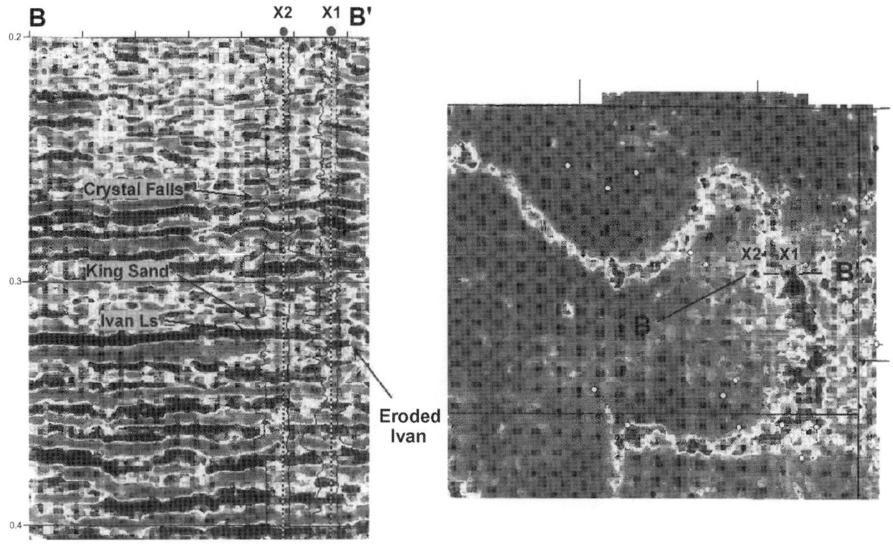

FIGURE 70. A point bar cut into the underlying Ivan limestone as picked out by varying seismic amplitudes on a horizon display, late Pennsylvanian to Early Permian, Baylor County, Texas (from Burnett, 1996). Reprinted with permission from the AAPG.

to pick the horizon everywhere else. Fluvial meander belts are commonly well differentiated on horizon slice amplitude displays (e.g., Brown et al., 1981; Rijks and Jauffred, 1991; Noah et al., 1992; Carter, 2003). A study in the Powderhorn field of Texas showed that the most distinct amplitude display images from seismic data could be derived from sand-prone lithofacies associations surrounded by muddy lithofacies (Zeng et al., 1996). These included coastal stream plain, delta, and delta plain depositional environments along with pinch-outs of back-barrier sandstones into lagoonal shales.

Carbonate sediments produce distinctive seismic facies with reefs and marginal reef environments commonly well imaged (Fontaine et al., 1987; Masaferro et al., 2003; Eberli et al., 2004). Deep-water marine deposits typically show sharp lithological contrasts between sand bodies and the encasing deep marine mudstones. These enable the sandstones to be picked on horizon slice amplitude and semblance displays (e.g., Varnai, 1998; Saller et al., 2004). *Semblance displays* (also known as *coherence cube*™ displays) are computed from seismic data by comparing the similarity of each seismic trace with its neighbors within a specific window of interest. Significant changes in the response corresponding to sand pinch-outs or faults are highlighted as edges (Bahorich and Farmer, 1985; Marfurt et al., 1998).

Geometrical patterns that allow depositional sedimentary environments to be recognized can sometimes be picked out by seismic facies analysis (Posamentier, 2004). *Seismic facies analysis* involves the analysis of seismic character to help predict the depositional environment. One method uses a computer-based neural network analysis of waveform character within a window of seismic data. A map is made showing the areal distribution of the waveform character classes, and this can be correlated with lithofacies variation.

Semblance and spectral decomposition methods were used to pick out individual macroforms in Pleistocene deltaic sediments in the Gulf of Mexico (Lopez et al., 1997). *Spectral decomposition* is a way of breaking down a seismic trace into its discrete component frequencies (Partyka et al., 1999). Certain stratigraphic features can be picked out because they are more sensitively tuned to specific frequencies although they may not be obvious in the seismic trace as a whole.

DETERMINING THE BASIN TOPOGRAPHY

Various maps can be constructed that allow the basin topography or bathymetry to be defined. The idea is to pick out sedimentary depocenters, dispersal patterns, and topographical features that may have influenced the distribution of sediments in the reservoir interval.

Sandstone percentage maps are contour maps that show the percentage thickness of sandstone within a gross rock interval. These can give a good indication of the sediment dispersal patterns and lateral pinch-out edges. *Gross sandstone thickness maps*, which are maps of the total thickness of sandstone within an interval, help determine the locations of sediment depocenters. These maps can also be used to infer depositional strike and depositional dip. Depositional strike is the dominant direction along which sedimentary bodies tend to be elongated. Depositional dip is the direction perpendicular to the depositional strike.

Hamilton et al. (2002) described the reservoir characterization of the Tertiary deltaic sediments in the Merecure unit A reservoir interval of the Budare field in

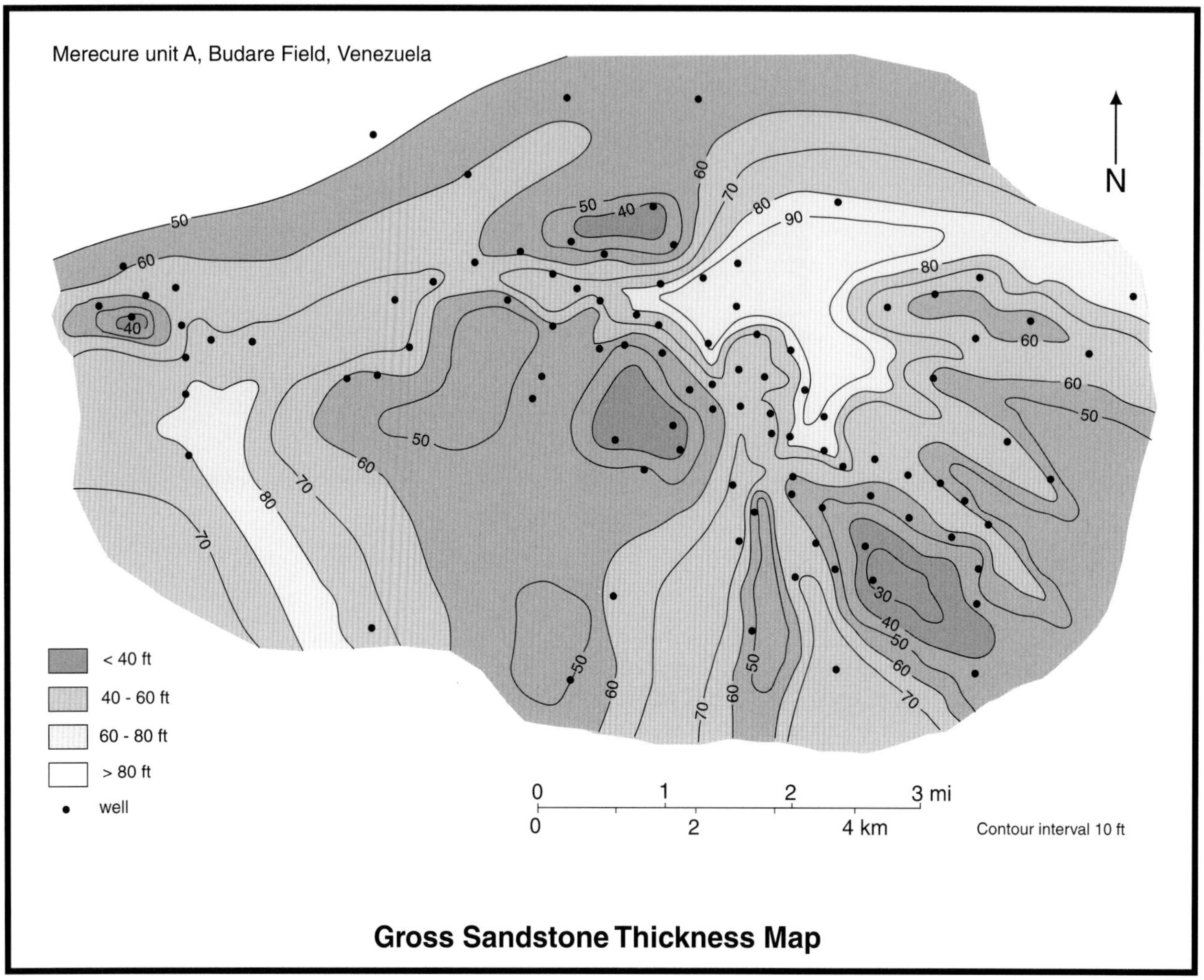

FIGURE 71. A gross sandstone thickness map can give an idea of the depositional dip and strike of the sedimentary system. In the Budare field of Venezuela, north–south strike elements correspond to distributary channels in the bottom part of the map. An east–west arcuate depositional element in the north of the map corresponds to a wave-dominated delta front (from Hamilton et al., 2002). Reprinted with permission from the AAPG.

Venezuela. The gross sandstone thickness map shows a dominant east–west grain defined by elongate, thick sandstone bodies. Narrower north–south lineaments are perpendicular to this main trend. The depositional environment is interpreted as a wave-dominated delta system. North–south-oriented features interpreted as distributary channels intersect at an oblique angle with arcuate to linear oriented trends interpreted as a marine reworked delta front (Figure 71).

Log facies maps give a sense of the internal sedimentary character of a reservoir interval (Shelton, 1971). For each well, a paper copy of the gamma-ray log is trimmed to the top and base of the reservoir unit of interest and pasted on a map. Computer applications are also available to help make these displays. Log facies maps give a visual impression of how the log facies varies over the field in terms of distribution, trends, and internal bedding characteristics. The various log patterns can be mapped across the field and then tied in to a lithofacies scheme (Figure 72). In the Budare field example from Venezuela (Hamilton et al., 2002), discrete zones of log character were mapped out. Blocky log profiles are related to distributary channel and aggradational mouth bar complexes, whereas multiple, thin serrated and subtly upward-coarsening log facies are interpreted as strand plain complexes flanking the delta front.

The end result is a series of *lithofacies maps* for each depositional sequence in the reservoir (Figure 73). These show the areal distribution of the various macroforms comprising a specific stratigraphic sequence.

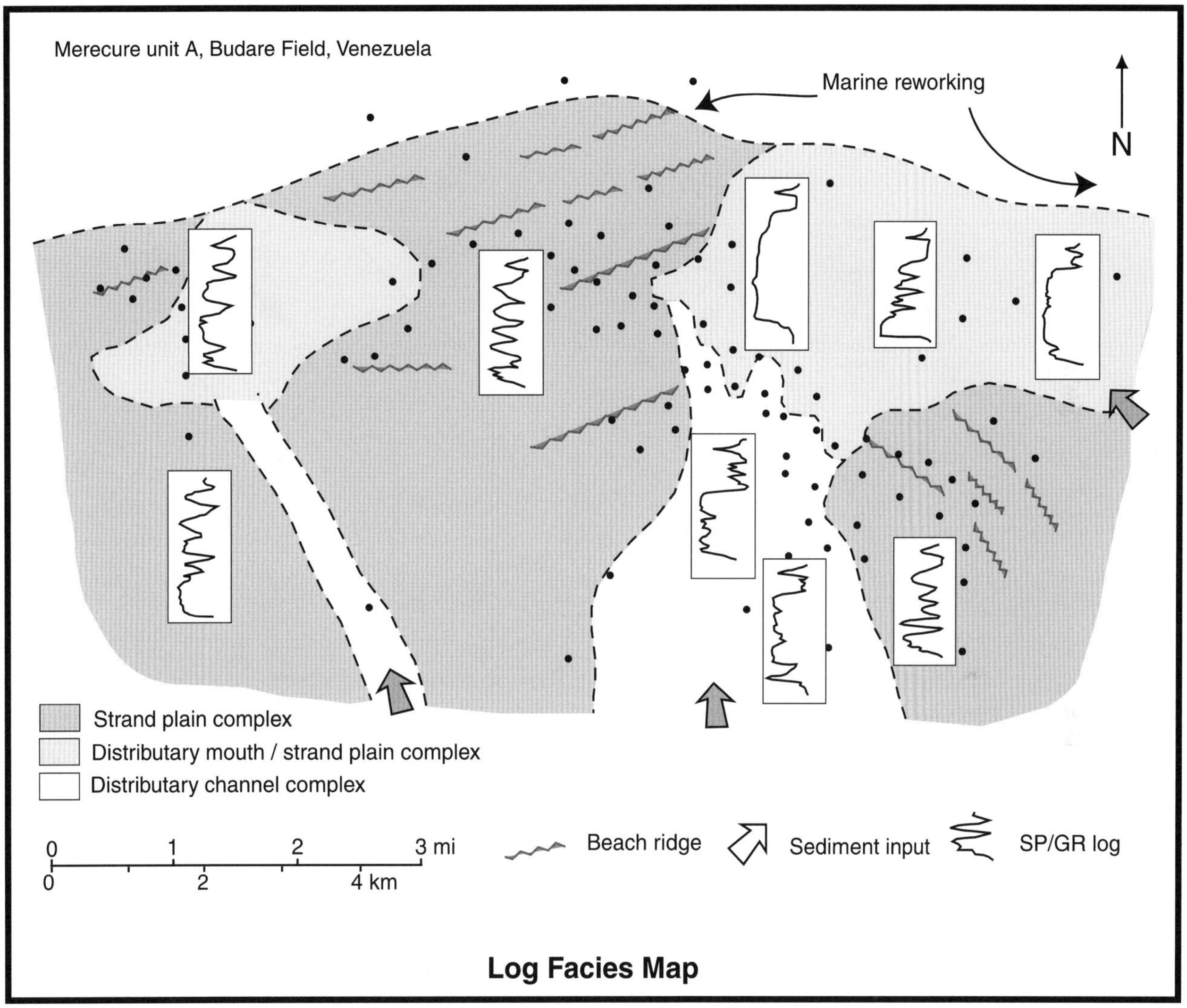

FIGURE 72. Log facies maps show the bedding configuration within the reservoir interval. The log patterns are then related to the lithofacies scheme and can be used to define lithofacies maps. This example is from the Budare field of Venezuela (from Hamilton et al., 2002). Reprinted with permission from the AAPG. SP = spontaneous potential log; GR = gamma-ray log.

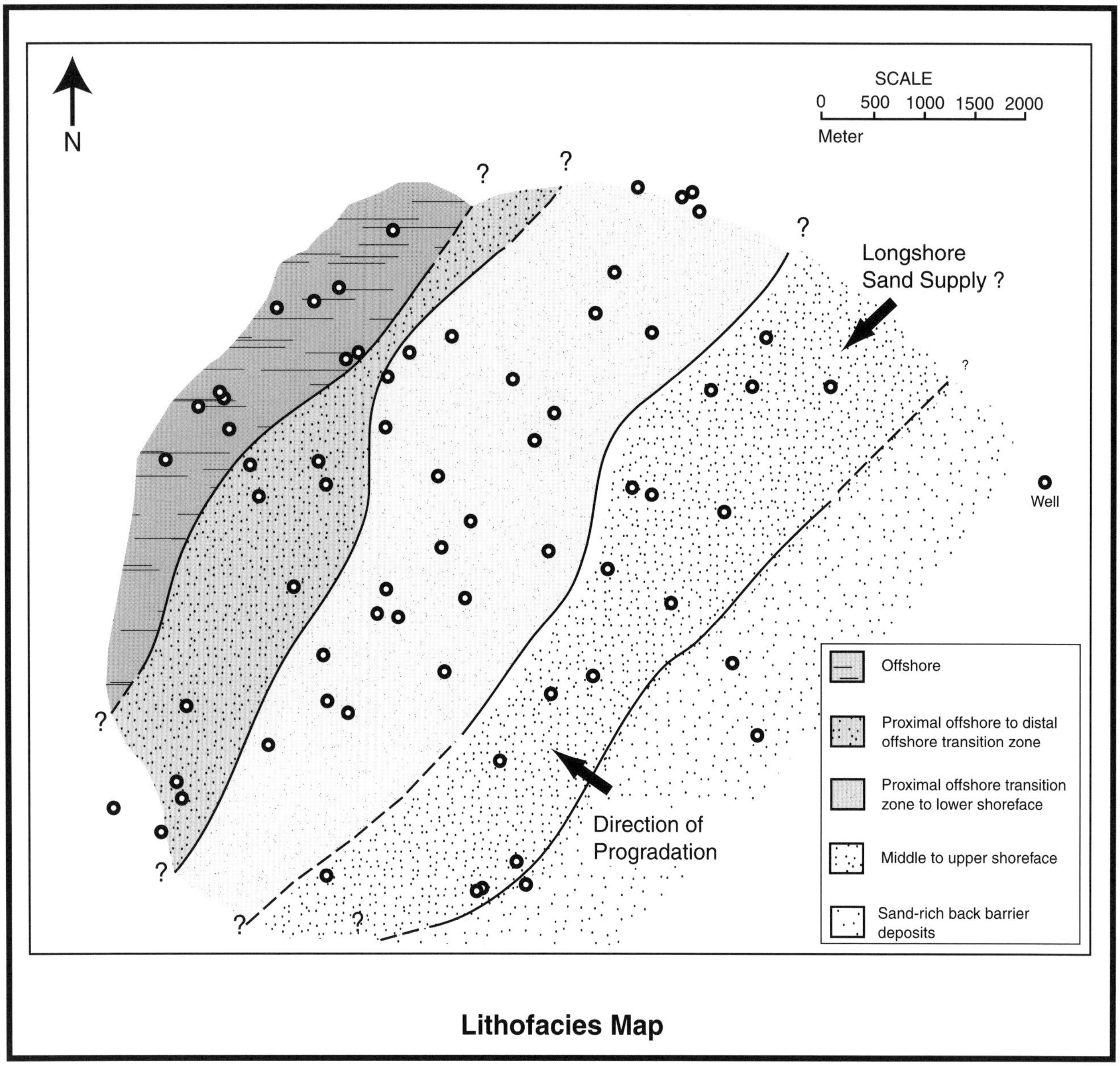

FIGURE 73. Lithofacies map for the upper Piper Sand interval of the Scott field, UK North Sea (from Guscott et al., 2003). Reprinted with permission from the Geological Society.

12

Shepherd, M., 2009, Analysis of rock properties, *in* M. Shepherd, Oil field production geology: AAPG Memoir 91, p. 99–105.

Analysis of Rock Properties

INTRODUCTION

Rock and fluid properties such as porosity, net to gross, and water saturation are required to build a geological model and to assess the in-place hydrocarbon volumes. An evaluation of permeability is the basis for understanding the productivity of a reservoir. Rock properties also represent the link between the static geological model and the dynamic reservoir engineering model that uses it. From the point of view of the reservoir engineer, the objective of a geological model is to adequately represent the distribution of rock and fluid properties for a reservoir simulation.

Predictive models for rock properties in the reservoir are made using data values from the existing well control. This is possible because there is a continuity of process influencing rock property distribution that allows predictions to be made. Rock property variation shows a significant correlation to lithofacies in siliciclastic sediments and rock texture in carbonates. Where rock properties are categorized by individual lithofacies on histograms, they commonly show discrete data distributions with shapes that are well known from classical statistics. Predictable areal and depth trends can sometimes be found for rock property variations in a reservoir.

ROCK PROPERTIES CATEGORIZED BY FACIES OR ROCK TYPES

Modern production geology characterizes reservoirs at the lithofacies level. Lithofacies that formed under high-energy conditions, such as fast moving water currents, will typically have good rock properties whereas lower energy lithofacies will tend to have poorer rock properties. Lithofacies will therefore correspond to packages with a distinct range of rock properties.

In certain reservoirs, lithofacies may not be the best packages for representing rock property variation. These are reservoirs that contain sediments where diagenesis and/or the pore geometry type are more of a controlling factor on storage and fluid flow than facies. Here the rock property groupings are made on the basis of the rock texture influencing fluid flow and are referred to as *rock types*. This categorization of rock properties is more common in carbonates than in clastic sediments (Lucia, 1995, 1999).

An example where rock types have been used is in the Tertiary carbonate reservoir of the Malampaya-Camago oil and gas accumulation, offshore Palawan Island in the Philippines (Grötsch and Mercadier, 1999). Lithofacies were defined, but these did not conform to discrete groups in terms of rock properties. Nonlinear porosity and permeability relationships are seen for instance. Individual lithofacies show a range in reservoir properties, which are controlled more by the degree of diagenetic overprinting than the primary depositional character. Five rock types were therefore defined, characterized by a combination of pore geometry, pore throat connectivity, and core-derived permeability cutoffs. Each rock type shows a characteristic pore geometry and distinctive values for porosity and permeability (Table 15).

STATISTICAL ANALYSIS OF ROCK PROPERTIES

Rock properties are analyzed according to lithofacies or rock types. Statistical analysis is used to understand the range and distribution of rock properties at this level (Table 16). It is often found that the rock properties cluster as a discrete group within a specific range of values showing one of the simple distribution shapes of classical statistics. This behavior helps the production geologist in the task of building a field-wide

DOI:10.1306/13161195M913372

Table 15. Rock types in the Tertiary carbonate reservoir of the Malampaya-Camago oil and gas accumulation, offshore Palawan Island in the Philippines.*

Rock Type	*Pore Geometry*	*Average Porosity (%)*	*Average Permeability (md)*
1	Connected intergranular and moldic	24.6	139
2	Dominantly moldic or finely intergranular	20.8	25
3	Poorly connected moldic	16.5	3.5
4	Unconnected moldic, poorly connected intergranular	7.4	0.5
5	Unconnected moldic, intergranular	4.0	0.04

**Data from Grötsch and Mercadier (1999).*

rock property model. The relationship between rock properties and lithofacies at the well scale can then be applied at the reservoir scale to build a rock property model conditioned to the lithofacies model. The range and distribution of the values from a statistical analysis of the well data will be replicated for the field-wide rock property model.

HISTOGRAMS

Histograms are used to graphically analyze the data distribution of rock properties. The data values are grouped in blocks of regular intervals from low to high. The number of values occurring in each group is the frequency and this is recorded on the vertical axis (Figure 74). The vertical axis can also be defined as the *relative frequency*, the number of data points in each group is expressed as a decimal fraction or percentage of the total number of data points. A plot that groups data with a relative frequency axis is known as a *relative frequency distribution.*

Histograms and relative frequency plots display how a specific rock property varies within the reservoir. They show the range of values for the data, indicate what are the most common values, and give a sense of what may be "bad" data points occurring as extreme values or outliers (Jensen et al., 1997).

Complex histogram shapes may be the result of mixing several different data populations. Sometimes, when the data are broken down further by a more detailed lithofacies classification, the simpler histogram shapes can become obvious (Chambers et al., 2000). For example, *bimodal distribution* of values on a histogram may be showing that the property under analysis is a combination of two overlapping subelements, such as two separate lithofacies (Figure 75).

Histograms help to recognize the ideal shape of the data distribution. If the tops of the bars on a histogram are joined with a continuous line, a curve will be formed. This *curve of the distribution* commonly resembles a simple shape such as the outline of a bell (Rowntree, 2003). Statisticians see these outlines as approximating to ideal distributions. The curve shown by these ideal distributions can be used to predict the probability that a given occurrence of a property will be found within the data population. The most common is a bell shape that shows a symmetrical distribution around the mean. This is referred to as a *normal or Gaussian distribution.* The peak of a normal distribution corresponds to the mean value.

A normal distribution shows an ideal shape that shows about 68% of the data values within a range of one *standard deviation* on either side of the mean (Figure 76). The standard deviation is a measure of the dispersion of the data around the mean. Just more than 95% of the data values occur within a range equal to two standard deviations on either side of the mean, and 99.7% of the data values are found within three standard deviations on either side of the mean. A normal distribution can

Table 16. Common statistical terms.

Statistical Term	*Definition*
Arithmetic mean	Measures the central tendency of a data set. Equal to the sum of the values divided by the number of values: $(x1 + x2 + x3 + \ldots + xn)/n$
Geometric mean	Equal to the nth root of the product of all the values, where n is the number of values: $\sqrt[n]{(x1 \times x2 \times x3 \times \ldots \times xn)}$.
Harmonic mean	Equal to the number of values divided by the sum of the reciprocal of the values: $n/(1/x1 + 1/x2 + 1/x3 + \ldots + 1/xn)$.
Mode	The value of the variable showing the most frequent occurrence.
Median	The value that splits the range into two parts, with half the values greater and the other half smaller than the median.
Standard deviation	A measure of the dispersion of data around the mean; given the symbol σ.

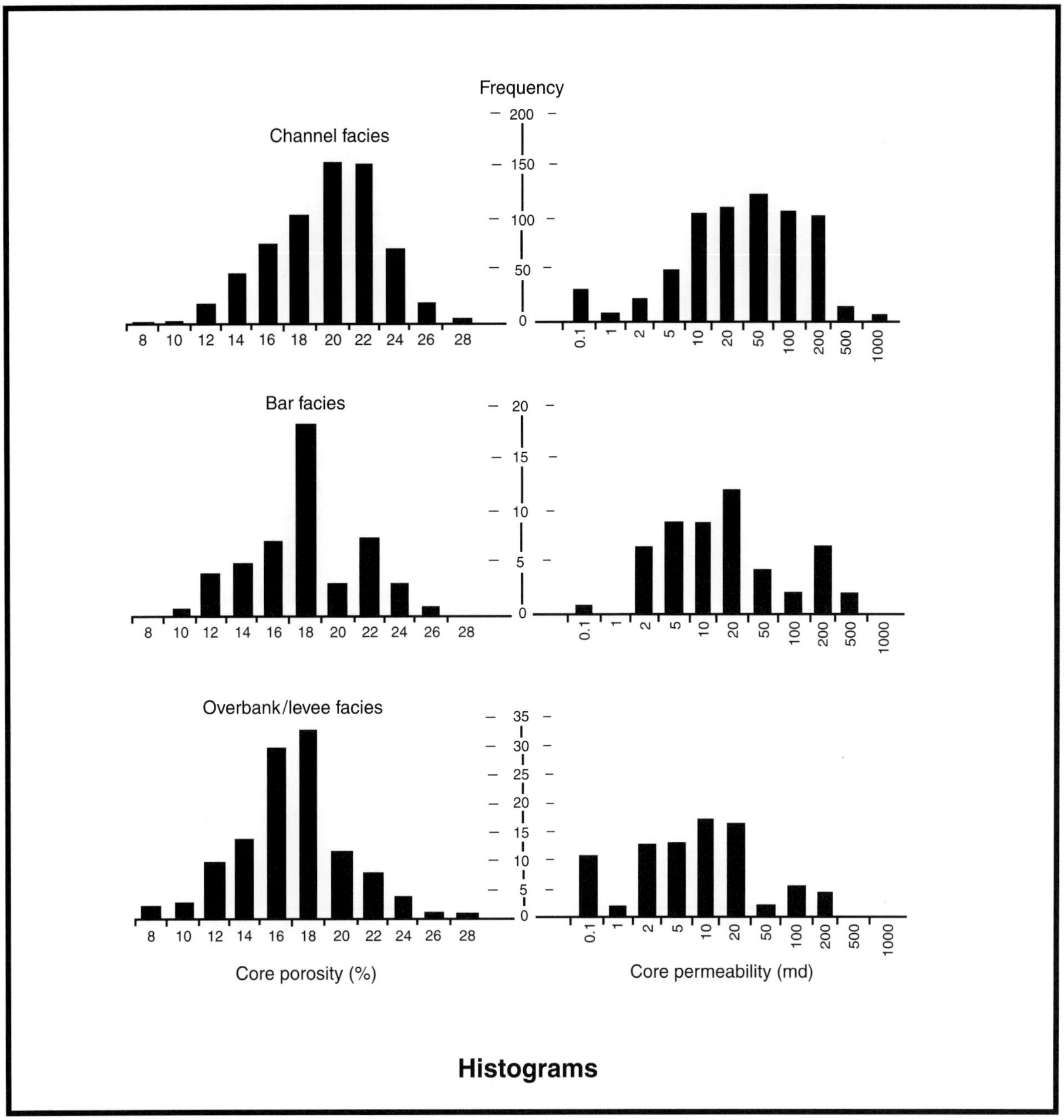

FIGURE 74. Histograms comparing the distribution of porosity and permeability according to lithofacies associations in the Frio sandstone D and E units, Rincon field, Texas (from McRae and Holtz, 1995). Reprinted with permission from the Gulf Coast Association of Geological Societies.

be completely defined by two parameters: the mean and the standard deviation (Rowntree, 2003).

Although reservoir measurements commonly form a normally distributed group, some may show a *lognormal distribution.* This is a probability distribution in which the logarithm of the variable is normally distributed. This type of curve shows a lopsided arrangement of variables on a continuous probability distribution with a small number of larger values and a larger number of small values (Figure 77). The mean, mode, and median values plot on different parts of the lognormal curve.

CHARACTERIZATION OF POROSITY

Porosity-depth Trends

Sediments show a gradual decrease in porosity with depth as a result of compaction and increasing cementation. Quartzose sandstones can show a reduction in porosity from depositional porosities of 35–40% to values of 15–25% at moderate reservoir depths (2000–3000 m; 6500–9500 ft). On occasions, anomalously high

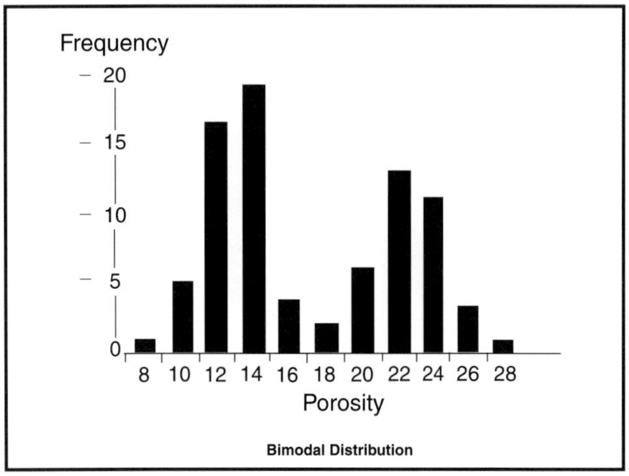

FIGURE 75. A histogram showing a bimodal distribution of values around two separate peaks. This may be the result of the combination of two separate data populations.

porosities are found in sandstone and carbonate reservoirs that do not lie on the expected porosity-depth trend.

An observed pattern, particularly in some of the deeper reservoirs, is to find enhanced reservoir porosities in the oil leg compared to the water leg. The porosity decreases markedly with depth toward the oil-water contact. This is a feature of many of the Chalk oil fields in the North Sea where the porosity contours frequently mimic the structural depth maps (Figure 78). For instance, in the Eldfisk field, Norwegian North Sea, the porosity in the chalk ranges from a maximum value of 45% near the top of the crestal wells to a downhole and off-structure value of 15–20% (Maliva and Dickson, 1992). Other types of carbonate reservoirs, and to a lesser extent some sandstone reservoirs, have been known to show these patterns (Wilson, 1977; Neilson et al., 1998).

One theory for this is that cementation resulting from late diagenesis can happen at roughly the same time as oil migration (Gluyas et al., 1993; Nedkvitne et al., 1993). Late diagenesis in carbonates results from grain contact dissolution and burial cementation (Tucker and Wright, 1990), whereas in sandstones, quartz is the major cement-forming phase at the temperatures of oil generation (Worden and Morad, 2000). A theory supported by many but not all geologists is that decreasing porosity and increasing cementation with depth in the rock hosting an oil column seems to reflect the long filling history of the field; a *race for space* between hydrocarbon filling and rock cementation (Gluyas et al., 1993). Once oil has displaced water from the reservoir rock, the net effect is to reduce the volume of solute available to form cement, to immobilize the circulation of solute material, and to inhibit the diffusion of the cementing phase (Worden et al., 1998). Although diagenesis will have almost stopped in the oil leg after

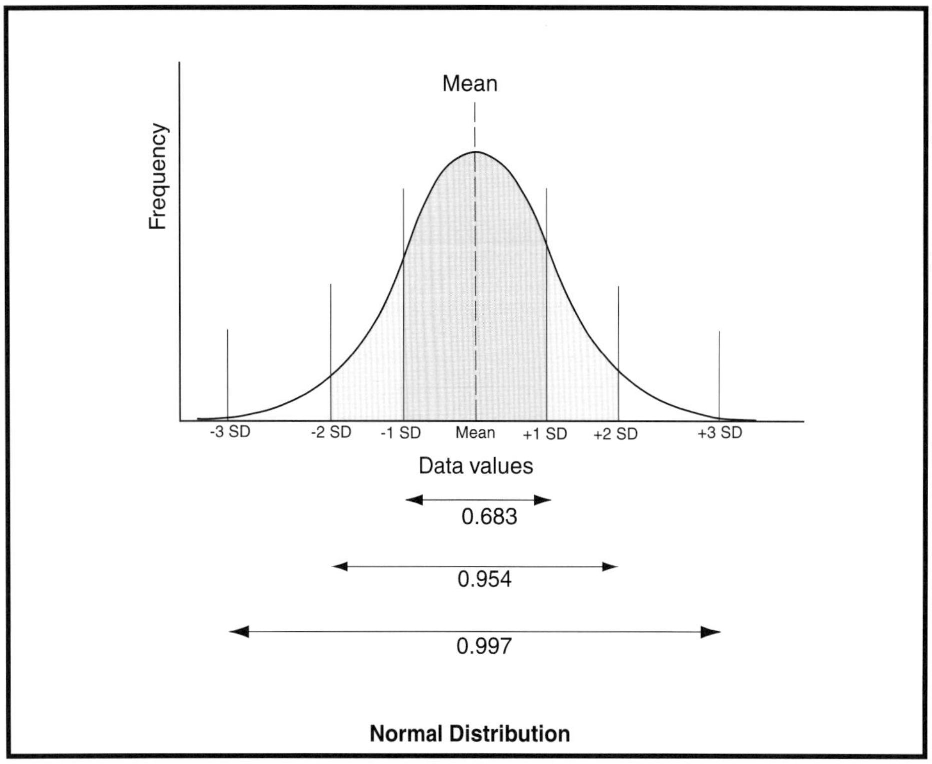

FIGURE 76. Many rock properties show a histogram shape that can be fitted with a normal distribution. This is a bell-shaped curve, which is symmetrical around the mean. 68.3% of the values are found within one standard deviation on either side of the mean; 95.4% of the data occur within a range equal to two standard deviations on either side of the mean, and 99.7% of the values are found within three standard deviations on either side of the mean. SD = standard deviation.

FIGURE 77. A lognormal distribution shows a lopsided curve with a large number of small variables and a small number of large variables.

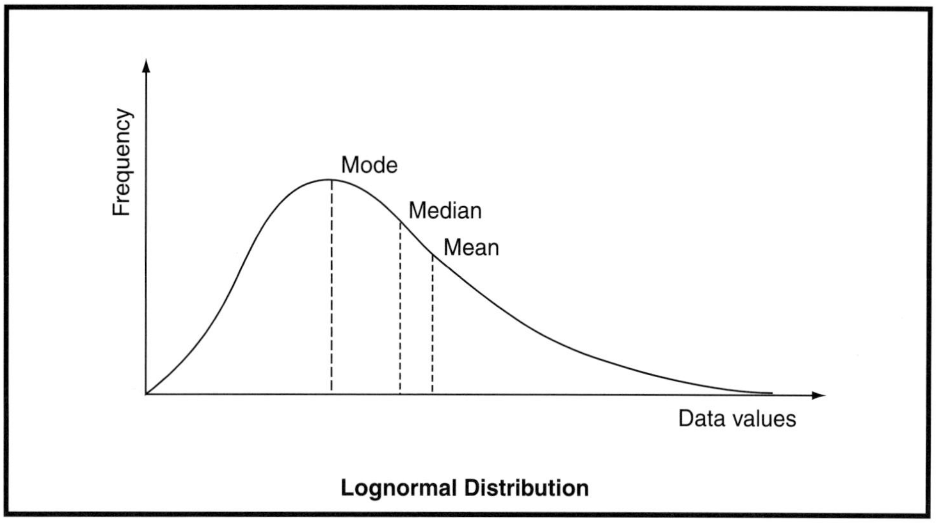

filling, diagenetic reactions may still proceed uninhibited in the water leg. For example, Webb (1974) recorded that the Cretaceous sandstones of Wyoming generally contain abundant authigenic kaolinite where water saturated, but little if any authigenic clay is found where the sandstones are hydrocarbon saturated. Sometimes, there can be a marked step change in porosity below an oil-water contact as a result of cementation processes carrying on after all the oil has been emplaced (Wilson, 1977).

Porosity Preservation by Grain-coating Minerals

Enhanced sandstone porosity can also be found where an early diagenetic mineral phase has extensively coated the sand grains. The grain-coating minerals are thought to retard later quartz cementation by blocking potential nucleation sites on the detrital quartz grains (Heald and Larese, 1974; Bloch et al., 2002). Common grain-coating mineral phases are chlorite (Pittman and Lumsden, 1968) and microquartz (Aase et al., 1996). An example of this is seen in the Upper Jurassic Norphlet Formation in Alabama (Schmoker and Schenk, 1994). Anomalously high porosities occur in Norphlet Formation sandstones despite depths of burial between 3000 and 7000 m (10,000 and 23,000 ft). Illite and chlorite are the common authigenic minerals within the sandstones. Chlorite-cemented sandstones have porosities that are 3 to 6 porosity units higher by comparison to illite cemented sandstone. No evidence has been found for any significant variation in porosity across oil-water contacts in this example.

FIGURE 78. The depth map and average porosity map show similar trends in the Chalk reservoir of the West Ekofisk field, Norwegian North Sea (from D'Heur, 1991). One theory for the systematic decrease of porosity with depth is that this results from a 'race for space' between oil filling and late stage cementation. Reprinted with permission from AAPG.

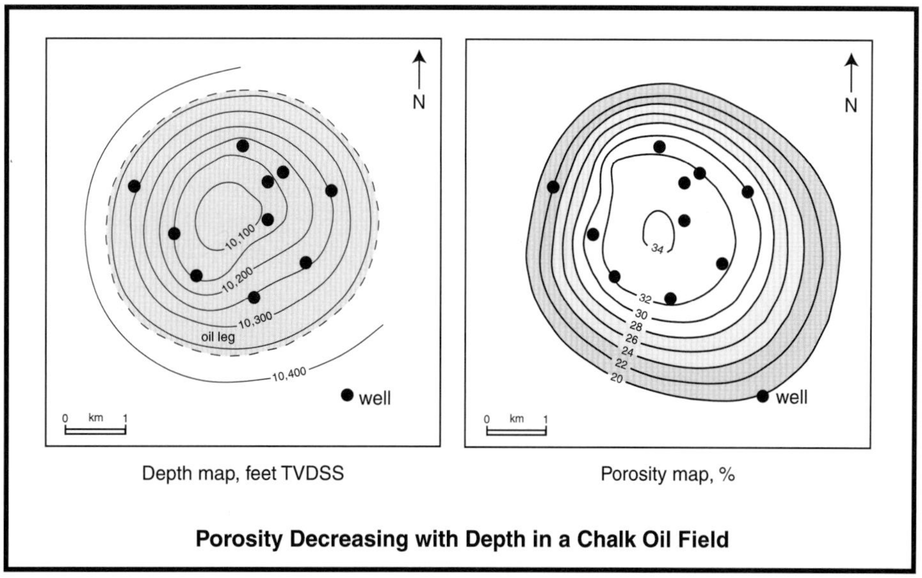

Porosity and Overpressure

High porosities can occur at depths within *overpressured* sandstones. This is where the pressure of the fluid in the pore space is higher than normal hydrostatic pressure. The fluid bears some of the weight of the overlying rock column and reduces the pressure on the grain-to-grain contacts (Bloch et al., 2002). The driving force for intergranular grain contact dissolution is smaller than in normally pressured sandstone, and high porosities can be preserved as a consequence. Sandstones at more than 5-km (3-mi) depth can still retain significant porosity; up to 35% in overpressured sediments in the central North Sea for instance.

Porosity from Seismic Data

Sometimes it is possible to get an indication of areal porosity variation from seismic data as a result of seismic inversion to produce an acoustic impedance data set. This involves removing the *seismic wavelet* from the seismic trace, that is, the shape of the seismic pulse. The seismic data are thus transformed by this process into a representation of the acoustic impedance (AI) character, the product of the rock density and seismic velocity. The seismically derived AI data values are correlated with porosity data from the wells at the same scale as the seismic data. Density and acoustic property values from core plug samples can be crossplotted with core porosity to calibrate the relationship. If a correlation is found between AI and porosity, this can provide a guide for mapping porosity between the wells.

The porosity estimated from an AI data set corresponds to the porosity of a volume of rock on a scale of tens of meters. The method gives an indication of porosity trends at this scale and can pick out 'sweet spots' within the reservoir. Geostatistical methods are available for creating porosity realizations from inversion data at a smaller scale (Haas and Dubrule, 1994; Rowbotham et al., 2003; Francis, 2006a, b).

CHARACTERIZATION OF PERMEABILITY

Permeability can be the most difficult to characterize of all the rock properties. The measurement of permeability is specific to a given volume of rock and is scale dependent. The scale of a core plug a few centimeters long is different in scale compared to the radius of investigation of a well test, which can be a hundreds of meters or more. The measurement of permeability at the various scales should be investigated as this information will give the geologist a sense of the heterogeneity within a reservoir (Haldorsen, 1986). It is often observed that permeabilities derived from well tests can be markedly different from the core-derived permeabilities from the same wells (e.g., Zheng et al., 2000). This is because heterogeneities at the well test scale, such as fractures, cross-bedding, and subseismic faults, will not have an influence at the core plug scale.

It is common practice to take permeability values from core plugs as the basic data for characterizing the permeability of larger volumes of the reservoir. This process involves *scaling up* from several core plug values to a single permeability value for the larger volume. Standard statistical methods are used for this. Jensen et al. (1987) stated that the aggregate permeability of a rock volume is a power average in which the exponent p can range between 1 and -1. However, it can be difficult to estimate p in practice. The more pragmatic approach is to use the more common averaging methods for permeability. Arithmetic, geometrical, or harmonic averages are used depending on the nature of the rock. Where the rock is more or less homogenous, the flow properties at the core plug scale are not much different from the larger volume of the rock, so arithmetic averaging will suffice. Such homogeneity of permeability measurements is rare. Where the permeabilities vary considerably along the flow path, geometric averaging is typically used (Warren and Price, 1961). Where the flow is orthogonal to the bedding plane in strongly laminated rocks, harmonic averaging is preferred.

Sometimes the measurement of core plugs at an interval of 1 per 30 cm is too sparse a data set to meaningfully characterize permeability in heterogenous rocks (Hurst, 1993). More detailed permeability profiles can be made using an instrument called a *probe permeameter*. The instrument measures the flow rate of gas as it passes from a probe into a porous rock sample. The permeability can be estimated from the flow rate and the gas pressure (Hurst and Goggin, 1995).

Vertical Permeability

Vertical permeability is the permeability perpendicular to the bedding planes. It can be a critical rock property controlling reservoir performance particularly in reservoirs with thick sandstone intervals. Fluid flowing through a sediment will be subject to two main forces. Flow along a bed will be influenced by the pressure differences driving the flow. Gravity will act vertically to pull the more dense fluids downwards. The relationship of vertical permeability to horizontal permeability dictates the way in which reservoirs are swept. This relationship is expressed as a decimal fraction of the vertical permeability to the horizontal permeability, the K_v/K_h *ratio*.

FIGURE 79. Porosity-permeability crossplots can be made from core data. An empirical formula is derived from a best-fit correlation. This allows permeability values to be estimated from log porosity in uncored intervals. This example, from the Rincon field in south Texas, shows that there is a strong facies control on the correlation (from McRae and Holtz, 1995). Reprinted with permission from the Gulf Coast Association of Geological Societies.

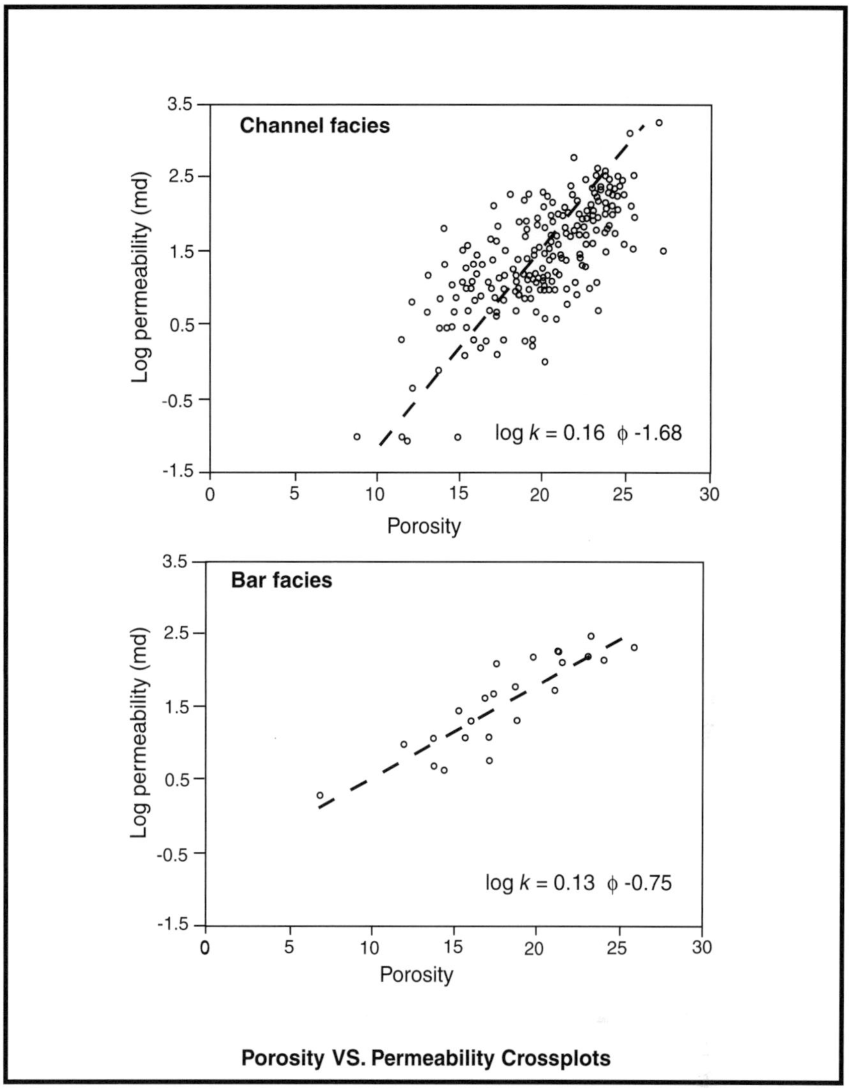

Geological Controls on Permeability

Permeability values show a significant relationship to the size of the pore throats connecting the various pores (e.g., Kopaska-Merkel et al., 1994). Pore throats act as chokes on the flow through the pore system. A correlation between grain size and permeability is commonly observed; the larger the grains, the larger the diameters of the pore throats are likely to be. Grain sorting will also have an effect on permeability; permeability will tend to increase with better sorting (Krumbein and Monk, 1943).

In many sandstones, there can be a strong relationship between porosity and permeability, particularly where interparticle pore throats are not occluded by clay cements or other processes (Kopaska-Merkel et al., 1994). Crossplots of porosity versus permeability are typically used as a basis for estimating permeability from wireline log porosity values in uncored wells (Nagel and Byerly, 1992). A regression line is fitted to the data, and an equation is derived relating estimated permeability to porosity (Figure 79). The porosity-permeability relationship is normally poorer for carbonates, although detailed textural analysis and subdivision into pore classes can help (Lucia, 1995, 1999; Lønøy, 2006).

A strong facies control on porosity-permeability relationships is frequently seen. For example, McRae and Holtz (1995) found markedly different porosity-permeability relationships for the channel and bar facies in the fluviodeltaic reservoir of the Rincon field, south Texas (Figure 79).

13

Shepherd, M., 2009, Structural geology: Faults, *in* M. Shepherd, Oil field production geology: AAPG Memoir 91, p. 107–122.

Structural Geology: Faults

INTRODUCTION

In structurally simple fields, the main control on production behavior is the distribution of lithofacies. In structurally complex fields, faults and fractures provide major elements influencing production performance. This chapter discusses the data used to establish the presence of faults and how faults are mapped for reservoir models. The reservoir structure can be analyzed at two different scales: the seismic scale and the well scale. The interpretation of faults and structure at the seismic scale is made by the seismic interpreter whereas the production geologist analyzes the structures from core and log data. Having established a fault framework for a field, it is important to know whether or not fluid flow communication occurs across the faults. Techniques are available to predict the likelihood of this. Sometimes sealing faults break down and open up to flow after a field has been producing for a few years. This reflects the change in the stress state of the reservoir as a result of pressure depletion.

SEISMIC INTERPRETATION OF FAULTS

The seismic data set is interpreted primarily using vertical time sections. These are displays that show a series of vertical seismic traces displayed side by side (see Figure 47). The peaks or the troughs are filled in with black shading or color. Continuous reflections stand out as an overlapping array of peaks or troughs. These create patterns on a seismic section that give a representation of the geological structure in the subsurface. The seismic interpreter will look for discontinuities in the seismic reflections likely to represent faulting. Various techniques can help in picking faults. The interpretation can be cross checked against *attribute maps* showing changes in *seismic dip* (magnitude of the time gradient), *azimuth* (direction of maximum dip), or abrupt changes in amplitude (Dalley et al., 1989; Hesthammer and Fossen, 1997). Another method is to use semblance data to detect edges in the data (see chapter 11, this publication).

STRUCTURAL CORE LOGGING

Structural features such as fault zones and fractures are commonly seen in cores. *Structural core logging* may be required if there is a high density of such features or where knowledge of the detailed fault or fracture pattern is important for reservoir development.

Core goniometry is a method for graphically depicting the structure in the core. The whole core is wrapped around with acetate film, and the structures and main bedding planes in the core are traced directly with felt tip marker pens. The unrolled film shows a 360° depiction of the structure comparable to the display shown by borehole image logs. Commercial rigs are also available, which take 360° photographs of the core for the same purpose.

Having established the structures in the core, it is important to know how they were originally oriented within the reservoir. Dipmeter data, scribed core, and paleomagnetic data have all been used to work out the spatial orientation of the core (Davison and Haszeldine, 1984; Bleakly, 1992).

Structural core logging provides a variety of useful information for the reservoir model. For example,

- the width of fault damage zones
- the orientation of the faults can be established and tied to the seismic interpretation
- the density and orientation of open fractures
- the composition and microstructure of material in the fault zone

DOI:10.1306/13161196M913372

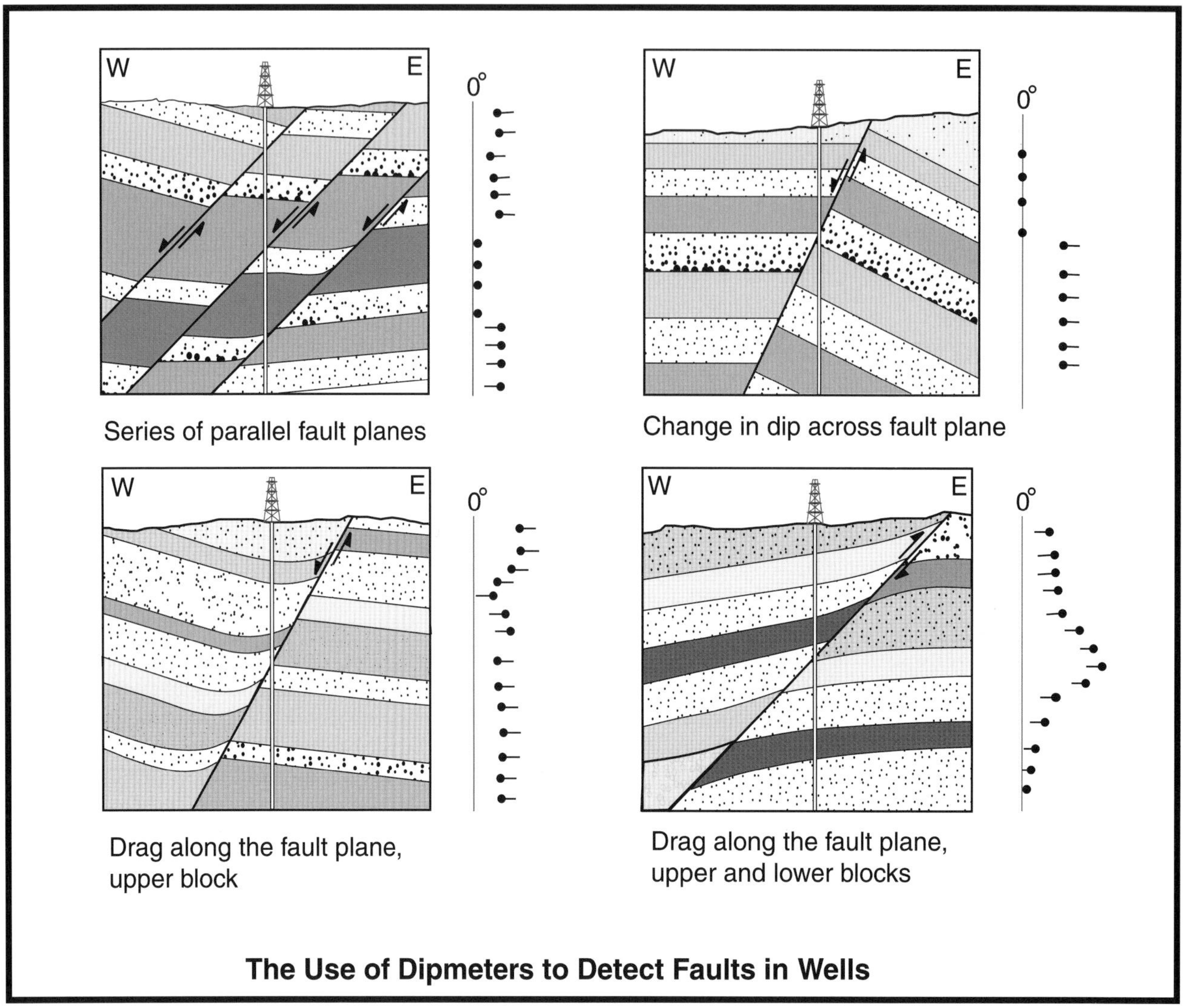

FIGURE 80. Dipmeter or image data can be used to pick likely fault planes in wells. Changes in dip amplitude or azimuth can indicate that a fault is present. Drag patterns may also be seen on the dip data above and below the fault intersection in a well (from Schlumberger, 1981). Courtesy of Schlumberger.

FAULT DETECTION METHODS

Dipmeter or borehole image data can be used to establish if and where any faults cut a well (Bengtson, 1981, 1982; Goetz, 1992; Adams and Dart, 1998). A sharp change in dip amplitude or azimuth on a dipmeter log can indicate that a fault is present. Drag patterns may also be seen on the dip data above and below the fault intersection in the well (Figure 80).

An anomalously thin reservoir section, perhaps in conjunction with the absence of a reasonably persistent marker horizon, may be caused by a normal fault cutting out part of the stratigraphic section in a well (Figure 81). The thickness of missing section can be estimated by comparison to nearby wells with unfaulted sections.

A fault-repeated section is sometimes seen in a well (Figure 82). Near-vertical or gently dipping wells cutting reverse faults will show a repeated pattern. A repeat section can also occur where a highly deviated well cuts through a normal fault at a shallower angle than the dip of the fault plane (Figure 82) (Mulvany, 1992).

WELL TESTS AND FAULTS

One method for locating faults is to check the results of reservoir engineering pressure transient analyses of well tests. The basis for these tests is that a production well, while it is flowing, will draw down the pressure for a considerable distance out into the surrounding reservoir. If the well is shut in and production is stopped, the pressure will build up as a result of the radial inflow of fluid toward the pressure sink in the immediate vicinity of the borehole. If a sealing fault or a feature likely to

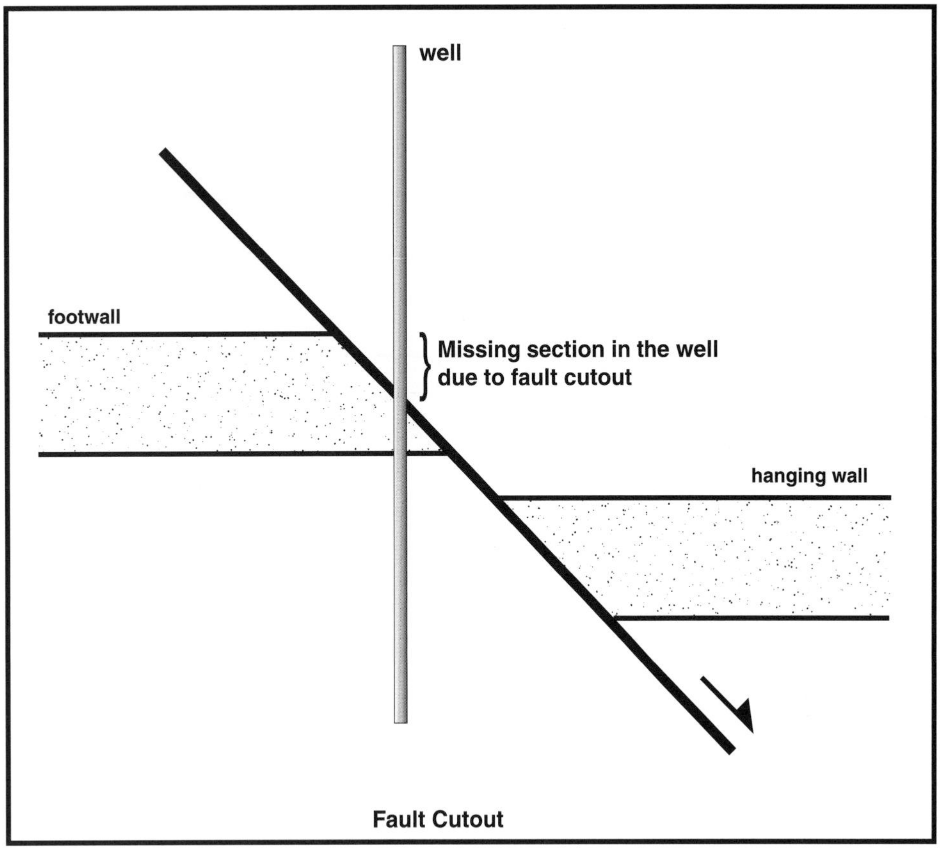

FIGURE 81. The stratigraphy in a well penetrating a normal fault will be incomplete due to fault cutout.

disrupt horizontal fluid inflow is present within the drainage radius of the well, then this can often be detected. The fault will disrupt the rate of pressure buildup once the catchment area for inflow of fluid increases outward with time and comes in contact with the fault plane (Figure 83). Analytical methods are available to make a rough estimate of how far away the fault is from the wellbore.

Care has to be taken that a feature such as a sand pinch-out or channel margin is not mistaken for a fault. It is a useful exercise for the reservoir engineer to have a working session with the seismic interpreter in order to compare test data for all the wells in the field with the interpreted fault pattern. An example of this is given by Márquez et al. (2001) for the LL-04 reservoir in the Tia Juana field, Venezuela. An integrated reservoir characterization study was made to identify reserve growth opportunities. Part of this study involved cross checking the seismic interpretation of faults with evidence of compartmentalization from engineering data. In places where inferred reservoir compartments and faults did not coincide, the seismic interpretation was rechecked to see if a fault had been missed. If no fault could be located, the geologists then investigated the possibility that stratigraphic pinch-outs could be the cause of compartmentalization.

MAPPING FAULTS

Structure maps show the contoured depth surface and a representation of any faults cutting the surface. The faults are drawn as *fault polygons* marking the *hanging wall* and *footwall fault cuts* for the interpreted surface. The hanging wall is the rock volume above the fault plane, and the footwall is the rock volume that lies beneath it (Figures 81, 82, 84).

Faults on structure maps should be checked for consistency. The fault polygons represent the length of the fault that can be picked from seismic data. Where the fault throw is less than the seismic resolution, the fault will not be mapped by the interpreter. The limits of the seismically mapped faults will therefore not represent the actual *fault tips* in the subsurface, the points at either end of the real fault where the fault displacement is zero. Estimates can be made of the extent of the actual fault tips for a seismically mapped fault. The ideal normal fault trace will have an elliptical shape with the maximum displacement in the center of the fault, decreasing gradually to zero at the fault tips (Barnett et al., 1987). If a linear length-to-displacement ratio is assumed, it is possible to use this geometry to extend the seismic fault traces to a feasible location of the fault tips in the subsurface (Pickering et al., 1997).

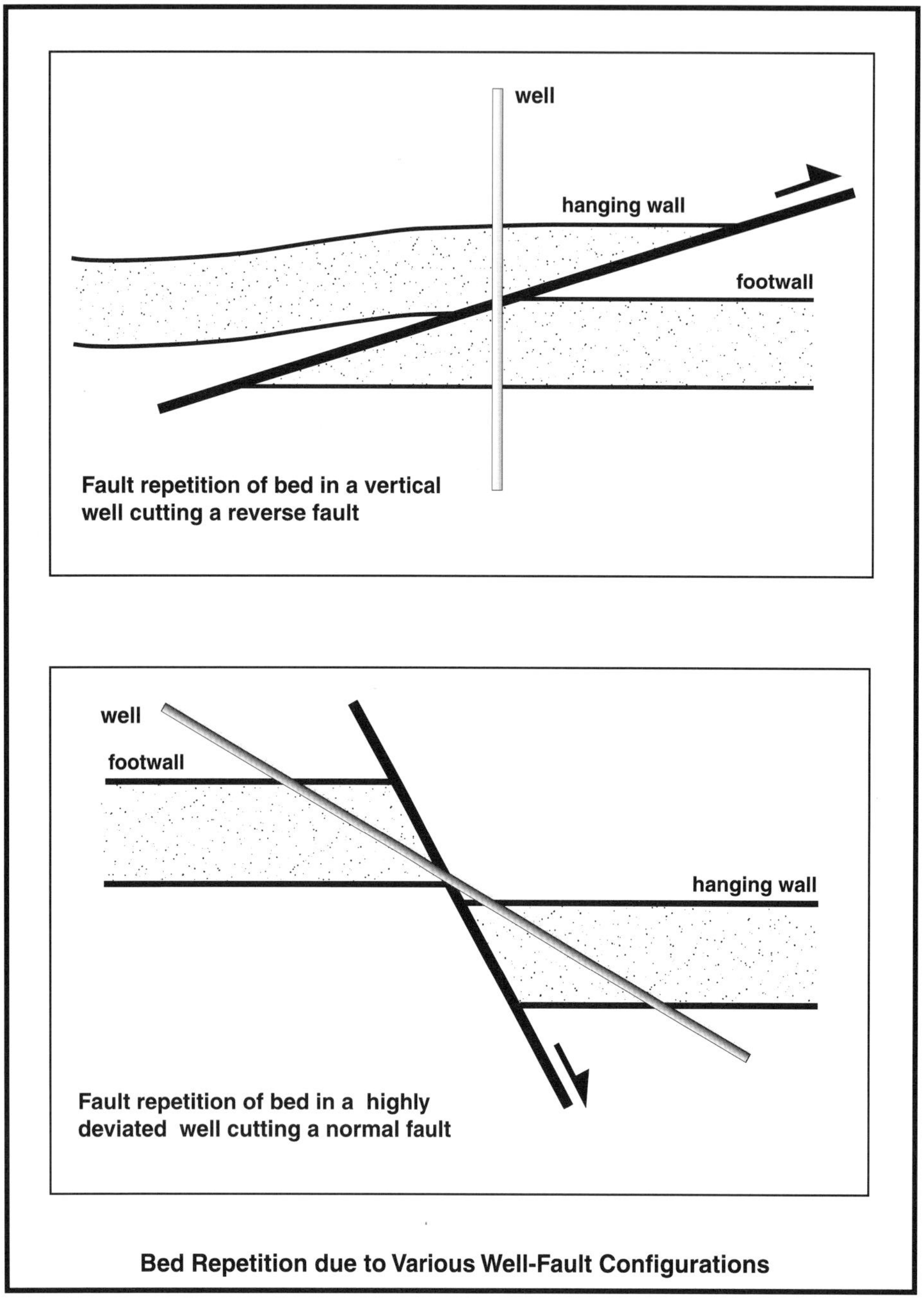

FIGURE 82. Repeated sections can be seen in a vertical well drilled through a reverse fault or with a highly deviated well penetrating a normal fault.

If the structure is computer mapped, the contours interpolated by the mapping algorithm around faults can sometimes be rather untidy. It is not unusual for a computer map to show spurious fault reversal along the length of the fault. Thus, it is important to check and edit the contour maps by hand where this has happened.

FAULT VALIDATION

Computer methods are available for validating the consistency of a reservoir fault framework. The most sophisticated of these will allow the geologist to examine the faulted model in 3-D and move the various fault blocks interactively back to the prefaulted undeformed state (Williams et al., 1997). If this can be achieved without any gaps appearing, then the fault model is valid in a geometric sense. However, if large gaps cannot be removed, then there are serious problems with the structural interpretation. Sometimes it can take several attempts at making a fault interpretation before a validated fault model is obtained.

Zamora Valcarce et al. (2006) used fault restoration to validate the El Portón field structure in Argentina prior to building a 3-D model of the field. The model was to be built to help plan the trajectories of new development wells. The idea behind validating the structural model

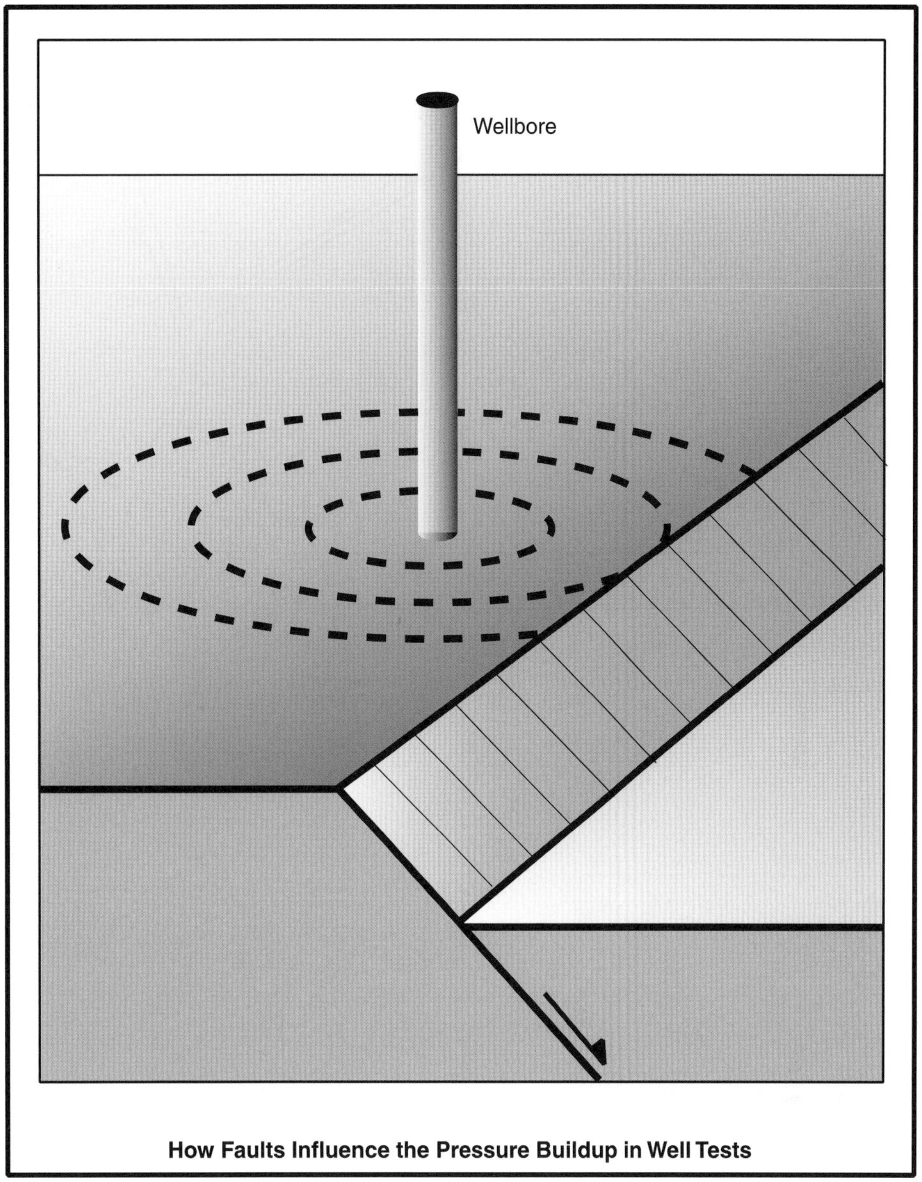

FIGURE 83. A pressure build-up test can be used to detect faults near the wellbore. The fault interferes with the way the pressure builds up with time, and this effect can be detected by pressure transient analysis.

was to give extra confidence that a planned well could be expected to intersect with the intended reservoir target given the structural complexities of the reservoir (Figure 85).

Fault restoration can also give insights into the structural history of an oil field. By determining the timing for episodes of faulting, uplift, and erosion, insights can be gained that allow the structural controls on reservoir development to be understood.

FAULT GEOMETRIES, LINKED FAULTS, AND RELAY RAMPS

What appears to be a simple large fault on seismic data may be more complex than it looks. The imaged fault may in reality comprise several closely spaced, overlapping faults, but because the seismic data cannot resolve the detail of the fault zone, it is shown as a single fault trace. These fault zones comprise linked fault segments with *relay ramps* in the overlapping areas between them (Figure 86) (Peacock and Sanderson, 1994; Needham et al., 1996).

It can be important to map relay ramps, as they can potentially provide pathways for fluid flow across a fault zone (Hesthammer and Fossen, 1997; Rotevatn et al., 2007). Identification of relay ramps can be difficult in practice as the gap between overlapping faults are small (e.g., tens of meters) and difficult to resolve. However, there are ways in which relay ramps can be recognized, despite the limits of seismic resolution:

1) Areas where fault traces show kinks on maps are commonly an expression of unresolved relay ramps (Fossen at al., 2005).

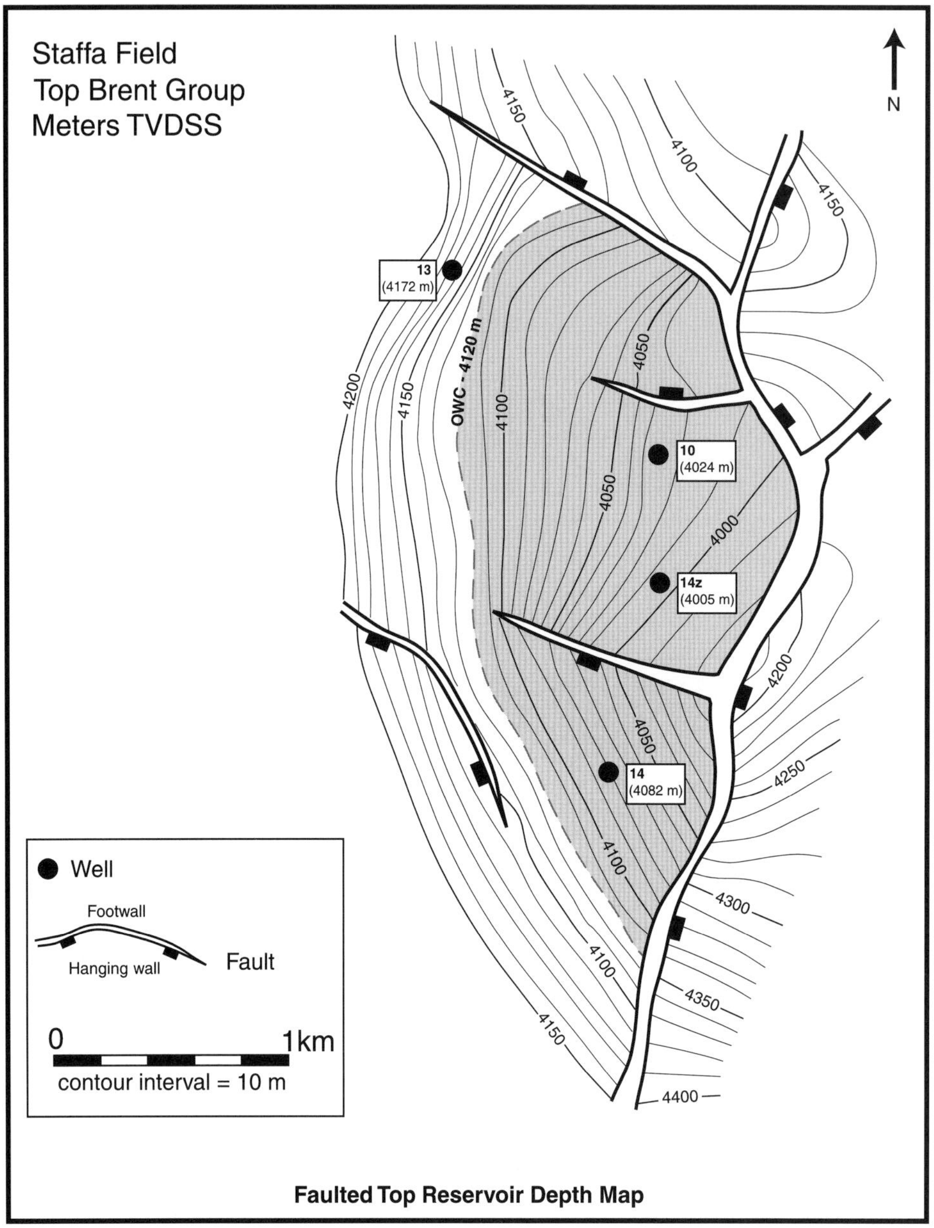

FIGURE 84. This faulted top reservoir map from the Staffa field in the UK North Sea is represented by a contoured surface and fault polygons. The fault polygons show the hanging wall and footwall fault cuts for the interpreted surface. The downthrown (hanging wall) side of the fault is indicated by a blocked out symbol (from Gluyas and Underhill, 2003). Reprinted with permission from the Geological Society.

2) Relay ramps may correspond to displacement minima along long faults (Needham et al., 1996).
3) Some of the longer faults may show anomalous length to displacement ratios. This can indicate that a relay ramp has been overlooked (Willemse et al., 1996).

FAULT DAMAGE ZONES

A large number of fractures, microfaults, and deformation bands can be found in a zone (up to 100 m [328 ft] or more wide) on either side of major fault planes (Aydin and Johnson, 1978; Jamison and Stearns, 1982; Antonellini and Aydin, 1995). These *damage zones* can be observed in outcrops and in cores from wells near large faults (Figure 87).

Clean, porous sandstones respond to localized strain by forming *deformation bands* (Figure 88). These are tabular zones where the grains are reorganized by grain sliding, rotation, and commonly fracturing in response to deformation processes including dilation, shearing, and compaction (Fossen et al., 2007). Deformation bands are frequently sheared with shear offsets on a millimeter to centimeter scale. By comparison to open fractures, which tend to enhance permeability, deformation bands have a much reduced permeability compared to the undeformed host sandstone (Antonellini and Aydin, 1994). Given that a damage zone can contain hundreds of deformation bands, then it is clear that even sand-sand

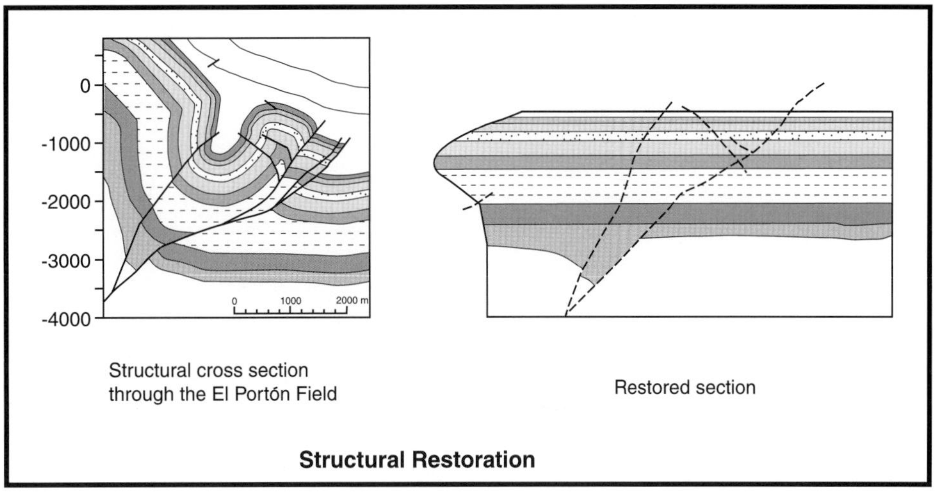

Figure 85. The structural framework of a reservoir can be shown to be valid if it can be taken apart and restored to its predeformed state without any gaps showing (from Zamora Valcarce et al., 2006). Reprinted with permission from AAPG.

contact faults with damage zones can have significantly reduced permeability across them.

Damage zones in impure sandstones (those with 15–40% clay) contain *phyllosilicate-framework fault rocks*. These are anastomozing zones where the rock has been disaggregated and the clays have been mixed in with the framework grains to produce a more homogenous mixture of clays than is present in the undeformed host rock. Faults affecting clay-rich sandstones with more than 40% clay content form clay smears (Fisher and Knipe, 1998).

The intensity of damage decreases away from the fault with the width of the damage zone roughly proportional to the throw of the fault (Knott, 1994; Knott et al., 1996). Field work on faulting in the Navajo Sandstone of Utah found that the summed width of the damage zones on either side of the fault core is approximately 2.5 times the total fault throw (Shipton and Cowie, 2001). Note that this observation is case specific for this locality. Large and rapid variations in damage zone thickness occur along many faults, and any estimate attempting to systematically relate damage zone thickness to fault throw is liable to a significant uncertainty as a result (Fossen and Bale, 2007).

A study on the Big Hole Fault in Utah based on core data showed a significant permeability reduction within the damage zone (Shipton et al., 2002). Probe permeameter measurements of permeability range from more than 2000 md in the undeformed host sandstone

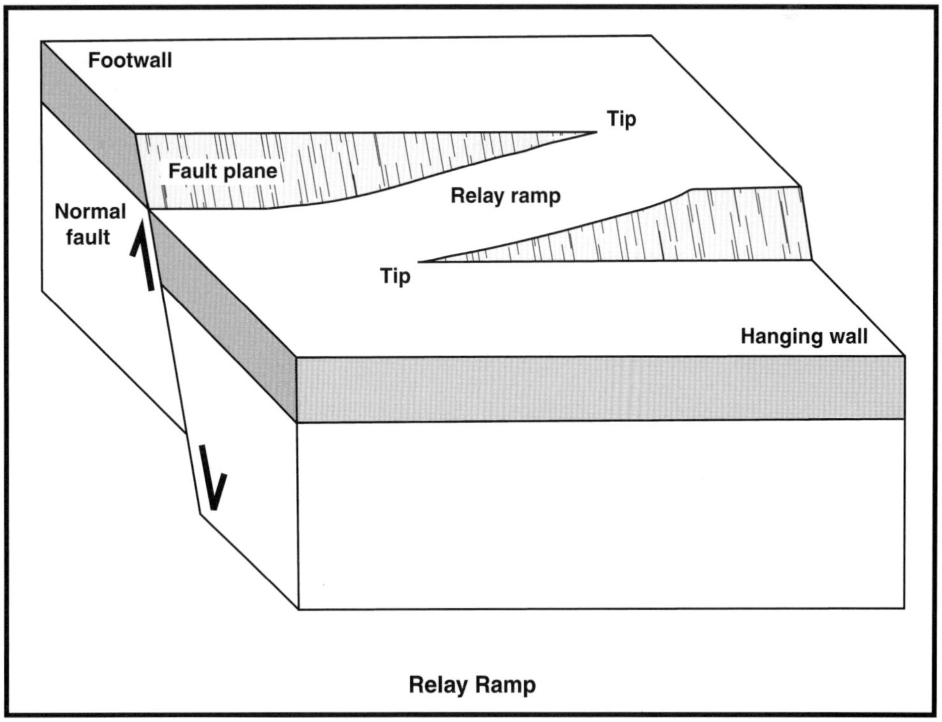

Figure 86. Relay ramps are found in the zone between two overlapping faults. They potentially provide pathways for fluid flow across a fault zone (from Peacock and Sanderson, 1994). Reprinted with permission from AAPG.

Figure 87. Fault damage zone from Moab, Utah. The outcrop is about 15 m (49 ft) high (photo courtesy of Angus MacLellan).

to less than 0.1 md in fault-damaged rocks near the fault. Whole-core tests showed that the permeability of individual deformation bands vary between 0.9 and 1.3 md. The transverse permeability modeled over 5–10-m (16–32-ft)-length scales across the fault zone was estimated as 30–40 md. This is approximately 1–4% of the permeability for the undeformed host rock.

The general consensus in the industry is that damage zones around faults probably baffle flow across them rather than acting as barriers to fluid movement (Sternlof

Figure 88. Deformation bands in the Aztec Sandstone, Valley of Fire, Nevada. Increased compaction compared to the undeformed rock causes the deformation bands to be more resistant to weathering and to stand out as ridges. Individual bands are approximately planar, showing distinct tips even where they are closely spaced (bottom left photo). Porosity loss resulting from granular rearrangement and clay accumulation in the bands results in lowered permeability (bottom right photo). DB = deformation band (from Sternlof et al., 2004). Reprinted with permission from AAPG.

et al., 2004; Fossen and Bale, 2007). The exception may be in deep reservoirs with high reservoir temperatures (more than 120°C). Here, accelerated quartz cementation at high temperature can decrease the pore throat diameters in the deformation bands to the extent that they become 100% water wet through capillary action. They thus become effective barriers to oil flow (Hesthammer et al., 2002).

Because of the abundance of low-permeability baffles and poorly connected volumes, production wells drilled in fault damage zones can significantly underperform. For example, wells drilled in fault-damaged zones in the North La Barge Shallow Unit of Wyoming are the poorest producers in the field (Miskimins, 2003). It is generally not a good idea to plan a new well trajectory too close to a large fault because of this.

FAULTS AND FLUID FLOW

Faults can have a significant impact on the fluid flow patterns within a reservoir. They can *juxtapose* one reservoir interval with another creating the potential for cross flow between the units. It is pragmatic to assume that all sand to sand juxtapositions allow fluid transfer across faults unless proven otherwise (James et al., 2004). Alternatively, juxtaposition of reservoir with nonreservoir rocks can cause the trapping of hydrocarbons against the fault. Deformation and cementation within the fault zone itself can create a zone of zero or very low permeability, which can cause the fault plane to act as a barrier to fluid flow. In some instances, fractures in the fault zone itself can act as conduits for fluid flow.

ALLAN DIAGRAMS

Allan diagrams or *fault juxtaposition diagrams* show the reservoir stratigraphy of both the hanging wall and footwall locations superimposed on the fault plane (Allan, 1989; Knipe, 1997). At a glance, the juxtaposition relationships of the various reservoir units across the fault can be seen (Figure 89). Allan diagrams are useful for the production geologist but are subject to the uncertainty in the input data used. The magnitude of vertical fault displacement estimated from seismic data is prone to error. Additionally, where fault drag is present but not picked up on seismic data, the vertical fault displacement can be overestimated. Complex fault zone architecture can also create large uncertainties in establishing fault juxtaposition relationships (Hesthammer and Fossen, 2000).

FAULT SEAL

Estimates can be made using Allan diagrams as to the probability that a fault will seal within a reservoir. In the first instance, *fault seal* can result from the juxtaposition of reservoir with nonreservoir rock. However, experience from many petroleum provinces has shown that faults can seal even where reservoir quality sand bodies are juxtaposed across a fault. The most common mechanism for sealing results from the incorporation of fine grained or dense material into the fault plane. Five different processes may cause this (Mitra, 1988; Fisher and Knipe, 1998):

1) *Clay smear*: Faults in clay-rich sediments are believed to form clay smears by the shearing of mudstone beds into the fault zone (Weber et al., 1978; Lehner and Pilaar, 1997).
2) *Cataclasis (shale gouge)*: Fault movement affecting clean sandstones will cause grain crushing and the breakage of rock in the fault plane, which will form a fault gouge (Lindsay et al., 1993).
3) *Diagenesis or cementation*: Fine grained fault rock and associated open fractures in fault zones can be prone to cementation. Fluids migrating up the fault zone can cause the mineralization of the host rock. It is a common observation to find carbonate-cemented intervals in wells drilled close to faults, whereas wells drilled farther away from the faults do not contain carbonate cements (e.g., Reynolds et al., 1998). This is an indication that the fault zones have acted as the locus for the fluids causing carbonate cementation.
4) *Pore volume collapse*: Ductile deformation during fault movement can cause poorly sorted sediments to mix and homogenize with a resultant decrease in porosity.
5) *Grain contact dissolution*: Fault zones can act as planes for intergranular grain contact dissolution and subsequent recementation of the dissolved material. This can be an important mechanism for fault sealing in carbonate rocks (Peacock et al., 1998).

When investigating fault seal, it is important to look at any faults in the core to determine which type of sealing mechanism may be present.

FAULT SEALING CHARACTERISTICS

Fine grained fault rock will have a higher capillary entry pressure compared to the undeformed host rock. Brown (2003) described how the seal behavior of water-wet fault fill defines three potential zones within a fault.

1) A fault can seal because the petroleum phase has insufficient buoyancy pressure to invade and displace water from the fine grained material within the fault rock; this has been termed *membrane sealing* (Watts, 1987).

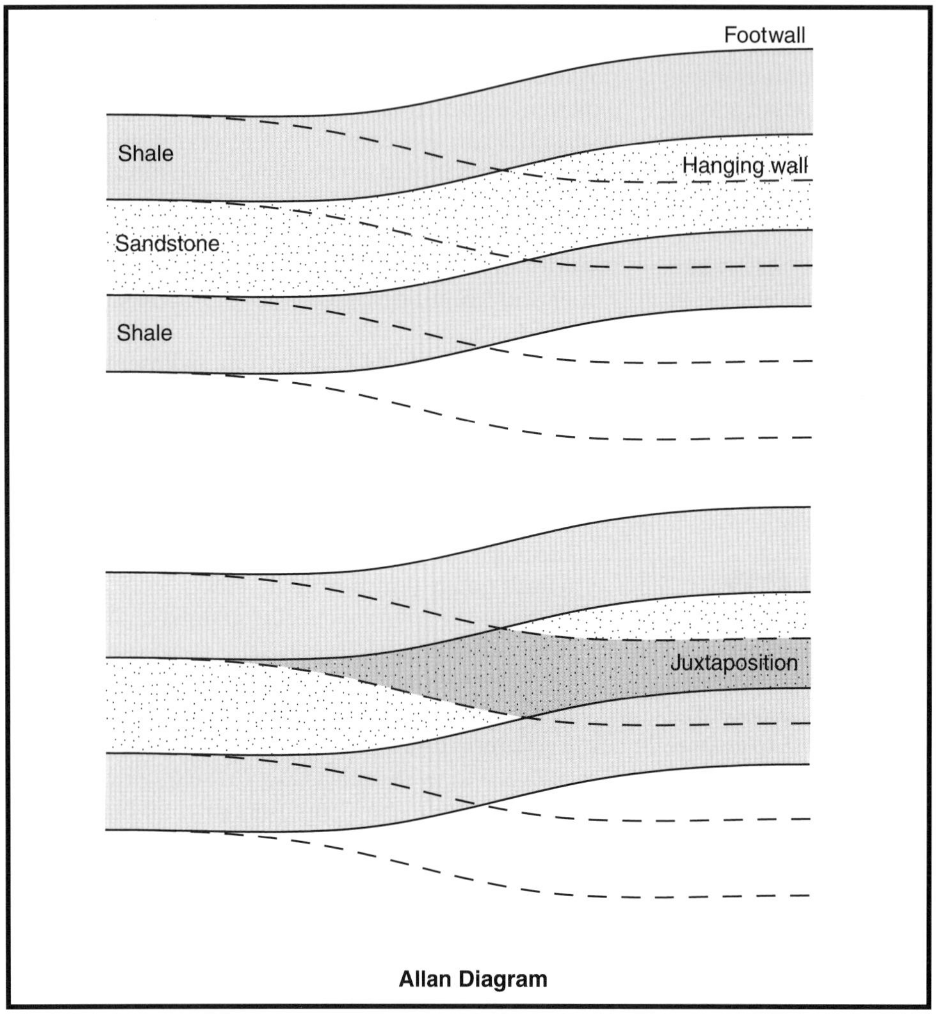

FIGURE 89. Allan diagrams show the reservoir stratigraphy of both the hanging wall and footwall blocks of a fault superimposed along the fault plane. At a glance, it can be seen where reservoir and nonreservoir lithologies are juxtaposed with potential for juxtaposition sealing.

2) Higher within the petroleum column, the buoyancy pressure can increase to the point at which the oil or gas can invade the fault rock and thus leak through it. However, the fault rock will have a very low permeability, and the rate of leakage can be trivial, even over geological time (Heum, 1996). The fault can then be considered to be effectively 'sealing' by *hydraulic resistance* (Watts, 1987).
3) Where an exceptionally thick petroleum column exists, even low-permeability fault rocks can leak at significant rates. This is the *zone of fault fill seal failure.*

FAULT SEAL PREDICTION

Where sealing faults are a key element controlling the fluid flow in a reservoir, they should be characterized for reservoir description and modeling (Fisher and Jolley, 2007). Much of the research to date has come about because of the particular importance of understanding fault behavior in deltaic reservoirs. In deltas deposited over thick and unstable mobile shale intervals, synsedimentary faults are a major element controlling reservoir continuity and size. The faults cut relatively unlithified sediments where the potential for clay smear along the fault planes is high and potentially predictable.

Algorithms are available for predicting the clay smear and shale gouge sealing potential of a fault. The basis for these algorithms is that the chances for clay smear to cause fault seal is controlled by the number and thickness of the shale beds displaced past a particular point on the fault. The thickness of the clay smear within the fault plane will decrease with distance from the source beds and with increasing throw of the fault (Yielding et al., 1997). The method involves taking the sand and shale distribution from a well close to the fault as a template for making the fault seal analysis.

The *clay smear potential* is calculated for a particular point on the fault plane as a function of the distance of that point from a shale bed acting as the source for the clay smear and the shale bed thickness (Bouvier et al., 1989; Fulljames et al., 1996) (Figure 90).

The *shale smear factor* (SSF) is dependent on the shale bed thickness and the fault throw but not on the smear distance (Lindsay et al., 1993) (Figure 90). Smaller values

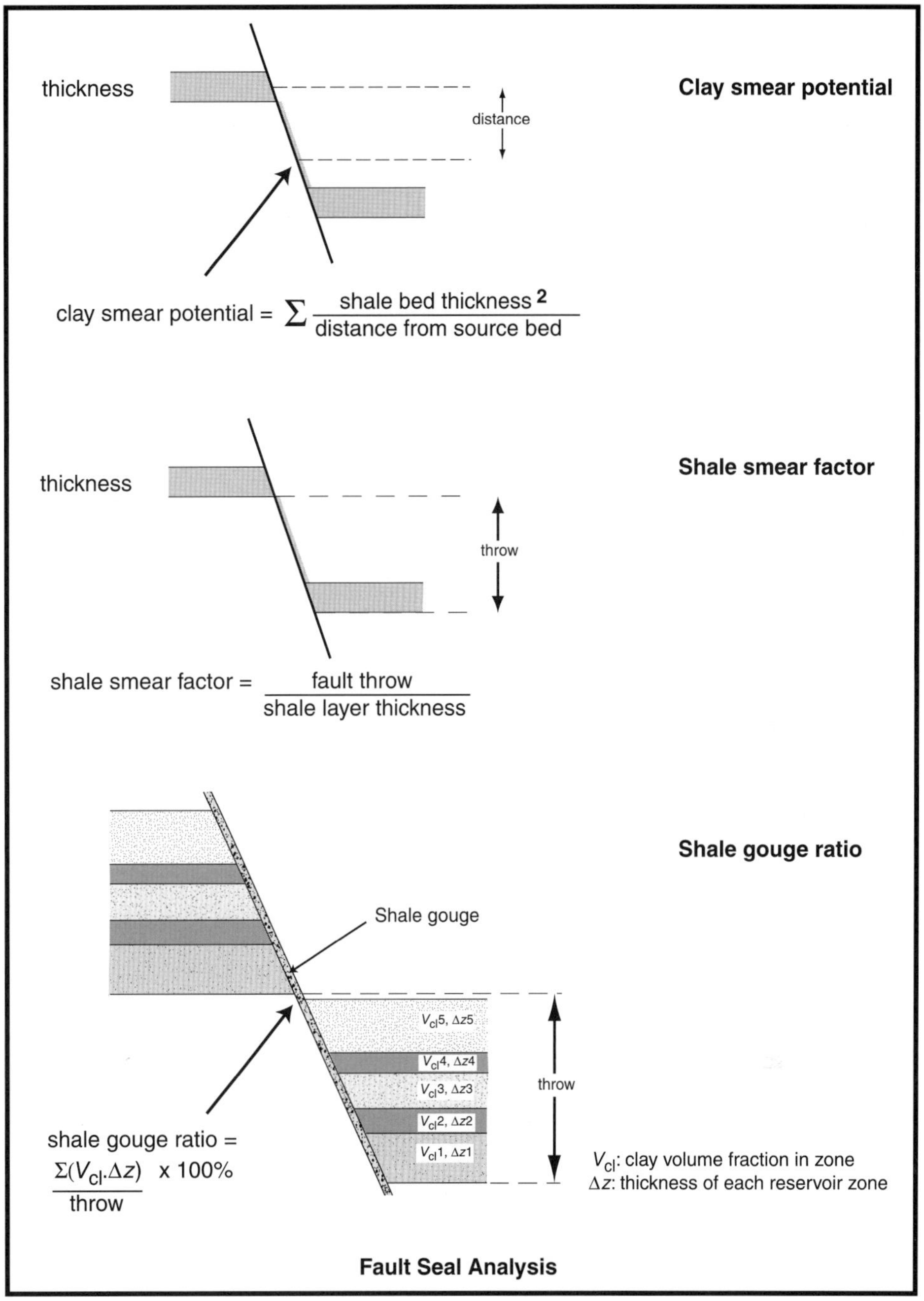

FIGURE 90. Fault seal analysis involves numerical methods of predicting the likelihood of fault seal (from Yielding et al., 1997). Reprinted with permission from AAPG.

of the SSF correspond to a more continuous development of smear on the fault plane. A large fault is likely to seal where the SSF is equal to or less than 4 (Færseth, 2006).

The *shale gouge ratio* works on the assumption that the sealing capacity is related directly to the percentage of shale beds or clay material within the slipped interval (Yielding et al., 1997). The shale gouge ratio is the proportion of the sealing lithology in the rock interval that has slipped past a given point on the fault (Figure 90). To calculate the shale gouge ratio, the proportion of shale and clay in a window equivalent to the throw is measured.

The prediction of fault seal is based on the assumption that if there is enough shale in the section undergoing faulting, then sealing is likely. There is often a continuous shale gouge or shale smear along fault planes where there is sufficient mudstone material available to be incorporated (Lindsey et al., 1993; Foxford et al., 1998). Nevertheless, a number of field studies show that fault zones can have a significant degree of complexity and variation in deformation style along their lengths (Childs et al., 1997; James et al., 1997). For example, Foxford et al. (1998) examined good exposures of the Moab fault in Utah. They found that the structure and content of the fault zone was so variable that it was impossible to predict the nature of the fault zone over even a 10-m (33-ft) distance. Doughty (2003) found that the

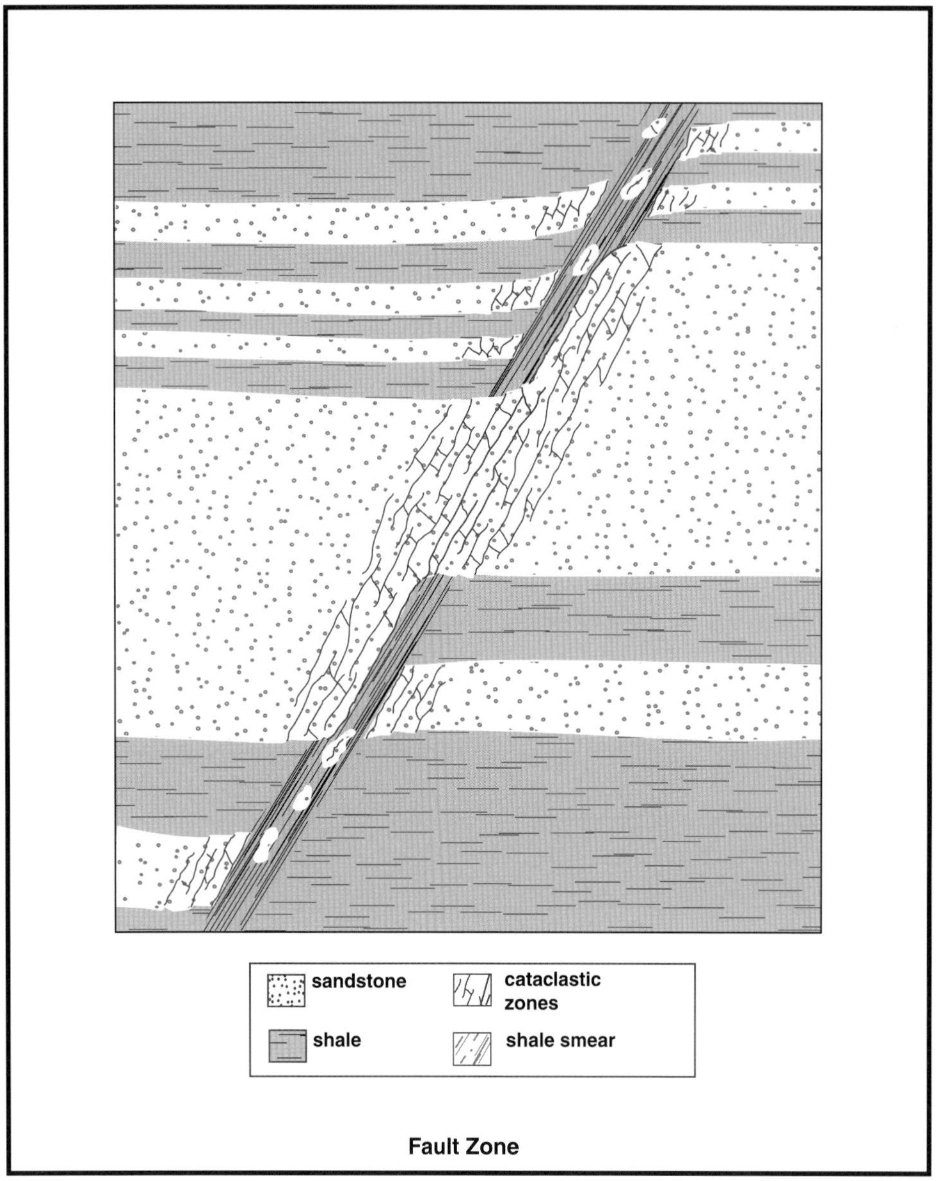

Figure 91. Schematic illustration showing the character of fault zones in siliciclastic strata based on outcrop and core observations from onshore and offshore Trinidad (from Gibson, 1994). Reprinted with permission from the AAPG.

clay smear along the Calabacillas fault in New Mexico showed numerous gaps particularly where minor faults within the fault zone complex cut out the shale smear associated with the major slip plane. The implication of these field studies is that fault seal can be predicted but is subject to chance factors affecting the reliability of the prediction. Because of this, any fault seal prediction should be calibrated against actual evidence that fault compartmentalization is present. Yielding et al. (1999) made a fault seal analysis for the Gullfaks field in the Norwegian North Sea. Areas of higher shale gouge ratios (>20%) were more likely to seal on the basis of pressure history and chemical tracer movement between wells.

Gibson (1994) provided a case history for fault seal analysis from the Columbus Basin, offshore Trinidad. Oil and gas fields occur in upper Miocene to Pleistocene deltaic sandstones of the Columbus Basin, located offshore to the southeast of the island of Trinidad. Numerous small faults dissect these reservoirs, and fault seal appears to be a critical feature controlling the size of these petroleum pools. Allan diagrams show that juxtaposition sealing is insufficient to explain the fault control on fluid contacts.

The sediments that form the reservoirs offshore are also exposed onshore along the east coast of Trinidad. Outcrops onshore and cores offshore provide control on the nature of the fault rock. In these outcrops, shale smears are found where shale beds have been displaced along the fault. The shale smears range in thickness from millimeter- to centimeter-thick shale partings to complex zones up to several meters thick (Figure 91).

Offshore, hydrocarbon columns up to 200 m (656 ft) thick are found within compartments interpreted as being sealed by clay smears along faults. The general

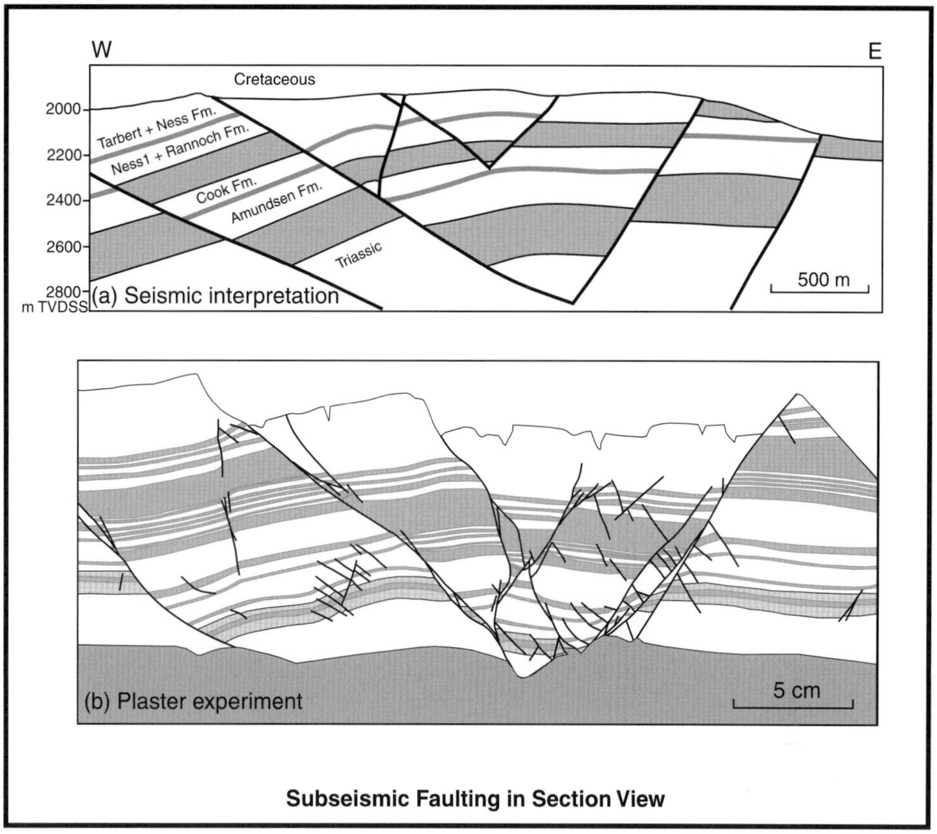

FIGURE 92. Comparison between (a) depth-converted seismic interpretation from the Gullfaks field, Norwegian North Sea, and (b) a plaster model deformed by plane strain extension. The plaster model shows that many small-scale faults are expected to exist in the Gullfaks structure but are below seismic resolution (from Fossen and Hesthammer, 1998). Reprinted with permission from the Geological Society.

observation is that the blanket of clay smear along faults only appears to be continuous and effective where the shale content of the displaced section exceeds 25%. The shale smear factor was estimated for faults from two of the fields in the basin. SSF values of between 1 and 4 were found for faults with throws more than 150 m (492 ft) that sealed the longest hydrocarbon columns. It was concluded that faults in this area could be modeled as sealing along their length provided the SSF did not exceed a value of 4.

SUBSEISMIC FAULTS

Only the faults that the geophysicist can pick from seismic data will be mapped, that is, those faults with vertical displacements down to the limit of seismic resolution. As mentioned in chapter 6, this can be about 20–40 m for reservoirs at moderate depths. However, a significant number of *subseismic faults* will probably be present with vertical displacements less than this (Figures 92, 93). Thus, the true degree of the structural complexity of a reservoir will be underrepresented.

It is possible to input subseismic faults into a reservoir model using *stochastic* methods (Munthe et al., 1993; Hollund et al., 2002). Stochastic modeling is described in more detail in chapter 19 of this publication. In summary, this is a computerized procedure for randomly inserting shapes representing geological features into a 3-D model while still honoring predefined rules and statistics controlling the global distribution of the data.

The first part of the method involves making an estimate of the number of subseismic faults by extrapolating from statistics on the length versus frequency of known seismic faults into the subseismic region. Fractal analysis has been used on the assumption that fault-size populations approximate to fractal distributions (Gauthier and Lake, 1993). Statistics are also compiled on fault orientations, length to throw ratios, and fault densities per square kilometer. A further step is to determine those areas of the field where subseismic faults are more likely to be present than elsewhere. One method is to predict the paleostrain regime of the reservoir at the time of faulting (Maerten et al., 2006). On this basis, a model will be made, which will include both the seismic and subseismic faults. Fault seal analysis can be applied to the subseismic faults in the model to determine whether they are sealing or not.

General experience with inserting subseismic faults into simulation models is that they will influence the flow behavior (Damsleth et al., 1998; England and Townsend, 1998; Ottesen et al., 2005). The critical feature seems to be whether the faults are sealing or not. Sealing faults can create an open framework of short baffles, which helps to improve sweep. The baffles increase the

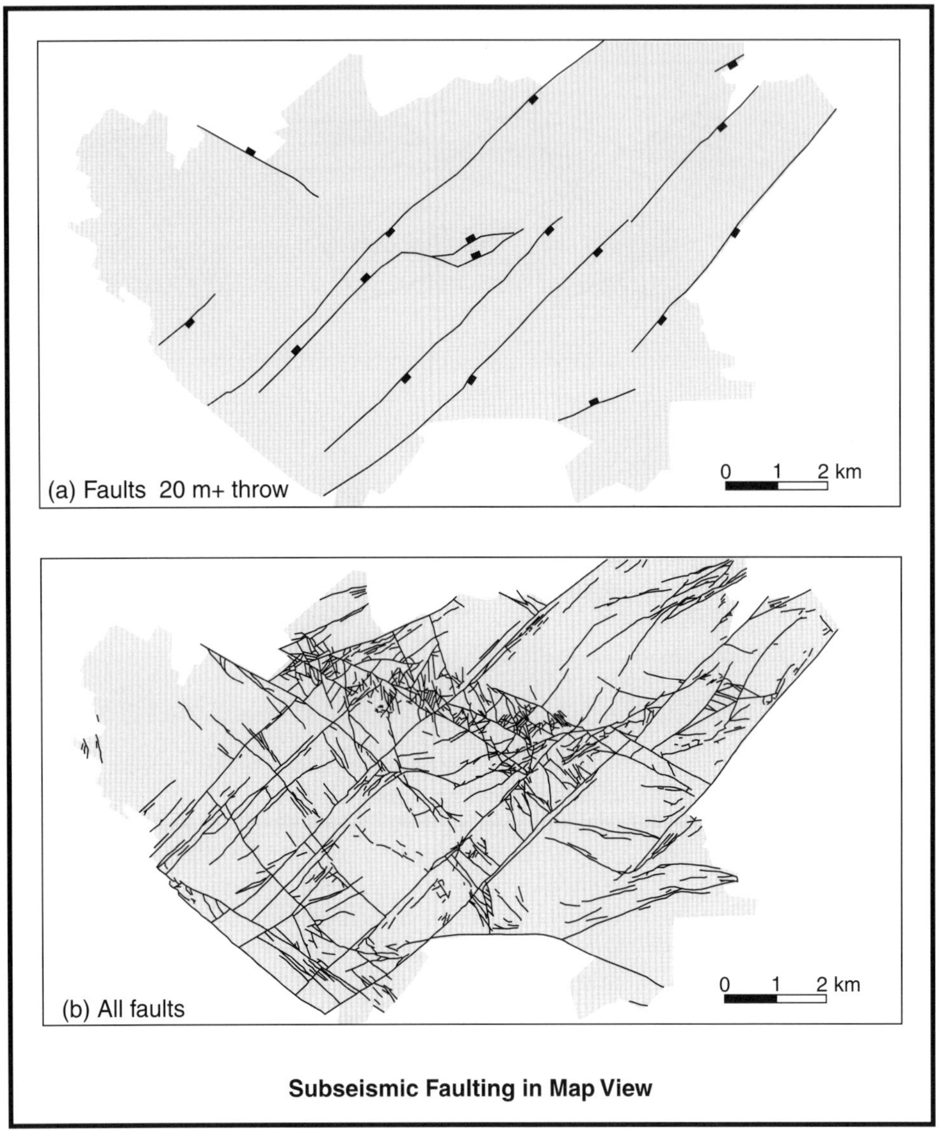

FIGURE 93. Fault maps of the East Pennine coalfield, United Kingdom. In map (a), only faults with throws of 20 m (64 ft) or more are shown. These are equivalent to faults that are detectable by seismic surveys at reservoir depths. In map (b), every mapped fault is shown, with fault throws of between 10 cm (4 in.) and 180 m (590 ft) (from Watterson et al., 1996). Reprinted with permission from the Journal of Structural Geology.

tortuosity of the flood front and delay water breakthrough. A large number of sealing subseismic faults in a reservoir will, on the other hand, create numerous dead ends, which will reduce the sweep efficiency of a waterflood. Nonsealing subseismic faults form cross-fault juxtapositions, which can improve vertical connectivity and enhance sweep.

GROWTH FAULTS

Growth faults are faults that were active at the same time as the sediments were being deposited (Figure 94). Many show a listric geometry with the fault soling out into shale horizons. They are common in areas with thick delta sequences. Growth faults can be recognized because sediments thicken into the hanging wall of a growth fault and the throw of the fault increases with depth. All the individual reservoir units may thicken up across a mapped growth fault. Alternatively, growth can be taken up by additional layers filling the accommodation space in the hanging wall (Hodgetts et al., 2001).

FAULTS AS FLOW CONDUITS

It is known that faults can conduct flow along the fault plane. Brittle rocks such as carbonates are more likely to contain conductive faults by comparison to shallow buried siliciclastic sediments, for example.

Specific examples of faults acting as fluid conduits have been described. Production wells located near faults showed rapid water breakthrough in the Fateh field, offshore Dubai. Well tests, production logs, radioactive tracer surveys, and interference tests indicate that aquifer influx is occurring along conductive faults within the reservoir (Trocchio, 1990). A campaign of horizontal drilling in the Prudhoe Bay field in Alaska showed that

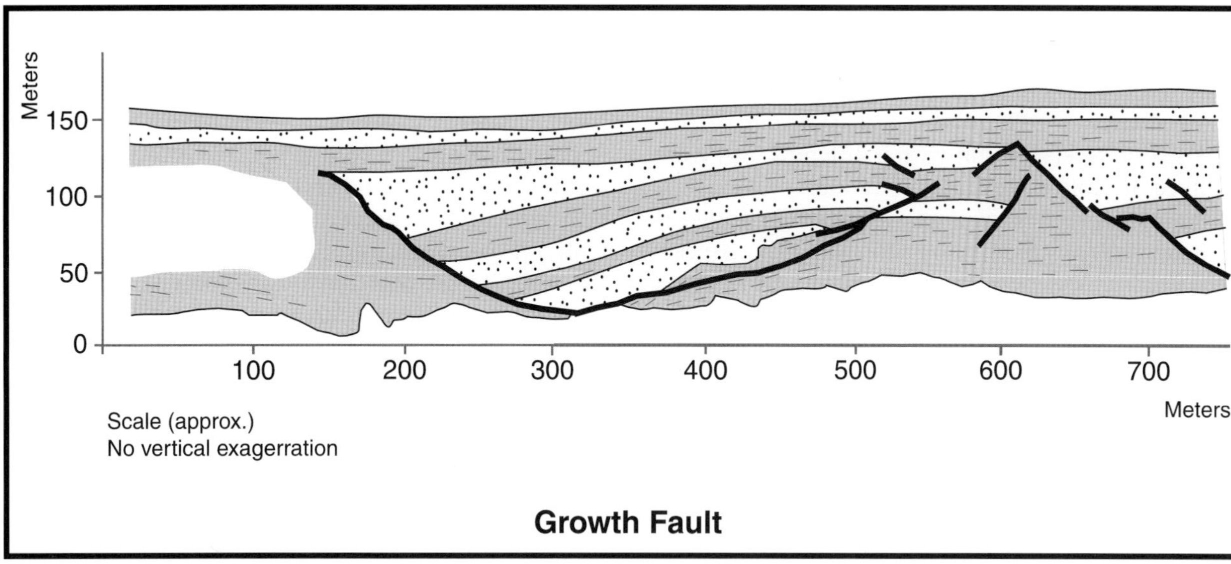

FIGURE 94. Reservoir intervals thicken markedly across growth faults. They are common in areas with thick delta sequences and mobile substrates such as shale or salt. This example is from Upper Triassic deltaic sediments exposed in the coastal cliffs of Svalbard (from Edwards, 1976). Reprinted with permission from the AAPG.

between 10 and 20% of the faults intersected by the wells were conductive to flow. These caused early water or gas production as a result of fault intersection with the water leg or the gas cap (Pucknell and Broman, 1994). Production, pressure, and production log data indicated that water flowing up faults had resulted in rapid water breakthrough in the crestal area of the Khafji field in the Arabian Gulf (Nishikiori and Hayashida, 1999).

STRESS CHANGES IN RESERVOIRS

In *stress-sensitive reservoirs*, fractures may dilate during injection and close during drawdown. These effects are most pronounced in low-permeability, overpressured, and naturally fractured reservoirs (Lorenz, 1999). Pressure depletion as a result of production will change the stress state of a reservoir (e.g., Hillis, 2001).

From a mechanical aspect, sandstone reservoirs are porous structures that form a load-bearing framework supporting the weight of the overburden. Reservoir depletion increases the *effective stress* on the grain framework; this is the difference between the total stress acting on all sides of the rock and the pore fluid pressure. The effective stress is applied at the grain to grain contacts. This leads to elastic deformation of the rock (recoverable on depletion reversal) and, with increasing stress, inelastic deformation. Inelastic deformation mechanisms include microcrack growth and closure, cement breakage, grain rotation, and sliding as well as deformation in clay, mica, and diagenetically altered feldspar grains (Bernabé et al., 1994; Schutjens et al., 1998, 2004; Wong and Baud, 1999). These mechanisms result in the compaction of the rock and a reduction in the porosity. Because the reservoir remains physically connected to the rock surrounding it, the overburden and underburden will also deform in response to reservoir depletion.

Compaction can lead to the reactivation of normal faults (Teufel et al., 1991; Goulty, 2003). In the Valhall and Ekofisk fields, offshore Norway, faults that were initially located in the crest of the field's anticlinal structure are thought to have spread out to the flanks as a result of reactivation induced by depletion and compaction of the Chalk reservoir. Casing failures have been attributed to shear along these spreading faults (Zoback and Zinke, 2002). Small earthquakes can be common around some producing oil and gas fields (Segall, 1989).

It is common to find that faults that were sealing over geological time in a reservoir start to leak after a few years of production. This may be noticed where a production anomaly occurs, such as newly drilled attic oil wells showing swept zones; a sudden, unexpected rapid rise in water or hydrocarbon production from production wells drilled close to faults; or an inexplicable source of pressure support appearing in the mid life of a producing well. In one example from the Endicott field in Alaska, a major sealing fault within the reservoir was known to act as a pressure barrier from early production data. Later on, it was established that radioactive tracer had crossed the fault from an injection well to a production well, and this indicated that the fault seal had broken down with production (Shaw et al., 1996).

Dincau (1998) analyzed fault breakdown with production in the South Marsh Island 66 field, offshore

Louisiana. The faults most likely to break down were those with limited predicted shale gouge and where the reservoir unit was fault juxtaposed against itself. Faults with an extensive predicted shale gouge and where they juxtapose one reservoir unit with a different unit were more likely to hold a pressure differential.

Examples of fault breakdown are often mentioned as a side issue in technical papers dealing with other aspects of field production. These include the Iagufu-Hedinia area of Papua New Guinea (Eisenberg et al., 1994), the Tia Juana field in Venezuela (Márquez et al., 2001), and the Veslefrikk field, offshore Norway (Pedersen et al., 1994). Fault breakdown is often attributed to the breaching of the capillary seal of the fault rock as a result of large differences in pressure across the fault. It is also possible that in some instances, fault breakdown is the result of fault reactivation induced by differential compaction between adjacent fault compartments, one significantly more depleted than the other. It is possible that the phenomena could be more common in depleting fields than is generally appreciated.

14

Shepherd, M., 2009, Structural geology: Fractures, *in* M. Shepherd, Oil field production geology: AAPG Memoir 91, p. 123–128.

Structural Geology: Fractures

INTRODUCTION

Fractures can lead to increased productivity from a field. As such, the characterization of fractures is important in reservoirs where they provide a significant contribution to flow. It is possible to model the distribution of fractures in a reservoir by determining how large-scale features such as folds and faults have influenced their development.

FRACTURES

Fractures are surfaces along which rocks have broken (Twiss and Moores, 1992) (Figure 95). They include joints, but faults can also be considered as fractures that show displacement. Open fractures can significantly increase the permeability of a reservoir rock (Stearns and Friedman, 1972). Where the matrix permeability is low or negligible, fractures can make the difference between a productive and nonproductive reservoir. In certain situations, such as extensively fractured basement rocks, all the porosity and permeability may reside in the fracture system. The storage and deliverability of these systems can be very impressive.

Nevertheless, the success of a fractured reservoir is usually the result of the assistance that the fracture permeability can give in producing from an otherwise low-permeability matrix between the fractures. For example, the Ekofisk field in the Norwegian North Sea produces oil from a fractured chalk reservoir. The matrix permeability of the chalk ranges between 0.1 and 10 md. Natural fractures increase the effective permeability in the wells by up to 50 md (Toublanc et al., 2005).

RECOGNIZING FRACTURES IN THE SUBSURFACE

The first step in a fracture study (Figure 94) is to examine the cores as they provide a direct observation of the distribution and properties of fractures in the subsurface (Mäkel, 2007). Key parameters include the fracture density, the population of open versus healed fractures, the fracture orientation, the fracture width, and any degree of diagenetic enhancement or degradation of fracture permeability. Care needs to be taken to differentiate between natural fractures and those induced by the coring process (Kulander et al., 1990). Breaks in the core not aligned parallel or perpendicular to the long axis of the core are more likely to be natural fractures.

Borehole image logs are an important source of data in fractured reservoirs (Haller and Porturas, 1998). These logs can show features such as fractures down to a few millimeters across. If fractures are visible, then it may be possible to determine if they are open or closed. Fracture orientation in the subsurface can also be established. A quantitative estimate of fracture density can be made from workstation analysis of image logs (Cheung and Heliot, 1990) (Figure 96).

Mud losses while drilling will occur if the drill bit penetrates a large open fracture. Even small losses can be a sign of fracture presence (Dyke et al., 1995). Mud losses can act as a spur to run image and production logs if noted (Ali-Ali and Stenger, 2001).

Full waveform array sonic logs can indicate the presence of fractures (Hsu et al., 1987). Acoustic waves travelling along the borehole wall are attenuated by open fractures, and these form chevron patterns on the log display (Hornby, 1995).

DOI:10.1306/13161197M913372

FIGURE 95. A fracture network in Devonian sandstones, Caithness, Scotland (photograph courtesy of Dominic McCormick).

Fractures can sometimes be picked out on resistivity or resistivity-based dipmeter logs where water-based mud has invaded the fractures. The fractures can show a very low-resistivity response within the hydrocarbon leg (Iverson, 1992).

The presence of open fractures can be established by production logs. A flowmeter log may show high flow rates over a very short interval (tens of centimeters). These are plausibly fractures if there are no indications of localized high-permeability intervals to explain the high

Workflow for Characterizing Fractured Reservoirs

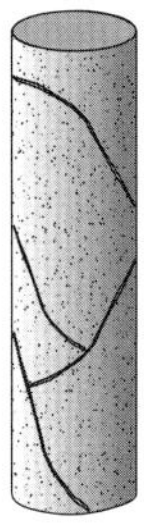

- Core fracture study
- Determine fracture presence from logs
  - image logs
  - full waveform sonic
- Establish fracture presence from well data
  - mud losses
  - production logs
  - well test analysis
- Assess fracture density
- Find out the fracture orientation(s)
  - fracture orientation vs. principal horizontal stresses
- Determine fracture aperture, porosity, and permeability
- Relate fracture density and location to specific large-scale features
  - folding
  - faults
  - seismic anisotropy
- Input to reservoir simulation model

FIGURE 96. Workflow for characterizing fractured reservoirs.

flow response. Such indications, coupled with image logs, can give vital information on open fracture trends within the reservoir.

Well tests can be used to infer the presence of fractures, particularly if the reservoir permeability from the test is greater than would otherwise be indicated by the matrix permeability from core plug data. Fracture conductivity and storage can be estimated from pressure transient analysis.

FRACTURE PROPERTIES, POROSITY, AND PERMEABILITY

For a fracture network to be effective in producing hydrocarbons, it must be extensively connected and conductive. A study of a fractured anticline in the foothills of Western Canada showed that *fracture density* was the most important factor controlling the development of good fracture connectivity (Jamison, 1997). Fracture spacing is influenced by the lithology, grain size, porosity, bed thickness, and stress history of the rock (Narr and Suppe, 1991). Methods exist for estimating the average fracture spacing in the subsurface from the borehole diameter and the fracture length (Narr and Lerche, 1984; Narr, 1996). For instance, in the Ekofisk field, Norwegian North Sea, fracture spacing was determined from core and image data. The fractures occur in swarms over a width of 1–3 m (3–9 ft) with a fracture intensity of two to five fractures per 30 cm (12 in.). The fracture swarms are spaced at intervals of about 5–30 m (16–98 ft) (Toublanc et al., 2005).

The *orientation of the fracture system* is an important parameter. A dominant horizontal stress direction is commonly present in the subsurface, although local features such as faults or salt diapirs can cause the azimuth to vary locally (Nordgård Bolås and Hermanrud, 2002). Much of the horizontal stress is imparted by present-day plate tectonic movement. A fracture set may be open perpendicular to the *minimal horizontal stress* yet in most cases, fracture sets will be closed in orientations perpendicular to the *maximum horizontal stress*. Occasionally, fractures in hard rock or mineralized fractures can stay open against the maximum horizontal stress (Dyke, 1992; Hillis, 1998).

Horizontal stress directions can be determined from evidence for *borehole breakout* in wells. These are triangular-shaped zones of enlargement resulting from stress failure, symmetrical around the borehole and oriented along the azimuth of the minimal horizontal stress. Borehole breakout can be established using caliper, dipmeter, or image log data (Bell and Gough, 1979; Bell, 1990). For example, Yassir and Zerwer (1997) analyzed the stress regimes in the Gulf Coast, offshore Louisiana, using wellbore breakout analysis (Figure 97). The bulk of the data set indicates a northeast–southwest orientation for the maximum horizontal stress, parallel to the strike direction of the Tertiary clastic wedge. Anomalous directions occur locally and have been attributed to stress perturbations associated with faults and salt structures.

Worldwide regional stress data sets have been incorporated into a *World Stress Map* by a research group based at the Institute of Geophysics at Karlsruhe University in Germany (Heidelberg Academy of Sciences and Humanity, 2004; Tingay et al., 2005). The maps can be browsed at http://www.world-stress-map.org/.

Fractures in a reservoir can be categorized according to *fracture sets*, a concept similar to lithofacies for sediments. These are designated on the basis of fracture type (extensional or shear), tectonic event, and orientation.

Fracture porosity for a volume of rock is the percentage or fraction of the void space in the fractures relative to its total volume. This is estimated from the average width of the *fracture aperture* and the fracture spacing. Image logs and core are the main source for this information. Fracture porosities are usually much lower than normal sandstone and carbonate porosities, typically less than 1% (Nelson, 1985). However, as the fracture porosity is continuous and pervasive over a very large volume, flow rates from fractures can be high.

Fracture permeability is proportional to the open aperture of the fractures (Aguilera, 1980). Narrow hairline fractures will tend to show only low flow rates, even though they may show good connectivity over large distances. Wider apertures are needed for decent flow rates. Fracture apertures can be determined from core and image logs (Luthi and Souhaité, 1990) or estimated from well test data.

FILLING IN THE GAPS

Filling in the gaps with a fracture distribution model is not easy, but it is possible. Common practice is to try and link fracture distribution in the wells to specific structural or lithological features that can be mapped at a large scale. If a relationship can be found, then this allows the fracture intensity and orientation of fractures to be predicted for the reservoir as a whole.

For onshore oil fields, satellite imaging of outcrops can yield analog data for reservoirs in the same area (Alpay, 1973; Hennings et al., 2000). The outcrops can be used to define the fracturing style, lithological control, orientation, fracture length, and spacing. Outcrops should be used carefully because of problems in interpretation resulting from weathering and differences in burial history (Cacas et al., 2001).

The crudest method is to interpolate fracture distribution between the wells from well data alone. A more sophisticated way of doing this is to constrain interpolation to *geomechanical properties* (Heffer et al., 1999). The degree to which a rock will fracture depends in the first instance on how brittle it is, the ductility as influenced

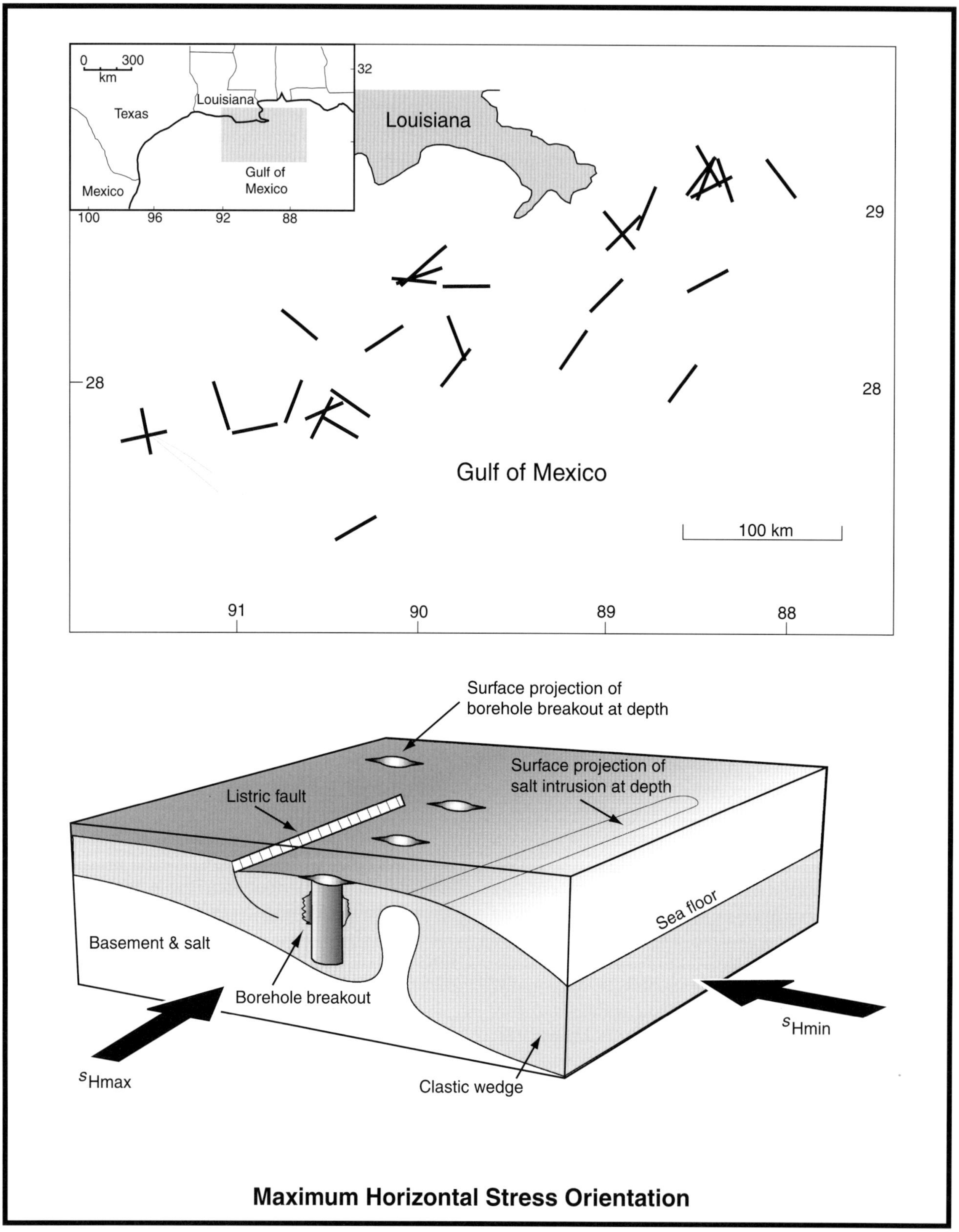

FIGURE 97. The map shows the present-day maximum horizontal stress orientation in the Gulf of Mexico, offshore Louisiana, as determined by wellbore breakout analysis (from Yassir and Zerwe, 1997). Reprinted with permission from the AAPG.

FIGURE 98. Fractures in lacustrine sediments of Devonian age, Caithness, Scotland, illustrating the concept of mechanical stratigraphy. The slightly coarser grained sediments are heavily fractured whereas there are few fractures within the more ductile, fine-grained sediments (photograph courtesy of Dominic McCormick).

by the volume of shale, the matrix porosity, and the bed thickness. If an outcrop analog is available, it may be possible to characterize the *mechanical stratigraphy* of the fractured rock (Corbett et al., 1987; Bertotti et al., 2007). This is not necessarily the same as the sedimentary stratigraphy. A mechanical layer can be defined as one or more stratigraphic units that fracture independently of the other units (Underwood et al., 1993). Fractures will tend to be stratabound within mechanical layers terminating against *mechanical layer boundaries* (Wennberg et al., 2007) (Figure 98).

If folds can be mapped, then a correlation between the intensity of fold curvature and fracture density can be derived (Antonellini and Aydin, 1995). *Curvature analysis* is used to determine zones of anomalously high strain so as to predict the intensity of fracturing (Lisle, 1994; Stewart and Podolski, 1998). The results of this analysis can be cross checked against the known intensity of fracturing from well and seismic data.

Toublanc et al. (2005) found a relationship between fracture and fault intensity in the Ekofisk field. They used the cumulative fault trace length observed per sampling area method of Dershowitz and Herda (1992) as a predictor (Figure 99).

Seismic facies analysis can be integrated with dip, azimuth, edge, and semblance analysis to identify fracture zones (Bloch et al., 2003). AVO analysis may indicate acoustic reservoir anisotropy caused by the presence of open fractures (Harvey, 1993).

Geostatistical techniques can be used to relate the fracture distribution to seismic, well imaging, and log data (Cacas et al., 2001). Ozgen et al. (2003) produced a "fracturability" map for fractured reservoirs in Mexico and Texas. The fracture index was estimated from lithology, porosity, and shale content and calibrated to fracture information from wells. This was combined with fold curvature analysis to produce a fracture connectivity model.

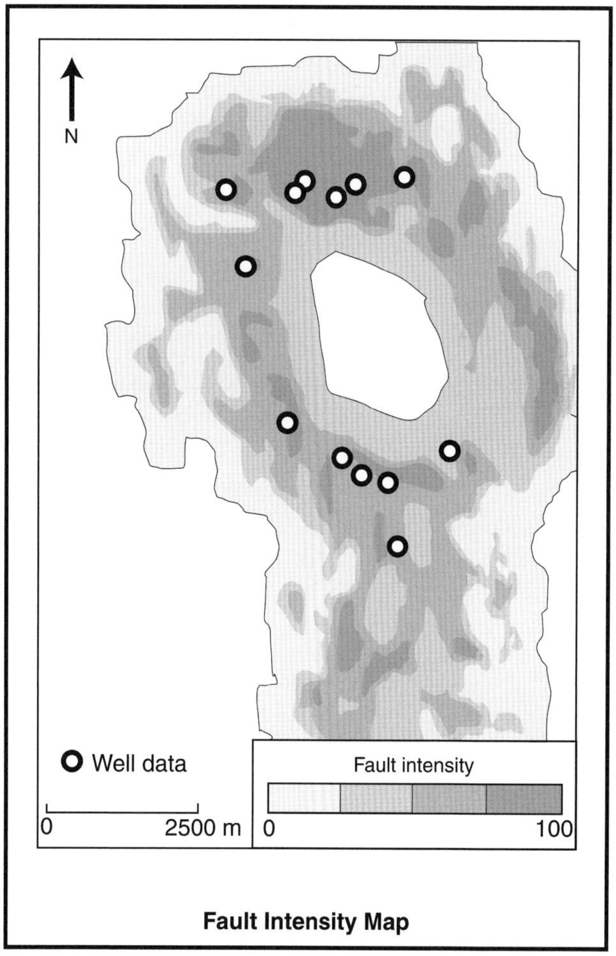

FIGURE 99. Fault intensity map at the top of the Ekofisk Formation and fracture data locations (borehole image data and cores), Ekofisk field, Norwegian North Sea (from Toublanc et al., 2005). Reprinted with the permission of the Geological Society.

The last stage in fracture characterization is to calibrate the fracture distribution to dynamic data. The contribution to flow by fracture permeability is determined by comparing the fracture intensity in the wells to the permeability calculated from well test data and production log profiles (Rawnsley and Wei, 2001).

RESERVOIR ENGINEERING MODELS OF FRACTURED RESERVOIRS

Geological modeling of a fractured reservoir can be used to provide input into a reservoir simulation. A common assumption in modeling fracture distribution is that the large faults are the main conduits for fluid flow. These are represented implicitly in models. The effects of smaller faults and the fracture network are typically included as an upscaled property (Walsh et al., 2002). Given that fractures generally increase the effective permeability of a fractured reservoir, the permeability in the grid is increased by a multiplier corresponding to the location of the fracture in the reservoir. Some specialized simulators designed for modeling fractured reservoirs can store the matrix and fracture properties separately (e.g., Cosentino et al., 2001). Examples of the geology to simulation workflow are given in technical papers on the Ekofisk field, offshore Norway (Toublanc et al., 2005), and an unnamed field, offshore Abu Dhabi (Lyon et al., 1998). An excellent review of the workflow involved in modeling fractured reservoirs is given by Mäkel (2007). Books on fractured reservoir characterization include Aguilera (1980), Nelson (1985), and Narr et al. (2006).

WELL PLANNING IN FRACTURED RESERVOIRS

Production wells in reservoirs where fracture production is important, should be drilled to intersect as many open fractures as possible. Horizontal wells are best suited for maximizing the potential of intersecting open horizontal fractures. Knowledge of the subsurface stress regime in the Keystone field of Texas was used to determine the likely orientation of open fractures. It was proposed that horizontal wells perpendicular to the strike of open fractures and in zones containing the highest remaining oil would maximize recovery from the reservoir (Major and Holtz, 1997).

Shepherd, M., 2009, Flow geology, *in* M. Shepherd, Oil field production geology: AAPG Memoir 91, p. 129–130.

Section 3

Flow Geology

GEOLOGY AND FLUID FLOW

The production geologist has the task to analyze and understand how the geological framework influences fluid flow within a reservoir. *Flow geology* is the term used for this in this publication. If the geologist intends to build a geological model on the computer, then it is important that the model should replicate the fluid-flow framework of the reservoir. Any reservoir simulation model built using such a geological model will have a better chance of matching and predicting the reservoir performance (see Section 4 of this publication). An understanding of the flow geology is the basis for locating the remaining hydrocarbons in a field.

The use of techniques for understanding the geological controls on fluid flow in an oil field predate the modern computer age of production geology. Many of the methods described here have been in use since at least the 1960s and they provide just as powerful a tool today as they have in the past. Nevertheless, the basic ideas are generally less well-known now than they have been previously. There are two possible reasons for this. Firstly, the modern computer applications used by production geologists do not provide integrated work flows for analyzing the flow geology. As such, younger geologists who use these programs extensively may not even be aware of this aspect of production geology. Secondly, there are a smaller number of technical papers dealing with the operational aspects of production geology by comparison to the more academic side of the subject. Several well-established practical methods are used by oil company geologists to understand reservoir geology that are rarely mentioned in technical papers, if at all. Some of these are described in this section and later on in Section 5 of this publication.

An excellent series of technical papers have been published by staff from the Bureau of Economic Geology, University of Texas at Austin, on the subject of flow geology and the use of this to locate the remaining hydrocarbons. These include studies on the Big Wells field of South Texas (Tyler et al., 1987), the Jackson field in Australia (Hamilton et al., 1998), and the Budare field in Venezuela (Hamilton et al., 2002).

DATA INTEGRATION

The method for coanalyzing geology and production data has been termed *data integration* (Bryant and Livera, 1991). The idea is that by combining production data with the geological interpretation, common patterns may be observed. The method is graphical and involves overlaying one set of data on the other. It is a two-way process. By overlaying production data onto geological maps and cross sections, data integration allows the likely flow character of the reservoir to be inferred from the geology. Features such as faults, lithofacies boundaries, or permeability fairways may act to control flow pathways. In turn, data integration allows the location of the main elements of the reservoir architecture to be inferred from flow patterns shown by the production data.

The work of understanding the flow geology of a reservoir typically involves building up a dossier of different observations from data integration. A "toolkit" of various techniques can be used to work out the flow geology of a reservoir; the main methods are given in this section. The observations are collated and are used to piece together an understanding of how the reservoir behaves as a whole. The work is painstaking, involves a large amount of data, and takes a long time to do properly. Nevertheless, the result can be extremely rewarding (Hartman and Paynter, 1979); the procedure is akin to detective work. It involves collecting clues to solve the bigger puzzle as to how the field behaves.

Not every production geologist carries out data integration analysis when they work on a reservoir. Arguably, if a geological scheme is not allied to an understanding of the flow performance of the reservoir, then

DOI:10.1306/13161198M913372

it is one that lacks focus. For example, when a geologist constructs a 3-D geological model of a reservoir, a lithofacies model is made on a computer, often with the individual macroforms represented as objects (see chapter 19, in this publication). There is an unstated assumption in making these computer models, that if the geologist honors the relative abundance, shape, dimensions, and rock properties of the various macroforms put into the 3-D geological model, then the connectivity of the reservoir will be directly replicated as a result. For some reservoirs this may be a valid assumption, but for many, it is not; local features such as cementation, mud drapes, and permeability contrasts at lithofacies boundaries commonly influence flow to a major extent. The effect of these may only be properly understood once a detailed data integration study has been made.

The understanding gained from integrating production data with the geology can be so fundamental that the previous geological model may be discarded and a new one, showing a better fit to the dynamic data, will take its place (Holtz and Hamilton, 1998). This is an important point. In the workflow followed by this book so far, the geological scheme has been based on a correlation of reservoir units that has been defined by a sequence-stratigraphic framework. Although there is a general correspondence between the various geological features controlling flow and the sequence-stratigraphic elements, there is not always a perfect fit. There will be numerous barriers and baffles to flow at a finer subdivision than that provided by the sequence-stratigraphic scheme. At this point, there is a need to refine the geological scheme to a *flow geology scheme*, a reservoir scheme that both honors the sequence-stratigraphic framework and is optimized to best represent the dynamic performance of the reservoir. The well correlation scheme needs to be redefined to match the flow geology. Correlation lines should be tied to the top and base of the individual hydraulic units.

DATA COLLATION

The first step in assessing how the geology and production interrelate is to collate the data. The history of every well needs to be collected within a database to evaluate fluid movement. This includes the following:

- The well production history: A plot of all the hydrocarbon and water production flow rates in each well versus time. Similar plots should be collated for water and gas injection data.
- The well status history: This includes the perforation history of the well together with any recompletions, plugs, patches, straddles, or reperforations.
- Postproduction formation tester pressure–depth plots.
- Production log results and interpretations.
- Well test and interference test data.
- Radioactive or chemical tracers.
- Any other relevant data that can be helpful for flow surveillance.

15

Shepherd, M., 2009, Production data and layering, *in* M. Shepherd, Oil field production geology: AAPG Memoir 91, p. 131–144.

Production Data and Layering

INTRODUCTION

Layered reservoirs frequently contain barriers to vertical flow; typically these barriers are shales. These can divide a reservoir up into discrete hydraulic units. Downhole variations in permeability strongly influence the flow profile of a well. Intervals of similar permeability can be defined as flow units. Baffles, barriers, and permeability variation result in distinctive flow patterns, which affect how a reservoir produces hydrocarbons. Various methods allow barriers and baffles to vertical flow to be recognized.

BARRIERS AND BAFFLES TO VERTICAL FLOW

Nonreservoir rock, such as shales, coals, cemented intervals, micritic carbonates, and bedded anhydrites, can act as extensive barriers and baffles to vertical flow within a reservoir. *Barriers* are laterally extensive and prevent flow communication vertically. *Baffles* are shorter ranging and can impede vertical flow but will not prevent it (Figure 100).

Shales are the major features causing flow layering in reservoirs. The areal extent of shales within a reservoir is generally related to the depositional environment. Shale barriers are more extensive in marine sediments than they are in continental sediments (Zeito, 1965; Richardson et al., 1978). Laterally extensive marine mudstones form marine flooding surfaces, which are correlatable on a basin scale. Shales in deep-water sediments commonly act as flow barriers. Shales are less widespread in fluvial channel systems and deltas (Le Blanc, 1977; Weber, 1982).

Discontinuous shales act as baffles to vertical flow. For fluid to move upward in a baffled reservoir, it has to take a tortuous route around the edges of the shale baffles. This situation can sometimes be favorable to production. For example, a rising oil-water contact will be obstructed by a series of discontinuous shales in a reservoir unit. The water will be forced to flow along a longer, more tortuous path through the reservoir, taking more time to reach the basal perforations in a production well and coming in contact with more oil. Oil production from the well will be water-free for a longer period of time than would be the case if the shale baffles were not present.

HYDRAULIC UNITS

A key part of reservoir characterization is therefore determined by shales. When examining cores and logs, it is important to determine from the vertical facies analysis whether a shale is likely to be extensive or laterally discontinuous (Bryant and Flint, 1993). Where shales are laterally extensive, they can split a reservoir up vertically into *hydraulic units* (Haldorsen and Lake, 1984). These have also been called *containers* (Hartmann and Beaumont, 1999). Hydraulic units comprise volumes of rock bounded at the top and base by geological features that act as permeability barriers to vertical flow. The reservoir interval between the permeability barriers is in hydraulic continuity vertically, with similar pressures from top to bottom (Figure 101). The shales at the top and base of the hydraulic unit may mark pressure discontinuities between the unit and the hydraulic units above and below.

VERTICAL FLOW BARRIER MAPS

Sometimes the key to understanding fluid flow in certain reservoirs is to determine whether the upper and lower surfaces of the individual reservoir units allow vertical flow across them or not. Are the units bounded top and bottom by shale barriers or are sand-sand contacts allowing fluid transfer? This is a common problem where a reservoir is made up of several stacked shoreface sand parasequences (Larue and Legarre, 2004).

DOI:10.1306/13161199M913372

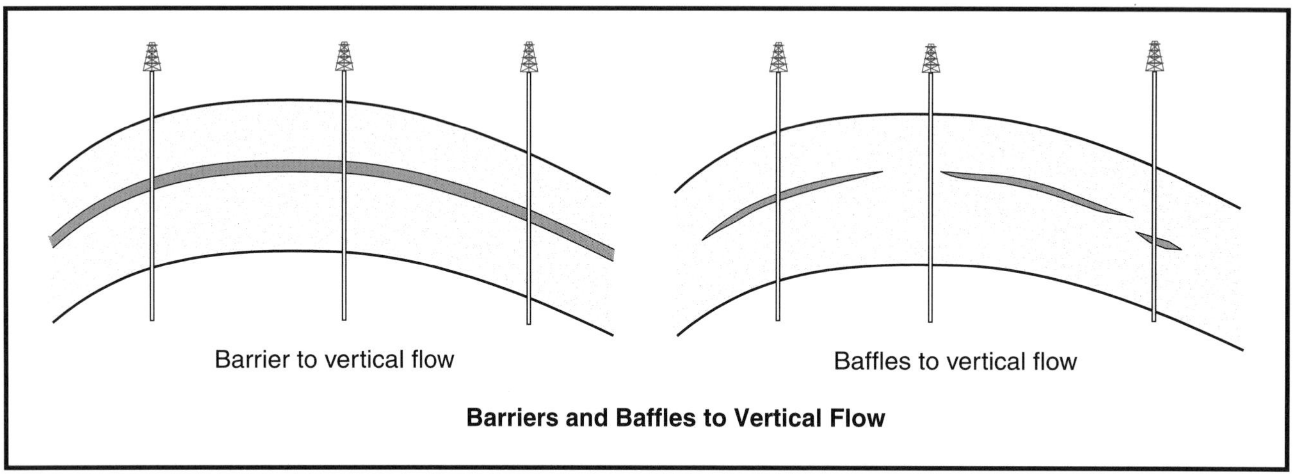

Figure 100. Barriers are laterally extensive and prevent flow communication vertically. Baffles are short ranging and impede vertical flow but do not prevent it.

Construction of *vertical flow barrier maps* can help (Cook et al., 1999). These maps describe the sealing nature of specific unit boundaries.

Two types of vertical flow barrier maps have been used. Deterministic maps are drawn to show the areas where the shale blankets are thought to be extensive and where they may have holes allowing vertical communication (Figure 102). For example, Tye et al. (1999) derived deterministic vertical flow barrier maps for the fluviodeltaic Ivishak Formation at the base of the Prudhoe Bay field in Alaska. These were used to understand how gravity drainage, waterflood, and EOR mechanisms operated so as to manage the reservoir effectively. An alternative is to draw probability maps of the likelihood of shale occurrence. This approach was used for the reservoir characterization of the deep marine sandstone reservoir of the Miller field in the North Sea (Garland et al., 1999).

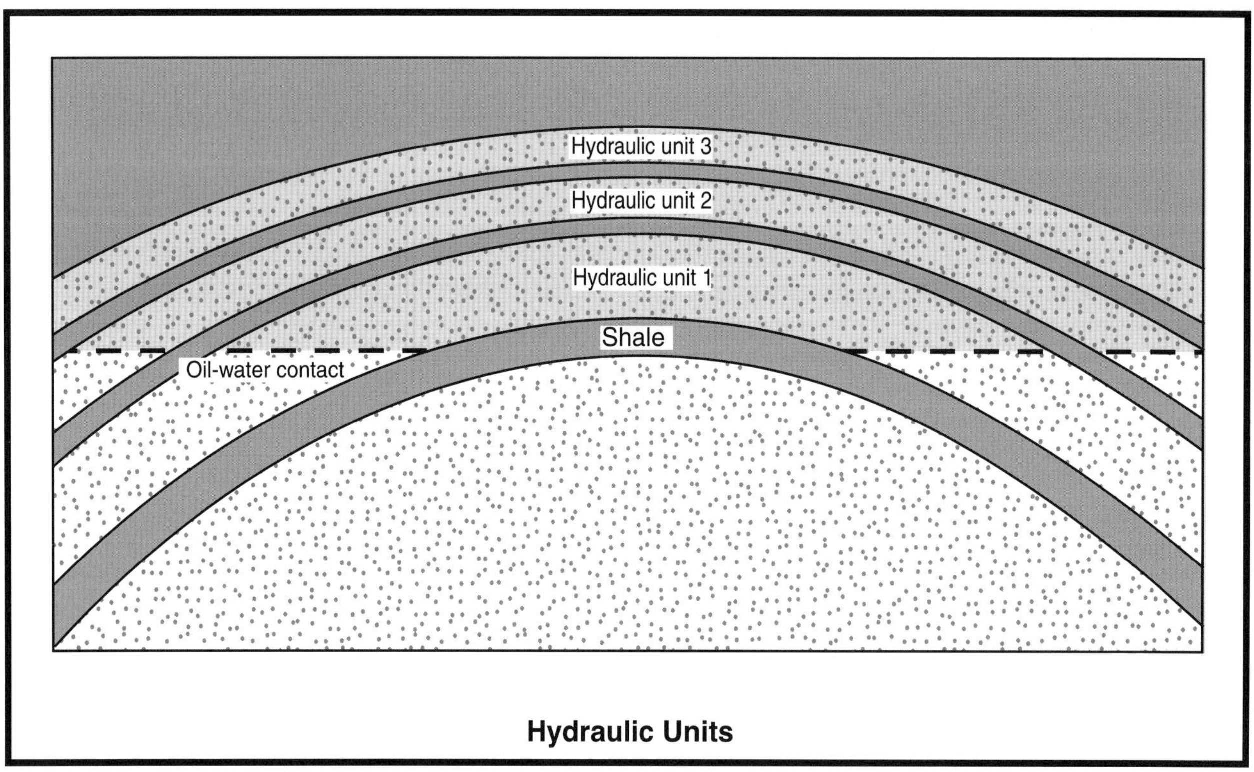

Figure 101. Hydraulic units are volumes of rock that are individually in hydraulic continuity but are separated from each other by permeability barriers to vertical flow.

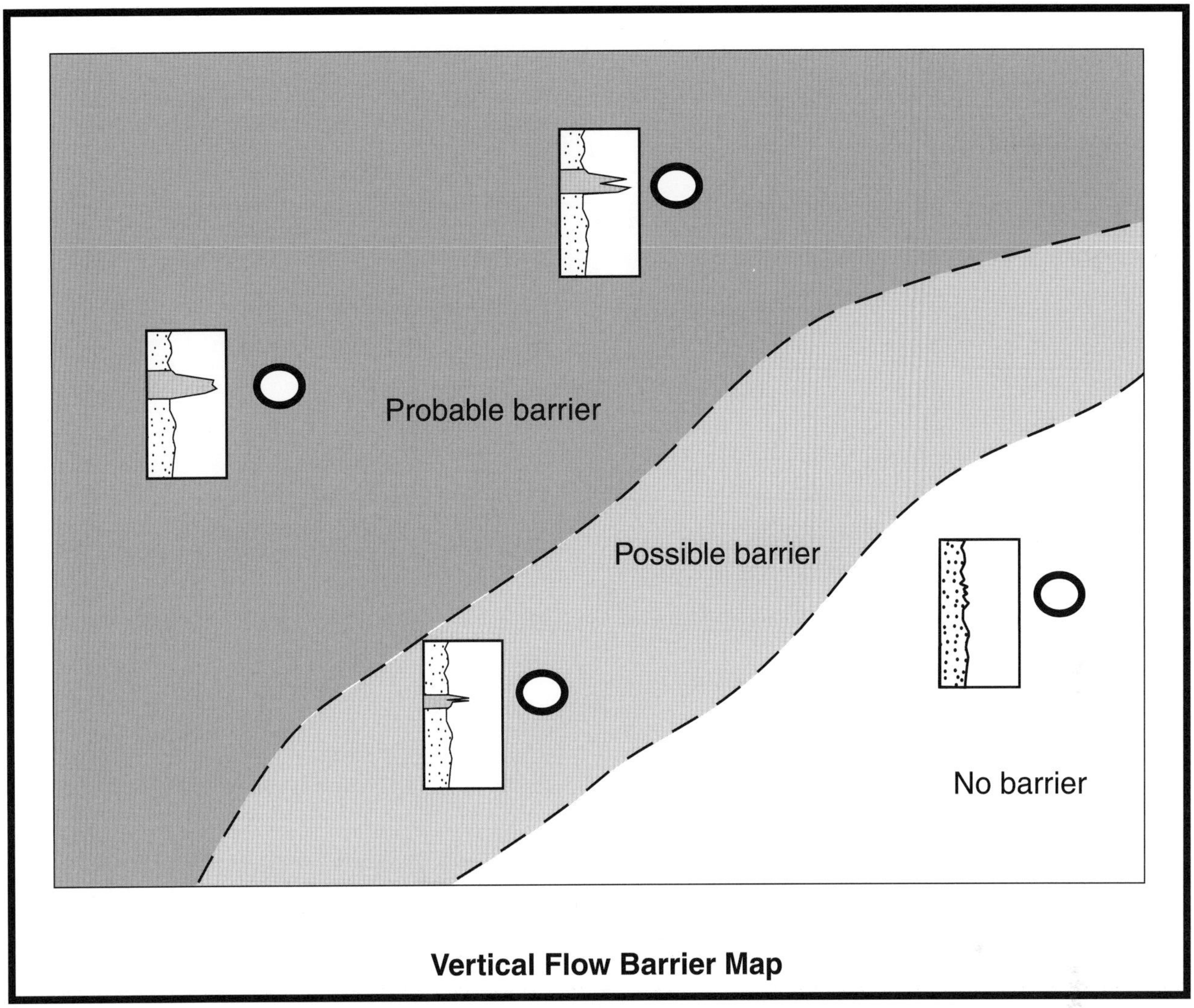

FIGURE 102. Vertical flow barriers can control the drainage patterns in a reservoir. The degree to which individual barriers are effective across the reservoir can be characterized by vertical flow barrier maps.

PERMEABILITY PROFILES

Permeability variation within a reservoir is the major control on the flow profile and sweep. Even if the entire reservoir interval has been perforated, the low-permeability intervals may be poorly drained (if at all). There will be a preference for water to displace oil through the higher permeability reservoir intervals.

It is useful to plot core plug permeability values onto copies of the core photographs. This gives the geologist an idea of how permeability and the sedimentology are related.

Where wells are extensively cored, plots of core permeability profiles with depth will indicate where the intervals of fast and slow flow are likely to occur. It is recommended that the permeability values are plotted on a linear instead of a logarithmic scale. As Dake (1994) pointed out, flow is linearly proportional to permeability and not to its logarithmic value. Only plots of the downhole permeability profile on a linear scale will show the relative flow capability within a reservoir interval.

FLOW UNITS

Flow units are specific parts of the reservoir that control the flow of fluids (Hearn et al., 1984). A flow unit has been defined as a mappable portion of the total reservoir rock within which the geological and petrophysical properties that affect fluid flow are internally consistent and predictably different from properties of other rock volumes (Ebanks et al., 1992).

A flow unit subdivision allows the downhole variation in horizontal permeability to be captured in a

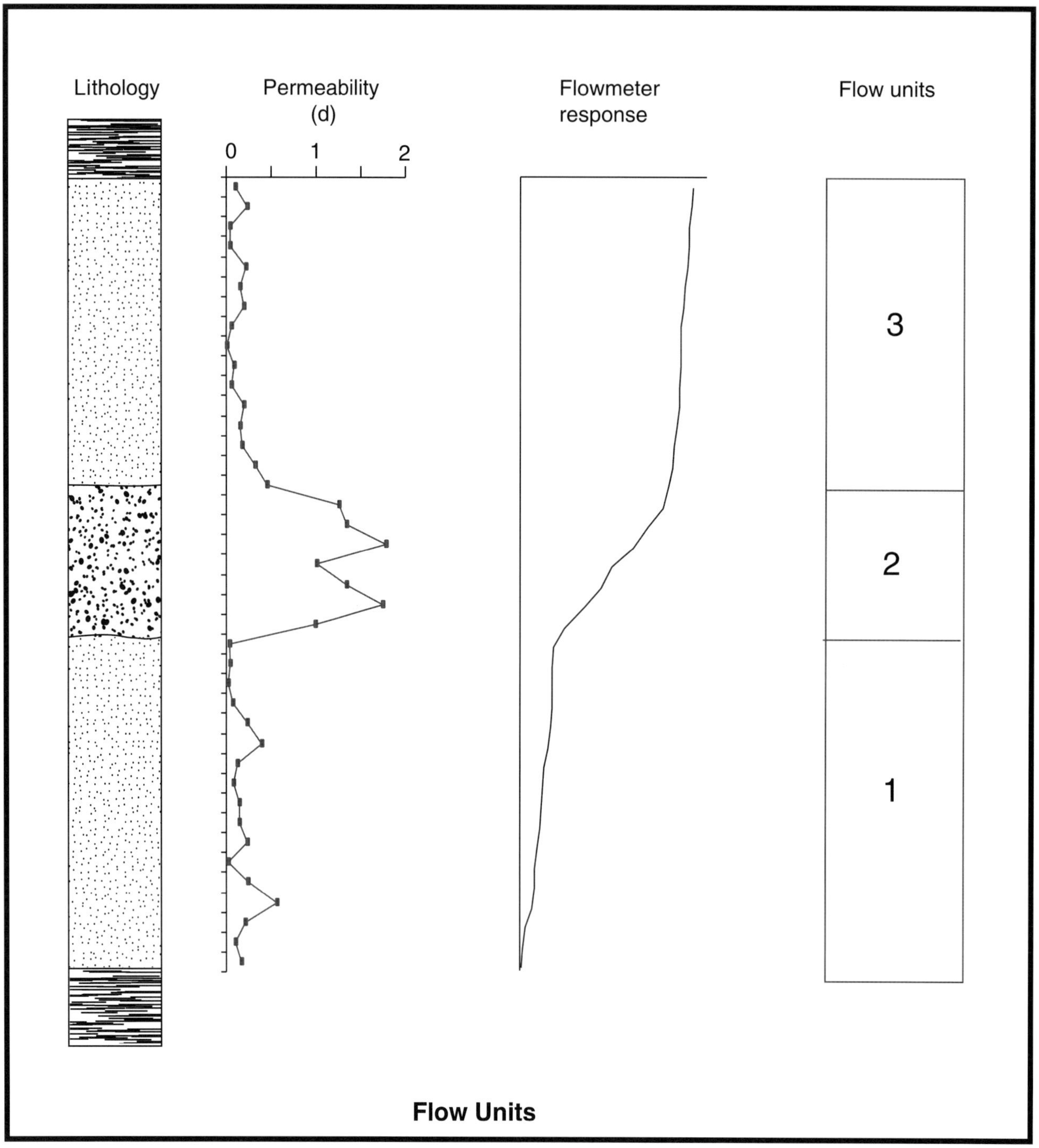

FIGURE 103. A reservoir interval can be subdivided into flow units according to its permeability and fluid flow character. The high-permeability flow unit 2 dominates the flow in this section.

meaningful way (Figure 103). Discretization of zones with similar permeabilities will give a better representation of sweep if the flow unit subdivision is to be exported to a reservoir simulation model. Flow units differ from hydraulic units in that they are bounded by significant changes in permeability, whereas hydraulic units are bounded by barriers to vertical flow.

CUMULATIVE PERMEABILITY PLOTS

Flowmeter logs can be used to determine flow units on the assumption that flow is directly controlled by permeability variations. One technique to verify this is to create a *cumulative permeability plot*. From the base

Table 17. Calculation for a cumulative permeability plot.

Depth	*Permeability (md)*	*Cumulative Permeability*
17,265	2	167 (142 + 10 + 13 + 2)
17,266	13	165 (142 + 10 + 13)
17,267	142	152 (142 + 10)
17,268	10	10

of the reservoir section or perforated interval, the permeability values are added up at incremental steps (see Table 17).

This calculation can be done quickly in a spreadsheet. The cumulative permeability plot can then be directly compared to the flowmeter log (Figure 104). The two curves should be roughly similar. Discrepancies between the two may indicate zones that are noncontributing, perhaps because of lower reservoir pressures, or, on the other hand, some zones may be flow enhanced because of the presence of fractures.

The flow unit is an invaluable term for describing reservoir behavior. Kuhn et al. (2003) defined *"fast" and 'slow' flow units* for the description of the reservoir performance of the Fulmar field in the UK North Sea. One set of fast flow units shows horizontal permeabilities greater than 600 md. Most of the movable oil in these units has been flushed. The slow flow units are fine-grained sandstones with low horizontal permeabilities (1 to 50 md). Fluid is considered to move very slowly in these units. Production logs indicate that these reservoir intervals are only slightly swept.

FLOW PATTERNS AND RESERVOIR LAYERING

The production performance of a reservoir will be influenced by barriers and baffles to a major extent. Certain production and flow characteristics allow barriers to vertical flow to be detected by the geologist.

BOTTOM WATER DRIVE

Where oil overlies a mobile aquifer, the aquifer moves upward to replace the oil as it is produced. This is called *bottom water drive* (Figure 105). It is a characteristic of clean reservoir sections where there are no significant permeability barriers to vertical flow. Pulsed neutron production logs, run at regular intervals in production wells, can show a history of a regular rise in the oil-water contact over time in reservoirs with bottom water drive.

The rise in the fluid contact may not always be regular. In reservoirs with very high vertical permeabilities, *water coning* may be a problem. If a well is produced too hard, viscous forces overcome gravity forces and water will be drawn upward into the producing perforations in a cone shape (Muskat and Wyckoff, 1934).

Following a regular rise in the oil-water contact over time in a bottom water drive reservoir, sometimes the oil-water contact can suddenly stop moving below a shale barrier. The producing oil-water contact is then observed from production logs to be static below this shale for a year or more. This indicates that the shale is acting as a widespread barrier to vertical flow.

EDGE WATER DRIVE

In strongly layered reservoirs, permeability barriers to vertical flow can suppress the bottom water drive while encouraging water to flow into the reservoir parallel to the beds. This is known as *edge water drive* (Figure 105). Edge water drive can be detected by production logs. A series of pulsed neutron logs can show the oil-water contact to be static under a shale for some time. Above the shale, the logs may show that some reservoir zones are swept by water whereas other intervals are at original oil saturations.

WATER OVERRUN (GAS UNDERRUN)

Laterally extensive shales can support an influx of water within an oil zone if the shale is extensive enough to connect with an injection well or an aquifer (Haldorsen et al., 1987). This is known as *water overrun* (Figure 106). Some zones of water overrun can be mapped for large distances between several wells in a field. A similar pattern is found with gas, but with the lighter gas underrunning the shales in an oil zone (Pucknell and Broman, 1994).

TAR MATS

Bitumen layers or *tar mats* can cause barriers to vertical flow. The bitumen forms in the pore space from the alteration of trapped or migrating oil. Permeability is reduced by the restriction or closing of pore throats by the bitumen. The permeability barriers formed may sometimes bear no relation to any preexisting facies or diagenetic or rock property characteristics of the reservoir (Lomando, 1992). A known pattern is for a tar mat to form at or just below an oil-water contact. Examples of this are found in the Burgan field in Kuwait and the Prudhoe Bay field in Alaska (North, 1985).

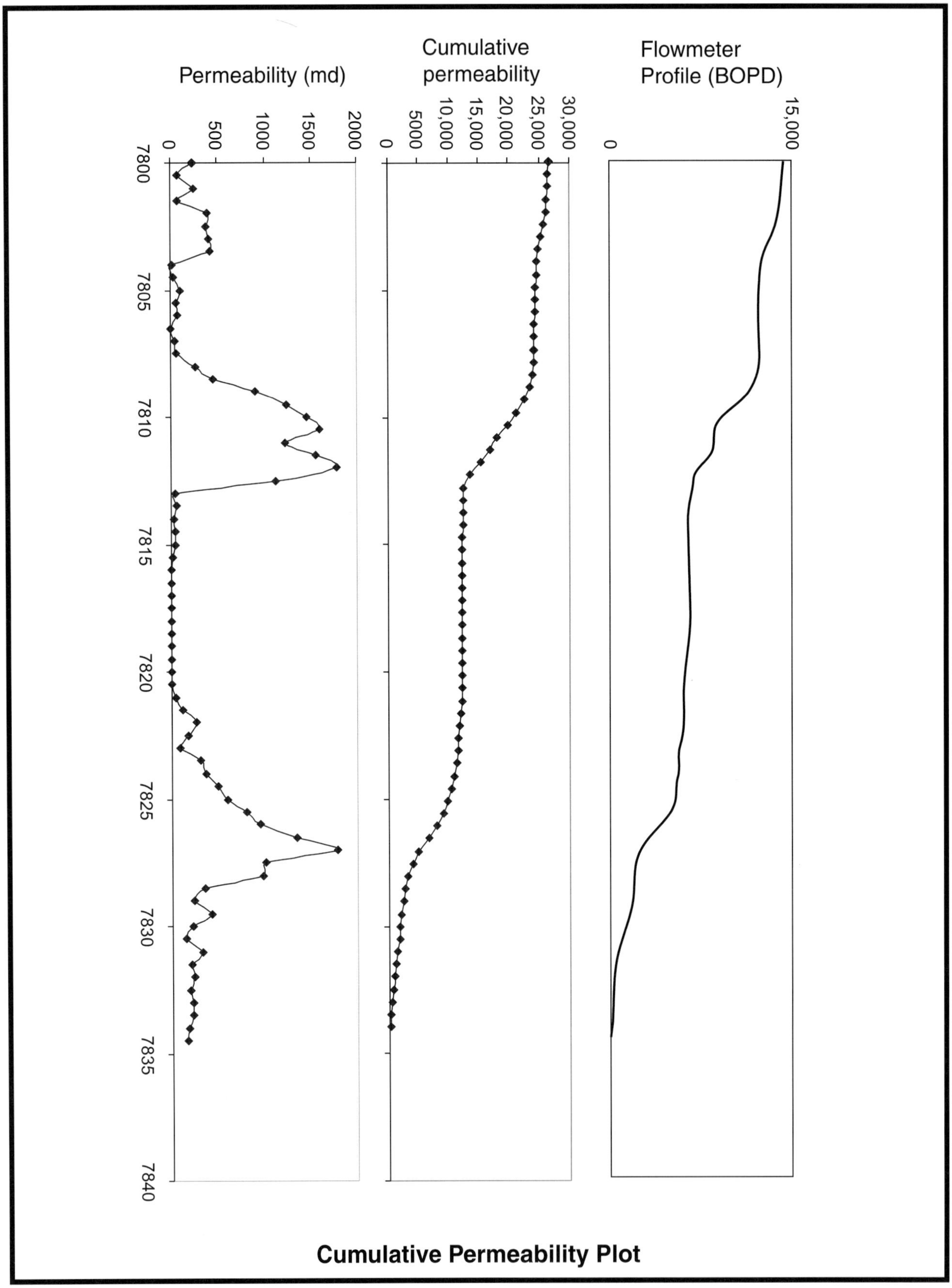

FIGURE 104. A cumulative permeability plot can be derived from permeability data in a well. The plot can be directly compared with flowmeter logs.

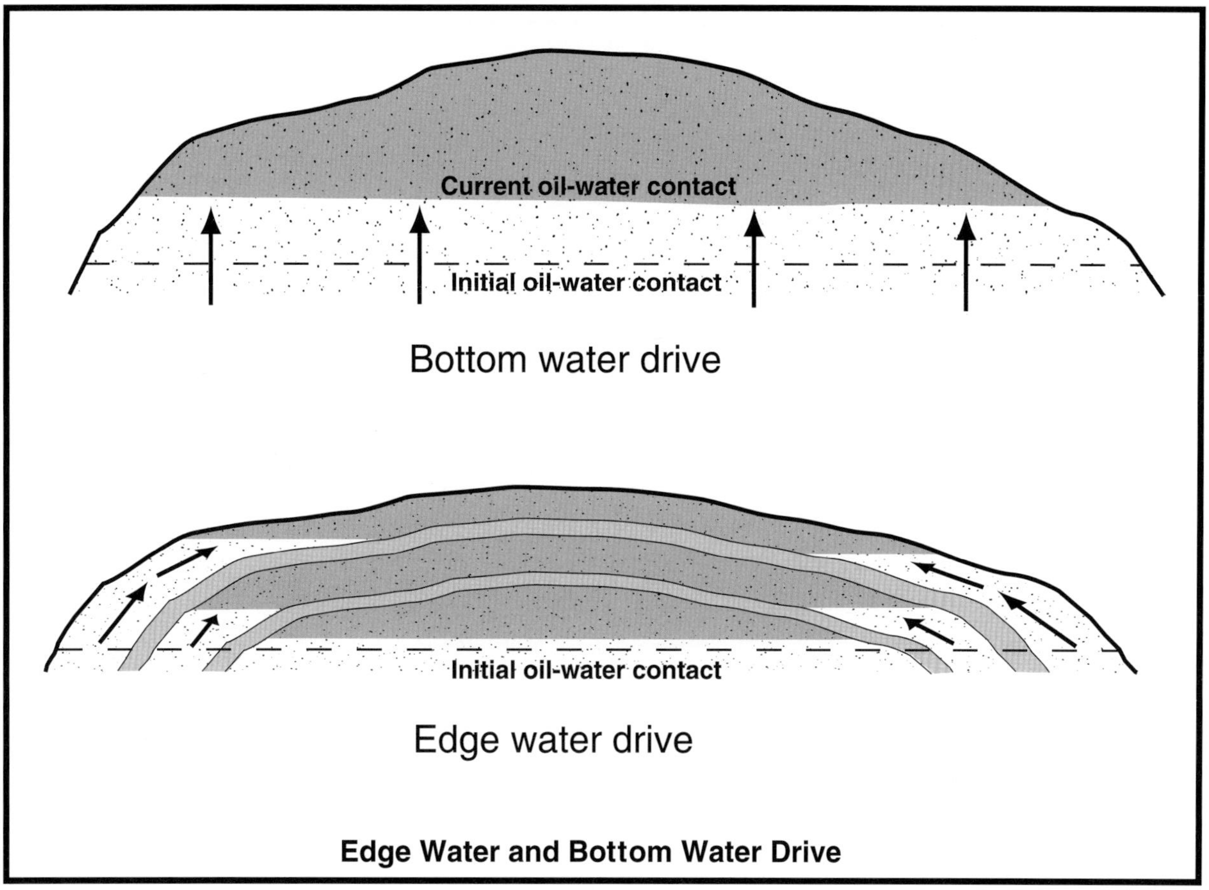

FIGURE 105. Bottom water drive occurs in clean reservoirs with few barriers to vertical flow. The oil-water contact rises in an even fashion with continuing production. In reservoirs with numerous barriers to vertical flow such as shales, water tends to encroach along the layers from the edges, hence the term edge water drive.

FIGURE 106. The pressure sink in the oil producer draws water up the structure along the top of a shale unit. The impermeable shale prevents the water from sinking through the less dense oil column.

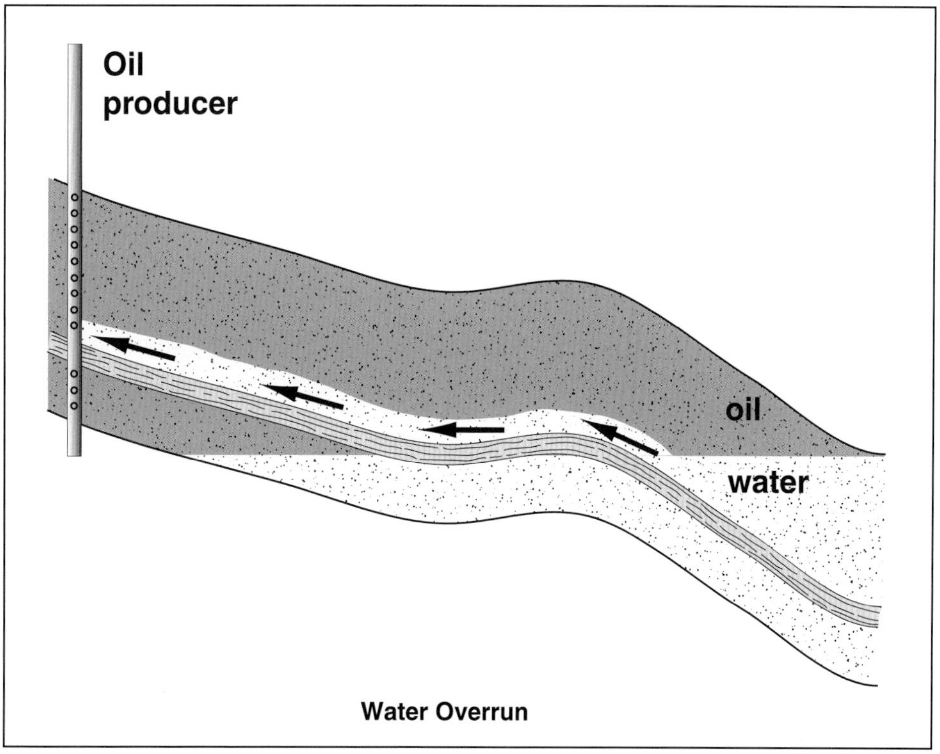

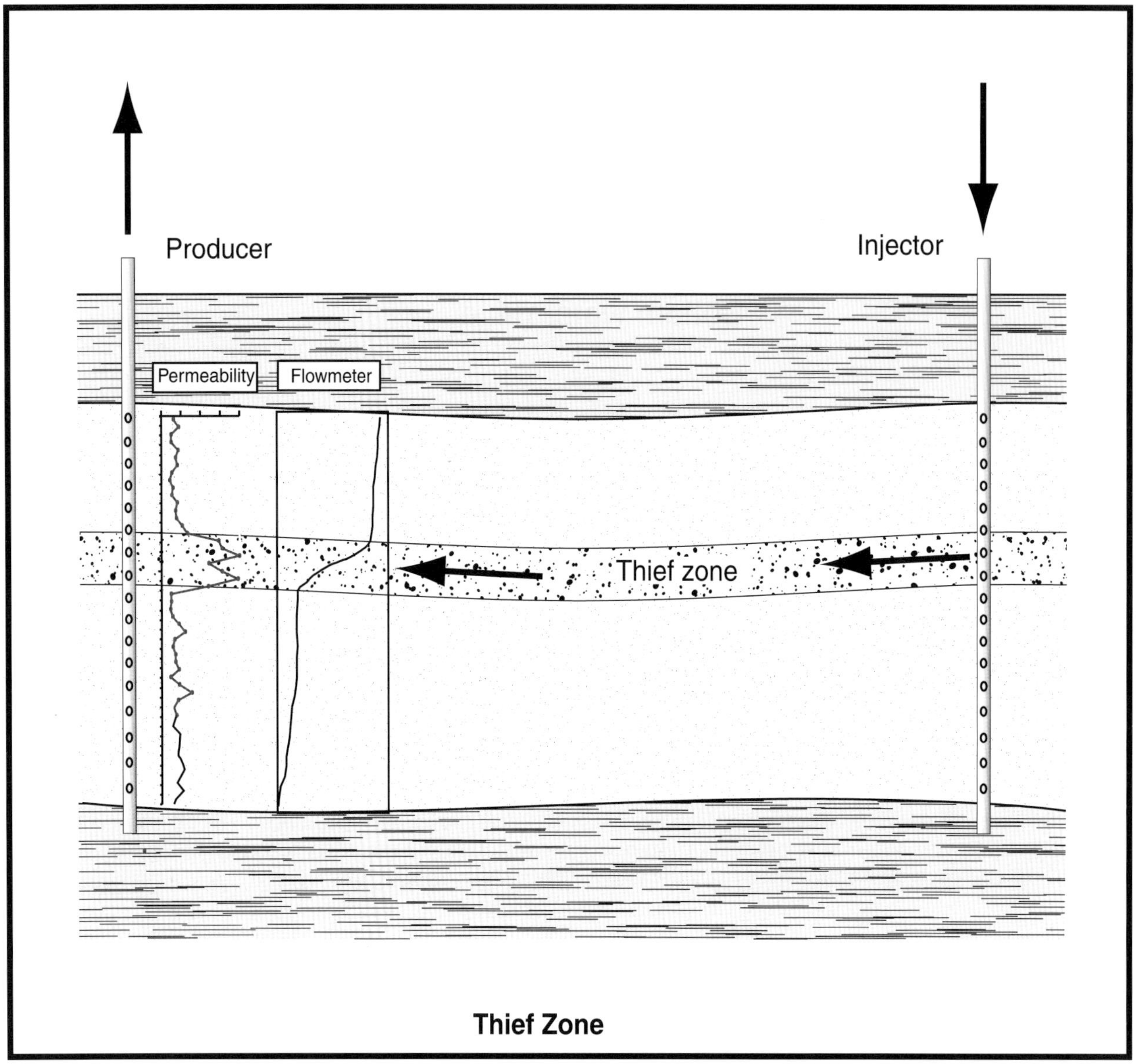

FIGURE 107. Thief zones can occur where very high-permeability reservoir intervals dominate the flow within the reservoir. Preferential production from such zones can result in rapid water breakthrough and loss of production.

THIEF ZONES

High-permeability sandstones can dominate the flow profile of a well as oil or gas will preferentially produce from these zones. Likewise, injection fluids will tend to enter high-permeability intervals. The uneven flow profile will result in an inefficient sweep profile for the reservoir (Figure 107). A *thief zone* has been defined as a relatively thin layer comprising 5% or less of the net pay thickness but taking more than 25% of the injected water in a given well (Felsenthal and Gangle, 1975).

Thief zones can cause the breakthrough of water (or injection fluids) very quickly. As only a small fraction of the reservoir ends up waterflooded, sweep may be poor in the remaining sediments. A large volume of water injection may circulate through a thief zone but will displace very little oil. Water production will increase rapidly, and the overall flow rates will decrease or stop altogether as the water loads up the well. Well intervention will be necessary to isolate the thief zone. If a new well is drilled and a thief zone is found, then it is a good idea to leave production from this zone until later, if practical. Likewise, it may be wise to avoid injecting into a thief zone. In the worst case situation, there may be good connectivity between the thief zone and the surrounding rock. In this instance, the thief zone can act as a conduit for flow within the reservoir even if it is not perforated in any of the injector and producer wells.

Thief zones occur in the Ghawar field in Saudi Arabia, the largest field in the world, where they are known as super-k intervals. Super-k zones have production or injection rates of at least 500 bbl of fluid/day/foot. The thief zones correspond to highly permeable limestone and dolomite intervals or open fractures (Meyer et al., 2000).

RESERVOIRS WITH UPWARD-INCREASING PERMEABILITY PROFILES

The waterflooding of an upward-increasing permeability profile in an oil zone will result in the water edging ahead in the high permeabilities at the top of the section. Eventually, the water will move downward under the effects of gravity and capillary action and in turn will displace the underlying oil (Gaucher and Lindley, 1960). The rate at which this happens will be governed mainly by the vertical permeability (Thomas and Bibby, 1991). This type of sweep can be efficient as a significant volume of the oil will ultimately be displaced by water. This pattern is commonly observed in shoreface and barrier bar sandstones where the sandstones tend to coarsen and become less silty upward (Figures 108, 109).

RESERVOIRS WITH UPWARD-DECREASING PERMEABILITY PROFILES

Oil zones in upward-decreasing permeability profiles are commonly poorly swept. During waterflooding, the water will move through the high permeabilities at the base of the flow unit. The lower permeability at the top has a tendency to be bypassed. This is a typical pattern in channel fills (Figure 109).

METHODS FOR DETECTING BARRIERS TO VERTICAL FLOW

A number of different observations can be made from production data that allow the geologist to detect where barriers to vertical flow exist within a reservoir (Figure 110).

PRODUCTION LOG DATA

Production logs, including pulsed neutron logs and flowmeter logs, can be used to determine flow layering in a reservoir.

Pulsed neutron logs, which detect hydrocarbon saturations behind the casing, may be run in wells several times during the lifetime of a well in order to monitor the rise in the oil-water contact as a result of production. The logs can show the oil-water contact rising steadily through a clean reservoir section, but then it can stop moving where impeded by a shale bed (Figure 110[a]).

Pulsed neutron logs can also pick out zones of water overrun above a shale. The log will show oil below the shale and an interval of water-swept rock immediately above (Figure 110[b]).

Hamilton et al. (1998) mapped out variations in the oil-water contact rise in the Jackson field, Australia, using production log data. In this way, they were able to define *oil-water contact rise domains* within the field. These are defined according to the magnitude of the rise in the oil-water contact and as to whether edge water or bottom water influx is the dominant drive mechanism. Log facies maps (see Figure 72) can help in predicting which drive mechanism operates. These will give the geologist an idea of the bedding patterns within a reservoir interval. Clean sandstones with few shales encourage bottom water influx. Interbedded sand-shale intervals are more likely to promote edge water influx patterns.

Flowmeter passes made with the well shut-in will show if *cross flow* is occurring between higher pressured and lower pressured hydraulic units in a layered reservoir (Figures 110[c], 111). If this is seen, it indicates that a vertical permeability barrier separates the zones of different pressure.

FORMATION TESTER DATA

Formation tester data from infill wells can show a reservoir separating into different pressure zones. This will happen because some hydraulic units may be producing more hydrocarbons and depleting faster or perhaps getting better injection support than the other hydraulic units. Formation tester data are presented as a pressure-depth plot with pressure on the horizontal axis and depth (TVD) along the vertical axis (Figures 110[d], 112). A pressure differential across a shale bed indicates that the shale bed is of sufficient extent to inhibit pressure equalization between the sandstone beds on either side.

WATER SHUT-OFF PERFORMANCE

In a mature field, water ingress into the perforated interval of a production well can cause the oil column to load with water, reducing the flow rate. Production engineers will respond by shutting off the zone of water production. For instance, this can be done by inserting a plug in the well above the perforations producing the water. The plug will be located opposite an impermeable

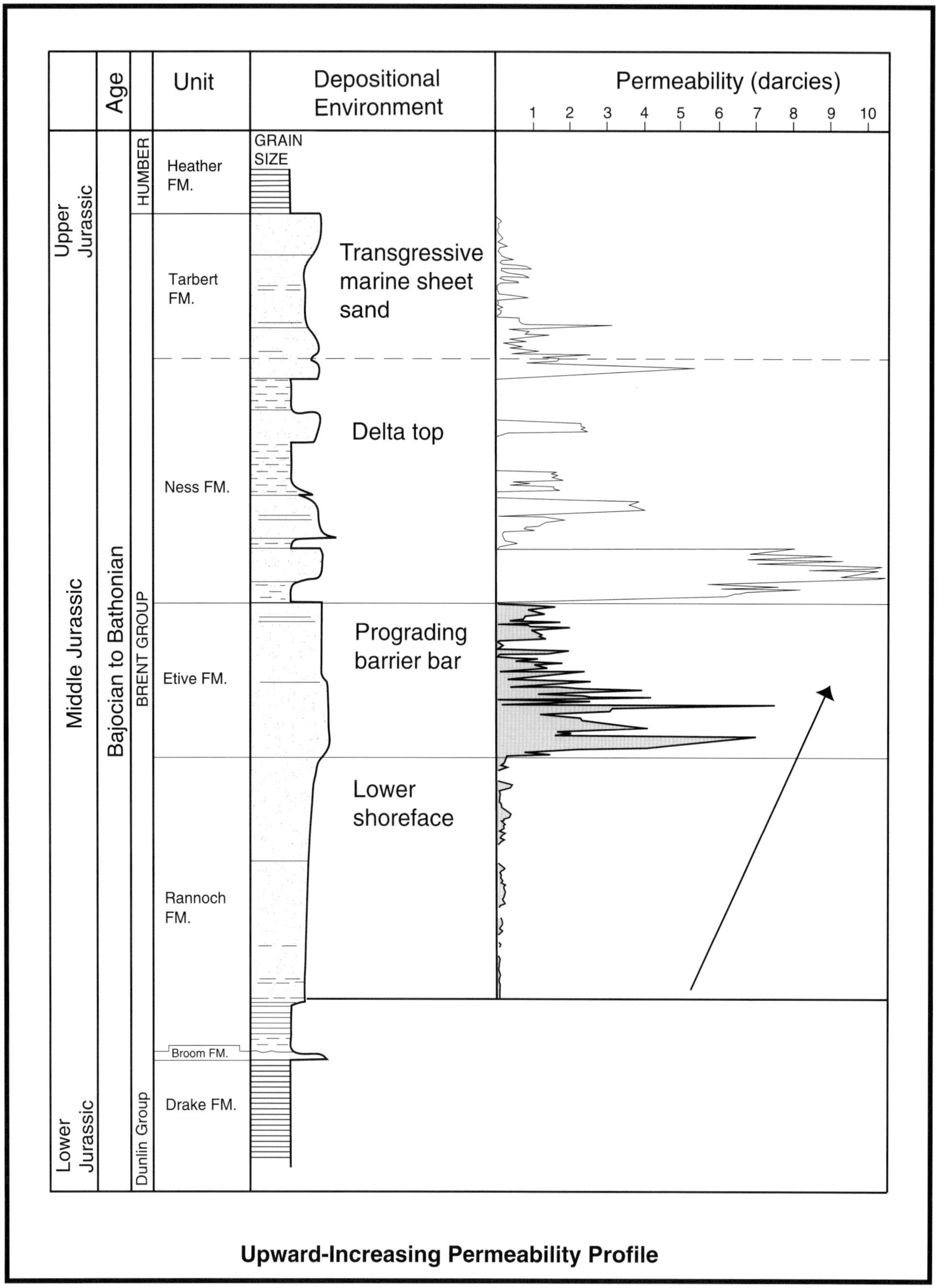

FIGURE 108. The barrier bar-shoreface interval of the Brent Group reservoir in the Thistle field, UK North Sea, shows an upward-increasing permeability profile. This pattern is favorable to a high sweep efficiency (from Williams and Milne, 1991). Reprinted with permission from the Geological Society.

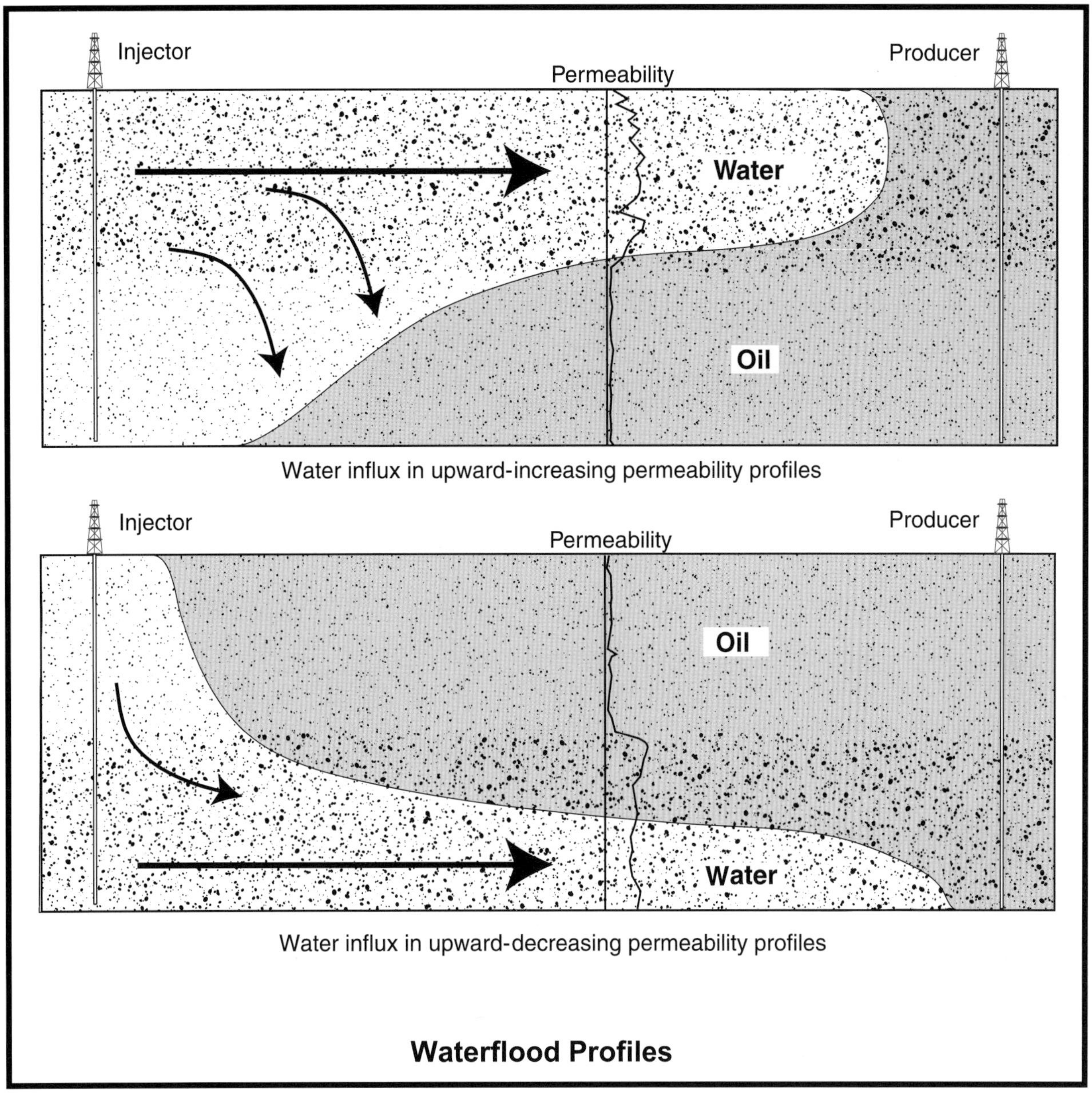

FIGURE 109. Sweep is more efficient in upward-increasing permeability profiles than in upward-decreasing permeability profiles.

zone in the wellbore, typically a shale. If this impermeable zone is laterally extensive, then the water shut-off operation will prove successful in isolating water production for a lengthy period of time. Both the plug and the shale in the reservoir behind the casing act together to isolate the open perforations in the well from a direct pathway for water influx from the swept zone. If the impermeable zone is not extensive, then water production will be unaffected in the well or will be reduced for only a short period of time.

Water shut-off performance was used by Hamilton et al. (1998) to evaluate the status of mudstones as permeability barriers within the Jackson field, Australia. They systematically analyzed the effect of each water shut-off operation in terms of how much the water cut was reduced and for how long the operation was successful in doing this. The result was a *water shut-off table*, which can be used to give a qualitative idea of the extent of the flow barriers (Figure 110[e]).

WATER AND OIL GEOCHEMISTRY

One method that may work in picking out barriers to vertical flow involves plotting the field-wide variation of the produced water chemistry.

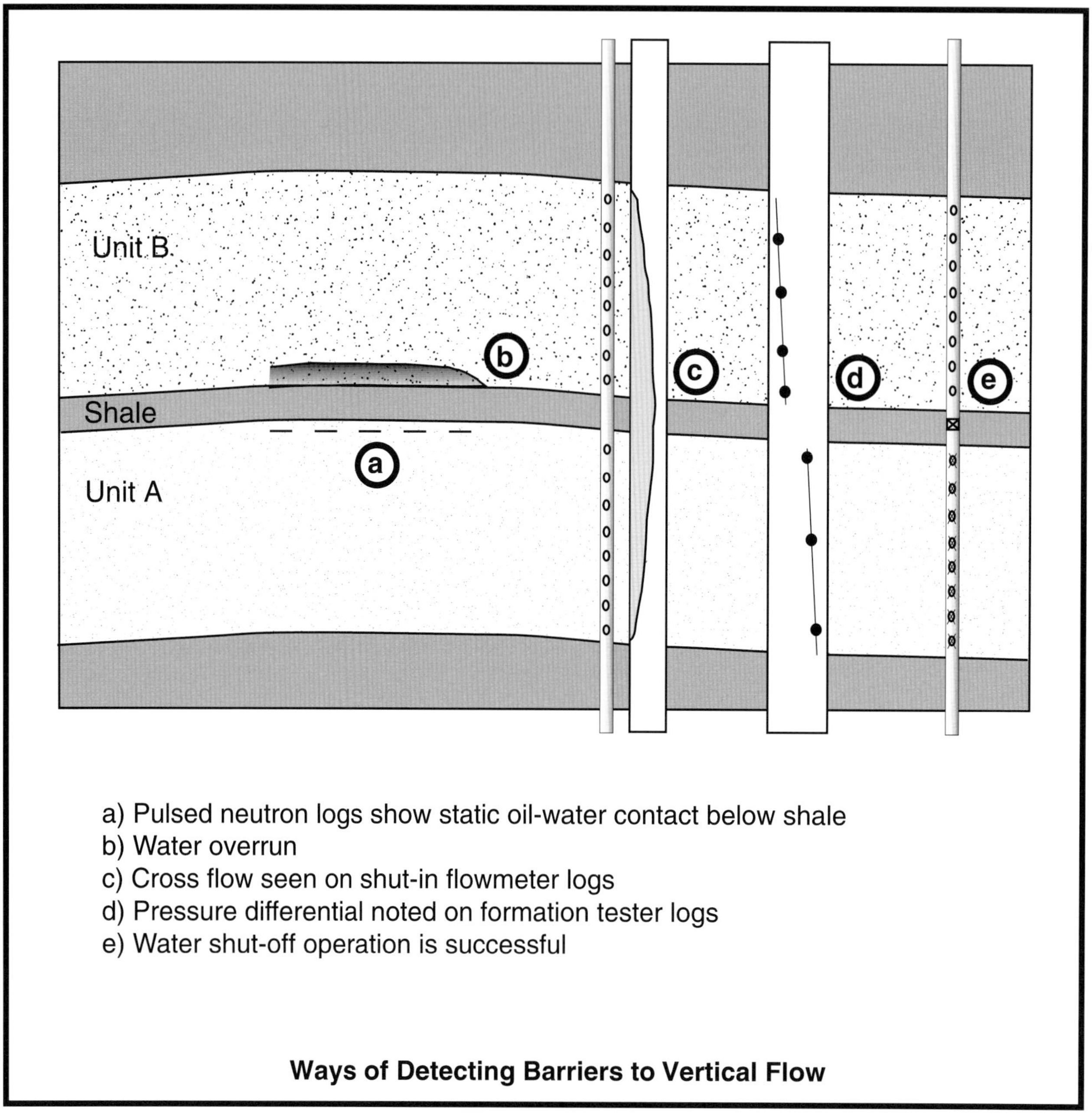

Figure 110. Different observations can be made from production data that allow the geologist to detect where barriers to vertical flow exist within the reservoir.

Different hydraulic units may show different produced water salinities. This can be a useful means of fingerprinting hydraulic units within a producing reservoir (Slentz, 1981).

Oil geochemistry can also be used to identify hydraulic units. An oil or gas field may be sourced from more than one source kitchen. This can result in the oil or gas geochemistry varying between hydraulic units.

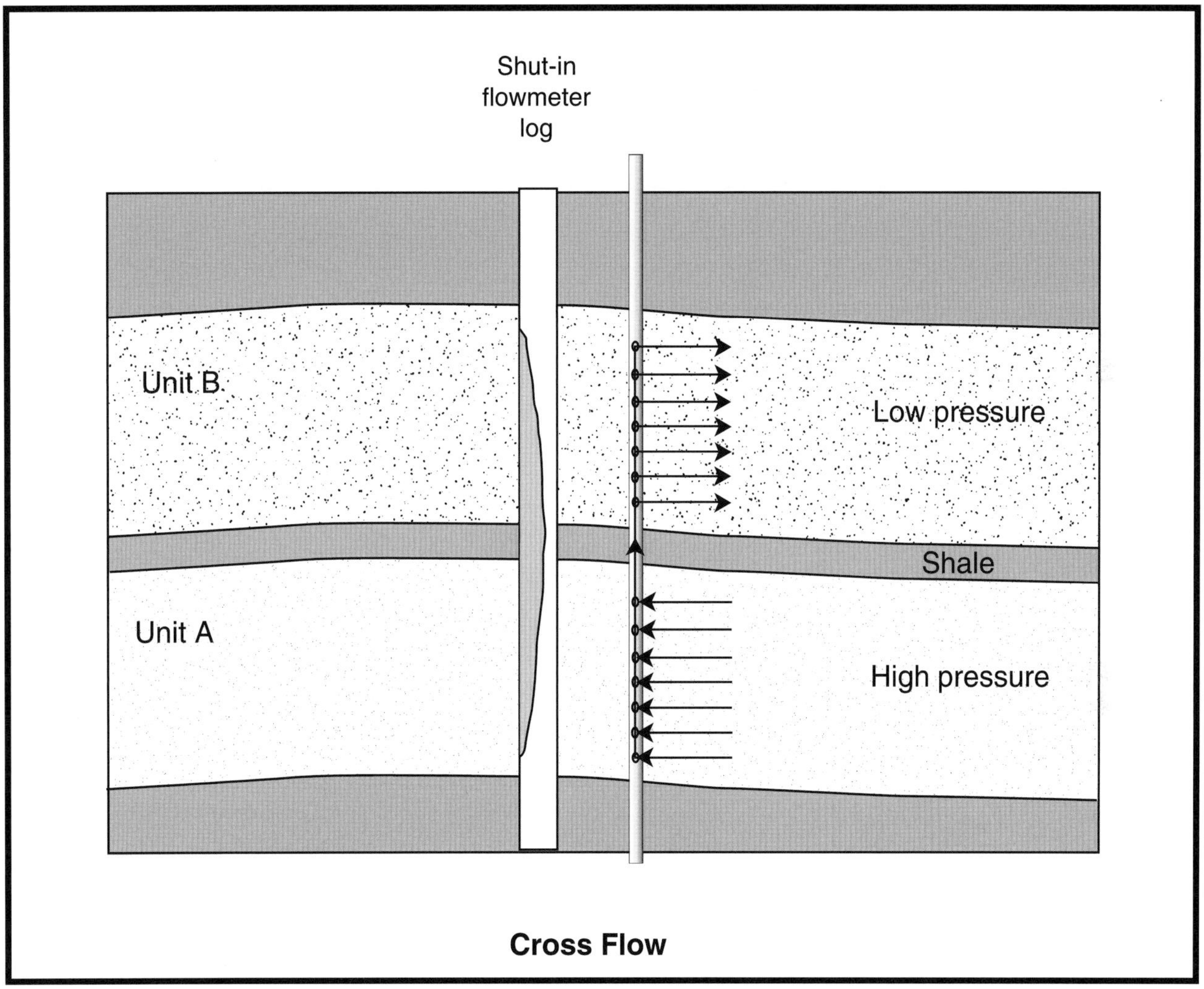

FIGURE 111. A shut-in flowmeter may detect cross flow of fluid from a high-pressured reservoir unit into a lower pressured unit. A permeability barrier separates the two zones.

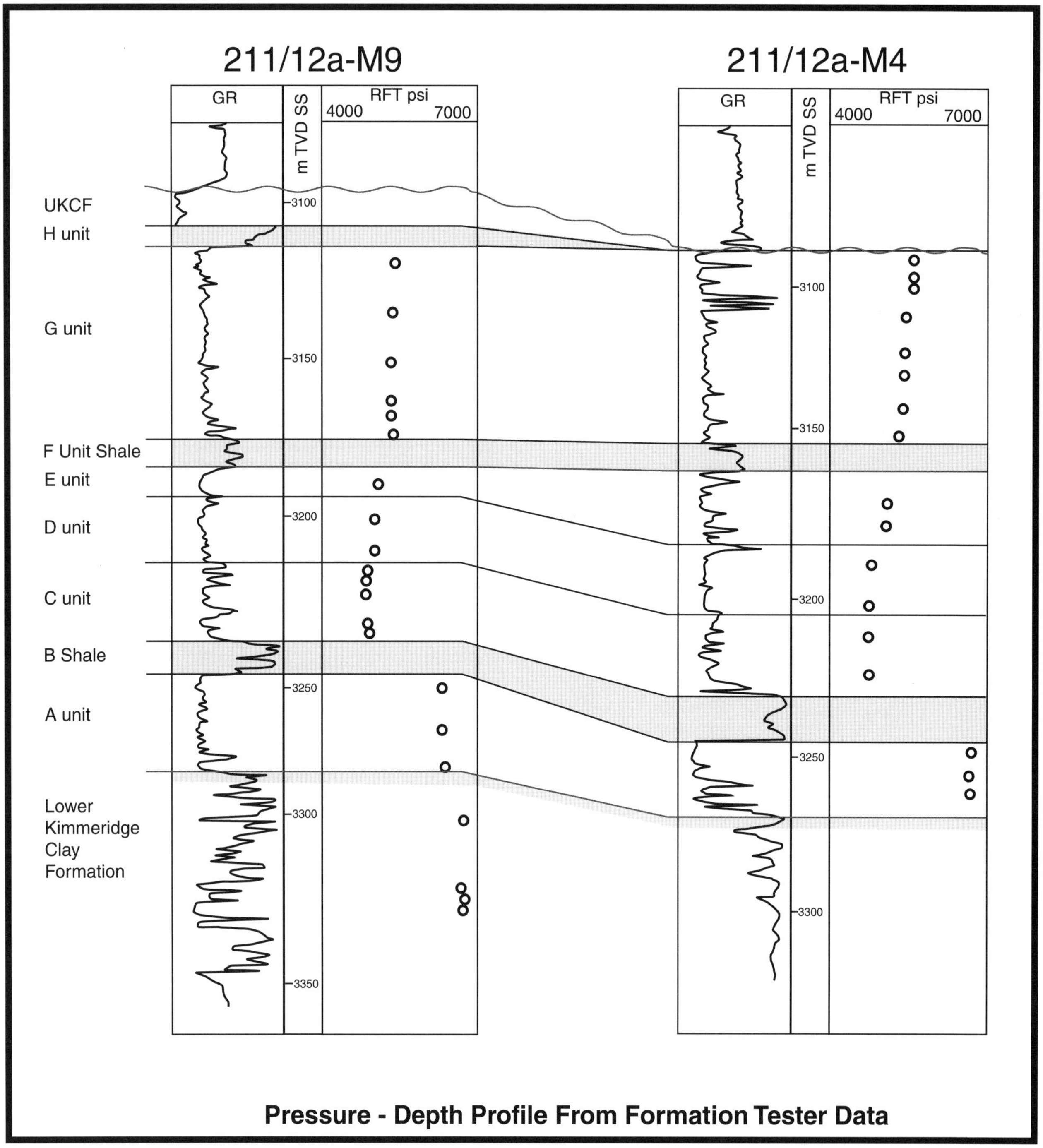

FIGURE 112. Formation tester data taken in wells that have been drilled postproduction provide invaluable data on how the reservoir splits up into hydraulic units showing different pressures. This example is from the Magnus field in the UK North Sea (from Morris et al., 1999). Reprinted with permission from the Geological Society. GR = Gamma Ray; RFT™ = Repeat Formation Tester; UKCF = Upper Kimmeridge Clay Formation.

16

Shepherd, M., 2009, Areal compartments, *in* M. Shepherd, Oil field production geology: AAPG Memoir 91, p. 145–147.

Areal Compartments

INTRODUCTION

Reservoirs may be segmented areally into compartments. These are regions of the reservoir that are effectively isolated from each other by barriers to lateral flow. The commonest bounding elements to areal compartments are sealing faults, although sedimentological pinch-outs, truncational erosion, and diagenetic features can also act to create compartments. The identification of areal compartments can be difficult. There are a number of features that indicate the presence of areal compartments, although they can be ambiguous in this respect. For example, two wells in a field may show different oil-water contacts and this observation could be used to infer the presence of two separate areal compartments. However, it may be that one of the wells has found a perched oil-water contact that is shallower than the other well, but with the two areas in pressure communication with each other (see Chapter 4, this publication). Careful judgement is required in defining area compartments in a field.

IDENTIFYING RESERVOIR COMPARTMENTALIZATION

Various observations may indicate that reservoir compartments are present:

1) Different hydrocarbon fluid contacts may be found in the various areal compartments (Figure 113). For example, the Hibernia field, offshore Newfoundland, contains more than 30 fault blocks, many of them forming distinct fault compartments with separate oil-water and gas-oil contacts (Sinclair et al., 1999).
2) Formation tester measurements taken at the field appraisal stage can show areal compartments at different initial pressures preproduction. For example, Smalley and Hale (1996) used this method to map out two separate compartments in the Ross field, UK North Sea.
3) A reservoir separating into patchwork areas of different pressures as a result of depletion can indicate compartmentalization. For example, individual well pressures were collated and used to define areal compartments in the Block 330 field in Eugene Island, South Addition, offshore Louisiana (Alexander and Handschy, 1998).
4) Mapping out initial PVT properties can sometimes indicate compartmentalization. PVT stands for pressure, volume, and temperature and refers to the physical properties shown by hydrocarbon fluids. For example, Leveille et al. (1997) noted differences in gas gravity and gas condensate yields between fault compartments in the Ganymede field in the southern North Sea.
5) Similarly, geochemical data can show intrafield variation in hydrocarbon and water composition (Slentz, 1981). Sahni (2003) used oil geochemistry to pick out compartmentalization in poorly connected sand bodies in a meandering fluvial reservoir. Oil geochemistry was also used to define fault compartments within the Ross field in the UK North Sea (Smalley and Hale, 1996).
6) Interference tests, pulse tests, or tracer data can be used to demonstrate that there is no or very poor connectivity between wells (see Chapter 6, this publication). This can be information of use in inferring the presence of areal compartments.
7) Production and injection profile data can sometimes give an indication of the degree of connectivity in a reservoir. If injection wells are present and were drilled after the production started, then production profiles for individual production wells can be examined to determine whether the start of injection support has had any obvious affect on

DOI:10.1306/13161200M913372

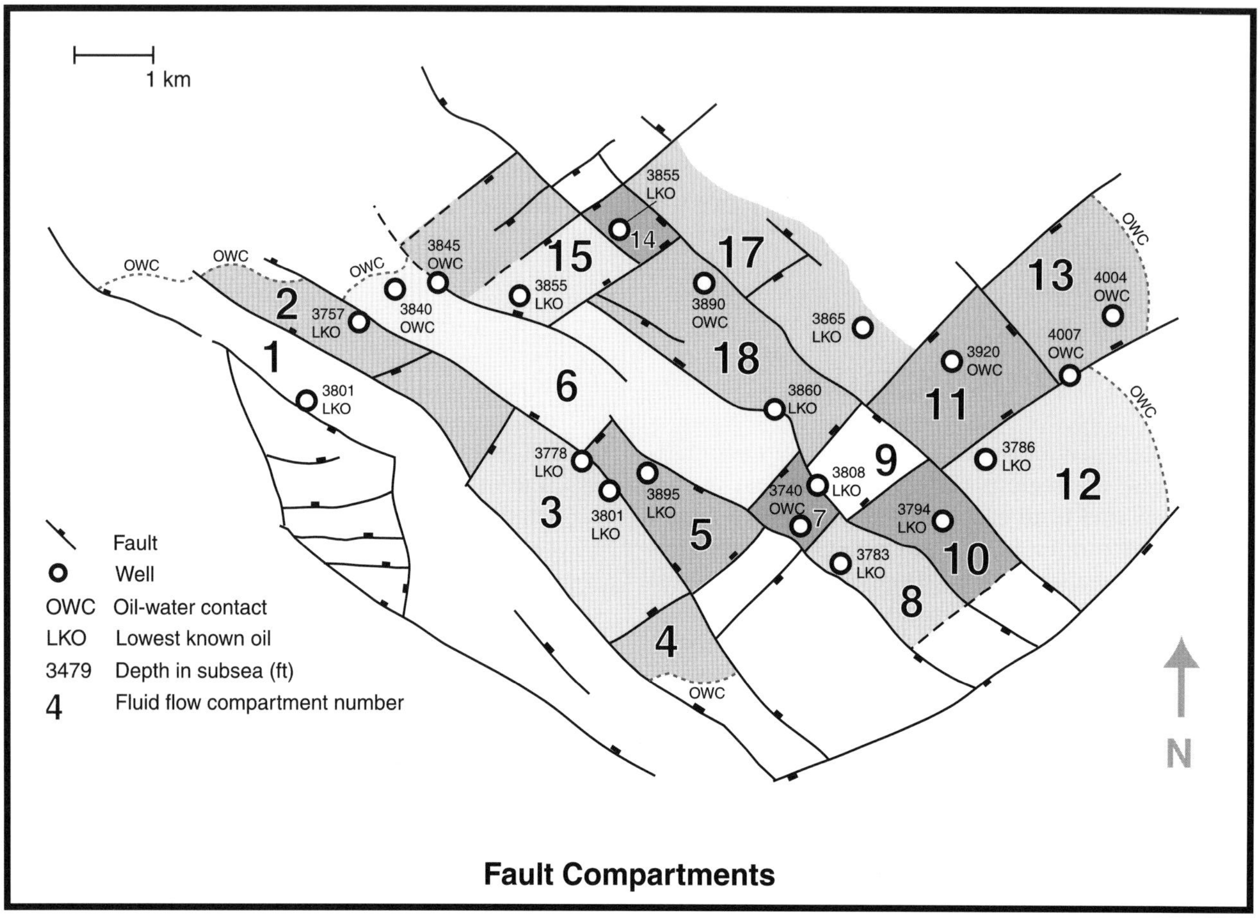

FIGURE 113. Fault compartments in the Merecure unit A, Budare field, Venezuela. Varying oil-water contacts and lowest known oil depths have been used to define compartmentalization (from Hamilton et al., 2002). Reprinted with permission from AAPG.

the oil rate. Similarly, a new well drilled in the same compartment as existing producers can cause a drop in production from the existing wells, once it is brought on stream.

In large fields, production wells in a common areal compartment can show very similar production profiles to each other. Trends on the production profiles will rise and fall mutually in each well. By contrast, wells in a neighboring compartment may show a totally different set of trends. One method to pick out areal compartments in a reservoir is to plot each well production profile on a map and to look for domains showing common production trends (Figure 114).

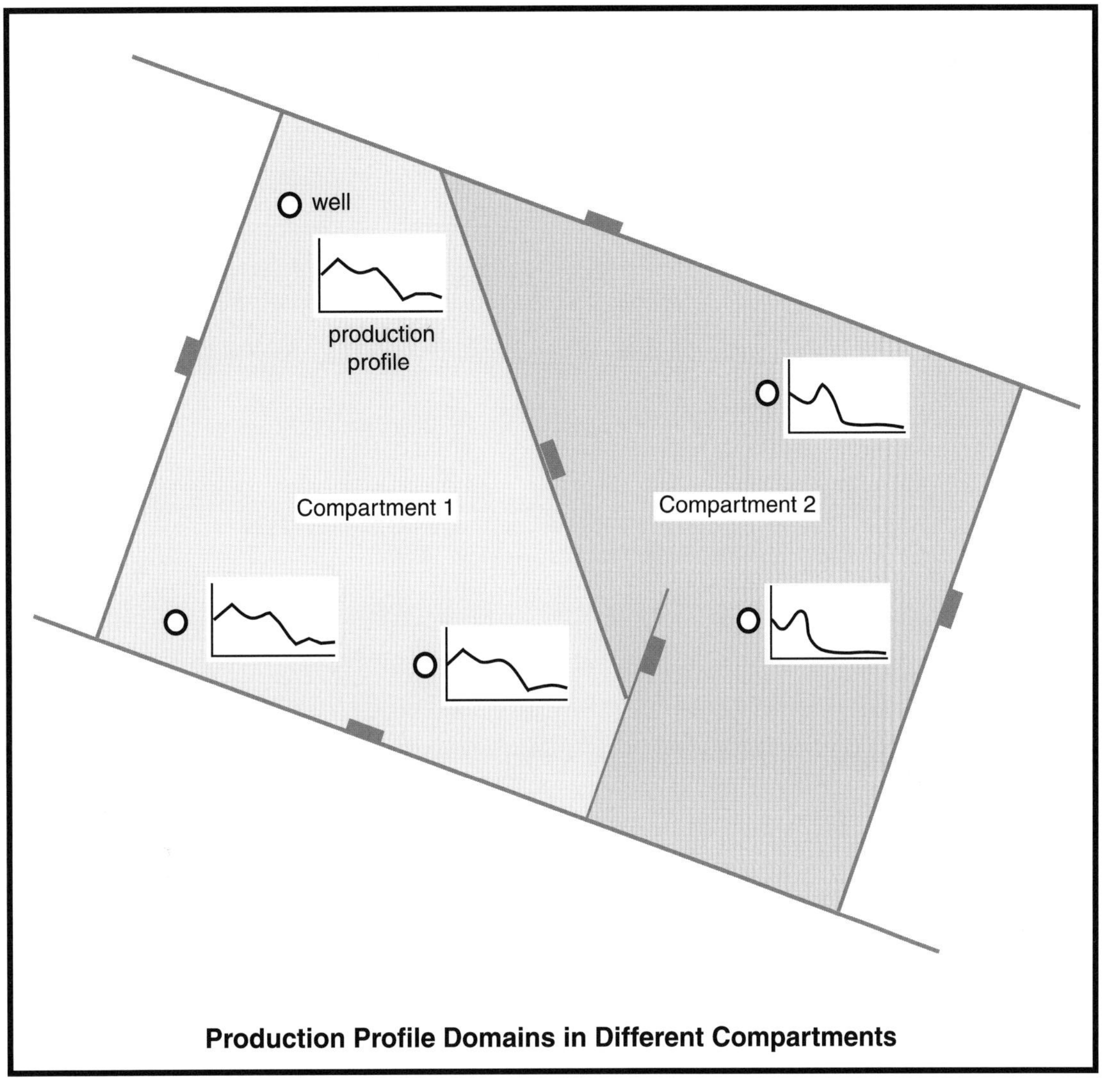

FIGURE 114. Areal compartments can sometimes be picked out by looking for domains with wells showing similar-looking production profiles.

17

Shepherd, M., 2009, Methods for analyzing areal sweep, *in* M. Shepherd, Oil field production geology: AAPG Memoir 91, p. 149–154.

Methods for Analyzing Areal Sweep

INTRODUCTION

Areal sweep is the fraction of an area of the reservoir pore volume that has been contacted by injected water. Areal sweep can be determined by several methods. Some of these are graphical, such as bubble maps, oil drainage maps, and water cut maps. In certain reservoirs, 4-D seismic and satellite-based interferometry surveys can be effective.

BUBBLE MAPS

One method for analyzing areal sweep involves compiling *bubble maps* of production data. The production (or injection) data values allocated to a well are represented by a circle, the diameter of which is proportional to the well flow rate. The maps show the areas of the field that are producing at high rates and those areas that produce very little.

The most useful map to make is a *cumulative production bubble map*. The size of the bubbles will be directly proportional to the total volume of oil (or gas) produced by each well. The map will indicate where the production sweet spots are in the reservoir and those areas where the production contribution is poor (Figure 115).

Cumulative production bubble plots can also be drawn for the total volume produced from individual hydraulic units. A given well may be producing from more than one hydraulic unit. To allocate a cumulative volume to a specific hydraulic unit, the geologist and reservoir engineer will need to determine the percentage of the total that has been contributed by the individual hydraulic units in a well. Alternatively, these can be estimated from historical flowmeter log data. If this is not available, allocations can be made according to permeability-thickness values or cumulative permeability plots (see Chapter 15, this publication).

One method of evaluating areal sweep involves taking the known oil production for a well and estimating the volume of porous rock around the well that would be needed to store this volume (Vining, 1997). *Oil drainage maps* can be made in this way (Figure 116). The *equivalent reservoir volumes* of rock that originally stored the produced oil can be estimated by reworking the volumetrics equation to determine the gross rock volume equivalent to the produced volume of oil (see Chapter 21, this publication). A volume of residual oil will be left behind on sweep, and this should be accounted for when the drainage volume is calculated.

The equivalent area of drainage can be shown in map view if an average thickness for the drainage volume is assumed. This area can be represented as a circle or it can be given a shape to fit a specific reservoir compartment. Overlapping drainage volumes are a sign that the wells are recovering significant volumes of oil from outside the immediate area of wells.

Atchley et al. (2006) used this method for the Innisfail field in Alberta, Canada, and called these *cumulative produced oil drainage area maps*. Individual drainage areas overlap with each other, and this indicates that the reservoir is connected over a large part of the field despite significant heterogeneity within the carbonate reservoir (Figure 116). A network of open fractures in conjunction with an intercrystalline pore network induced by dolomitization is thought to be responsible for the extensive connectivity.

Water cut maps can also provide information on the flow geology. The water cut of a well is the percentage of water flowing relative to the total flow. In mature fields, most of the wells will be flowing at high water cuts. Nevertheless, there may still remain a small number of wells producing with low water cuts. An area of low water cut may be the result of only one well producing from a large volume of contactable oil. This may happen because there is a significant volume of oil present between the production well and the source of water encroachment. It is worth investigating whether more wells are needed to support production from the area and to ensure the timely drainage of the contactable

DOI:10.1306/13161201M913372

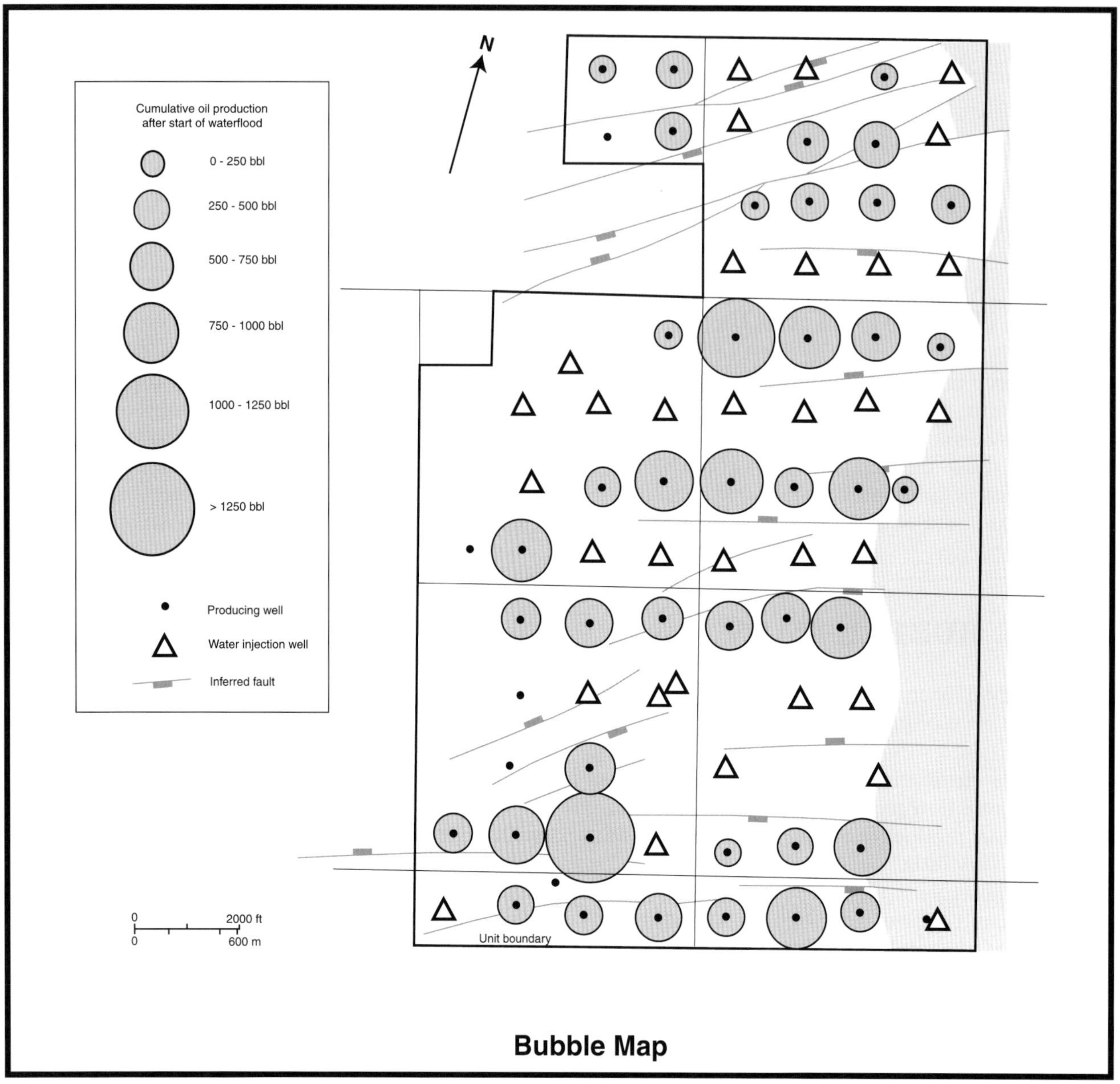

FIGURE 115. Bubble map showing the cumulative oil production since the implementation of a waterflood project in the Three Bar field, Permian Basin, Texas (from Montgomery, 1998). Reprinted with permission from the AAPG.

volume. Alternatively, a volume of oil in an areal compartment producing from a single well may be totally isolated from any source of water influx affecting wells elsewhere in the field.

Water cut maps were used by Galloway (1986) to locate volumes of poorly drained oil within the Greta sandstone of the Frio Formation in Texas. Bypassed oil is thought to be trapped in isolated tidal inlet channels cutting a more laterally extensive and better swept barrier bar system.

A *time series of water cut maps* shows the developing pattern of water influx over several years (Figure 117). These can highlight preferential pathways acting to focus sweep over the lifetime of a producing reservoir (Tyler and Ambrose, 1986). The maps are effective in edge water drive reservoirs. If the maps are overlain on lithofacies maps or fault maps, it may be possible to identify the geological features that are responsible for the water influx patterns. For example, Tyler and Ambrose (1986) noted that water ingress in the North Markham-North Bay City field in Texas has preferentially developed along the beach ridge axes of a strand-plain beach system.

A *time series of pressure maps* can be drawn to show the areal pressure variation in wells across the field. These can be used to pick out a patchwork of pressure domains

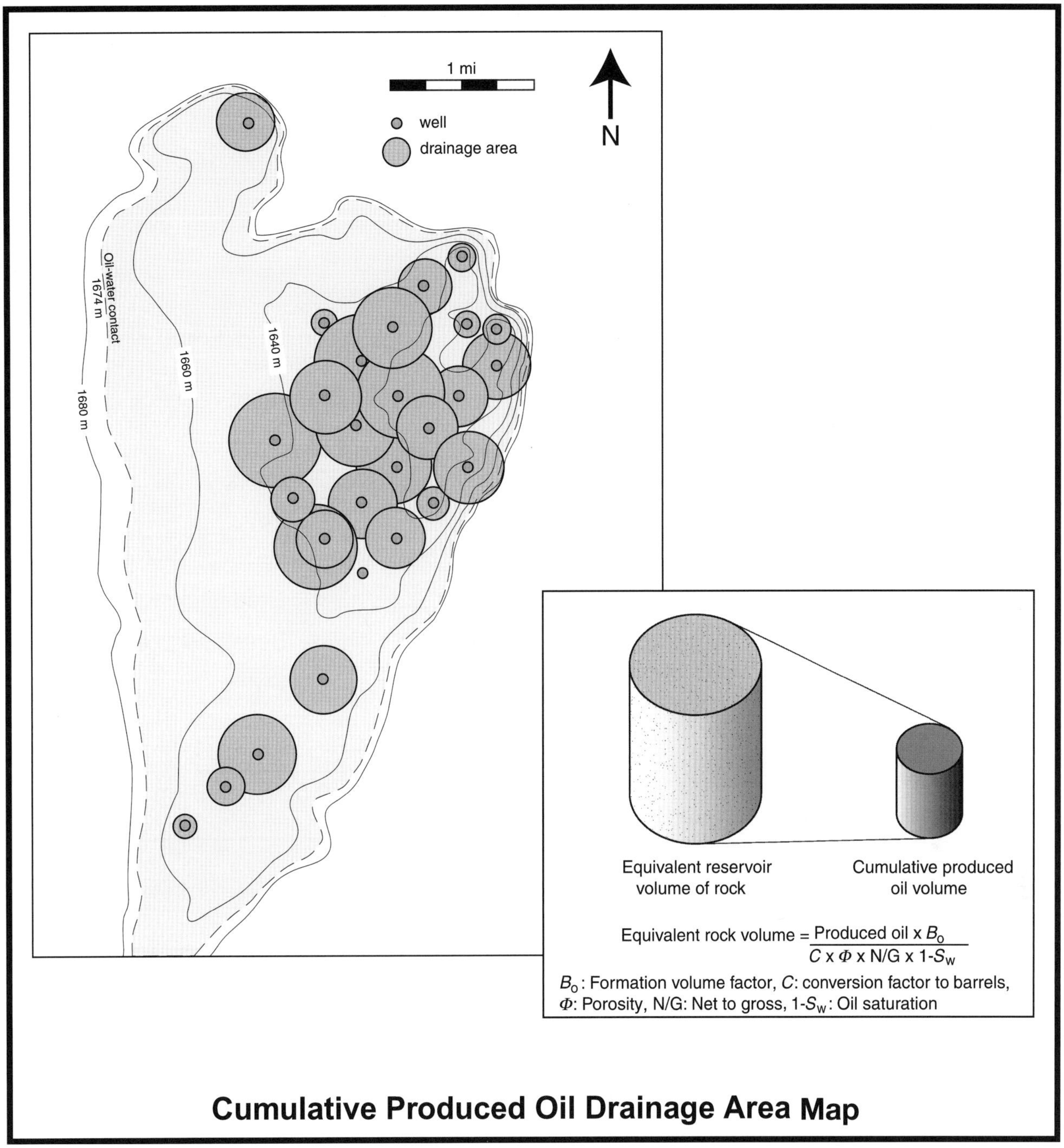

FIGURE 116. The cumulative production of an oil well in barrels can be converted to the equivalent reservoir volume of rock that produced the oil. Assuming a thickness for the drainage volume allows a drainage area around the well to be estimated. The cumulative produced oil drainage area map shows the drainage areas for all the production wells in a field, as in this example from the Innisfail field in Alberta, Canada (from Atchley et al., 2006). Reprinted with permission from the AAPG.

developing with time. High pressures may correspond to areas receiving good pressure support, or, alternatively, they may indicate relatively undepleted, isolated volumes. Low pressures correspond to areas receiving poor pressure support or undergoing rapid depletion.

4-D SEISMIC SURVEYS

4-D seismic surveys can be used to determine volumes of poor areal sweep (Figure 118). The analysis of 4-D seismic data, sometimes called *time-lapse seismic* data,

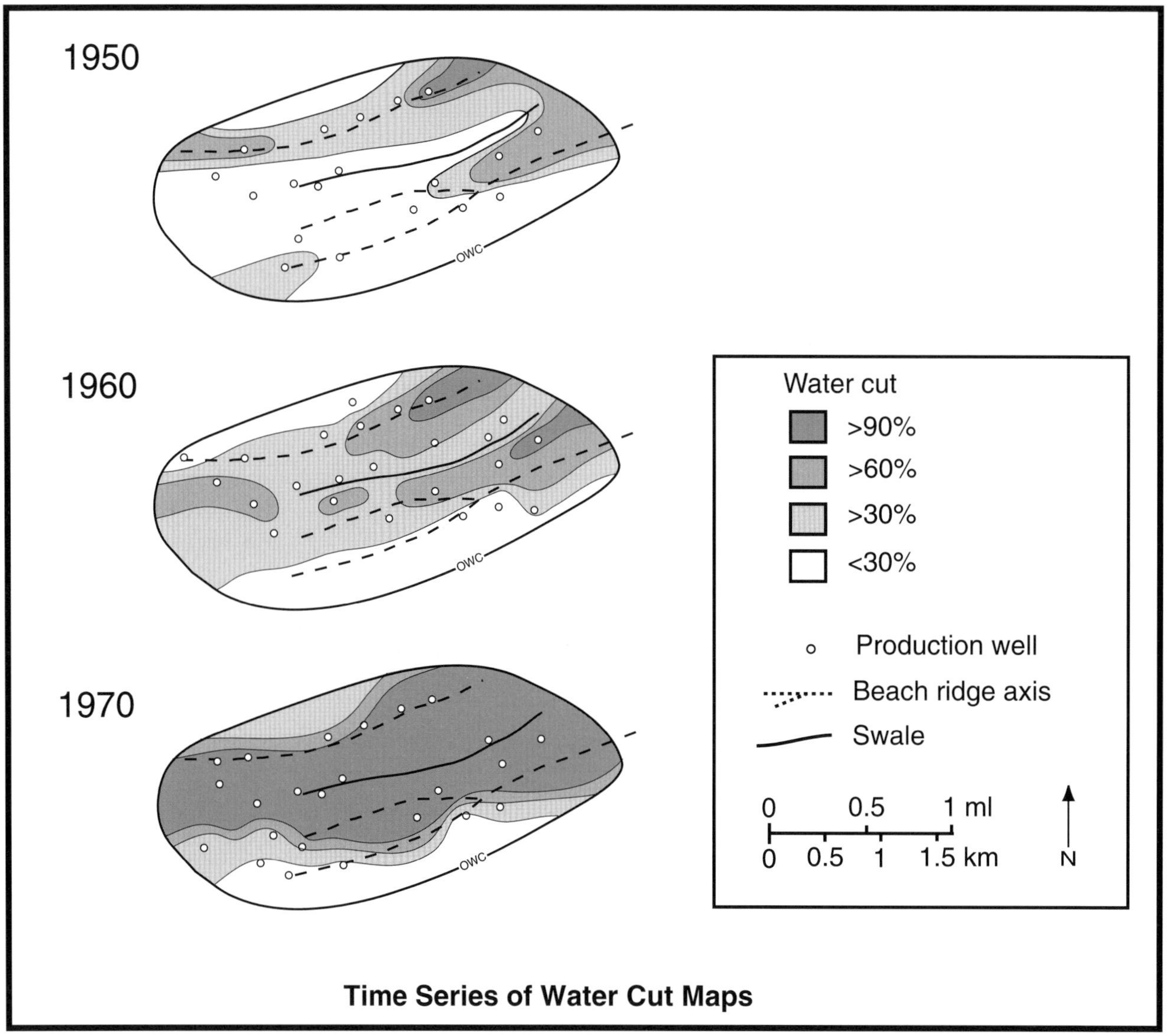

FIGURE 117. Time series of water cut maps from the West Cornelius reservoir, North Markham-North Bay City field, Texas. In this strand-plain reservoir, east-northeast–west-southwest-oriented beach ridge macroforms are fairways for water ingress. Tidal mud flat deposits south of the field restrict water influx from this direction (from Tyler and Ambrose, 1986). Reprinted with permission from the AAPG.

involves comparing different vintages of 3-D seismic surveys acquired a few years apart. Seismic amplitude changes and time shifts can result from compaction, pressure differences, and/or areal changes in fluid type as a result of sweep. The changes tend to be small. The method works if the various 3-D surveys can be compared directly with each other without any major problems resulting from differing navigation, acquisition, velocity analysis, and processing parameters.

The results of a 4-D seismic survey can be spectacular. Optimal conditions include thick reservoirs, high porosities, a less consolidated rock, and shallow reservoir depths. A high-density contrast between the various fluids moving through the field also helps. This includes reservoirs undergoing steam injection for heavy oil, water flooding of light oil, gas injection, and oil production with an expanding gas cap (Brown, 2004).

The 4-D seismic survey is becoming an increasingly powerful tool for imaging the areal sweep efficiency of hydrocarbons in fields where the method is effective. The largest economic benefit of the 4-D seismic method comes from the way in which it can pick out unexpected areas of sweep within the reservoir (de Waal and Calvert, 2003). The volume of petroleum fluids in these '4-D surprises' can be significant enough to target with new wells. The added value for individual fields can be in the

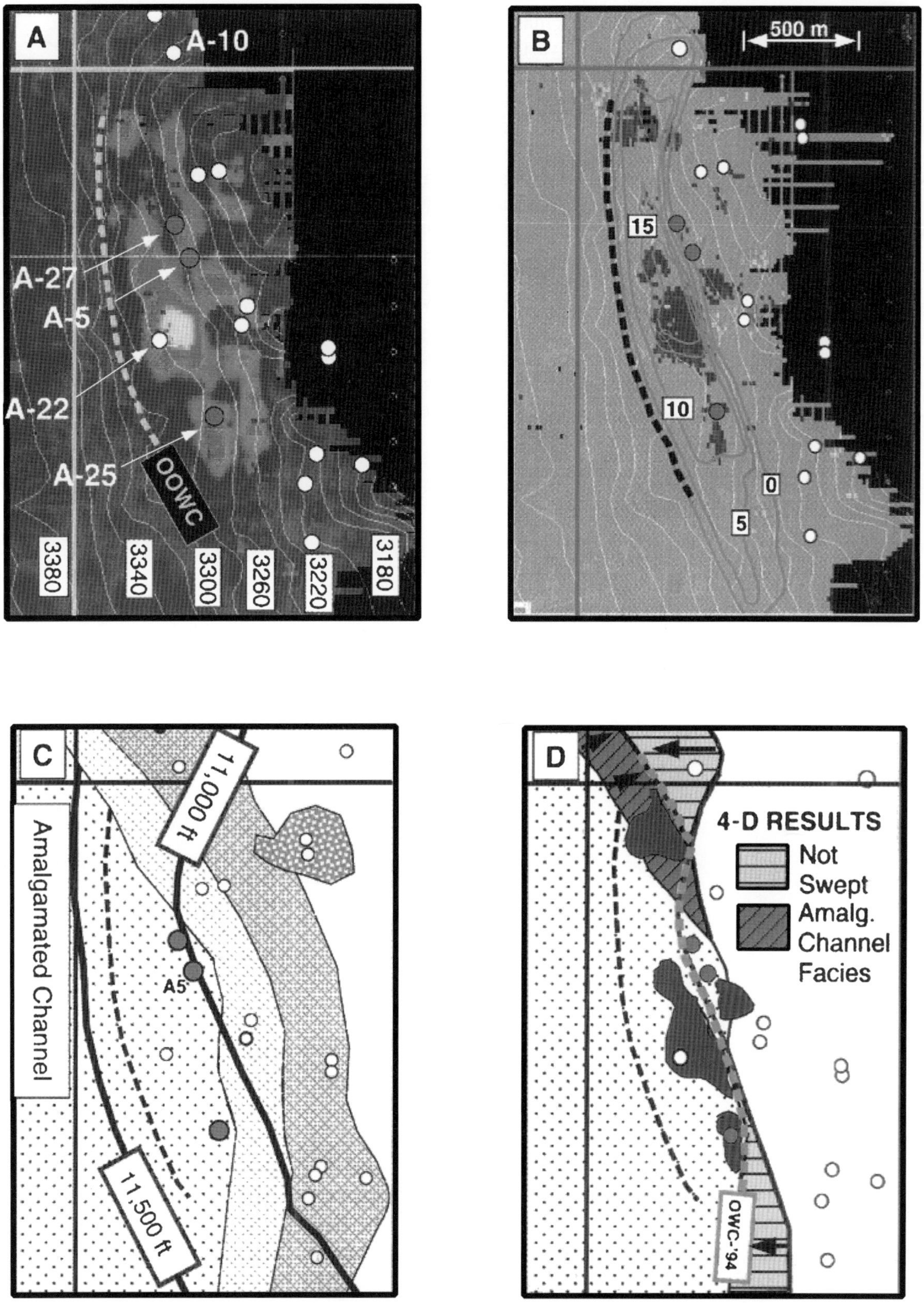

FIGURE 118. Resiults of a 4-D seismic survey from the South Timbalier Block 295 field, Gulf of Mexico. (A) 1988 amplitude map of the K40 sand unit. (B) Seismic amplitude differences between the 1994 and 1988 surveys. (C) Facies belts in a turbidite channel complex within the K40 sand; from west to east, amalgamated channel, channel margin, and levee with a localized overbank splay in the northeast corner. (D) Sweep map based on 4-D seismic data. Drainage is assumed to occur only in the amalgamated channel facies belt, with unswept oil within the channel margin and levee facies belts (from Hoover et al., 1999). Reprinted with permission of the AAPG. OWC = oil-water contact; OOWC = original oil-water contact.

range of tens to hundreds of millions of dollars, making the value of information from 4-D seismic data vastly in excess of the cost of taking the survey. For example, the results of a 4-D seismic survey indicated that a sand-rich turbidite channel in the south center of the Nelson field (UK central North Sea) was unswept. An infill well was drilled and established that the unswept volumes were indeed present, confirming the value of the 4-D seismic data (McInally et al., 2003).

4-D seismic data can aid reservoir management in other ways. Water injection wells can be optimally located to enhance the sweep of volumes of "slow" oil shown by the 4-D response. On occasion, a 4-D seismic response can also indicate where injectors are badly located in small compartments, serving only to pressure up an isolated volume (Staples et al., 2002).

SATELLITE-BASED RADAR INTERFEROMETRY

Current satellite and radar technology makes it possible to detect and measure very small movements of the Earth's surface; under ideal conditions down to a resolution of millimeters between repeat acquisitions. The *radar interferometry* technique allowed the mapping of the effect of cyclic steam stimulation on sweep within the Cold Lake heavy oil field in Alberta, Canada. Injection of steam into the reservoir at 415–470 m (1361–1542 ft) below the ground causes the land surface to heave up to 36 cm (14 in.) and then to subside at the end of each cycle. The effect of this could be clearly determined from satellite images (Stancliffe and van der Kooij, 2001).

18

Shepherd, M., 2009, Drainage cells, *in* M. Shepherd, Oil field production geology: AAPG Memoir 91, p. 155–156.

Drainage Cells

INTRODUCTION

Hydraulic units act as discrete flow layers within a reservoir and are bounded by permeability barriers. Additionally, fields may be segmented into compartments areally, commonly by sealing faults. Lateral and vertical sealing elements together act as bounding surfaces to individual volumes of the reservoir that behave as discrete *drainage cells* (Shepherd, 2007) (Figure 119). Drainage cells behave in a self-contained manner, although internally they may contain baffles to flow. This concept of a drainage cell is similar to what some geologists would describe as a compartment. For instance, Larue and Hovadik (2006) defined compartments as 'non-connected parts of the reservoir'. In this book, the word compartment has been used in the sense of an areal compartment only.

THE SIGNIFICANCE OF DRAINAGE CELLS

Drainage cells can be considered as acting as mini reservoirs within the framework of the reservoir as a whole. For instance, a detailed flow geology analysis of an oil field can change the perception that the field is behaving as a single "tank" of oil to one where the field is deemed to be producing from several independent drainage cells.

The number of drainage cells in a field will depend on the number of vertical layers and the degree to which the field is segmented into individual areal compartments. Some unfaulted oil and gas fields are relatively simple with only a few drainage cells present. Other reservoirs may be very complex with many hydraulic units and numerous areal compartments. The easiest drainage cells to define are those bounded by intersecting faults. These are often found to be boxlike in shape. The most difficult drainage cells to define are those where the lateral edges are stratigraphic or diagenetic in nature.

The recognition of drainage cells is important throughout the reservoir life cycle. One of the critical features to recognize at the appraisal stage is the number of drainage cells present. If a potential field development contains numerous drainage cells, then it will require numerous production wells to ensure that each individual cell is swept. The expense of drilling these wells will erode the overall value of the project, perhaps to the extent that the project becomes uneconomic. However, field developments with few drainage cells require fewer wells to recover the hydrocarbons and are more likely to be profitable.

For fields at the mature stage of development, the overall performance of the field is the combined performance of each drainage cell. Some drainage cells will produce large volumes of hydrocarbons with the recovery expected to be high. Other drainage cells will be underperforming with low ultimate recoveries likely. Perhaps some drainage cells have no active production wells. Hydrocarbons isolated in one drainage cell will not be recovered by a well in an adjacent drainage cell. Underperforming drainage cells are where the remaining hydrocarbons are most likely to be located in a mature field.

The recognition of the number and location of drainage cells in a field can lead to a breakthrough in the understanding of a reservoir. Drainage cell definition is the way toward framing a reservoir such that the remaining oil can be localized to specific cells. For instance, in a field with 12 drainage cells, it may be that only five of them contain any significant quantities of unrecovered mobile oil.

DOI:10.1306/13161202M913372

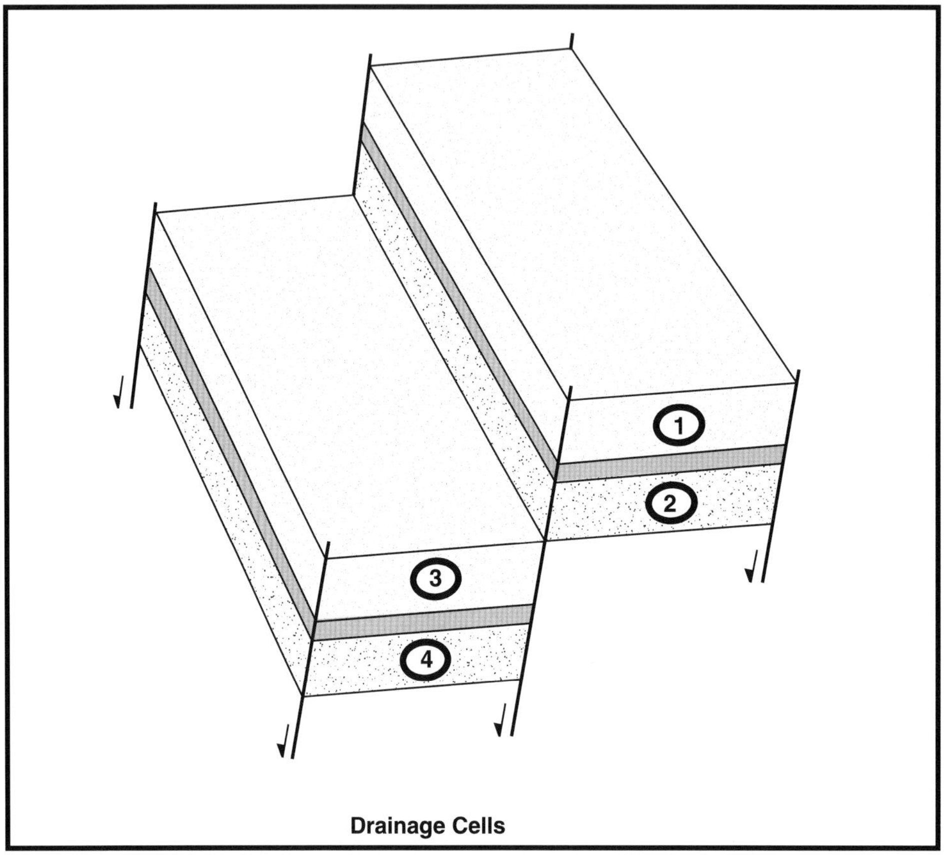

FIGURE 119. Drainage cells are individual segments of the reservoir, which are bounded by vertical and lateral permeability barriers. A producing field may comprise numerous drainage cells; four individual cells are shown in this figure.

Shepherd, M., 2009, Reservoir modeling and geostatistics, *in* M. Shepherd, Oil field production geology: AAPG Memoir 91, p. 157.

Section 4

Reservoir Modeling and Geostatistics

3-D GEOLOGICAL MODELS

An important aspect of modern production geology involves the building of 3-D geological models, particularly for large fields. These models have many practical uses. They help the geologist and the rest of the subsurface team to locate the remaining hydrocarbons in mature producing fields. The trajectories for new wells can be planned to find the optimal well path relative to the reservoir geology.

Geostatistical methods are used to build these 3-D models. Although these were originally developed as tools to analyze the spatial distribution of rocks and rock properties, there is often not enough data available from oil fields to do this properly. More typically, the tools are used in production geology as a means of making a 3-D model that replicates the geological scheme. A lithofacies model can be produced, which honors the statistical distribution of rock properties for the individual lithofacies.

3-D geological models are used to estimate the inplace hydrocarbon volumes. It is important to understand the uncertainty involved in estimating these volumes. This can be a difficult job to do reliably, although there are several methods available that allow the geologist to do this in a rigorous way.

One of the main uses of 3-D models is to provide a framework for the reservoir engineer's simulation model. The simulation model is used to estimate reserves and to help with managing the field, and is the basis for economic forecasting within an oil company.

DOI:10.1306/13161203M913372

19

Shepherd, M., 2009, Geostatistical methods, *in* M. Shepherd, Oil field production geology: AAPG Memoir 91, p. 159–174.

Geostatistical Methods

INTRODUCTION

Geostatistics involves specialized methods for analyzing and representing the spatial variation in geological phenomena. Production geologists use geostatistics as a means to build facies and rock property grids for 3-D geocellular models on a computer. Geostatistical methods allow models to be built that are heterogeneous and have "realistically" complex flow pathways when they are used in a reservoir simulation.

3-D geological models built using geostatistics should be based on a well thought out conceptual geological scheme. Geostatistical methods provide a tool to help the geologist to realize the geological scheme as a 3-D model (Figure 120).

GEOSTATISTICAL REALIZATIONS

The easiest reservoirs to make predictions for are those where the reservoir geometry is layer-cake; thickness changes are gradual, and rock properties tend not to vary substantially between the wells. Because of this, the interwell geology and rock property variation can be represented with a reasonable degree of confidence in a geological model by the simple mapping applications described in Chapter 8 of this publication. Alternatively, maps can be made by kriging, a geostatistical application described later in this chapter. When the gaps are filled in using simple mapping techniques, the geological framework is said to be *deterministic*. The use of the word deterministic suggests that there is a single, unique way to describe the geology between the wells. Although there is never enough data to do this with total confidence in the result, the assumption can be made that a deterministic model of a layer-cake reservoir will be close enough to reality for practical purposes.

With the more complex jigsaw-puzzle reservoirs and the even more complex labyrinthine geometries, making a computer model of the geology between the wells becomes a difficult task. Not so long ago, geologists did not try too hard to solve this problem. The best that could be done was to contour rock properties in 2-D using a mapping program on a computer. A simple algorithm would be used to interpolate between the well values. The resulting contour maps could only represent a very smoothed out version of the geology. If the geologist wanted to show a representation of a more complex sedimentological scheme, they digitized hand-drawn contours of the net-to-gross maps to honor the appropriate depositional geometry. Admittedly, this is a rather unsophisticated method of modeling a reservoir, but this was all that could be done at the time.

Production geology has made progress since then, and a number of computer-based geostatistical methods for modeling heterogeneous reservoirs are now available. The computer produces a geostatistical representation or simulation of the reservoir geology. It will not be a totally accurate model of what is in the reservoir, but, with the very limited control available, it is not possible to know if it is right or wrong. The methodology is designed to honor the available statistical data, and there is a degree of heterogeneity imparted to the model that would be difficult to achieve using 2-D mapping software. The *realization* is not unique. In practice there could be many possible realizations that can be made that will fit the data and the underlying statistics. These are all just as possible as each other or *equiprobable*. This computerized process is known as *stochastic modeling* (Haldorsen and Damsleth, 1990). *Stochastic* means governed by chance or probability.

Two main types of stochastic modeling procedures are used by geologists. The first method is object modeling, in which the computer inserts graphical objects representing macroforms into a background lithology such as shale. The second method involves grid-based

DOI:10.1306/13161204M913372

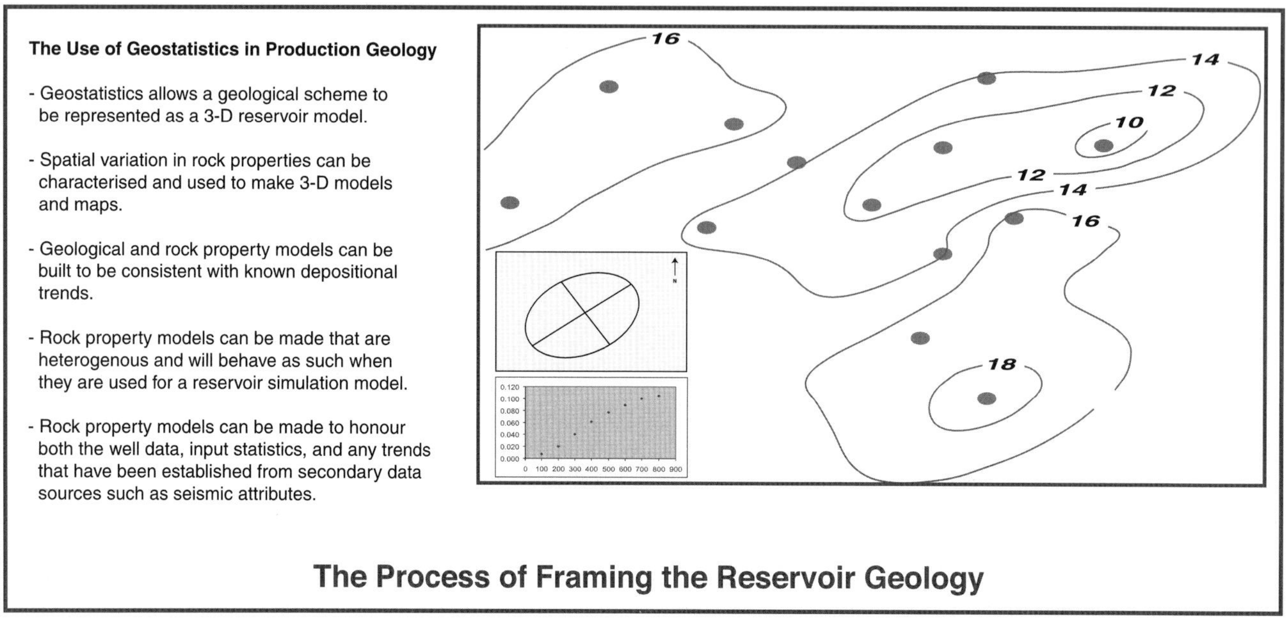

FIGURE 120. Geostatistical methods provide a tool to help the geologist realize.

algorithms. The computer assigns individual values to all the cells within a 3-D grid using stochastic methods.

OBJECT MODELING

An *object model* is a 3-D computer model of interlocking sedimentary bodies that resembles a feasible 3-D representation of the sedimentological scheme for the reservoir (Figure 121). It is created by a computer algorithm that stochastically inserts geometrical shapes (objects) representing the various macroforms into a background volume such as mudstone. Specific geometrical shapes can be assigned to each of the sediment bodies. For example, a half-pipe shape, a cylinder cut in half along its length, can be used to represent a channel (Clemetsen et al., 1990; Damsleth et al., 1992).

Geometrical parameters are defined for the objects using a series of menus in the computer program. For example, where a channel belt is being modeled, the channels in the model can be given an orientation corresponding to the depositional strike of the system. Other parameters can also be specified, such as channel sinuosity, wavelength, and amplitude. The geologist can also define a list of replacement rules for the erosional relationships of objects to each other (Figure 122). For instance, a rule may be made such that channel bodies will always cross cut crevasse splays.

The model is constructed so as to try and replicate the global percentage terms of the facies as recorded in the wells (an assumption is made in this approach that the facies percentages in the wells are representative of the reservoir as a whole). For example, the geologist may want to model a delta front depositional environment. The computer application will prompt the user to assign the facies percentages. The global facies percentages in the wells could be:

Distributary channel 37.2%
Mouth bar 20.8%
Background shale 42.0%

The object modeling software will try and match these percentages as closely as possible.

The range in size, length, and width of the various objects to be inserted in the model are predefined by the geologist. The program will prompt the user to type in these numbers. A reservoir analog is required to provide the range in ratios between the thickness, width, and length of the various types of geobodies to be modeled. Alternatively, ratios of thickness to width and width to length can be provided as input. Statistics are available for outcrop analogs from fieldwork conducted by researchers and students. The larger oil companies have in-house databases available for reservoir analogs. Other companies can subscribe to academic Web sites where these data are made available. Data can also be found in technical journals. An example of sediment body measurements is shown in Figure 123 and Table 18 (Reynolds, 1999).

A correlation between the thickness of a sediment body and its width is commonly found, and this width-to-thickness relationship is called the *aspect ratio* (Gluyas and Swarbrick, 2004). The assumption is that the thicker the sediment body, the larger the scale of depositional process responsible for creating the body will be. Much

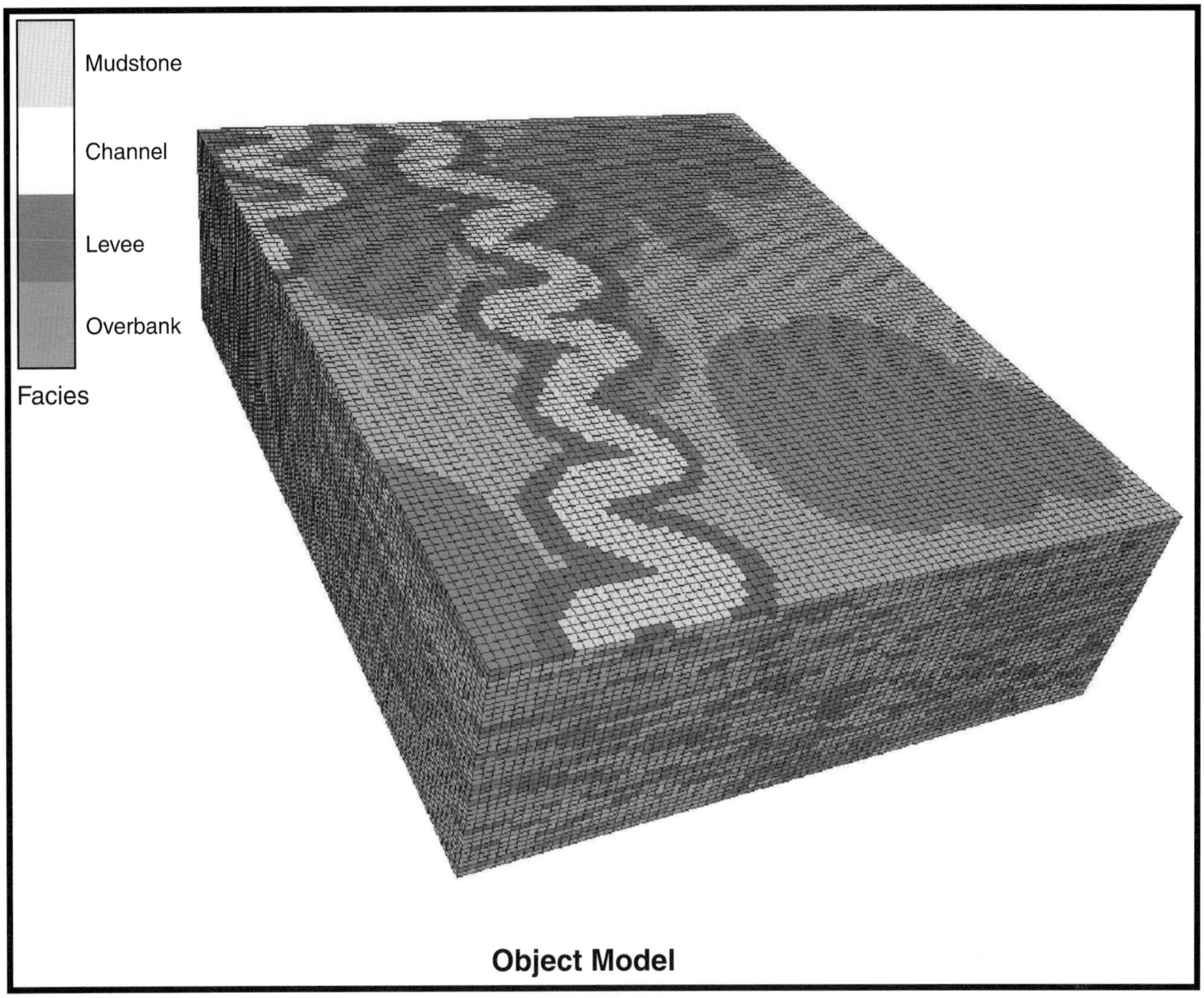

Figure 121. An object model is a 3-D graphical representation of lithofacies. It is created by a computer algorithm that stochastically inserts shapes representing macroforms into a background volume. The diagram shows a channel and associated overbank deposits within a background of mudstone.

research has been carried out on outcrops to determine aspect ratios, and these are available in several technical papers, e.g., Bryant and Flint (1993). Thickness values can be measured directly from the wells (on the basis that the sediment body is not eroded), and the widths of the various macroforms are estimated using the aspect ratio. Length-to-width values have also been measured in the field, but these are much more difficult to obtain. The full lengths of sedimentary bodies are not generally seen in outcrops (Geehan and Pearce, 1994).

Modern depositional analogs can be used to provide sedimentary body dimensions (Tye, 2004). As the sedimentary bodies are easily distinguished from each other, the length, width, and orientation of individual macroforms can be readily measured.

The relative abundance of the various lithofacies within a reservoir interval will vary both vertically and areally. Vertical and areal trends can be used to influence where specific objects are placed by the computer within the grid. *Vertical proportion curves* summarize in a single diagram the vertical distribution of facies within a unit (Eschard et al., 1998; Doligez et al., 1999). A geocellular grid can be populated by facies using the vertical proportion curve as a control. The advantage of this is that where a specific lithofacies is more common at a specific horizon, the localized concentration will be captured by the vertical proportion curve (Figure 124).

It is also possible to influence the areal distribution of facies in an object model by using a *probability trend.* This is represented on a contour map showing the probability of encountering a given lithofacies on a scale between 0 and 1. If the geologist has data to suggest that a particular feature is present, for instance, a geobody seen on a seismic amplitude display, then this is a way of influencing the object model to include the feature.

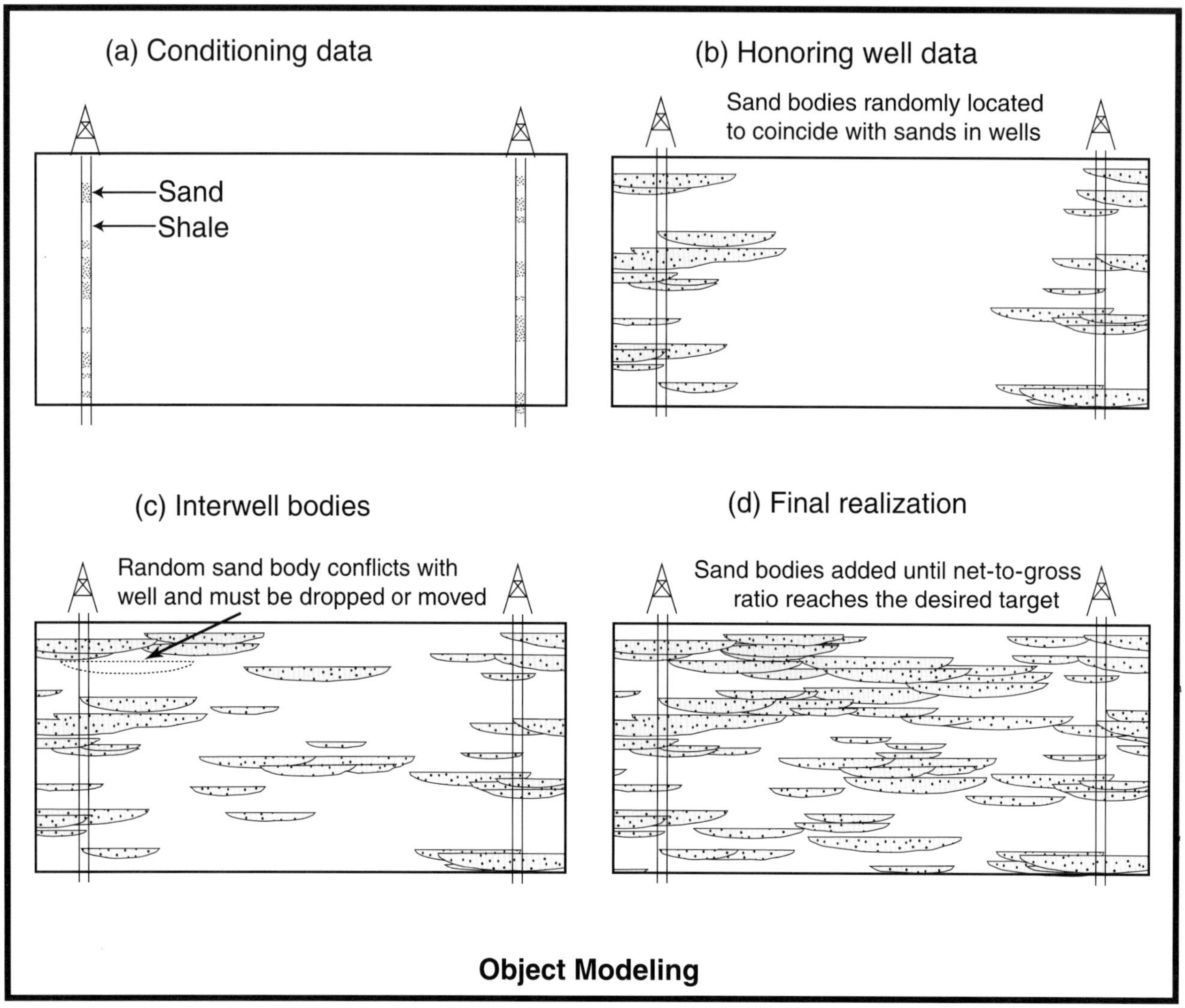

FIGURE 122. Object modeling input (from Srivastava, 1994). Reprinted with permission from the AAPG.

Having set up the computer application with background information on how to model the reservoir, the geologist then starts the program. The computer will conduct various iterations in an attempt to fit the various objects into the grid. Simultaneously, it will make an effort to honor the facies found in wells and the predefined settings while attempting to pack in all the facies objects neatly.

The steps are as follows (Srivastava, 1994):

1) Fill the reservoir model with a background lithofacies.
2) Randomly select a grid node.
3) Randomly select one of the lithofacies shapes and draw this with an appropriate width, length, thickness, and orientation.
4) Check to see if this shape conflicts with any shape or well value that is already there. If not, then keep the shape; if there is a conflict, reject the shape and try again.
5) Check to see if the global proportions of the various shapes have been reached; if not, go back to step 2.

The computer will make a large number of iterations to pack everything in while honoring all the input data. The end result is a geologically constrained means of filling in the gaps between the wells.

Typically, object modeling is used for sedimentary environments where the objects are smaller in dimensions than the interwell spacing. This configuration gives the object-based algorithm a practical chance of achieving a reasonable model. The program works more elegantly in this instance as it is not very easy for the computer to fit in objects that are big enough to extend across two or more wells (Dubrule, 1998). In fields with many

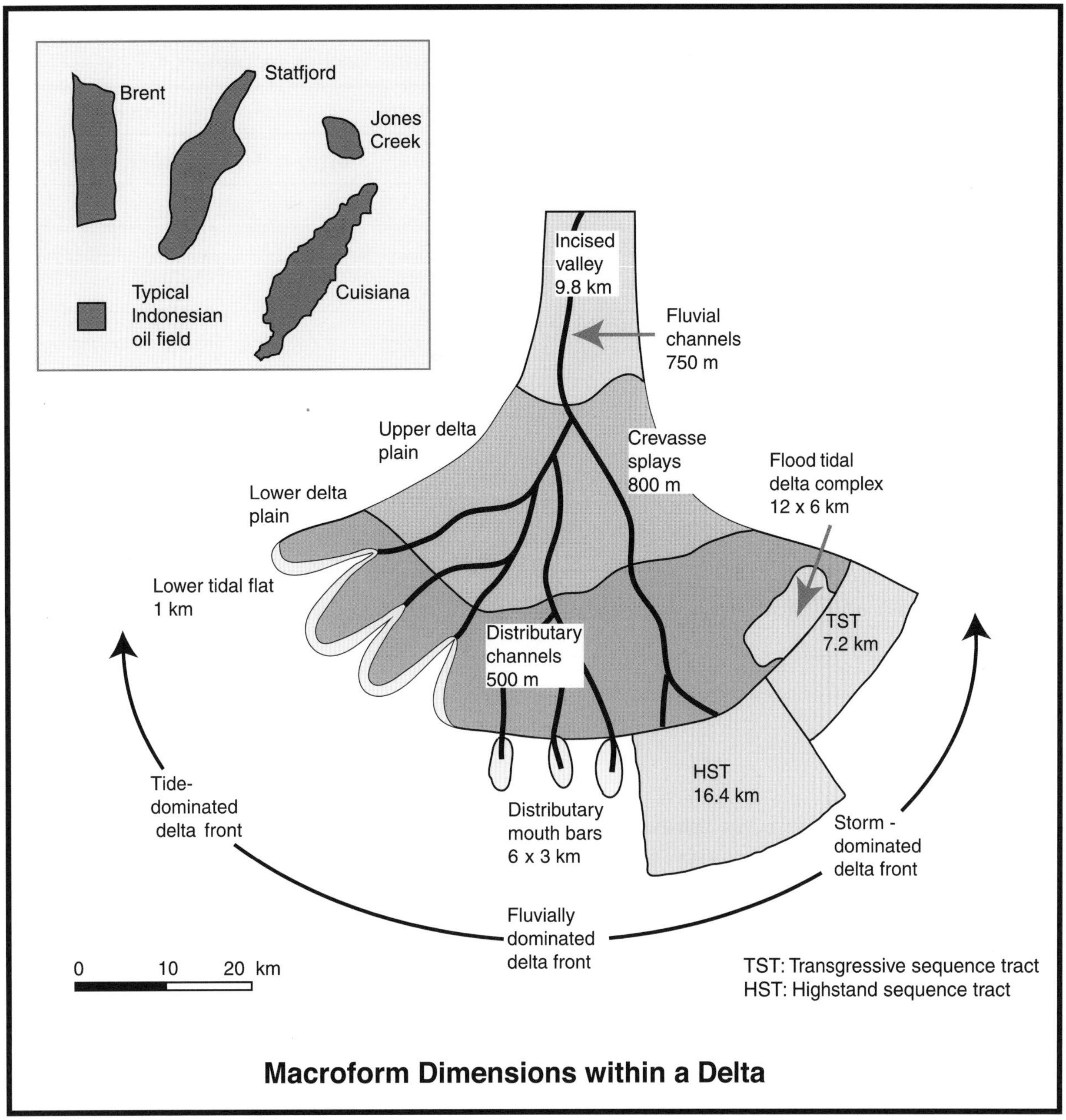

FIGURE 123. A schematic delta showing a range of sand body types at their average dimensions, together with several oil and gas fields at the same scale. The delta front is divided into three segments that are storm-, fluvial-, and tidal-dominated, respectively. The delta and its divisions are not to scale (from Reynolds, 1999). Reprinted with permission from the AAPG.

wells and where the objects are large, the program may struggle to achieve any match at all.

Object modeling is often used for fluviodeltaic reservoirs and any reservoirs tending toward jigsaw-puzzle or labyrinthine geometries (Dubrule, 2003). One advantage of object modeling is that the realizations produced will show sharp boundaries between the objects representing the macroforms. This character will influence the flow geology response where the model is used for a reservoir simulation.

VARIOGRAMS AND SIMULATION MODELS

An alternative to object modeling is to use one of several variogram-based geostatistical methods to create a *simulation*. A simulation is the process by which the input data, the variogram, and once in progress, any already simulated cell values are used to assign a value for individual grid nodes (cells) and subsequently to fill

Table 18. Statistics of dimensional data for deltaic sandstone bodies in meters.*

Sand Body Type	*Width*			*Length*			*Thickness*			
	Mean	*Max.*	*Min.*	*Mean*	*Max.*	*Min.*	*Mean*	*Max.*	*Min.*	*N*
Incised valleys	9843	63,000	500				30.3	152	2	91
Fluvial channels	755	1400	57				9	24	2.5	6
Distributary channels	518	5900	20				7.8	40	1	268
All types of systems tracts	25,365	106,000	1600	93,166	190,000	47,000	19.1	49	2.7	67
Highstand systems tract	16,425	43,000	16,000							36
Transgressive systems tract	7150	20,000	3300							5
Distributary mouth bars	2868	14,000	1100	6477	9600	2400	9.7	42	1.2	26
Flood tidal delta complex	6201	13,700	1700	12,300	25,700	2900	6.7	23	1.8	13
Crevasse channels	58	400	5				2.4	17	0.2	44
Crevasse splays	787	7700	18	5577	11,700	160	1.4	12	0.3	84
Lower tidal flat	994	1550	400				4.6	9	2	14
Tidal creeks	813	1550	161				5.2	18	1	15
Tidal inlet	1850	2550	700	4300	4300		4.8	7	3	3
Estuary mouth shoal	2400	2900	1700	3750	4700	2200	10	35	10	4
Chenier	3650	6400	900	21,758	38,600	49,000	5.8	7	4.6	2
All sands	5094	106,000	5	35,313	190,000	160				671

**From Reynolds (1999). Reprinted with permission from the AAPG. N = number.*

out the entire grid. The parameter value is picked at random from a computed probability distribution and then assigned to the grid node. The process is repeated until all the grid nodes have values.

VARIOGRAMS

A variogram is a geostatistical technique for describing and quantifying the spatial order shown by rocks and rock properties. Spatial order means that any given attribute of the rocks in the reservoir can be found to occur over a specific volume and its properties will vary in intensity spatially in a way that can be measured. The closer two data points are located to each other, the more similar the data values are likely to be. The farther apart the data points, the less similar they are, until a point is reached where the difference between the values is uncorrelated, i.e., unpredictable. The degree of predictability shown by rock properties occurs at a given length scale and this may be over centimeters, meters, or kilometers. The length scale may also vary according to azimuth. Rock properties tend to be more predictable for longer distances along depositional strike than they are along depositional dip.

Some geologists have difficulty in understanding what it is that variograms show about the reservoir geology. It is easier to recognize the significance of variograms once it is realized that they measure something that is real. The spatial distribution of geological properties is not a difficult concept to grasp. Here is a thought experiment: You find a large nugget of gold in a field. Are you more likely to find another piece of gold at a distance of 10 m (33 ft) or 10 km (6 mi) from where you are standing? Most people would start looking closer instead of farther away from the original find; in this way, they are thinking geostatistically. They may not know it, but they are making a prediction based on an intuitive feel for the spatial order of natural phenomena.

Here is another thought experiment: If a geologist wants to find out how a specific rock property varies spatially, what can they do to measure this variation? One way would be to crossplot the distance between pairs of wells and the difference in rock property values. This is almost what a variogram shows. The method is slightly more complex than this but not by that much.

The importance of variograms is that if there is a quantifiable degree of order to the spatial distribution of geological phenomena, then this provides a meaningful way to predict reservoir properties at any point in the gaps between the data (Chambers et al., 2000). The construction of variograms in geostatistics is frequently a stepping stone for their use in applications such as kriging, which are specifically designed to estimate values in the gaps between wells. Thus, a single value can be estimated at a specific location or many values can be derived for the nodes of a regular grid.

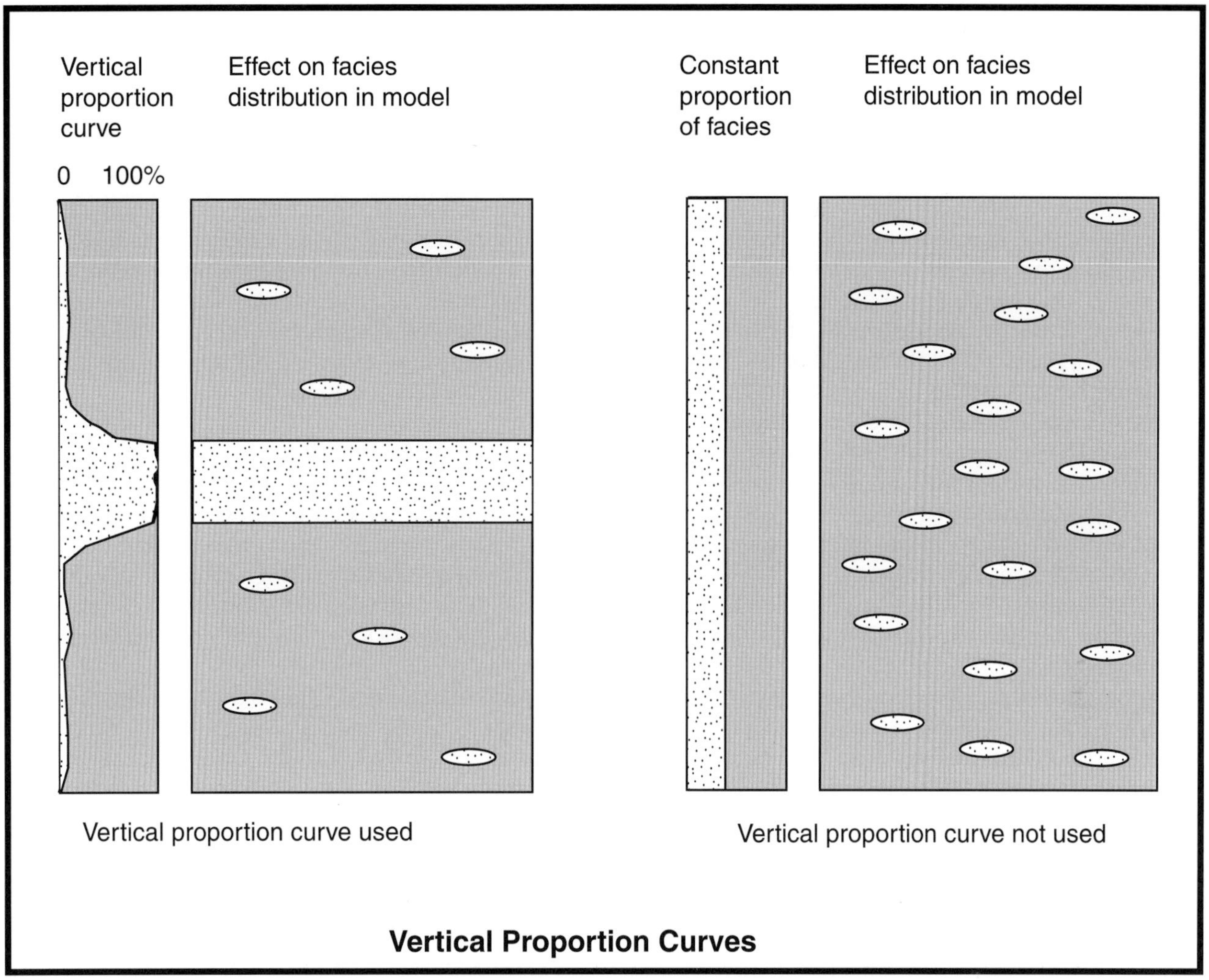

FIGURE 124. Vertical proportion curves show the vertical distribution of lithofacies within a reservoir interval. They can be used to ensure that the trends they show are honored in the modeled rock property distribution.

Geological data require processing before the data can be used to make a variogram. A basic assumption for variograms and geostatistical mapping applications is that the mean and variance of distribution of reservoir properties derived from the well locations represents the entire area of the field. This has been called *stationarity*. Not all geological data sets show stationarity; some may show an underlying trend that causes a systematic trend in the mean areally. This is called *drift* (Clark, 1979). Where variograms are to be used for estimation purposes, it is necessary to remove the drift before they are made so that the stationarity assumption applies. The geostatistical computer application will provide an analytical package to help do this.

The first step in making a variogram is to take pairs of data points corresponding to well values, then the difference between the two values is calculated and the distance between the two points is measured (Figure 125). It is also important to note the azimuth along which the two wells line up; a variogram is defined for a specific orientation in the subsurface. A well pair will only be selected if they are aligned along a common azimuth defining the variogram. An example of data selected for calculating a variogram could be for two wells, 500 m (1640 ft) apart, oriented along a northerly azimuth, with a difference in porosity between the wells of 0.143 – 0.121 = 0.022. Many such pairs are used to construct the variogram, with the distance between the wells and the difference in rock property value measured for each pair.

It can happen that the geological data values are too irregularly distributed and sparse to construct a meaningful variogram. Geostatisticians solve this problem by defining data pairs as usable if they are roughly a given distance apart and more or less oriented along a given direction. This is done by binning the data (Figure 126). The data are grouped according to the *lag distance* between the data points; for instance, all the well pairs that are 500 m (1640 ft) apart or all the well pairs that are 600 m (1968 ft) apart. In practice, there will be a *lag tolerance* specified to capture data close to the lag distance. This will be plus or minus 50% of the distance between the lags. Where, for instance, the lags are defined in increments of a 100 m (328 ft), the lag distance of 500 m (1640 ft) will capture well pairs that are between 450 and 550 m (1476 and 1804 ft) apart. Each data pair is thus assigned to a lag distance.

The data values are also binned between a range of azimuths to get usable pairs (Figure 126). For example, data can be defined as pairs if they occur within a range of orientations that have an *angular tolerance* of 22.5° on either side of the orientation of interest. The angle tolerance is restricted to a specific bandwidth for the larger lag distances (Deutsch and Journel, 1992).

Having binned all the data pairs into lags in this way, the next step is to calculate the variogram for the data values. The variogram (also referred to as the semivariance) is the sum of the squared differences of the data points falling within a specific range of distances (the lag) divided by twice the number of pairs found for that lag (Chambers et al., 2000).

The *experimental* (or *sample*) *variogram* is a graph of the raw data showing the lag distance plotted against the variogram value (Figure 127). It provides a representation of how a rock property varies over distance.

If there is a spatial relationship to the data, then the variogram should increase with increasing lag distance for the first few values. This shows that the rock property values are becoming more and more different from each other as the distances between the well pairs increase. The lag distance will eventually reach a point where there is no further increase in the variogram; in the ideal situation, the variogram value will reach a plateau after this. The variogram value where this happens is called the sill; the corresponding lag distance to the sill is the *range*. No relationship is found between the variogram and lag distance beyond the range.

The significance of the range where a variogram is used as input to stochastic modeling is that large ranges will mean that points far apart are correlated. The result is a model with a significant degree of continuity and smooth-looking variation. Small ranges mean less continuity between widely spaced points, and this will give a more heterogeneous result and rougher looking grids. According to Dubrule (2003), the range imparts a wavelength to the data areas. A contour map derived from a property grid will show areas of highs and low values. The range will influence the width of the areas of highs and lows on these maps.

Variograms can also be made to characterize spatial variation in the vertical plane. A vertical variogram of core porosity, for instance, will represent the range in thickness of clusters of high and low porosity values.

When some variograms are plotted, it is found that the values do not intersect the origin of the graph; instead they show a positive value at the intercept with the vertical axis. The intercept value for the variogram is referred to as the *nugget value*. This behavior can result from significant variation in data over distances shorter than the sample spacing, or alternatively it may be the result of measurement errors.

The pattern shown by the individual values on an experimental variogram plot will normally show some scatter, but an overall trend should be obvious if the analysis is meaningful. Once the experimental variogram has been constructed, then a variogram model is defined using a computer. This involves fitting a smooth line to the raw data. This is a necessary step as the variogram will later be used as input for other geostatistical applications, and these require a mathematically described shape for the method to work.

Three basic shapes are the most frequently used for variogram models: the *spherical*, the *exponential*, and the *Gaussian models* (Figure 128). The three variogram types vary significantly in their slope near the origin, and this character has a direct influence on the appearance of maps and grids derived using variogram-based algorithms. A steep gradient near the origin means that closely spaced samples will show a lot of variation. A low gradient at the origin will produce smoother looking maps and grids.

The Gaussian model shows the lowest gradient near the origin, and this model gives the lowest variance between closely spaced samples. Geostatistical maps and simulations based on a Gaussian model tend to be smooth looking. The Gaussian model type is preferred when mapping very continuous phenomena such as topography or thickness variation.

The exponential model will have the highest gradient near the origin, and will produce a more heterogeneous-looking effect over short distances in simulations. The spherical model is intermediate between the Gaussian and exponential models with respect to the line gradient near the origin. It is the most frequently used model type (Clark, 1979) because many geological parameters seem to fit this pattern.

The variogram is directional in nature. A variogram aligned along depositional strike for a property may be different from one along depositional dip. Rock properties can be more correlatable for longer distances along depositional strike than along depositional dip. Variograms should also be made for the vertical direction. It is the nature of the availability of data in the oil field that the vertical orientation will be sampled on a decimeter scale, whereas horizontal sampling is on a length scale of hundreds of meters between wells. Because of this, vertical variograms can be produced with much more confidence by comparison to horizontal variograms.

If several variograms are made for the same data set but with different azimuths, then this can give a sense of the *degree of anisotropy* of the data. If the plots are similar in all directions, then the data are isotropic. If the graphs are dissimilar, then the data are *anisotropic*. Three essential directions should be selected for analyzing anisotropy with experimental variograms: one along depositional

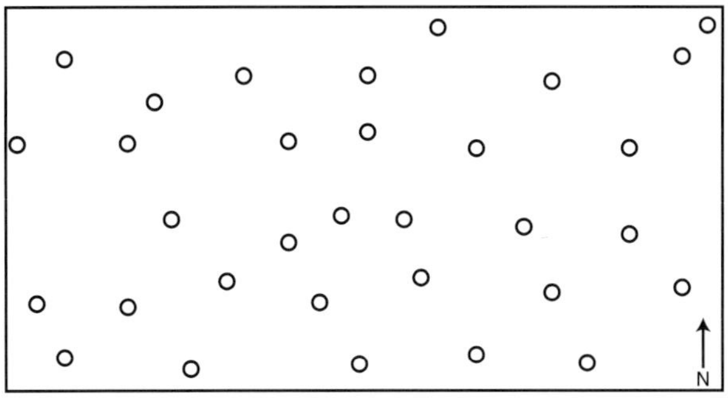

1) Raw data

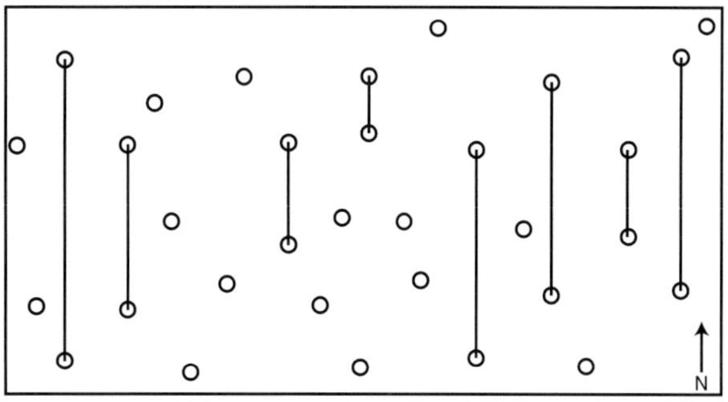

2) Well pairs selected for analysis, e.g., along a north-south azimuth

Well pair	Distance apart	Binned to a lag	Assigned Lag value	Difference In value
1	605 m	550-650 m	600 m	0.122
2	315 m	250-350 m	300 m	0.078
3	195 m	150-250 m	200 m	0.050
4	110 m	50-150 m	100 m	0.013
5	415 m	350-450 m	400 m	0.071
6	455 m	450-550 m	500 m	0.104
7	160 m	150-250 m	200 m	0.042
8	440 m	350-450 m	400 m	0.117
and so on ...				

3) Data tabulated

Lag value	Variogram
100 m	0.007
200 m	0.020
300 m	0.040
400 m	0.061
500 m	0.077
600 m	0.089
700 m	0.100
800 m	0.105

4) Variogram value calculated for each lag value

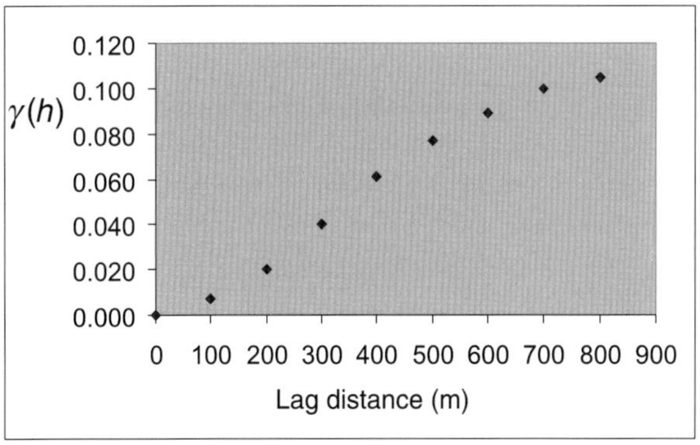

5) Plot the experimental variogram

$$\gamma(h) = \frac{1}{2N(h)} \sum_i (z_i - z_{i+h})^2$$

$\gamma(h)$: variogram

$N(h)$: number of well pairs for the lag value of h

$(z_i - z_{i+h})$: difference between rock property values for each well pair

Construction of a Variogram

FIGURE 125. Steps in the construction of a variogram.

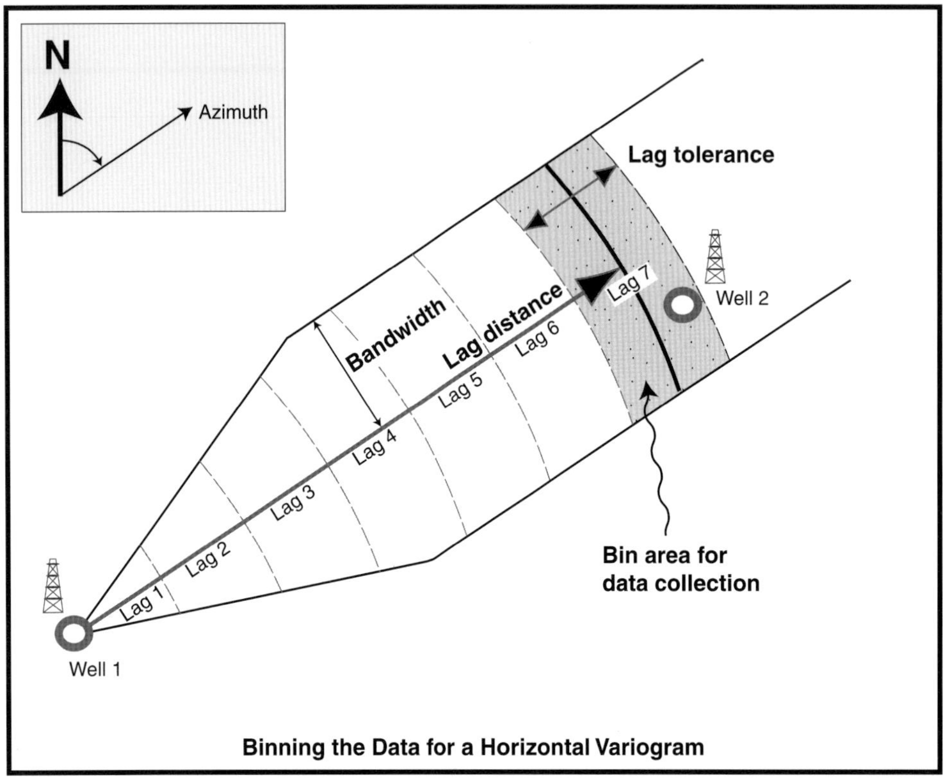

FIGURE 126. Because the available data for making a horizontal variogram in an oil field can be sparse, it is necessary to bin the data to derive a meaningful variogram (modified from Deutsch and Journel, 1992). Copyright Oxford University Press®, reprinted with permission.

strike if known, one along depositional dip, and one vertically. According to Olea (1994), variograms should be made for as many different directions as possible with four directions being the minimum.

Where geometric anisotropy has been established, then a single model can be used to fit all the experimental variograms by making the range a function of the azimuth (Figure 129). An ellipse is fitted to the data, and the model is defined by the size and orientation of the *major and minor axes* of the ellipse (Olea, 1994).

Many geostatistical methods were originally developed for the mining industry. Mining data sets comprise hundreds of densely spaced boreholes, ideal for geostatistical analysis and for producing high-quality predictive models for ore grade variation. Well data are sparser in oil fields, especially offshore, and this limits

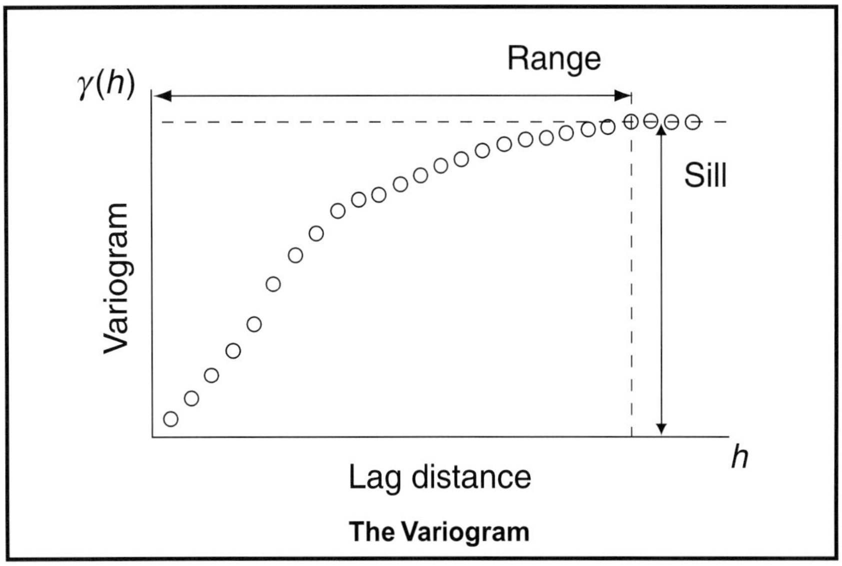

FIGURE 127. The variogram is a tool for describing the spatial variation in rock properties. It shows how (half) the average variation in rock property values between samples increases with the distance between the points at which the measurements were taken.

the use of geostatistics as an analytical tool. A common problem for fields with a small to moderate number of wells is that there is not enough data to construct any horizontal variograms that can be trusted. There are practical ways of finding a proxy for horizontal variograms in a field where there are not many data points. Dubrule (1998) suggested choosing a variogram model by experimenting with the inputs used. The parameters of the variogram are varied until a realization is obtained that looks like a geological model the geologist can feel comfortable with. Another way is to use analog data where usable variograms have been derived.

Variograms can be used to model the spatial character of two types of data. *Continuous data* includes data sets such as porosity that cover a large area and vary areally in value. *Discrete data* can be characterized numerically by one where the property of interest is present and zero where it is absent. Lithofacies is an example of discrete data. The *indicator variogram* measures for each lag distance, the probability that two discrete values such as facies at this distance apart are different (Dubrule, 2003).

KRIGING

Kriging is a method for estimating the value of a surface at an unsampled location in a statistically rigorous manner so as to minimize the error involved in the prediction. The estimate is made by interpolating a value between the wells where the influence of individual well values on the estimate is weighted using the variogram model (Davis, 1986).

Values can be calculated by kriging not only for a single point but also for all the node points on a grid. This allows the method to be used as a contour mapping technique, which many geologists consider to be an improvement over conventional mapping methods, e.g., inverse-weighted distance algorithms. Because the variogram model is used, a degree of anisotropy can be imparted to the shape of the contours, and the resulting map can look geologically realistic (Chambers et al., 2000).

Kriging is such a rigorous method for creating maps that, in the areas beyond the influence of the data points, the surface will smooth out to the mean value of the data. This will happen if the variogram range is less than the average interwell distance. The maps produced may look aesthetically unpleasant to the geologist, but, in a strict sense, they will be statistically valid given the input data. The appearance of kriged maps is therefore sensitive to the ranges used in the variogram model. A short range will create a local adjustment of the mean value creating the ugly-looking maps mentioned (Figure 130b). If a long range is used, each point has a large influence radius and the maps start to look more reasonable from a geological view point (Figure 130c). Kriging uses the variogram model to produce an optimal set of weights for interpolating between data points, and the weights themselves can change according to the degree of anisotropy (Figure 130d).

Kriged maps have been described by Yarus and Chambers (2006) as preserving only low-frequency information. Because there are only a small number of widely spaced data points used to make the maps, the resultant areal variation will be much smoother than is likely to be the case. Without a densely sampled data set, there is no exact way to replicate the "high-frequency" local variation that probably exists in reality within the reservoir. Other applications are available to make facies and rock property models with a semblance of heterogeneity. The simulation algorithms, to be described shortly, are designed to create grids with a rough texture, which approximate to "high-frequency" local variation.

It can be useful to create kriged maps before using the more sophisticated geostatistical mapping methods. Kriging makes the underlying trends more visible, such that any anomalous values in the data will show up (Hirsche et al., 1998).

COKRIGING

Cokriging is a method of incorporating secondary data into a kriged map or a model (Xu et al., 1992).

Kriging uses weighting derived from the variogram to interpolate between well control points. Cokriging uses additional data, typically high-density data sets such as seismic data, to help fill the gaps while still honoring the primary data. *Collocated cokriging* is a commonly used type of cokriging, which in its simplest form, uses a correlation coefficient of the secondary variable with the primary variogram. One of the benefits of cokriging is that it provides a general framework for integrating seismic and rock property data (Hirsche et al., 1997). Established ways in which cokriging is used in practice include combining seismic impedance values with well porosity values to create a porosity map (Doyen, 1988) or creating permeability maps that mimic the contour patterns of a porosity map.

SEQUENTIAL INDICATOR SIMULATION

Sequential indicator simulation is a simulation method that is suitable for integer-coded discrete variables; rock types or lithofacies for example (Figure 131). The method uses an integer variogram model as the basis for

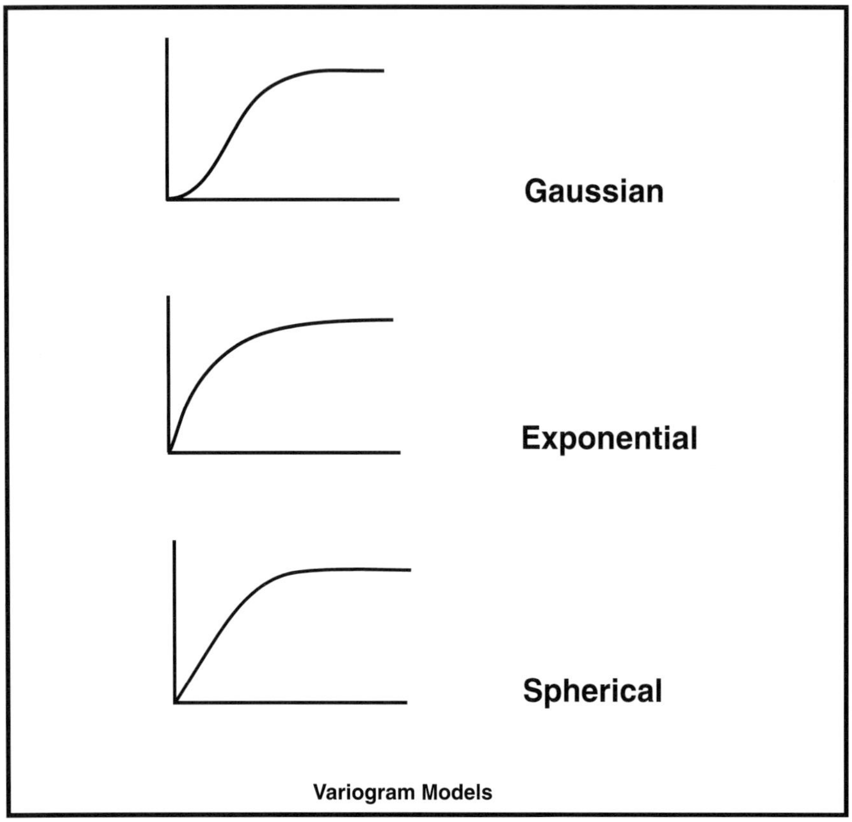

FIGURE 128. The three most commonly used shapes for variogram models are shown. The appropriate model selected will depend on the best fit to the experimental variogram.

populating the grid (Journel and Gomez-Hernandez, 1989). The 'sequential' part of the algorithm name derives from the way the model is built. That is, the model is built step by step, one grid node at a time, with the grid becoming progressively filled with values.

The computer program works as follows (Srivastava 1994):

1) The program selects a grid node at random where no value has yet been assigned.
2) Using all known values at the well points and from those previously assigned by the program, it computes a probability distribution of each of the categories likely to be present at that point. The probability functions are calculated by kriging.

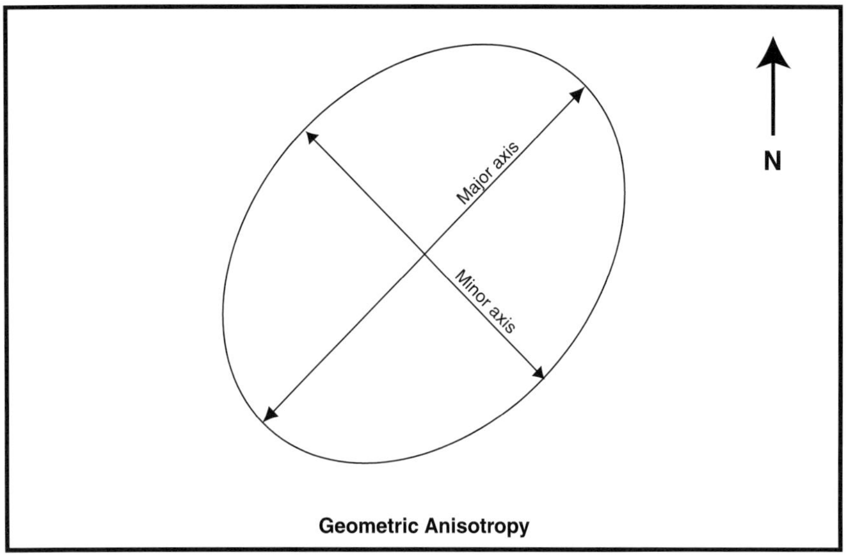

FIGURE 129. Where geometric anisotropy has been established, a single model can be used to fit all the experimental variograms by making the range a function of the azimuth. An ellipse is fitted to the data, and the model is defined by the size and orientation of the major and minor axes of the ellipse.

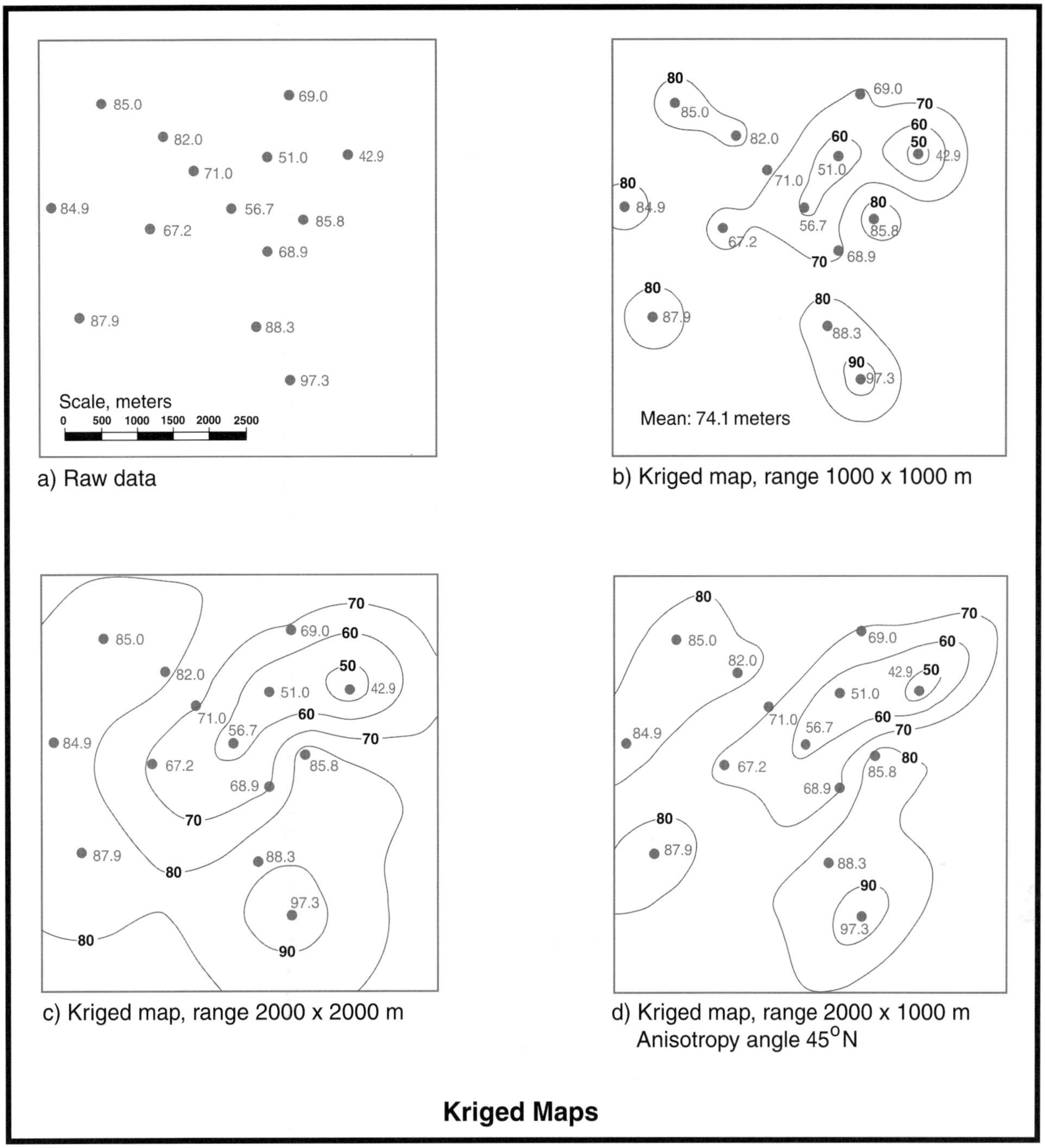

FIGURE 130. Kriging is a geostatistical technique that can be used to produce contour maps. Variograms provide the basis for kriging, and this allows a degree of directionality to be imparted to the contour distribution.

3) The program selects a value at random from the probability distribution function.
4) The program then assigns that value to the grid node.
5) The procedure is repeated from step 1 until the entire grid has been filled.

The following input data are required:

- The indicator variogram for each property (facies)
- The well data

The program is designed to create a property grid with a heterogeneous character while still honoring all the input data. The well values will be correct, the histogram of all the values filled out in the grid will closely reproduce the histogram of the input data, and the original variogram model can be backed out using data from the grid.

Where sequential indicator simulation is used to model facies, at step 2 the program will estimate the probability of occurrence for each facies type at a specific

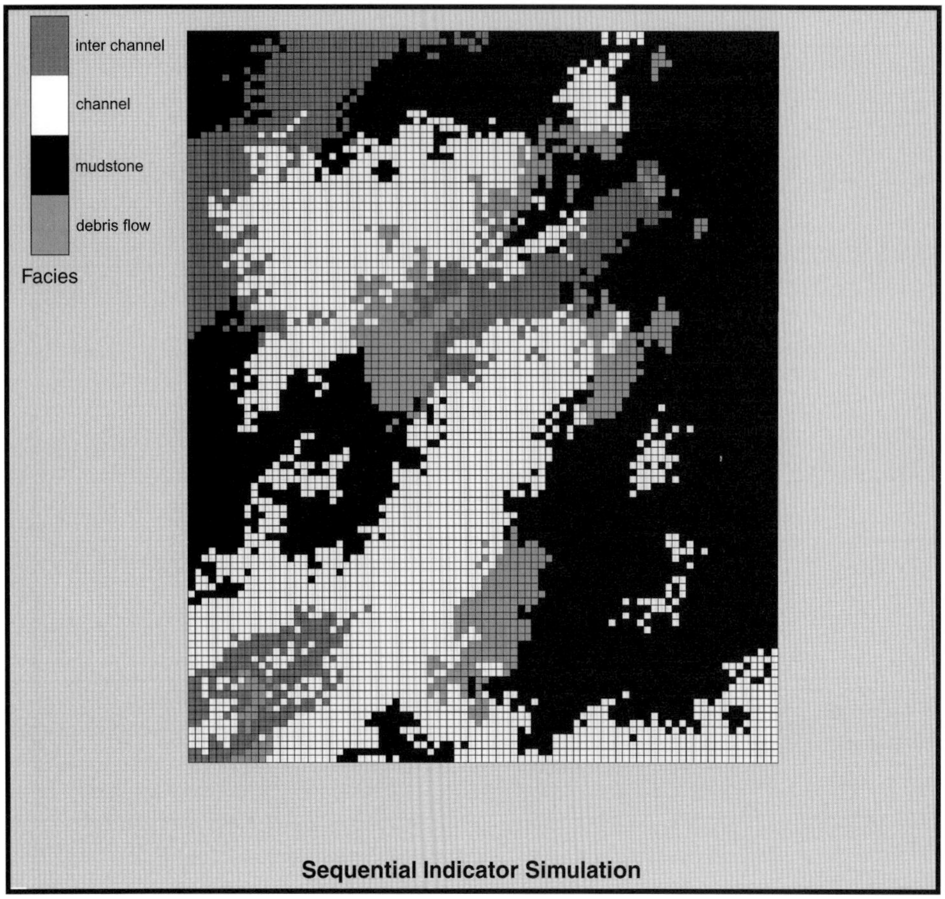

FIGURE 131. Facies model made using sequential indicator simulation.

node point, e.g. 60% shale, 40% sandstone. The program will then randomly select one of the facies types and will assign it to the grid node. The higher the probability that a specific facies type is present, the greater will be the chances that it will be selected. The sequential indicator simulation tends to produce a fuzzy, not very geological-looking picture of the facies distribution. This type of algorithm is not set up to include statistical data such as sediment body aspect ratios for instance (Dubrule, 1998). However, probability trend maps can be used to influence the areal distribution of lithofacies.

One advantage of sequential indicator simulation is that it produces facies models that are rough looking; according to Yarus and Chambers (2006), they created a high-frequency component that is a proxy for heterogeneity.

ROCK PROPERTY MODELING

Rock properties in siliciclastic rocks tend to be facies controlled. Thus, if the geologist is modeling rock properties, it is meaningful to analyze rock properties on a facies by facies basis and to control the distribution of values in the model by conditioning to lithofacies.

The variogram-based algorithms have to date been the most popular methods for modeling rock properties (Dubrule, 1998). 3-D modeling software provides data analysis tools for the user to look at histograms showing the distribution of rock properties such as porosity. Ideally, the sample data set should show a regular distribution on histograms. The data should also be checked for extreme values as these can have a major impact on the results of spatial correlation if they are left in the data (Hirsche et al., 1998).

SEQUENTIAL GAUSSIAN SIMULATION

The *sequential Gaussian simulation* algorithm is suitable for continuous variables such as rock properties (Deutsch and Journel, 1992). The method works in a similar way to the sequential indicator simulation method mentioned above and also uses the variogram as a basis for modeling. One difference is that it uses an alternative method of assigning values at step 2. The program calculates the mean and standard deviation for a cell value by kriging. A probability distribution is then derived, which has a mean and standard deviation equivalent to the kriged estimate. The value assigned to the cell is then selected at random from within this distribution. The method can produce local variation and will reproduce input histograms (Figure 132). A sequential Gaussian simulation algorithm can assign rock property grids for the whole grid. By preference, the algorithm should be used to fill out the grid on a facies by facies basis. The program will honor the input data for each facies class within the grid.

Sequential Gaussian simulation is commonly used for the stochastic interpolation of rock properties such as porosity and permeability in computer models. The program starts off with a *seed value* randomly selected by the computer or, if desired, a specific seed value selected by the geologist. This is where the first stochastic value will be posted by the computer. When the seed value for a series of runs is varied, each output realization will be different. If the seed value is the same for each run, then the output model is identical every time.

ADAPTING MODELS TO EXISTING TRENDS

Extra information can be used to fill in the gaps between the wells in a 3-D model. It may be possible to pick out the shape of individual genetic bodies from seismic data. Sometimes production data may show that there is connectivity between two parts of the reservoir and the geologist may want to ensure that this connectivity is honored.

The typical method of doing this is to define maps showing the probability that a given lithofacies is present in the area of interest. Both object and simulation modeling methods can use these maps as underlying trends to bias the interpolation of facies within the model. Some modeling packages allow the user to edit probabilistic models interactively to come up with a model that satisfies seismic geomorphological inputs. Another method of incorporating geomorphology information involves *multiple-point geostatistical simulation* (Strebelle and Payrazyan, 2002; Liu et al., 2004). The method combines the ideas behind the simulation and object-based approaches. The model is built one pixel at a time in a sequential manner but at the same time honors the reproduction of facies shapes and patterns in a manner similar to object modeling. A *training image* is used to represent the sedimentary bodies. This is becoming a more popular tool to represent the geology in 3-D models.

GETTING FAMILIAR WITH HOW THE VARIOUS GEOSTATISTICAL APPLICATIONS WORK

One of the best ways of understanding how the various geostatistical applications work is to take a small data set (e.g., a porosity data set with 10–15 well values) and to experiment with different settings to see what effect this has on the appearance of the grid.

Here are a few things to try:

1) Krige the data with a spherical model, with no nugget value. Try varying the range, e.g., 500/1000/2000 m (1640/3280/6560 ft), to see what effect this has on the grid. As the range increases, the width of the highs and lows increases, with the maximum width close to the range value.
2) As above, but this time make the major range axis 2000 m (6560 ft) and the minor range axis 1000 m (3280 ft), and give the azimuth of the major axis a value of 90°. The resultant grid will show a series of highs and lows elongated along the specified azimuth.
3) This time try the sequential Gaussian simulation algorithm on the same data set, no nugget, and a constant range (e.g., 2000 m; 6560 ft). Try in turn using an exponential model, a spherical model, and a Gaussian model (use a nugget of 0.01 to avoid the latter model crashing). The exponential model produces the roughest looking grid; the spherical model will be smoother, while the Gaussian model is the smoothest of the lot.
4) Try the sequential Gaussian simulation algorithm one more time. This time use a spherical model and a constant range (e.g., 2000 m; 6560 ft). Increase the value of the nugget by steps of 0.2 up to a value of 0.99. The grid will become noisier and noisier as the nugget value increases. The underlying grid trends will become more and more indistinct until they virtually disappear.

From this, it will be seen that the settings used for the various geostatistical applications can dictate how smooth or rough the grid trends appear, the width of the high and low patches, and will influence how much random variation is present within the grid.

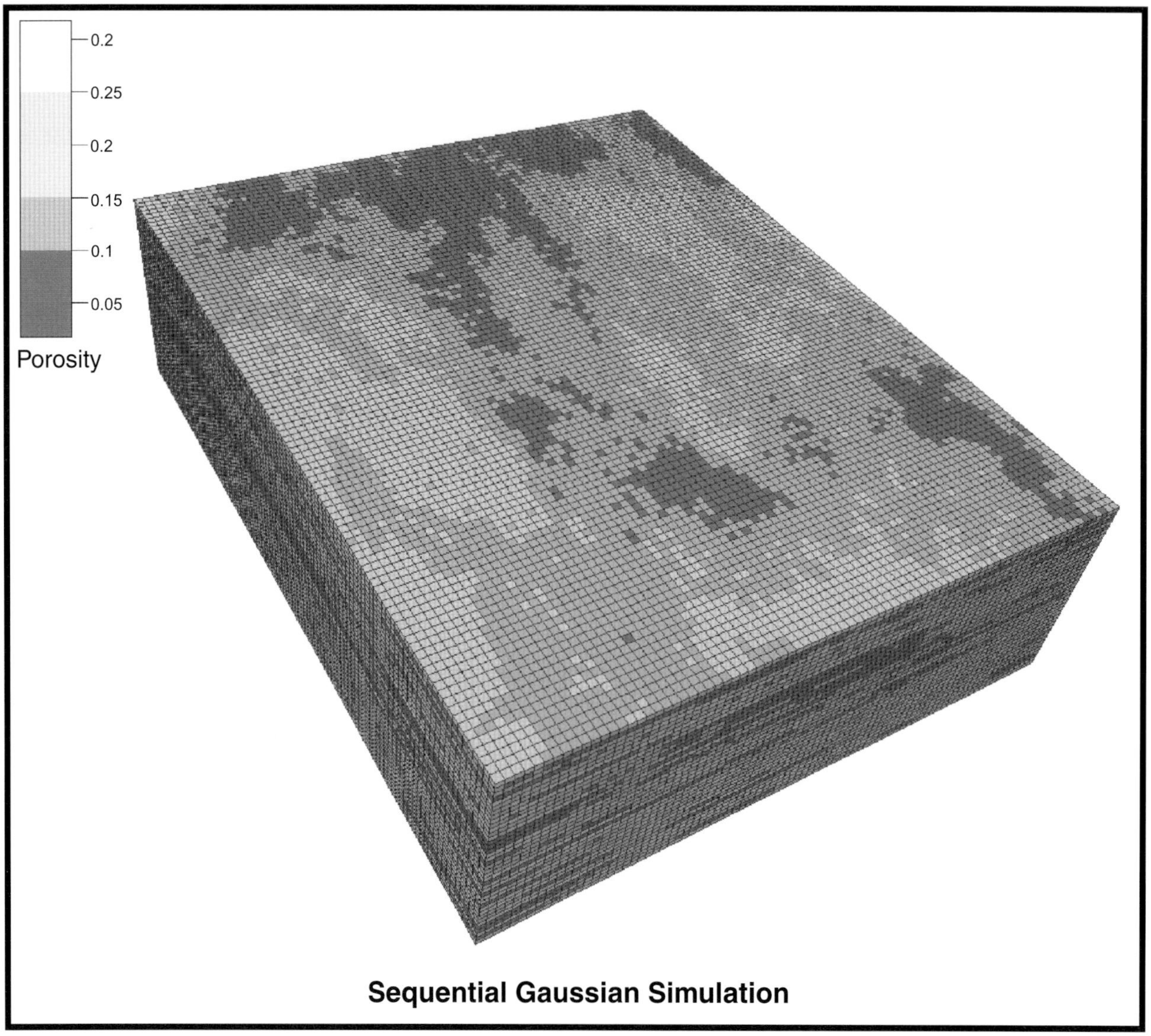

FIGURE 132. Porosity grid created using sequential Gaussian simulation.

20

Shepherd, M., 2009, 3-D geocellular modeling, *in* M. Shepherd, Oil field production geology: AAPG Memoir 91, p. 175–188.

3-D Geocellular Modeling

INTRODUCTION

Production geologists are increasingly using computer-based, 3-D *geocellular modeling* packages to represent the reservoir geology. The models replicate the 3-D structure of the reservoir, with the stratigraphic envelope, reservoir sublayers, and faults all represented in three dimensions. The reservoir volume is divided into a 3-D mesh of cells, a typical geocellular model having hundreds of thousands to millions of cells in it. Typical cell dimensions in models are about 0.5–2 m (2–6 ft) thick and 50 × 50 m (164 × 164 ft) areally.

For each cell, a lithofacies type and rock properties such as porosity can be assigned. Geostatistical applications allow the entire 3-D grid to be populated with values extrapolated from well control. A cell gets a single value for each reservoir property only. It is also possible to calculate new attributes within a model using appropriate equations. Thus, for instance, the oil in place for each cell can be calculated using the rock and fluid properties assigned to it. The computer can then calculate the oil in place for the entire reservoir by summing up the values in all the cells.

WHY 3-D MODELS ARE REPLACING A 2-D REPRESENTATION OF THE RESERVOIR

Formerly, the main method of representing the reservoir geology was to map the structure and rock properties in two dimensions using mapping software. The use of 2-D mapping is a crude method of capturing the reservoir geology. It is impractical to represent small-scale heterogeneity with 2-D isochore maps as it would be necessary to produce a large number of maps to do so. Similarly, the rock property maps are grossly simplified. The mapping algorithms use an averaging method to interpolate values between the wells, and, as such, any small- to medium-scale reservoir variation is mostly lost. By contrast, the cellular nature of 3-D models allows a heterogeneous reservoir to be given a chopped up rough-looking character between the wells.

THE GRAPHICAL BENEFITS OF 3-D MODELS

Once a 3-D model has been built, it is possible to display the various reservoir surfaces, faults, and wells on the computer screen in 3-D graphics. These can be selected, moved, and rotated in 3-D using the computer mouse cursor. There is something remarkable about the ability to visualize the structure of the reservoir so easily on a computer. It is an excellent tool for showing non-geologists in the company what the reservoir looks like. 3-D reservoir models aid communication between the disciplines. Those who are not inducted into the mystery of geological jargon or who do not share the geologist's innate spatial ability to look at 2-D images and see them in their 3-D context can still visualize and understand the major subsurface issues by looking at these models. Many of the larger companies have a dedicated *visualization room* where 3-D models can be inspected on a large display screen. These have a particular benefit for well planning, often reducing the time taken for this by a significant amount. Many packages also allow interactive well planning with instant visualization of the chosen well path (Figure 133).

Any item of data can be displayed on the screen by 3-D modeling software, including isochores, facies, and rock property grids. Well correlations can also be brought up on screen along with the various horizon tops picked by the geologist. Seismic data can be imported and shown in 2-D and 3-D. Statistical applications are provided for the analysis of rock properties and for applying arithmetical and logical operations on the data.

DOI:10.1306/13161205M913372

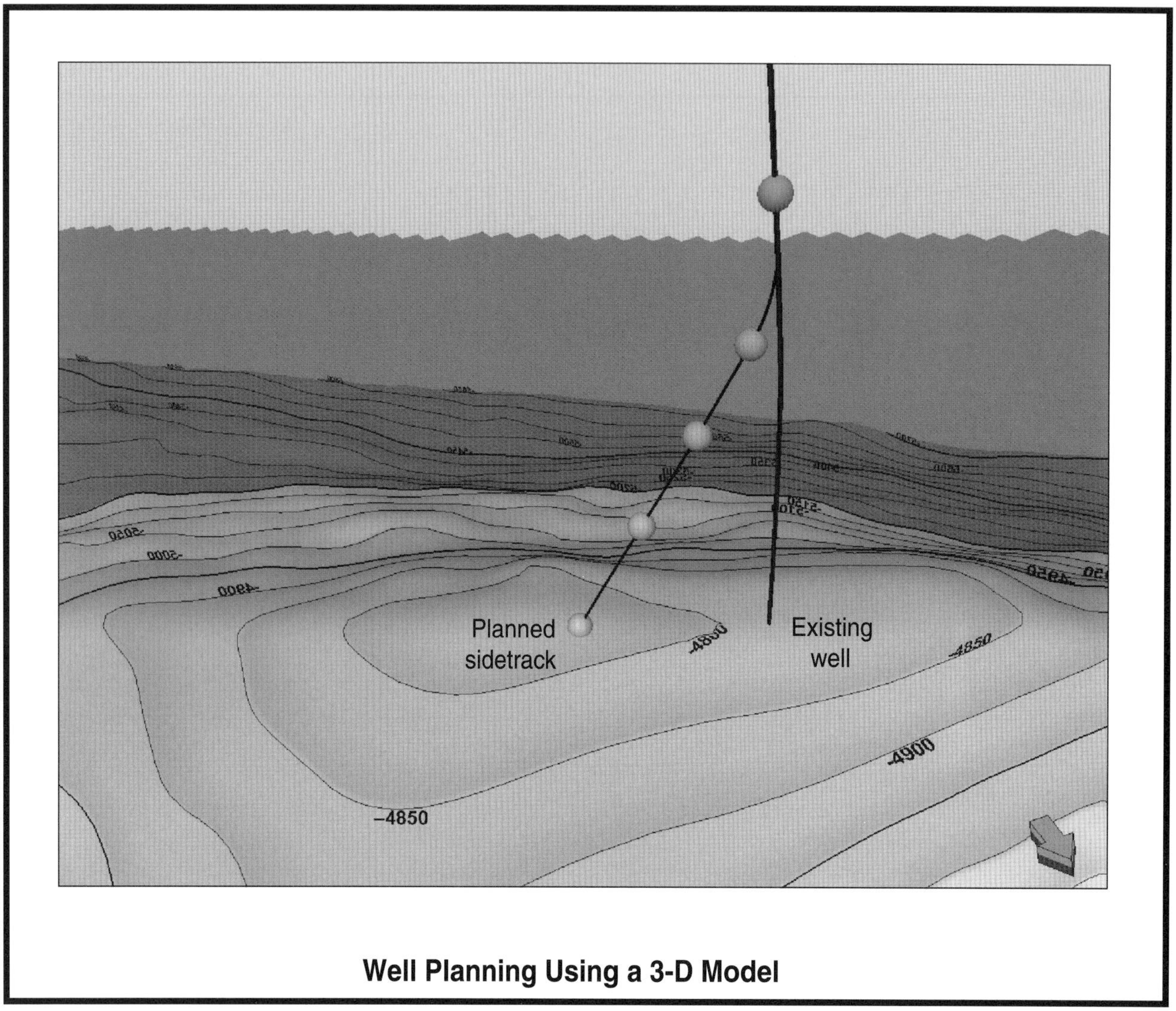

FIGURE 133. Some 3-D modeling packages include a well-planning module, which allows new well locations to be planned interactively.

Commercial modeling packages include, amongst others, GOCAD™, Irap RMS™, and Petrel™ (the illustrations showing the 3-D models in this book have been made using Petrel™ software, courtesy of Schlumberger).

WHAT 3-D GEOCELLULAR MODELING PROGRAMS CAN DO

Here is a checklist of some of the various facilities offered by these programs (Figure 134):

- 3-D visualization of faults and structure maps
- 3-D visualization of well paths
- 3-D representation of well tops and associated depths
- 2-D representation of depth, isochore, facies, and rock property maps
- Spreadsheet of well tops
- Cross sections
- Import or export data to other applications
- Screen capture of graphics for reports
- Interactive well correlation
- Interactive editing of well tops
- Interactive facies classification for wells
- Seismic interpretation of an imported seismic volume
- Depth conversion
- Fault modeling
- Data analysis of facies and rock property data sets
- Arithmetic and logical operations on rock properties
- Facies modeling
- Petrophysical modeling
- Object modeling
- Derivation of variograms
- Kriging and other geostatistical techniques
- Simulation modeling

3-D Reservoir Modeling Workflow

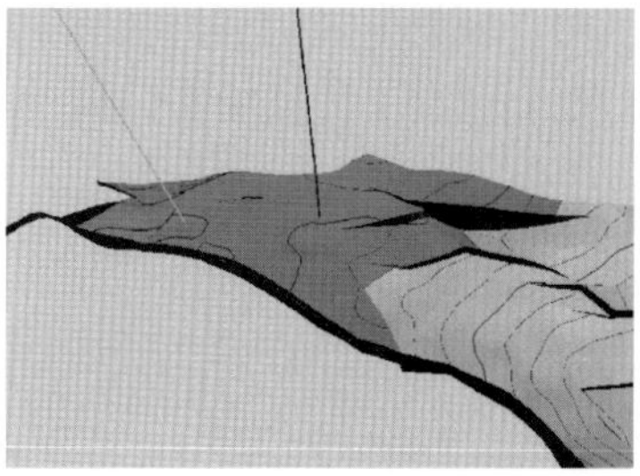

- Preplanning
    - object of the model
    - decide the key parameters
- Establish scope of work
- Collate and check the input data
- Build the structure
    - put the faults in
    - create the grid
    - make horizons and layers
- Populate the grid with a facies model
- Populate the grid with rock properties
- Quality control the model
- Calculate hydrocarbon volumes
- Make a modeling report
- Upscale (if required) for the reservoir simulation model

FIGURE 134. 3-D reservoir modeling workflow.

- Upscaling for reservoir simulation models
- Volumetrics
- Well design

BEFORE STARTING

The geologist will need to do some thinking before starting the modeling project. First of all, what is the logic behind building the model? There should be clear objectives for the model as this will ultimately influence how it is made. A model built to evaluate hydrocarbons in place will not need to be as complex as one built for a reservoir simulation. Maybe all that is required is to know if a specific drainage cell in the reservoir can yield better recoveries, in which case, a sector model may be more appropriate and timely to build than a full field model.

MODEL COMPLEXITY

The geologist will now have to consider how complex the model should be made.

Experienced 3-D modelers will advise the geologist to try and keep their models as simple as possible without losing the key elements of the heterogeneity that control the way the reservoir behaves. These models will be updated at regular intervals, and the simpler the model, the easier it is to modify it. A second recommendation is that if a particular geological feature cannot be drawn on a piece of paper, then there is probably not enough information to model it (Dubrule, 2003). The feature is either too complex to draw or the geology is not properly understood. Therefore, if you can't draw it, don't model it.

CAPTURING THE RESERVOIR CHARACTERISTICS IN 3-D MODELS

It is good practice to have lithofacies maps drawn before reservoir modeling starts. A valid 3-D geological model should replicate the geological scheme and should not be a substitute for it. It is also important to have a comprehensive view of the reservoir flow geology before modeling starts. The random use of geostatistics can produce impressive-looking geological models but there is a chance that they may not replicate important heterogeneities such as thief zones, mudstone barriers, and sand pinch-outs that are critical for reservoir simulation studies (Larue and Legarre, 2004). If the geological model is to be used for a reservoir simulation, then the geologist should have a checklist of the features that are required to be in the upscaled model, once it is complete.

KEY GEOLOGICAL PARAMETERS FOR MODELING

A *prescreening review* should be made. There is a need to decide on the area of interest to be modeled, the cell size, the number of reservoir units, the number of lithofacies, and what the key faults are that are required for the model.

The *area of interest* will be defined by the objectives for the model. There is no point in modeling the whole field if the model is only going to be used to help maximize the reserves in a particular sector of the reservoir. On the other hand, it may be that the reservoir engineer is interested in the aquifer performance around the field, and the aquifer may need to be modeled in addition to the hydrocarbon volumes.

If the reservoir is not too complex, it may be advantageous to make the *cell size* such that the geological model can be imported straight into a reservoir simulation without the need for upscaling. This can save an enormous amount of effort later on, particularly if there is a rush to get the model finished. On the other hand, if the reservoir is known to be complex, then the cell size should be at the scale of the smallest component of heterogeneity likely to influence flow. Some modelers recommend that the grid cell size should be at least half the width of the narrowest macroform (Figure 135). The cell size should also be such that there is no more than one well per cell at the very least and ideally with several grid blocks between the wells if the model is to be used for a reservoir simulation. Reservoir engineers also prefer to use geological models where there are at least two cells between closely spaced parallel faults.

The number of *reservoir units* will need to be decided. There should not be too many because the more units there are, the more work is needed to complete the model. If there are any laterally continuous shales influencing flow in the reservoir, it may be worth considering representing them explicitly as a layer in the model. Alternatively, shales at unit boundaries can be modeled in the reservoir simulation using transmissibility barriers. This was the method chosen by Cook et al. (1999) for a reservoir simulation of the Meren field in the Niger Delta, offshore Nigeria. Multipliers ranging from 0.1 to 0.0001 times the transmissibility between cells on either side of the surface marking the shale were used to represent the flooding shales between marine parasequences. In this way, the degree to which a specific shale acts as a barrier or baffle can be represented. Tuning of the multipliers in the simulation model controlled the history matching of production logs and the degree of water and gas encroachment.

Similarly, some geologists think that it is bad practice to model a large number of facies. In the early days, when geological modeling packages started to be used, it was common to see models with a dozen or more facies types. The argument could be made that by replicating all the facies, a true representation of the complexity of the reservoir architecture could be obtained. However, the more facies that are modeled, the more cumbersome the model will be. The end result can be a snake pit of facies types that is very difficult to make sense of visually and is near impossible to replicate or edit once new data values become available. These days, many geologists prefer to keep the modeled facies to somewhere between three and five groups per reservoir layer. This has been called the *KISS principle*; Keep It Simple and Sensible. It may be necessary for the geologist to modify a complex facies scheme into a smaller number of groups for modeling purposes. In this instance, lithofacies that are similar in their rock properties and geometrical shape should be lumped together as a single group, if this is practical.

Likewise, there is a need to determine which faults should be used in the model. For a model with the aim of calculating hydrocarbon volumes, the modeling of many small faults may not make much difference to the final result. However, small faults in the hydrocarbon zone will have an impact on the local fluid flow and may need to be represented if the geocellular model is to be carried through to a reservoir simulation.

SCOPE OF WORK

Once the geologist has a reasonable idea of what they want to do, then a *scope of work* should be defined for the project. This will include the objectives and a workflow. Managers will also insist on a timetable giving details as to how long the modeling is likely to take; especially if integrated subsurface models are required to justify wells and the rig to drill them has already been booked for a given date.

It is a good idea at this stage to allocate extra time to make revisions to the model later on. It is a common experience to find that certain starting assumptions have to be made about the reservoir to make progress with a model. However, the act of building the model can in itself be a learning process by which new insights into the geology of the reservoir are gained. If these show the initial assumptions to be inappropriate, then the model will have to be modified.

INPUT DATA

The first stage in working on a model is to compile the input data. There are variety of data types and sources of data that may be included in the input list:

- Top reservoir structure map data supplied by a geophysicist

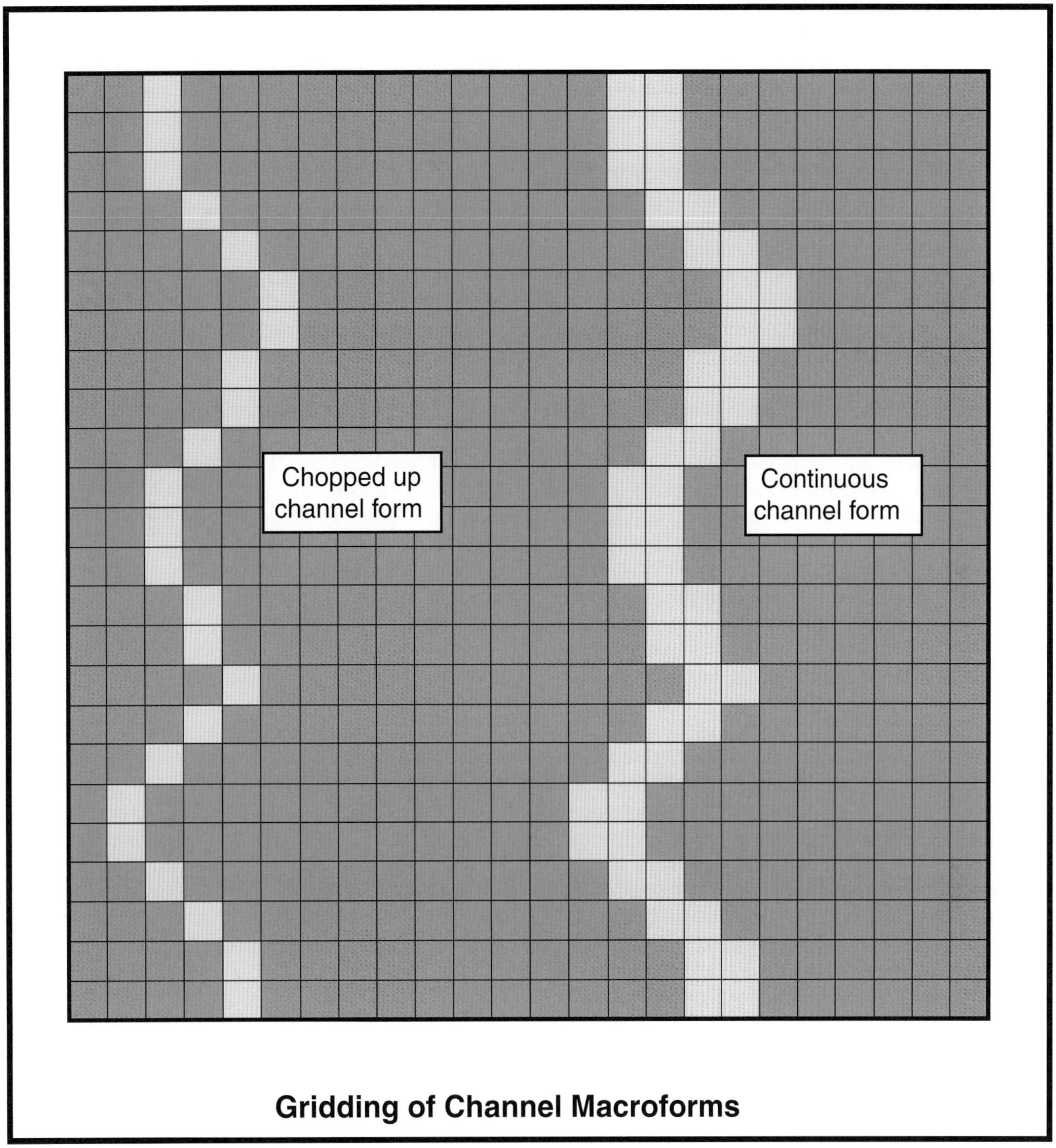

Figure 135. The grid cell size should be chosen at half the width of the narrowest macroform. This ensures that there will be neighboring connections in a reservoir simulation model that will adequately replicate the flow continuity of the macroforms.

- Fault polygon or stick data supplied by a geophysicist
- Isochore map data supplied by a geologist
- Directional survey data for all wells supplied by the well database
- Wireline log data for all wells supplied by a petrophysicist
- Petrophysical log data for all wells (especially porosity and permeability curves) supplied by a petrophysicist
- Formation and flow unit top data supplied by a geologist
- List of fault cut data in the wells supplied by a geologist

QUALITY CONTROL

All the data to be used for the model should undergo a quality control check for erroneous values. Large data sets are prone to errors and they do need to be validated. Although the work involved is painstaking and tedious, the time taken to validate the input data for a 3-D model is well spent. It will save the geologist a lot of embarrassment and hard work later should mistakes come to light once the building of the model is at an advanced stage. A quick and easy check is to visualize the data on the computer screen (Tinker, 1996). Erroneous values will often stand out from the general cluster of data in three dimensions and should be obvious to spot.

The surface locations of the wells should be checked against the master database. A typical problem is to find that different wells are referenced to different cartographic projection systems. Deviation surveys in wells are a well-known source of error, and many production geologists have spent hours checking this type of data. Deviation survey errors can often show up by looking at the well path visually in 3-D. The well tops should lie on the appropriate surface. Drill floor elevations for all the wells should be verified because these will be subtracted to ensure that all TVDs are referenced to a flat datum.

The lithofacies in the wells should be depth shifted from core depths to log depths. A visual comparison of the lithofacies versus the log response will give a sense check on this. The wireline logs used to build the model should be examined. If the log values on the petrophysical log look unlikely, it is worth getting the petrophysicist to recheck the original interpretation. Each log type should have a consistent naming policy instead of, for example, having one set of wells with the gamma-ray logs named "Gamma" and the other set named "GR." Rock property inputs should be checked. Some petrophysical programs assign unusual numerical values to null points, −999.25 for instance. These need to be removed. Some petrophysical software programs will create the porosity output with a few negative values. These should be removed or assigned a zero value. Likewise, any water saturation values greater than 1 should be removed or assigned a value of 1.

At this stage, the geologist should tabulate a list of all the tops for the various reservoir units in the wells. This should be consistent with the same set of tops used by the geophysicist to create any depth-converted structure maps. A list of fault cuts in the wells will also be required.

BUILDING THE STRUCTURE

The first stage in making the model is to create the 3-D structure of the reservoir by constructing surfaces and faults. The faults are created by interactively fitting 3-D surfaces on the computer screen to the depth-converted fault polygons or fault sticks made by the geophysicist in the seismic interpretation (Figure 136). The faults created in this way will need to honor the location of fault cuts in the wells. Care is needed to connect branching or crossing faults in a consistent manner.

CREATING A GRID

Once the fault modeling is complete, a 3-D grid can then be created. This will wrap around the fault structure within the defined area of interest. At this stage, the *X-Y* grid increment is specified for the grid cells in the horizontal plane. Larue and Legarre (2004) gave details of the grid dimensions that were assigned for a 3-D model of the Meren field, offshore Nigeria. The area of interest is 15 × 4 km (9 × 2 mi). The model cells are 25 × 25 m (82 × 82 ft) in area by approximately 0.75 m (2.5 ft) thick. Overall, the geological model contains more than 9 million cells. Note the much larger areal dimensions of these cells compared to their thickness. This is because the reservoir geology is likely to vary much more in the vertical plane than in the horizontal plane.

The *orientation of the grid* can be important. If the model is to be used for a reservoir simulation, this should be discussed with the reservoir engineer. They may want to ensure that the long axis of the grid is oriented in a particular direction so as to optimize the model performance. This may be along the expected flow direction between injection wells and the production wells or relative to the main direction of aquifer influx (Figure 137). Alternatively, they may wish the grid to be parallel to any horizontal well paths. Some modeling packages allow control lines to be defined so that the grid can be locally aligned in specific directions.

HORIZONS AND SUBLAYERS

The next stage involves inserting the major reservoir horizons into the grid. The program will fit the horizons to the fault model and should also ensure that the well tops tie the surfaces. The isochores can be inserted above, below, and between the major horizons to create the individual reservoir zones. The program will compensate for any gaps or overlaps that may result from this process.

The structure should be reviewed at this stage before going further. Some basic checks involve visualizing the model to see if the well tops lie on top of the appropriate surface and that they also lie on the well tracks. Horizontal wells should intersect the stratigraphy only where they are supposed to. The direction of throw for the modeled faults should be consistent unless they have been interpreted as scissor faults. The grid should be

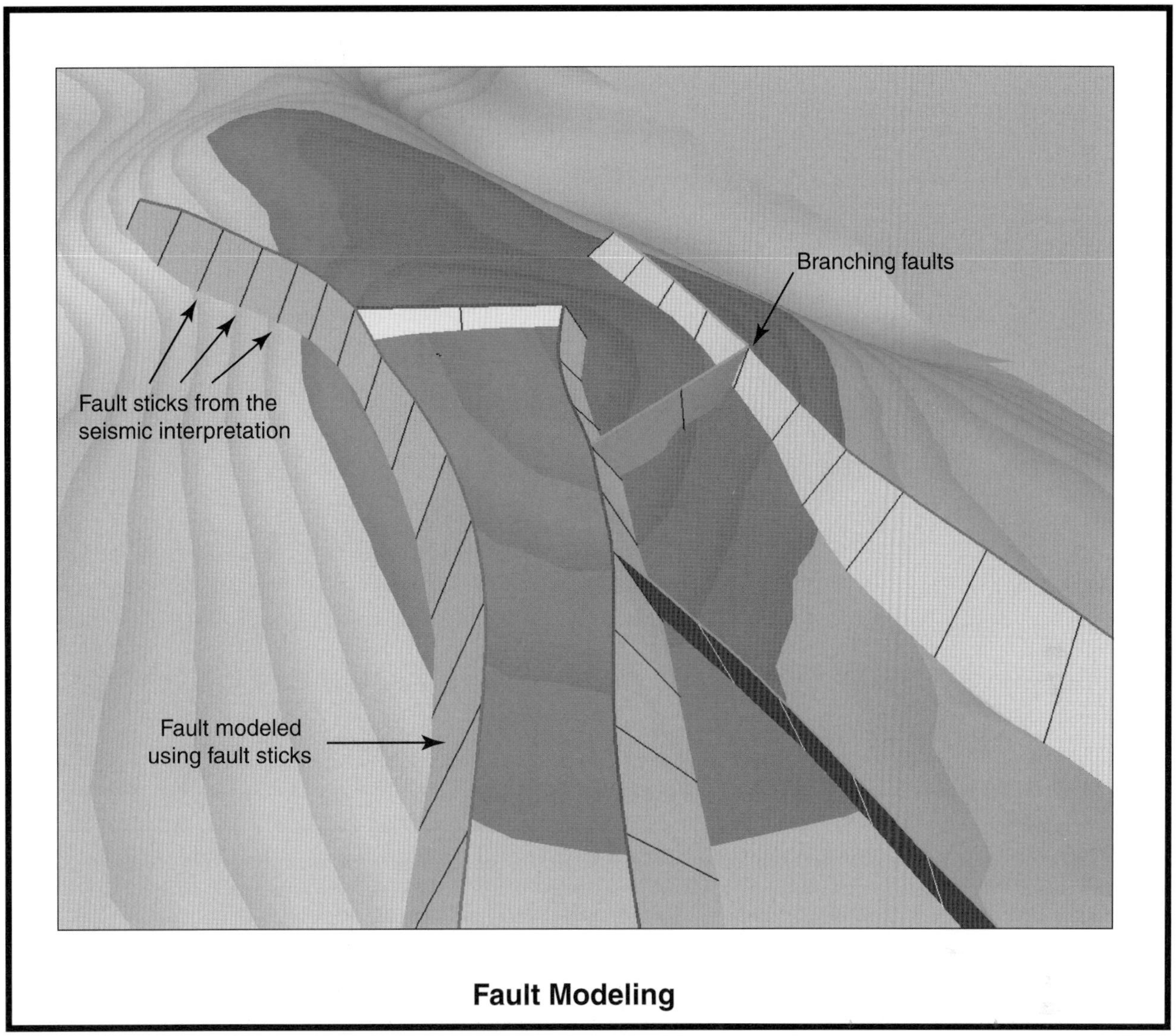

FIGURE 136. Modeled faults in a 3-D geological model. The faults are created by interactively fitting surfaces on the computer screen to fault sticks or polygons from the seismic interpretation.

checked to ensure that it is regular in form with no awkward twists, turns, and overhangs. If the model is to be used for a reservoir simulation, it may be worth exporting the grid at this stage to see if the reservoir engineering software will accept it. Isochore maps should be produced from the model to see if they look as expected.

The zones are then divided into several sublayers a meter or so thick in order to capture the small-scale vertical heterogeneity. In practice, it is best to keep the thickness of the grid cells at, or less than, the thickest of the smallest macroforms to be modeled. If it is important to capture permeability layering, then this is better represented with thinner cells. If the cells are too thick, then the subtleties of the permeability layering may become averaged out such that the potential effect on flow is lost. The number of sublayers per zone can be a compromise between minimizing the number of cells in the 3-D model and the need to capture the vertical heterogeneity at the appropriate level of detail.

The program will subdivide the zones into sublayers according to a user-specified geometry, for instance *proportional*, *top parallel*, or *base parallel* (Figure 138). The type of geometry used will have a significant impact on the appearance of the grid where the zones are wedging in thickness. A proportional geometry will divide the wedge into zones of equal thickness, all decreasing proportionally in thickness toward the thin end of the wedge. This is the preferred option if the model is to be used for a reservoir simulation. It creates the relatively simple cell-to-cell geometry that makes the simulation model easier to run.

A top parallel configuration will replicate an onlap geometry. Here, the upper layers are areally more extensive than the lower layers. A base parallel geometry

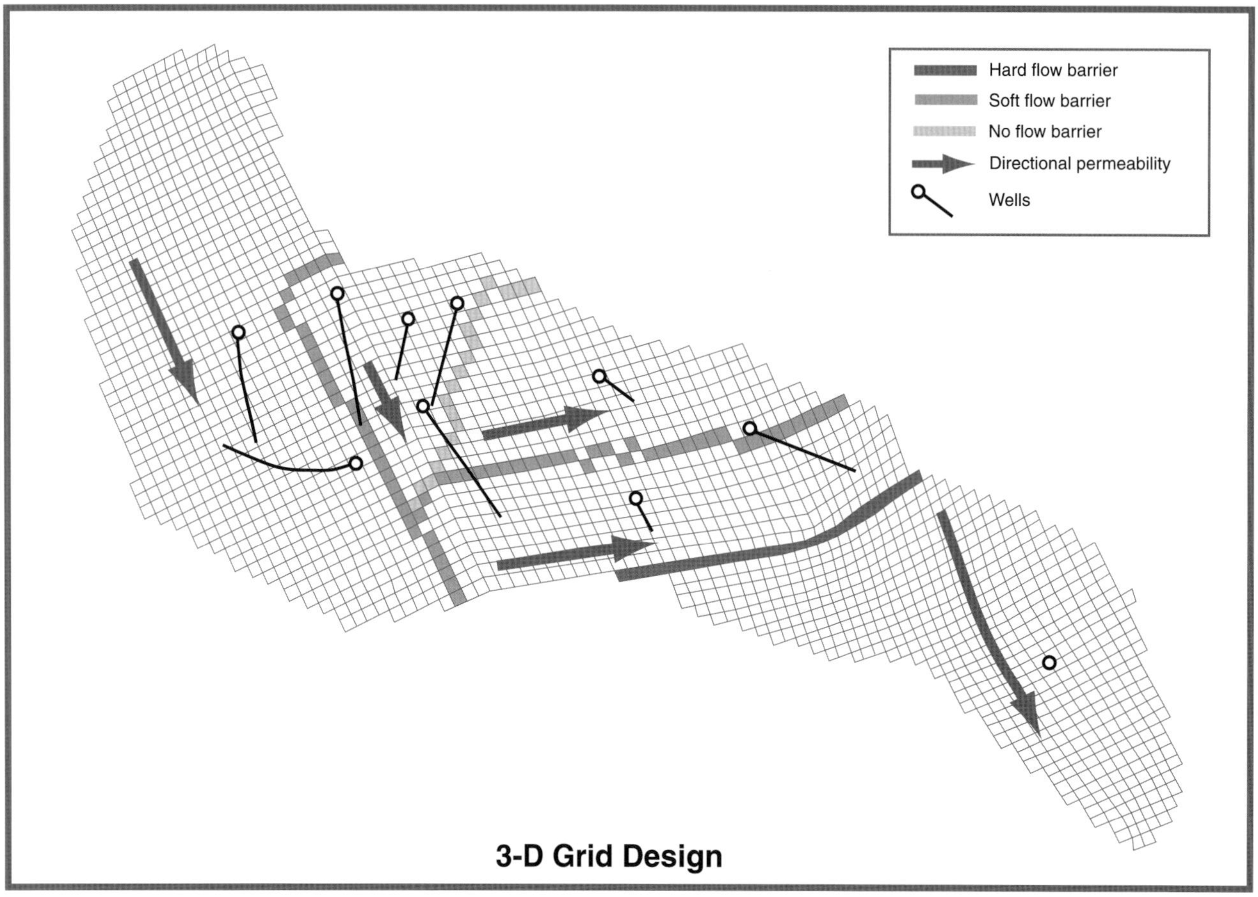

FIGURE 137. 3-D grid design for the MacCulloch field, UK North Sea. The grid has been designed to accommodate flow barriers and directional permeability in the various sectors of the field. Grid cell dimensions are 50 × 50 m (164 × 164 ft) (from Scorer et al., 2005). Reprinted with permission from the Geological Society.

can be chosen if erosional truncation is a dominant feature affecting the reservoir. In this instance, the upper sublayers are sequentially truncated by the overlying surface.

BLOCKING OUT FACIES AND ROCK PROPERTIES

The geometrical framework for the 3-D geocellular model will now be in place. The next stage is to populate each cell in the grid with a single facies type and a single value for each rock property. A typical model will have a representation of the facies model, porosity, permeability, net to gross, and water saturation properties. Before this can be done, facies and rock property values have to be assigned to the cells intersected by the wells.

Each cell will have a unique value assigned to it as a result of *blocking out* the well data (Figure 139). Blocking out involves taking a representative value of the rock properties within each individual cell intersected by a well. For example, the cells in the reservoir model may be 1 m (3 ft) thick whereas the data density assigned by the petrophysical analysis in the wells may consist of one value every 0.15 m (0.5 ft). To assign a single value to the cell, the rock property data are averaged. In the case of facies, an intersected cell may be given either the most frequent facies present or the specific facies type opposite the midpoint of the cell.

The results of blocking out the facies and rock properties should be checked before going further with the model. It is important that the main heterogeneities that influence flow are preserved after blocking out the logs. This can include features such as laterally extensive shales, which act as permeability barriers and high-permeability thief zones.

BLOCKING OUT NET TO GROSS AND POROSITY

A net-to-gross parameter is usually required for rock property modeling unless the layering is fine enough to represent both net and non-net intervals separately. Care

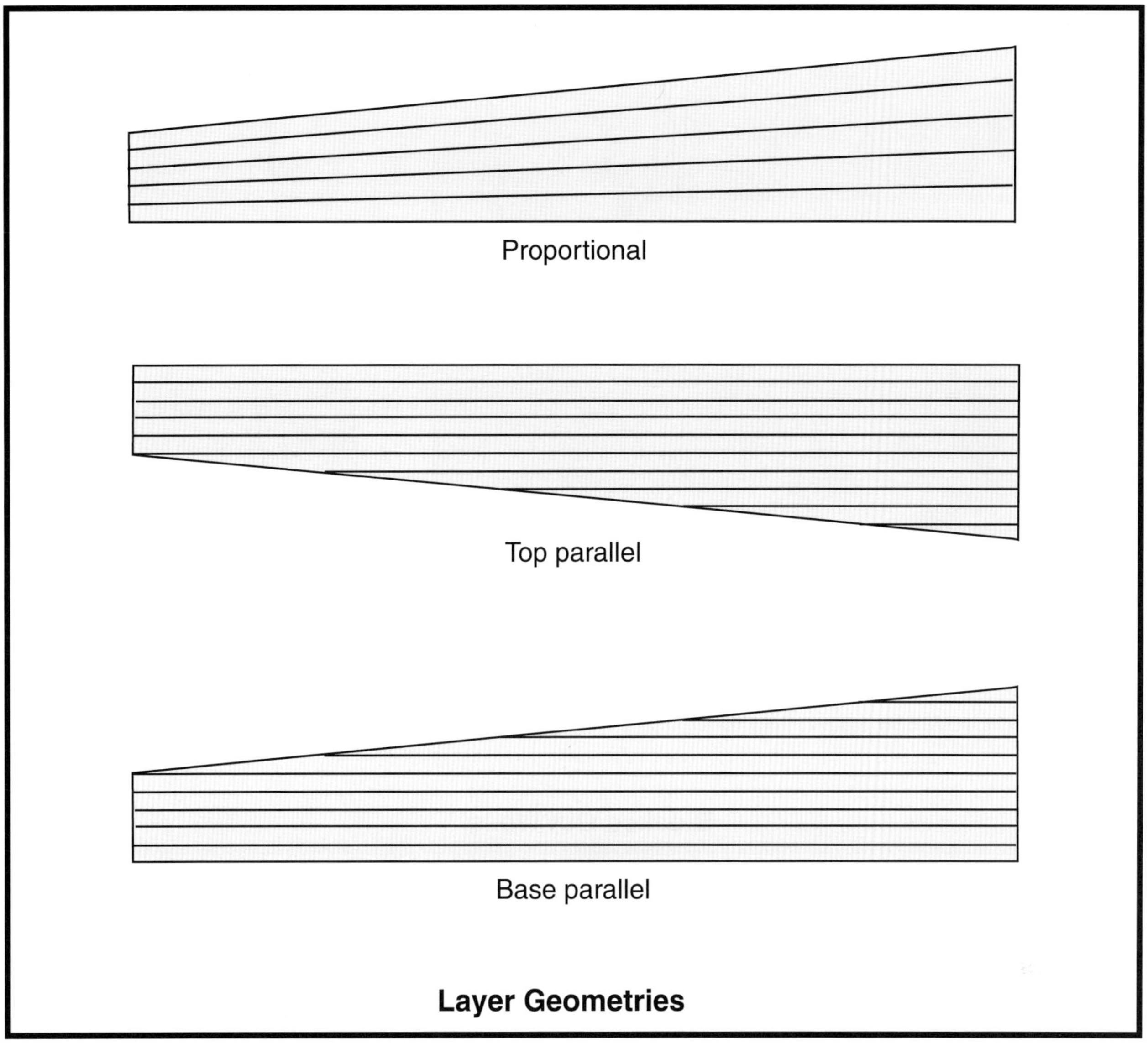

FIGURE 138. Proportional, top parallel, and base parallel layer geometries.

has to be taken in blocking out rock properties where a net-to-gross parameter is being modeled (Figure 140). Values in non-net rock should not be used for calculating the average value of rock properties for a cell.

Take, for example, a specific cell 1 m (3 ft) thick with six porosity values. Three values are measured in non-net rock and three values are net. The porosity values in the non-net rock may be zero or have very low values. If these are included when averaging the porosity for the cell, then the overall porosity value assigned to the cell will be unrealistically low. For this reason, porosity values in the non-net intervals should be set to undefined (but not zero). The result is that the assigned scaled-up porosities for each cell are representative of the net rock (and thus the rock that is likely to produce).

BLOCKING OUT PERMEABILITY

The geologist should be aware of the scaling issues when creating a 3-D property grid for permeability. It is common practice to use a log transform equation to calculate an estimate of permeability from the porosity log (see Chapter 12, this publication). Core plug data will have been used to make this equation and as such it will be specific to rock volumes at the decimeter scale. The transform should therefore be applied to the individual log porosity values in the wells before blocking out. It is inappropriate to apply the transform to the scaled up porosity values.

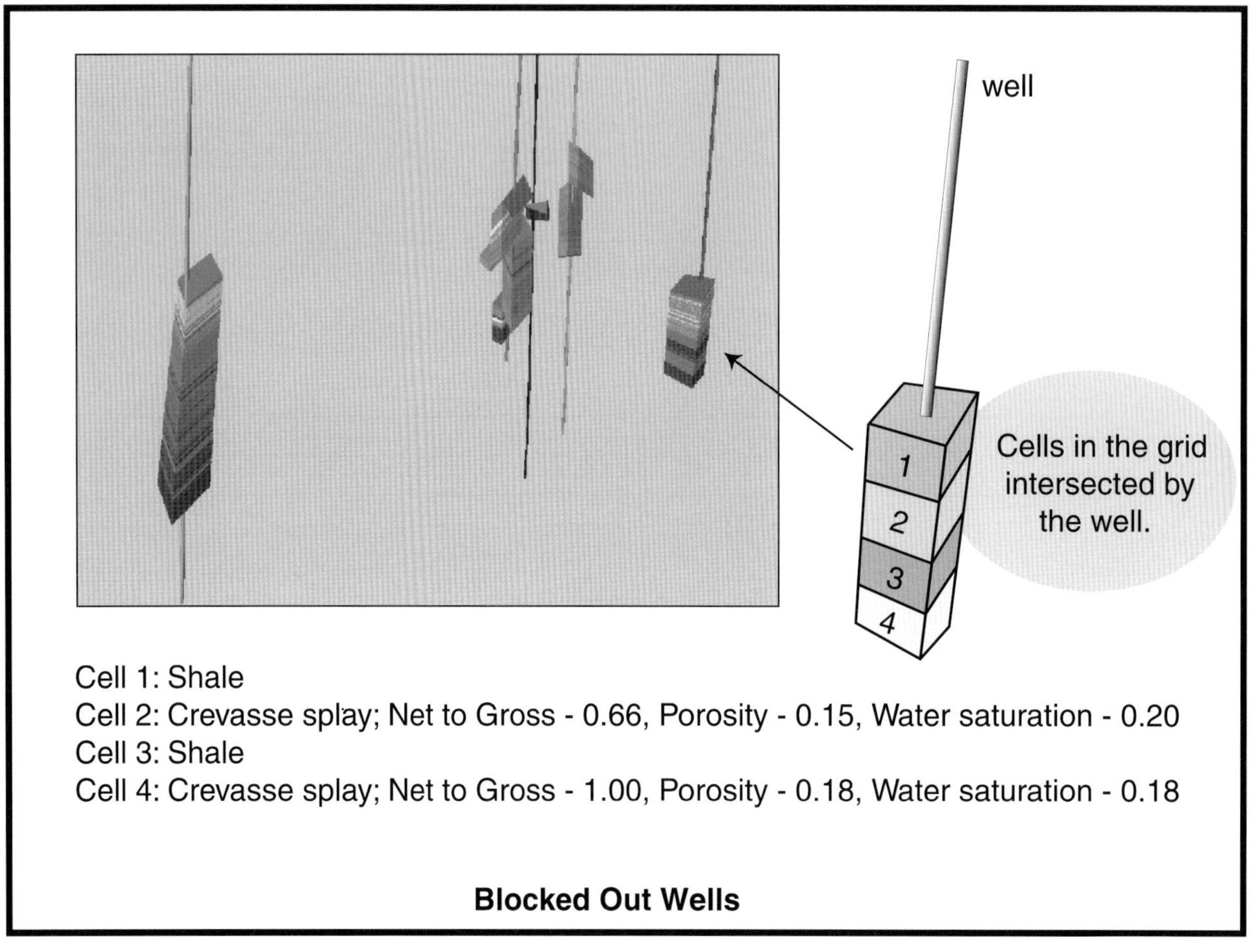

FIGURE 139. Blocked out wells in a 3-D model.

The scaling up of permeability well values should be looked at closely where a net-to-gross parameter is used. Permeabilities should be undefined in non-net intervals to ensure proper averaging. The procedure for averaging permeability values in blocking out a cell varies from company to company. A typical procedure is to use a geometric average for scaling up horizontal permeability. Some companies prefer to use arithmetic averages for blocking out horizontal permeabilities in wells. This is on the basis that it is the higher permeability rock that will dominate flow within a layer and arithmetic averaging gives a better chance of preserving the high-end permeabilities that influence the flow character.

QUALITY CONTROL

At this stage, the cells in the model intersected by the wells will have been populated with lithofacies and rock properties. A visualization of the model will show that the wells are represented as a stack of colored boxes with nothing to be seen in the large gaps in between.

FACIES AND ROCK PROPERTY MODELING

Having blocked out the lithofacies within the wells, the next stage is to populate all the cells in the model with lithofacies. For jigsaw-puzzle and labyrinth reservoirs, object-based modeling is a commonly used method of generating facies models. For layer-cake models, sequential indicator simulation or assigning a single deterministic lithofacies may be more appropriate.

Stochastic modeling methods will honor the global proportion of the facies seen within the wells. That is, if say the overall statistics for all the facies in the wells indicate that 26% of the reservoir interval comprises crevasse splays, then the stochastic modeling algorithm

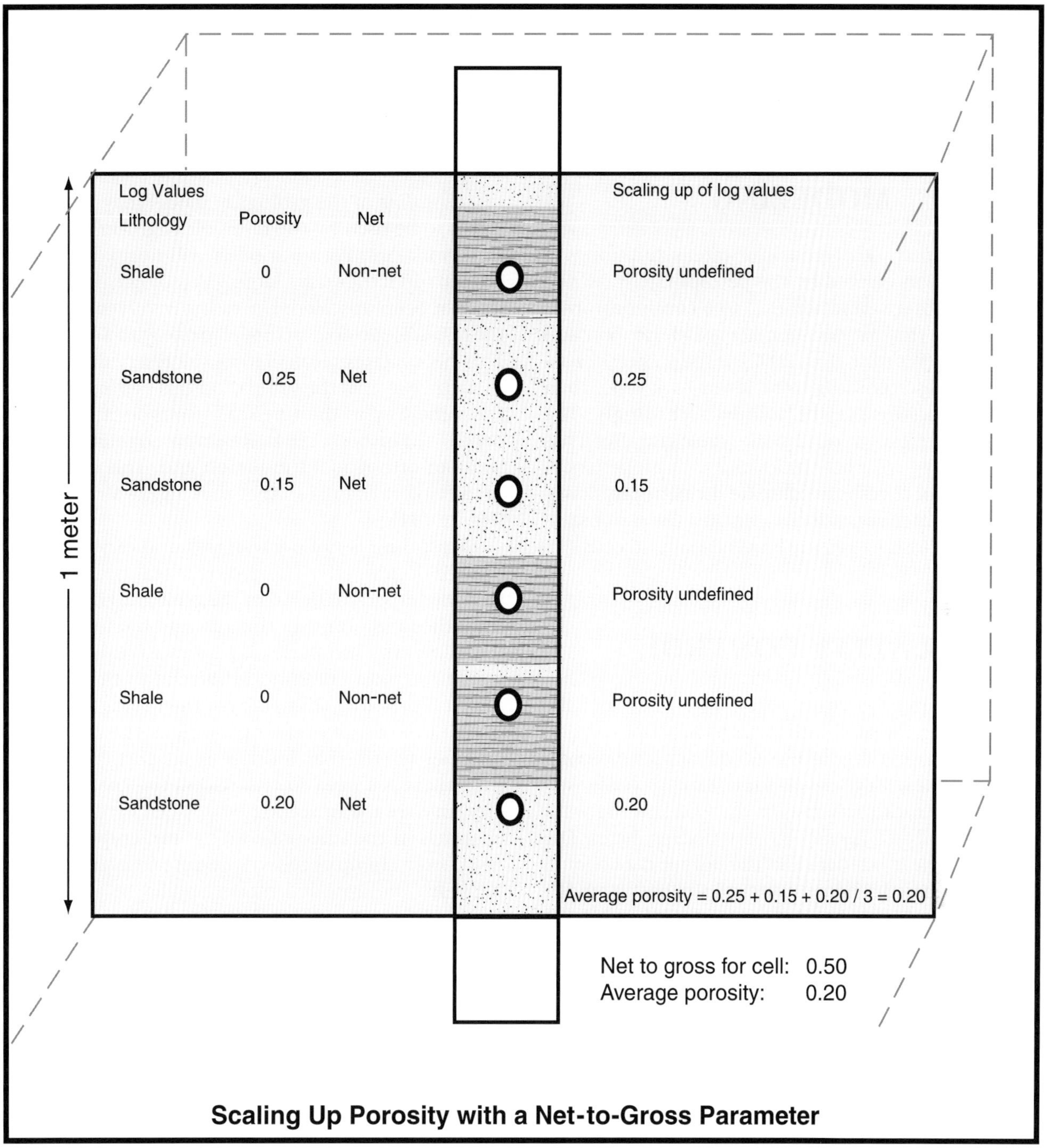

FIGURE 140. Where a net-to-gross parameter is used, porosity values in the non-net rock should not be used in calculating the average porosity.

will try and replicate this proportion. However, there are instances where there may be a bias in the global facies proportions. It may be that most of the wells have been drilled in the crest and there are only one or two wells on the flanks. A bias can also happen in models with horizontal wells. The intent in drilling horizontal wells is to target the sandy reservoir units, and their inclusion in the global facies breakdown can boost the percentage of sand-prone intervals at the expense of the poorer quality facies. It may be worth omitting horizontal wells for calculating global facies proportions because of this bias. However, they should still be blocked out and used for creating the facies and rock property grids.

Sequential Gaussian simulation is the preferred method for creating porosity, net to gross, and permeability grids. These grids should be conditioned to facies. It is a common practice to cokrige permeability values with porosity as a secondary variable.

WATER SATURATION MODELING

A typical method of modeling water saturation is to use an equation derived by the petrophysicist, which relates the water saturation (S_W) to a combination of height above the oil-water contact (gas-water contact) and porosity (Worthington, 2002). If the S_W-height relationship varies according to facies or rock type, then separate equations may be defined accordingly. The equation is typically applied to the whole grid after it has been populated with lithofacies and rock properties. A quality control check is to compare the petrophysicists' interpreted water saturation log against the water saturation values assigned to the blocked-out grid cells.

QUALITY CONTROL

Another quality control check is required once the grid has been populated with facies and rock properties. The facies proportions in the model should be compared with those in the wells for consistency. The statistics for the rock properties in the grid and the wells should be analyzed for any obvious discrepancies. The statistics of the input data and the output data should then be compared. Theoretically, all stochastic modeling algorithms should be able to reproduce the statistics of the input data (Isaaks and Srivastava, 1989). If a poor correlation between the input and output data is observed and there is no obvious reason for it, then there is a problem with the model, which merits further investigation (Al-Khalifa, 2004). If everything is satisfactory, then volumetrics can be run to determine the in-place hydrocarbon volumes (see Chapter 21, this publication).

MODELING REPORT

Time should be taken to write a detailed modeling report on the completion of the 3-D geological model. This is essential as the chances are that the 3-D model will have a working life of several years. Other geologists will use the model, and they will want to make modifications to it. It is very difficult to use another geologist's model unless everything that was done in that model is clearly and fully documented. Every step in the construction of the model needs to be itemized in detail. It can be useful to include screen dumps from the computer showing the input parameters for the model at each stage.

UPSCALING

Upscaling of the geological model may be required if it is to be used by the reservoir engineer for simulation work. Simulation models will normally take too long to run if they use a large 3-D geological model that has not been upscaled. Upscaling involves converting the fine-scale geological model to one with a coarser grid. This necessitates recalculating the reservoir properties in such a way that the simulation grid performs in much the same way as the fine-scale reservoir model would behave (Mansoori, 1994). Reservoir engineers will also want an orthogonal simulation grid. This can involve modifying the faults so that they zigzag along the grid (Figure 141).

Upscaling is not a trivial task and can take a long time to do properly. The geologist will need to ensure that the important detail in the geological model is not smoothed out or lost in the upscaled model.

When the lithofacies are upscaled, it is normal to assign the most common lithofacies to the single upscaled block. It is advisable for the geologist to check on the final result of the upscaling. If the model has a channel system that influences production behavior, then there is a need to ensure that the channel geometry survives the upscaling process. Linear features like channels in the fine-scale model can end up as a disconnected clump of cells in the upscaled version. If this happens, the upscaled model should be modified to preserve the connectivity. Features that cause barriers to vertical flow should be preserved in the upscaled model as implicit features, or at least as a surface that can have a transmissibility barrier applied to it in the simulation model.

Most rock properties are relatively simple to upscale and are averaged out. Water saturations are usually calculated for the cells in the upscaled model rather than averaged up from the geological model. The same equation used to define S_W for the fine-scale model can be used for the upscaled model.

Special care has to be taken in upscaling permeability. A simple method is to upscale horizontally using arithmetic averaging and to upscale vertically using geometric averaging (Laudeman, 1992). A more sophisticated technique uses an algorithm that serves to keep the flow properties the same for the coarse grid as for the fine grid. For a given pressure gradient, the average flow velocity for the fine grid cells is calculated, and this value is given to the coarse grid cell.

Here is an example of what happened in upscaling a 3-D geocellular model for a North Sea field. The model

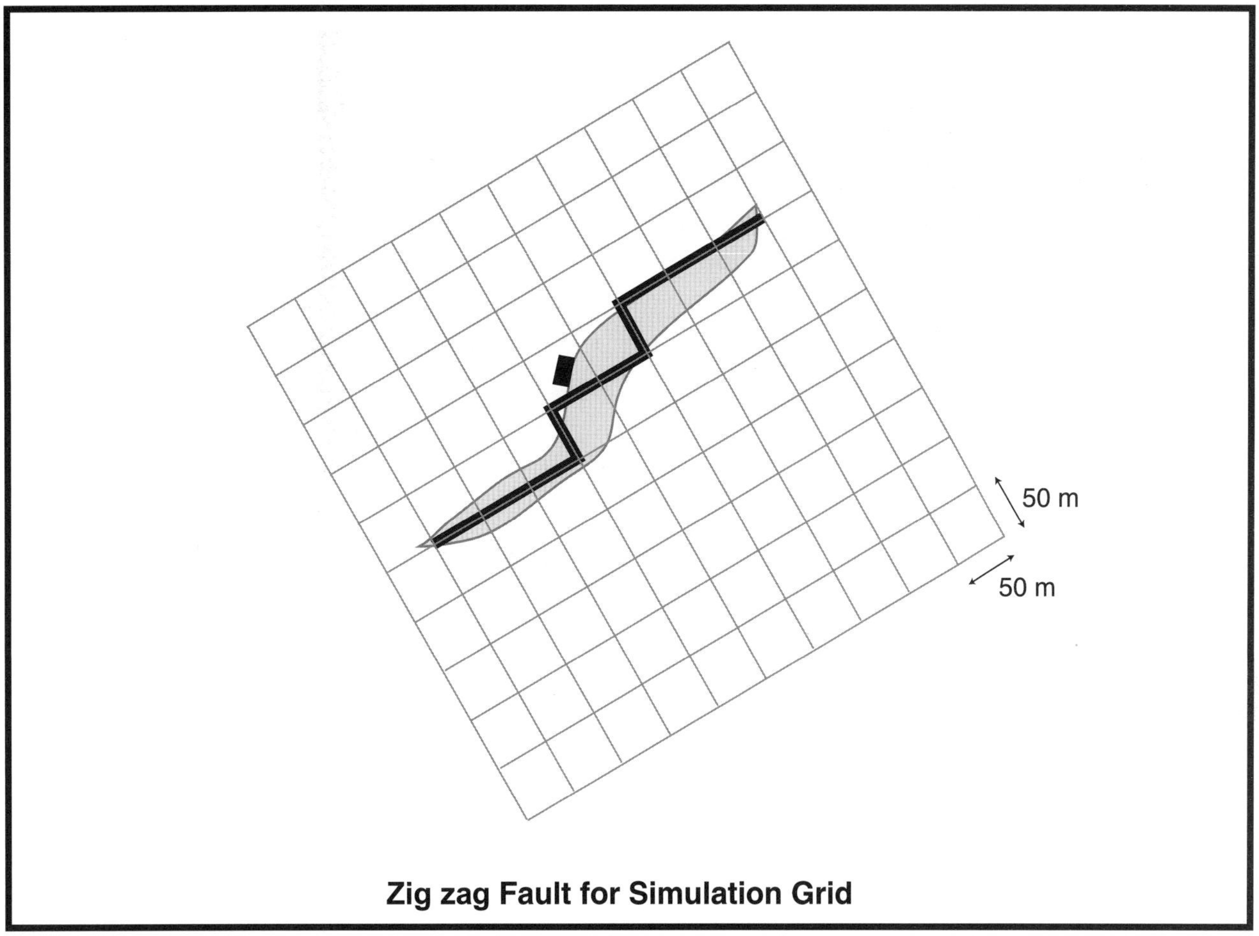

Figure 141. Modification of faults so that a regular orthogonal simulation grid is created.

has 15 horizons, 14 zones, and 100 layers that have been gridded with cells 100 × 100 m (328 × 328 ft) in the *x-y* direction. This is a rather coarse grid, and this was agreed with the reservoir engineer at the prescreening stage to avoid upscaling if possible. The field is a low-complexity shoreface reservoir with a tendency to a layer-cake geometry. Given the lack of any significant fine-scale heterogeneity that mattered in terms of flow, there was no special need to produce a finer scaled grid.

Nevertheless, the model has a total of 1,107,000 cells. Not all of these are active cells as far as the reservoir simulation is concerned, i.e., many contain shale only. In the reservoir simulation model, these have been *blanked off*, and this results in 127,494 active cells. To run a simulation on this particular model took 6 hr given the immense number of calculations that occurs for each cell in a sequence of incremental time steps.

The reservoir engineer then decided to reduce the run time for simulation by upscaling after all. This was reasonable because the first run had resulted in a fairly close match to the oil production, water cut, and pressure history of the field. The geological model was providing an approximate solution to the engineering field performance. The reservoir is not complex in nature. Therefore, it was felt that there would be no detrimental loss of essential detail that would cause the match to deteriorate. The result is a simplified model that could be run in 30 min. After some minor modifications to the vertical transmissibility locally within the model, a reasonable history match was obtained. The shorter run time is much more convenient for the reservoir engineer as they may be required to perform numerous sensitivities and profiles by rerunning the simulation model repeatedly with slightly different parameters.

THE PREDICTIVE VALUE OF GEOLOGICAL MODELS

Production geologists make predictions from their geological schemes and models. The geologist can predict how the reservoir is likely to behave based on its geometrical configuration and the distribution of rock properties. They can also predict where there is a chance of finding bypassed oil and where these volumes are big enough to justify well locations. The effectiveness of a geological model is therefore in its *predictive value.*

With this in mind, the new geological model should be screened to try and establish what its predictive value is. This involves visually inspecting the model and making inferences as too how such a model could behave as a reservoir. It may be possible to get an idea on the following reservoir characteristics from this:

- Location of flow pathways and connected volumes
- Flow continuity across macrofacies boundaries
- Strength of the aquifer (or other drive mechanisms)
- Degree of heterogeneity (whether creating numerous uneconomic oil volumes or minor heterogeneity with large drainage cells)
- Thief zone behavior
- Location of sand-sand fault juxtapositions
- Effect of diagenetic cements on connectivity
- Number of drainage cells

Shepherd, M., 2009, Volumetrics, *in* M. Shepherd, Oil field production geology: AAPG Memoir 91, p. 189–193.

21

Volumetrics

INTRODUCTION

The geologist is responsible for estimating the in-place hydrocarbon volumes. These can be calculated using 3-D models where they have been made for the larger fields. For smaller fields, volumes can also be calculated using 2-D mapping software or from simple graphical methods. The reservoir engineer will estimate field reserves using the geologists' in-place volumes as a framework for the calculation.

UNITS OF MEASUREMENT

The classical unit of measurement for oil is the *barrel*, abbreviated as *bbl* (Table 19). A barrel holds 42 U.S. gallons, 35 imperial gallons, or 0.159 m^3 (5.6 ft^3). Some companies express oil volumes in meters cubed. It is also common for oil to be defined in terms of *metric tonnes*, which is a weight measure rather than a volume.

Gas is often measured in cubic feet; the common abbreviations are *Bcf* (billion cubic feet) and *Tcf* (trillion or 10^9 cubic feet). Gas flow rates are typically stated in *MMCF* (million cubic feet) per day. Companies using the metric system will quote gas volumes in meters cubed.

OIL IN PLACE

The estimated volume of oil in place is normally defined as *STOIIP* or *stock tank oil initially in place*. The *stock tank* is a storage tank on the surface for oil. The significance of this for the definition of oil in place means that the oil volumes are measured at surface conditions; it is the volumes at the surface which are commercially important and sold. The convention for this is that stock tank conditions are defined at a temperature of 60°F (15°C) and one atmosphere pressure (Archer and Wall, 1986). The initially in place term refers to the volume of oil that was originally in place before production started from the field.

The formula for STOIIP is as follows:

$$\begin{aligned}\text{STOIIP} = \text{GRV} &\times \text{conversion factor}\\ &\times \text{net to gross} \times \text{porosity}\\ &\times \text{oil saturation} \times 1/B_o\end{aligned}$$

GRV is the gross rock volume of the hydrocarbon-bearing interval, typically expressed either in cubic meters or acre-feet. An *acre-foot* is a common unit of measurement for GRV in the United States and is a unit volume of 1 ac in area by 1 ft high.

A conversion factor converts the GRV to barrels (bbls). The GRV is multiplied by 6.29 where the GRV is measured in cubic meters or by 7758 where it is measured in acre-feet.

Oil saturation is equal to 1 minus the water saturation ($S_o = 1 - S_w$).

B_o refers to the *formation volume factor* and is also called the *shrinkage factor*. This parameter, defined at initial reservoir conditions, converts the volume of oil in the reservoir to the volume of oil at the surface under standard pressure and temperature. The surface volume is smaller mainly because of the oil shrinkage that occurs as gas separates out of the solution on the way up the well. The pressure and temperature will also decrease uphole, and the oil volumes will change to some extent as a result of this (Archer and Wall 1986).

GAS IN PLACE

The calculation of *gas in place* is similar to that for oil in place. It is normally defined as *GIIP* or gas initially in place.

$$\begin{aligned}\text{GIIP} = \text{GRV} &\times \text{correction factor} \times \text{net to gross}\\ &\times \text{porosity} \times \text{gas saturation} \times 1/B_g\end{aligned}$$

The correction factor corrects the GRV to cubic feet if this is appropriate.

DOI:10.1306/13161206M913372

Table 19. Common volumetric terms and abbreviations.

Term	*What it Means*
HIIP	Hydrocarbons initially in place
STOIIP	Stock tank oil initially in place (a stock tank is a surface storage vessel)
GIIP	Gas initially in place
bbl	Barrel
stb	Stock tank barrel
M	Thousand (10^3)
MM	Million (10^6)
Billion	1000 million (10^9)
BOPD	Barrel of oil per day
Bcf	Billion cubic feet (gas volume)
Tcf	Trillion cubic feet (gas volume)
GRV	Gross rock volume
NRV	Net rock volume
NPV	Net pore volume
HPV	Hydrocarbon pore volume
S_o	Oil saturation
S_g	Gas saturation
S_w	Water saturation
B_o	Formation volume factor for oil
B_g	Formation volume factor for gas
BOE	Barrel of oil equivalent

Gas saturation is equal to 1 minus the water saturation ($S_g = 1 - S_w$).

B_g is the formation volume factor for gas at initial conditions.

BARRELS OF OIL EQUIVALENT

Sometimes oil company reports will quote reserves as *barrels of oil equivalent (BOE)*. A barrel of oil equivalent is either a barrel of oil or a volume of gas, commonly 6000 ft^3 (the number varies slightly between oil companies), which is considered to have the same energy level as that released by burning a barrel of oil. Investors analyzing the assets of different oil companies, some with numerous oil fields and others producing large volumes of gas, can look at the reserves figures in BOEs and make a meaningful comparison between the companies.

THE CALCULATION OF RESERVOIR VOLUMES

There are three main methods used to calculate reservoir volumes. The simplest and most basic uses a graphical method, the *area-depth method* (Figure 142). Hand-drawn contour maps are made for the top and base of the reservoir, including contours for any fluid contacts. The area enclosed by each of the contours is measured by using a hand-held instrument known as a *planimeter*. This measures the area enclosed by each of the depth contours. These values are plotted against depth to produce an *area-depth plot*. The GRV for the hydrocarbon accumulation is calculated by measuring the area enclosed by the curves and lines representing top reservoir, base reservoir, and fluid contacts. An average net to gross, porosity, and hydrocarbon saturation can be used to calculate the in-place volumes; an example is given below:

$$\begin{aligned}\text{STOIIP} &= \text{GRV} \times \text{conversion factor} \\ &\quad \times \text{net to gross} \times \text{porosity} \\ &\quad \times \text{oil saturation} \times 1/B_o \\ &= 32 \times 10^6\ \text{m}^3 \times 6.29 \times 0.62 \\ &\quad \times 0.18 \times 0.72 \times 1/1.28 \\ &= 12.6\ \text{MM stb}\end{aligned}$$

GRV (from the area-depth method)	32×10^6 m^3
Conversion factor	6.29
Average net to gross	0.62
Average porosity	0.18
Average oil saturation	0.72
Formation volume factor	1.28

More rigorous estimates of hydrocarbon volumes are made using either 2-D mapping or 3-D geological modeling software. These have custom-designed applications for deriving volumetrics. Computer methods of estimating hydrocarbon volumes have largely supplanted the older graphical methods.

VALIDATING VOLUMETRICS

It is easy to make a mistake when calculating volumetrics, and the importance of the numbers is such that a mistake is not wanted. One method for validating volumetrics involves noting the intermediate steps in the calculation of the oil in place/gas in place value. The computer application used to calculate the hydrocarbon volumes should also be set up to output the gross rock volume (GRV), the net rock volume (NRV), the net pore volume (NPV), and the hydrocarbon pore volume (HPV) for the hydrocarbon leg.

These intermediate volumes can be related to the average rock properties of an oil zone as follows:

NRV = GRV × average net to gross
NPV = NRV × average porosity

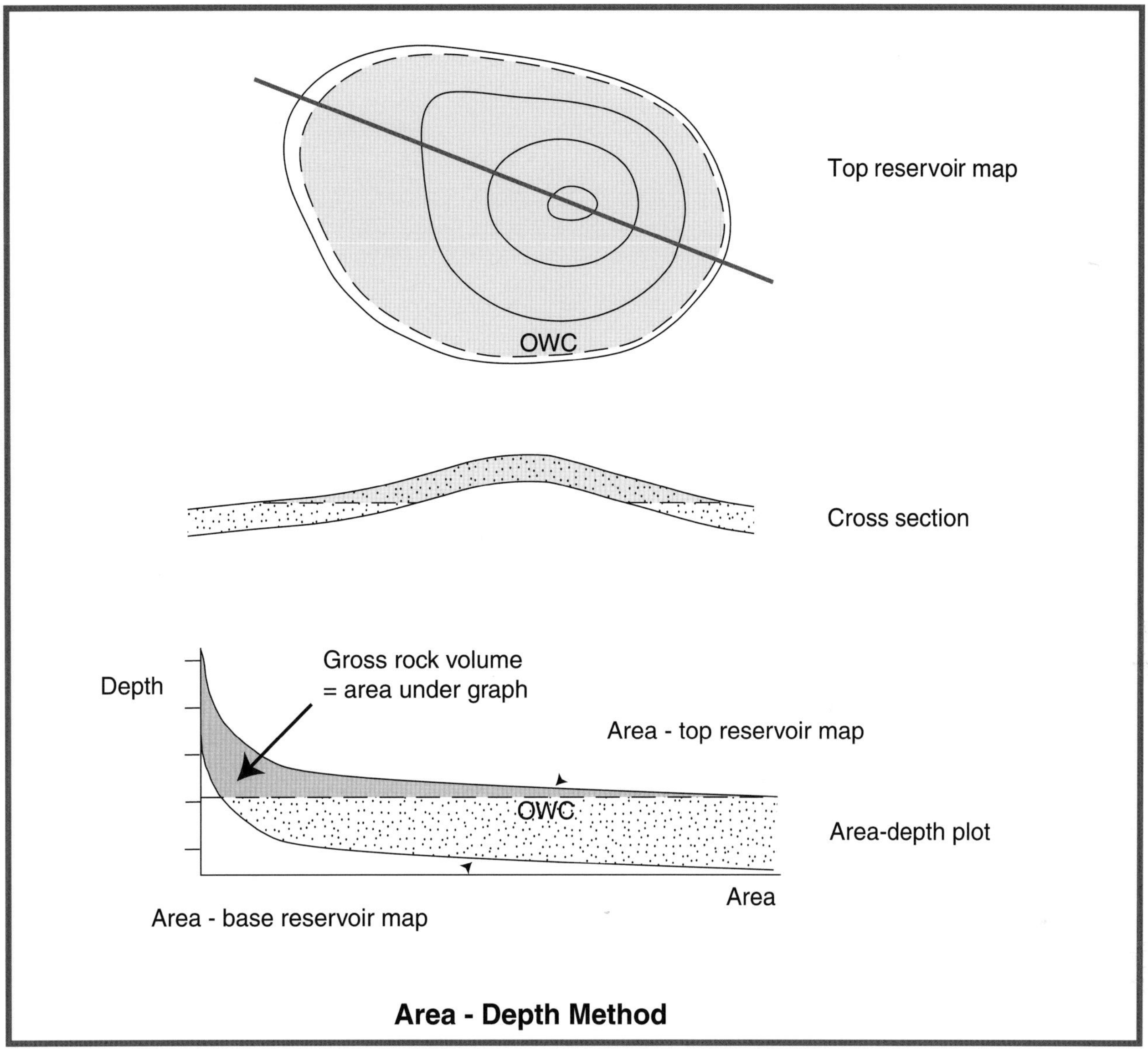

FIGURE 142. The area-depth method. OWC = oil-water contact.

HPV = NPV × average oil saturation
STOIIP = (HPV (oil) × $1/B_o$)

The average grid rock and fluid properties can then be back-calculated for the hydrocarbon leg:

Net to gross = NRV/GRV
Porosity = NPV/NRV
Oil saturation = HPV/NPV
Formation volume factor = HPV/STOIIP

Comparing these numbers to the well values can sometimes make it immediately obvious that a mistake has been made.

A simple calculation can be made to verify the order of magnitude of the volumetrics calculation. The reservoir volume in map view can be fitted with a simple geometric shape that allows the gross rock volume to be calculated very approximately. The simplest is a rectangular area. The area is multiplied by the average reservoir thickness to give a *slab volume.* Average rock properties are used to derive the volumetrics. If a large discrepancy between this simple calculation and the computer-derived version is found, then there is a need to go back and check for mistakes.

It is also useful to know the value for the reservoir *yield.* The yield is the average volume of hydrocarbons that a given volume of rock can hold. For instance, an excellent reservoir can have an average yield of about 1 MMbbl of oil per 1 MMm^3 of rock. A calculation multiplying the slab volume by the yield gives a quick check that the more detailed volumetric calculation is roughly correct.

The distribution of oil in a reservoir or a reservoir interval can be visualized by making a *hydrocarbon pore*

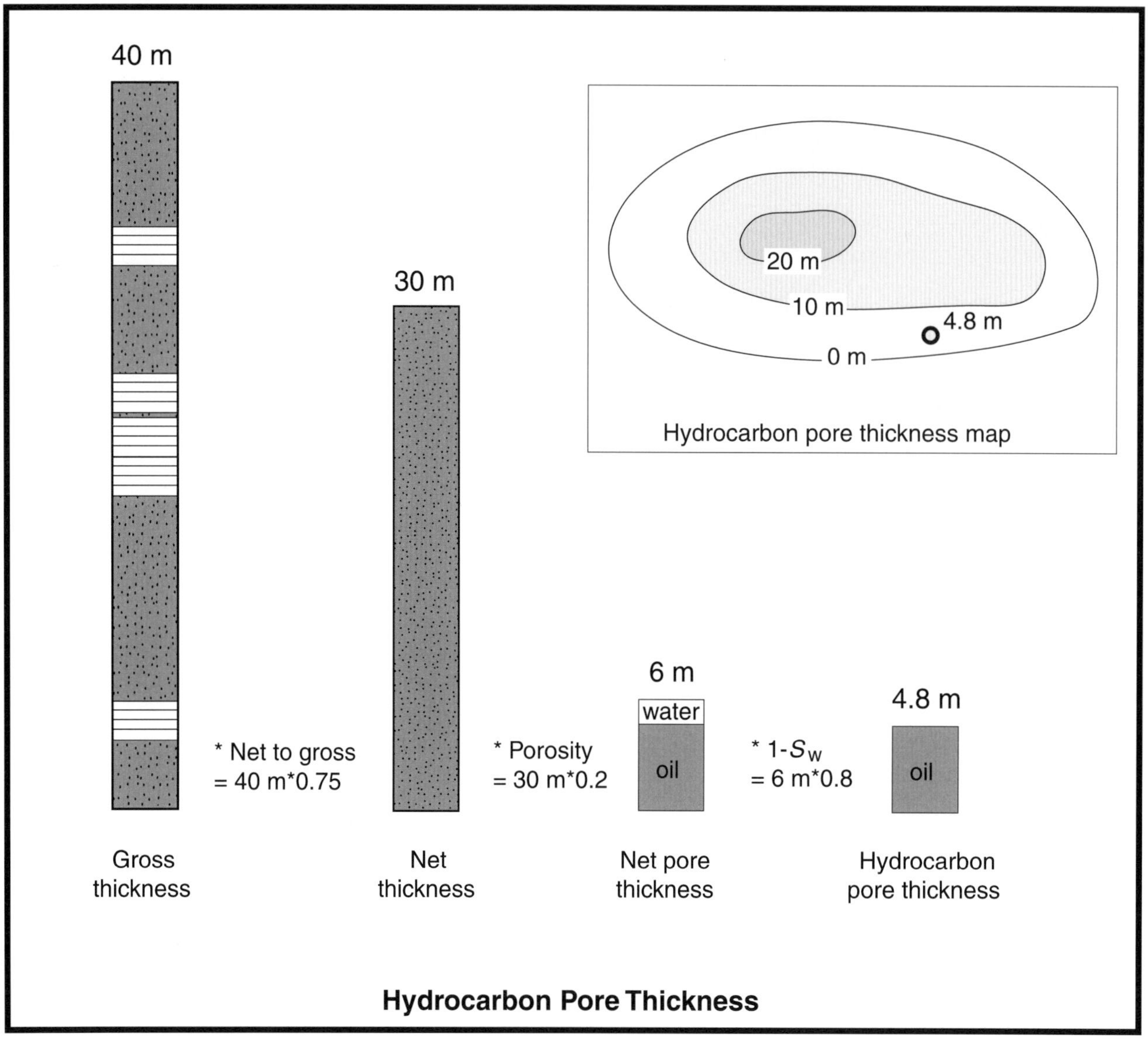

FIGURE 143. Hydrocarbon pore thickness is equal to the gross thickness × net to gross × porosity × the oil saturation. A hydrocarbon pore thickness map is a contour map showing the distribution of the hydrocarbons in a reservoir. S_w = water saturation.

thickness map (Figure 143). The *hydrocarbon pore thickness* for a particular reservoir location is equal to the gross isochore thickness in true vertical thickness × net to gross × porosity × the oil saturation. One way of thinking about what a hydrocarbon pore thickness map illustrates is that it is a map of the depth of a lake that could be made from the oil if the reservoir rock was not there.

RESERVES

Reserves are the volumes of oil and gas that a company estimates will be produced from a field from a given date to the end of field life. A 'given date' in this definition effectively refers to the date of the most recent reserves estimate, and this may be preproduction or at some phase during the production history of the field. Any hydrocarbons already produced from the field will not be categorized as reserves. The term *estimated ultimate recovery* is used for the combined volume of cumulative production to a given date plus the reserves estimate.

It is the responsibility of the reservoir engineers to estimate reserves as the methodology to do so is within their remit, e.g., from computer simulation. Reserves are important because they not only signify the present value of the company, but they also give an indication as to the likely performance of the company in the future. Oil companies will quote the *reserves replacement ratio,* which is a measure of how much reserves have been added to replace those that have been produced over a year. If the number is larger than 100%, the company is growing, if it is less than 100%, it is shrinking.

The estimate of reserves is prone to even more uncertainty than is involved in the estimate of the hydrocarbons in place. This is because in addition to the geological uncertainty, reserves estimation also has to account for reservoir engineering uncertainty and the economic uncertainty. Even the definition of reserves categories varies from company to company. Oil and gas companies trading on the New York Stock Exchange are required to submit their reserves estimates to the U.S. Government Securities and Exchange Commission (SEC), who have their own definitions (www.sec.gov). The definitions given by the Society of Petroleum Engineers (SPE) and the World Petroleum Council (WPC) are also commonly used (see the SPE Web site, www.spe.org).

22

Shepherd, M., 2009, Reservoir uncertainty, *in* M. Shepherd, Oil field production geology: AAPG Memoir 91, p. 195–200.

Reservoir Uncertainty

INTRODUCTION

It was mentioned in Chapter 9 of this publication that the sampling ratio of wellbore to total reservoir volume for an offshore field can be in the order of magnitude of one to three million. The corollary to this statement is that there is an enormous amount that is unknown and mostly inferred about reservoirs, e.g. 2,999,999/3,000,000ths of the volume.

Nevertheless, the nature of the production geologist's job is to make predictions about the geology in the gaps between the wells and to have an idea as to how reasonable that prediction is (North, 1996). Any prediction about the subsurface will be an estimate with uncertainty involved.

Decisions to spend money on subsurface work such as drilling wells will be influenced by this uncertainty. Tools are available to help the geologist assess the range of uncertainty in the reservoir volumes and also to guide decision making for subsurface activities.

MEASUREMENT UNCERTAINTY

Reservoirs are somewhat data-sparse volumes to make predictions about. If this fact of reservoir life alone makes the work of the production geologist rather challenging, consider then that the available 'hard' data will not be that precise either. Much of the data gained from oil and gas fields are liable to *measurement uncertainty*, and prone to a large variety of possible interpretations. This includes seismic and wireline log data.

GROSS ROCK VOLUME UNCERTAINTY

At the typical depths at which reservoirs are found, the frequency content of conventional 3-D seismic data can give a minimum vertical resolution for features, only if they are greater than about 20–40 m (66–132 ft) in height at moderate reservoir depths. Additionally, variation in the signal to noise ratio, migration accuracy, and depth conversion can introduce uncertainty in the location, geometry, and even the very existence of some of the features in the subsurface (Stewart and Holt, 2004).

There will be structural uncertainty resulting from the depth conversion calculation. This uncertainty will increase with distance from the well control. As the top reservoir surface defines the upper envelope for the reservoir, the gross rock volume (GRV) between the top of the reservoir and the fluid contacts will be sensitive to the depth conversion uncertainty. An indication that the depth conversion is not precise is shown by the mismatch between the seismic prediction of the top reservoir and the actual top reservoir depth seen in wells when they are drilled during the development of the field (Bahar et al., 2003).

It is also common for fluid contacts not to be known precisely for reservoir compartments. This can occur even in mature producing fields with numerous wells. The combination of structural uncertainty with fluid contact uncertainty is that the GRV estimate typically provides the largest uncertainty of the input parameters used for calculating volumetrics.

WIRELINE LOG DATA UNCERTAINTY

Wireline log data also allow a large latitude of possible interpretations. Log porosity values in uncored wells are calibrated to core porosity data. A large uncertainty can occur when assessing the best fit for the core porosity-log porosity correlation in reservoirs. When a best fit is found, it is then used to assign porosity values from logs to noncored intervals. Henriquez and Jourdan (1995) found a range of 5–10% difference in the calculation of hydrocarbon pore volume as a result of the different possible fits that could be used to

DOI:10.1306/13161207M913372

calibrate log porosity in a North Sea field. Determining net to gross can be a major problem particularly in thin-bedded reservoirs. Flowmeter measurements can sometimes show hydrocarbons flowing in intervals that are supposedly nonpay (Henriquez and Jourdan, 1995). The derivation of a water saturation (S_w) model provides a major concern for the petrophysicist particularly in complex reservoirs with thin beds. The S_w model is often grossly simplified for ease of calculation of volumetrics.

PREDICTIVE UNCERTAINTY

Predictive uncertainty is uncertainty about predicting the lateral configuration of structure, lithofacies, and rock properties. The ability to make a prediction about a reservoir decreases in confidence as the geological complexity increases.

The geologist is sometimes asked to make a prediction about reservoir connectivity, where this can be a very difficult problem to judge. For instance, a company has just drilled a very expensive well in deep water in the Gulf of Mexico. The well has discovered oil in a reservoir interval interpreted as channelized turbidites. Any subsequent appraisal and development wells will be very expensive in this new oil pool, but the wells can be justified if there is a large enough connected volume between the individual channels. The problem is that it is very difficult to establish an appropriate model for channelized turbidite sandstones with any degree of confidence, even if several wells have been drilled. Variables include the geometry, width, and stacking patterns of the channels. These in turn will affect the connectivity and the optimal number of production wells required to ensure the adequate drainage of the pool. The problem is complex and will stretch the ability of a geologist to make a realistic prediction. One approach is to make an extensive review of the characteristics of turbidite systems that allow predictions to be made with the minimum of information. However, the decision to develop the field, if it is sanctioned, may be the result of the management taking a chance on a successful field development in the absence of a reliable predictive reservoir model.

WHEN UNCERTAINTY MATTERS

What matters in the oil field environment is the bottom line of whether any particular project will be economic or not. Thus, the focus of the work done by the subsurface team should be targeted toward adding value to the operation, and this includes consideration of the uncertainty. Reservoir uncertainty needs to be translated into financial uncertainty for any oil field project that the geologist is involved in.

VOLUMETRIC UNCERTAINTY

In practice, uncertainty for the production geologist primarily concerns the consideration of the range of uncertainty of the volume of an oil or gas field. When a specific volume of petroleum is identified as a single deterministic value, it will fall into one of two categories; it will be either large enough to make money or too small to make any money. In practice, reservoir uncertainty is such that a given volume of hydrocarbons is more logically represented as being somewhere within a range of likely volumes. At some point within this range, there is a threshold volume that is economically attractive to an oil company. This volume will vary from oil company to oil company as different economic criteria (*metrics*) will be used to define the success case. The degree of risk will also vary from company to company. A cautious conservative company will be particularly keen to ensure that the low side case is economic. A less risk-adverse company may sanction a project if the most likely case (P_{50} value) is economic and the chances of an upside are considerable. Perhaps an oil company will decide to develop a risky project on the basis that if it subsequently proves profitable, then it unlocks several similar hydrocarbon accumulations within the company portfolio as projects for development.

HOW VOLUMETRIC UNCERTAINTY IS REPRESENTED

Volumetric uncertainty is represented by a range of values. Both the high side and low side cases should have a reasonable chance of happening, although the two values should also be far enough apart to capture the range in uncertainty involved. This is not always easy to do. There is a tendency to put too narrow a range on the reservoir uncertainty. Human beings think they are much more accurate in estimating the bounds of large quantities than is the case in reality. However, they have a better chance of capturing uncertainty using standard techniques designed for the purpose (Capen, 1976). Methods for doing this are provided below.

Deterministic Method

An old and fairly basic technique used to evaluate volumetric uncertainty is to calculate for each input property such as porosity, GRV, net to gross, and hydrocarbon

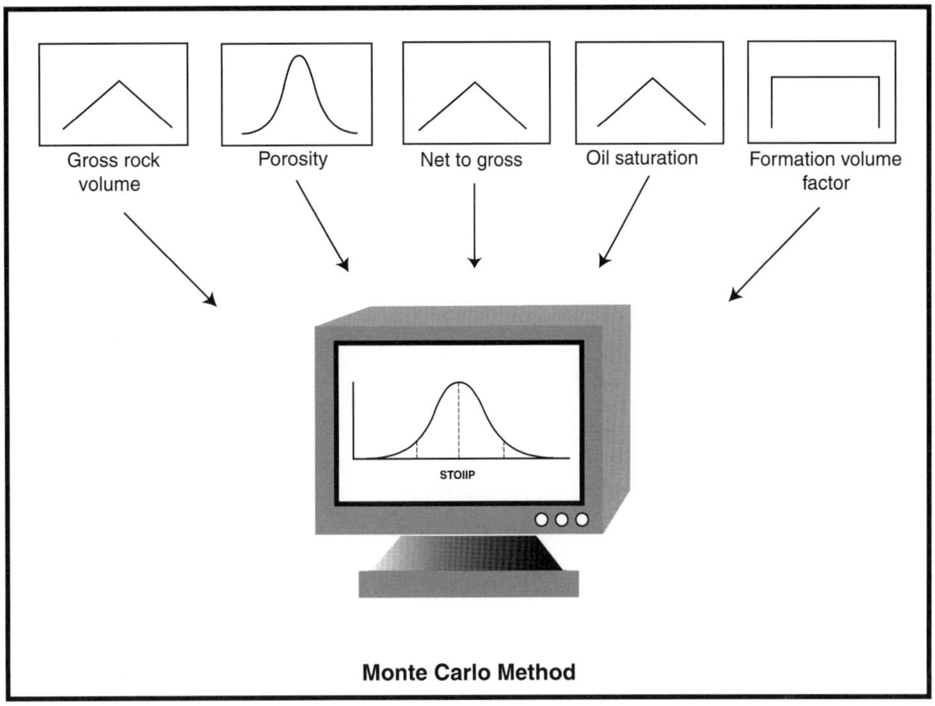

FIGURE 144. The Monte Carlo method can be used to calculate a distribution range for the oil in place. STOIIP = stock tank oil initially in place.

saturation, a *low, most likely*, and *high case*. For example, porosity might be given a range of values of 0.10–0.13–0.16. The properties are then combined for each case to derive the minimum, median, and maximum values for the parameter of interest, e.g., oil in place. Although it is simple to calculate these values, there is the problem that combining a minimum value with another minimum is likely to be unduly pessimistic (Dubrule and Damsleth, 2001).

Monte Carlo Method

The Monte Carlo method involves the use of a computer program or spreadsheet add-on to evaluate volumetric uncertainty (Figure 144). The user will define a specific range and distribution for each of the input properties used to calculate the hydrocarbons in place (porosity, GRV, net to gross, and hydrocarbon saturation). The computer then calculates a hydrocarbon volume by taking values at random from somewhere within the distribution range for each property used in the volumetric calculation. Many such calculations of the hydrocarbon volume are made (hundreds to thousands of them), and the idea is that after a large number of trials have been conducted, the range of values calculated will approximate to a distribution of all possible combinations. The resulting distribution curve is taken as representing the probability range of the hydrocarbon volume in place.

The results of all these hundreds of trials can be plotted as a *relative frequency distribution plot* or as an *expectation curve* showing the cumulative frequency as a function of the data values (Figure 145). These plots are used by geologists to show the range in possible values for hydrocarbons in place. The expectation curve can be used to define the P_{90} (low estimate), P_{50} (medium estimate), and P_{10} (high estimate) values.

Scenario Method

A tool that can be used for appraisal work is the *scenario method* (Taylor, 1996). At the appraisal stage, there may not be much information to decide on an appropriate subsurface model for the reservoir. It may be that several geological models are just as feasible as each other based on what little information is available. In this instance, the geologist could investigate not just one geological model but several scenarios. Each scenario is analyzed separately with the economic sensitivity estimated for each one. The less robust scenarios may give rise to concerns for the project feasibility, and there may be a case for obtaining extra data to constrain the geological model better and reduce the risk on development. This may include shooting a new seismic survey or drilling an appraisal well.

USING 3-D GEOLOGICAL MODELS TO ASSESS VOLUMETRIC UNCERTAINTY

The modern method of assessing volumetric uncertainty is to use the capability of 3-D modeling packages.

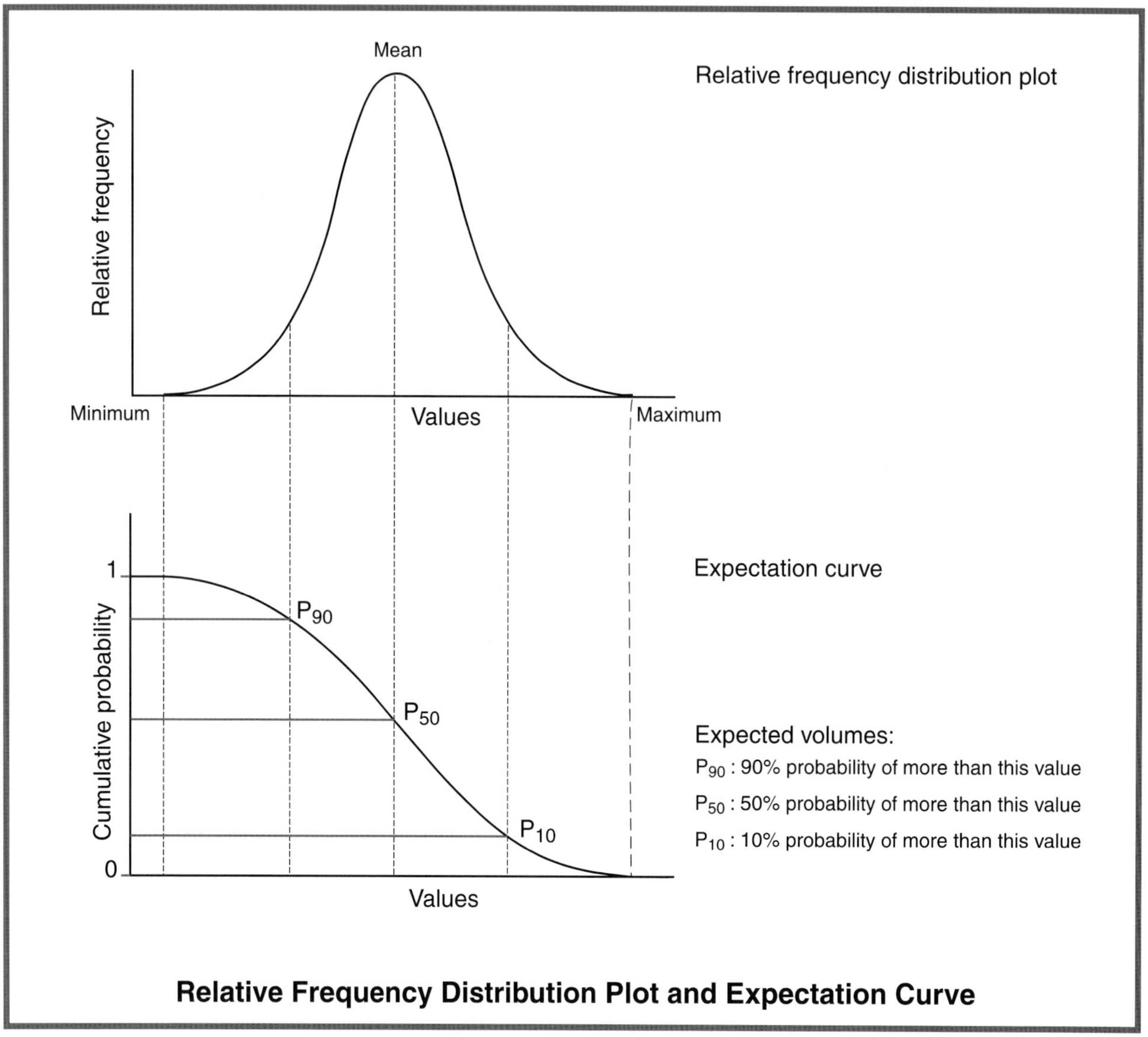

FIGURE 145. Relative frequency plot and expectation curve derived from a Monte Carlo simulation.

A base case model is defined and then the input parameters are varied in the model to analyze the effect on the volumetrics (Abrahamsen et al., 1992). Automated workflows can be set up within the computer to create multirealizations.

The uncertainty of the depth map surface at the top reservoir is a function of the measurement uncertainty primarily associated with the depth conversion as distance increases away from the well control (Thore et al., 2002). If this uncertainty is estimated, then the uncertainty range can be sampled stochastically as input to producing a large number of realizations of the depth surface. For example, this can be useful where the range in the volumes of oil to be found in a low lying fault compartment is sensitive to the depth at the top reservoir. Some methods also allow for the degree of confidence in picking a seismic reflector at the top reservoir level where the quality of the seismic pick can vary.

The uncertainty in fluid contacts can also be modeled. Some compartments within the field may show hydrocarbon-down-to or water-up-to depths with the location of the actual hydrocarbon-water contact somewhere in between (see Figure 10). A Monte Carlo simulation can be used to generate contacts for individual realizations. If a saturation-height function has been used to estimate water saturations, scripts can be written to regenerate water saturation grids for each fluid contact for each individual realization.

Multiple realizations of lithofacies and rock properties can be created. For example, this can be done by changing the seed number in a sequential Gaussian simulation of a rock property grid. For each realization,

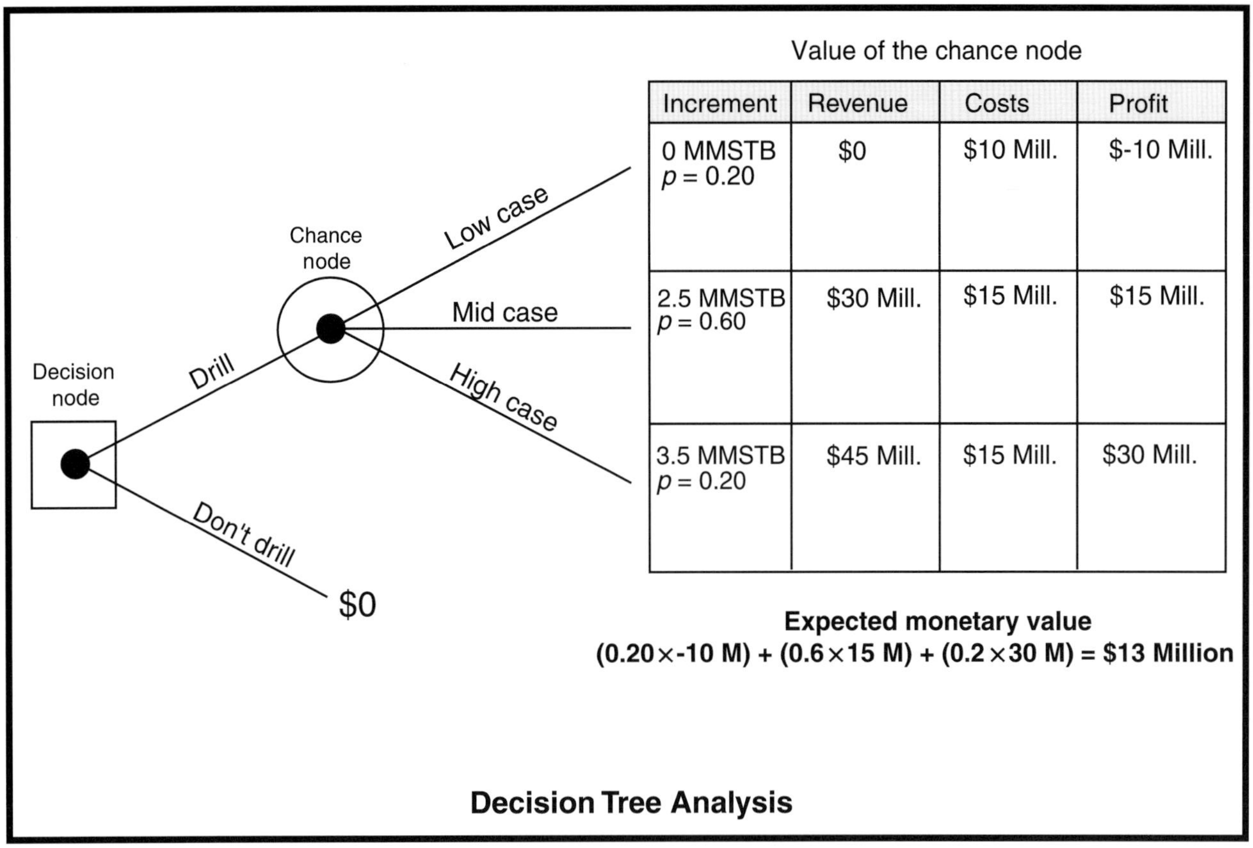

FIGURE 146. Decision tree analysis for an infill well. This is an effective tool for evaluating any oil field decision. In the illustrated example, the expected monetary value for drilling a new well is \$13 million. This return could justify a decision to drill the well. MMSTB = million stock tank values.

the GRV, porosity, net to gross, water saturation, and hydrocarbons in place can be calculated. A range in hydrocarbon volumes can be generated in this way.

UNCERTAINTY AND DECISION MAKING

Subsurface operations involve making some very expensive decisions. However, given the paucity of data on which to base an understanding of the subsurface, the nature of any action made will be based on a perceived balance of probability that the operation will be successful and make money.

DECISION TREES

A decision tree is a diagrammatic technique for evaluating the monetary return of several possible actions (Newendorp, 1996) (Figure 146). The various actions are split into branches, each eventually leading to a possible outcome. *Decision nodes* are rectangular and *chance nodes* are circular. The revenue and the probability of occurrence of each outcome are calculated. The value of a chance node is the sum of the net profit of each outcome multiplied by its probability. This is termed the *expected monetary value (EMV).*

CUMULATIVE PROBABILITY CURVES

If it is important for an infill well to encounter a given lithofacies in order to ensure high production rates, then the question to ask may be, "What is the probability that two wells drilled 500 m (1640 ft) apart will encounter the same sand body?" or "If this sand body has to be more than 10 m (33 ft) thick to provide an economic flow rate, what are the chances of this happening?" Analog data sets can be used to help address these questions, with the data plotted on a *cumulative probability curve* (Capen, 1992). The data are sorted and ranked sequentially from low values to high

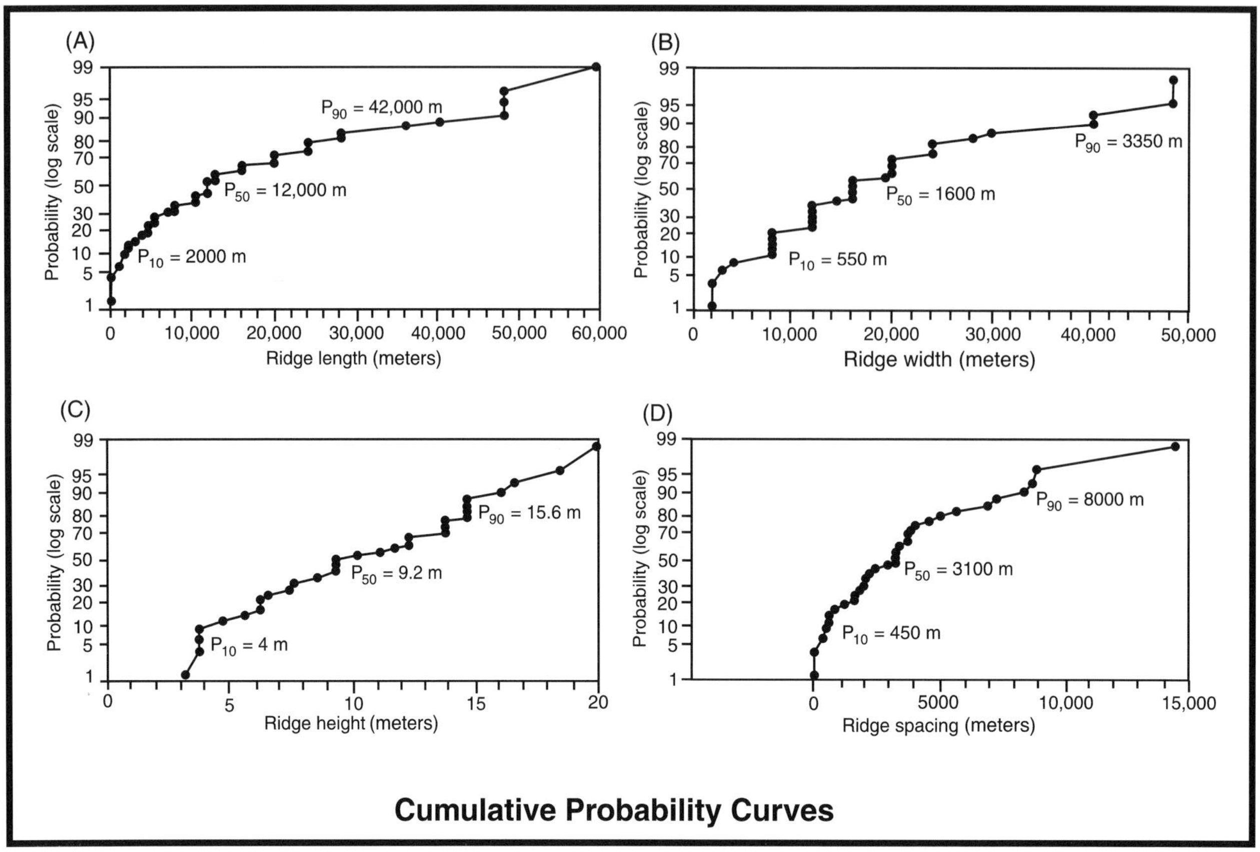

FIGURE 147. Cumulative probability curves showing the probability of occurrence of specific tidal ridge dimensions based on data from modern tidal ridge examples (modified from Wood, 2004). Modified with permission from AAPG.

values. The cumulative frequency for each point is then calculated according to the formula:

$$\text{Cumulative frequency} = (\text{ranking number}) / (1 + \text{total number of points})$$

The values are converted to percentages by multiplying by 100 and are presented on a logarithmic scale. The cumulative frequency is then crossplotted against the data values. Logarithmic graph paper is used on the assumption that many geological properties show a lognormal distribution (Capen, 1992).

Wood (2004) used cumulative probability curves to show the range in the geometrical dimensions of tidal sand bodies worldwide (Figure 147). Cumulative probability curves were also used by Shanley (2004) for fluvial sand body widths to estimate the well spacing required to optimally recover the remaining gas in the Jonah field, Wyoming. The present operation in the field has been to drill infill wells to a spacing of 40 ac (402 m; 1319 ft). However, cumulative probability curves indicated that only 14% of the fluvial sand bodies were likely to have widths equal to or greater than this. Given corroborative production evidence for poor connectivity between individual sand bodies, a smaller well spacing would be required to increase gas recovery.

23

Shepherd, M., 2009, Interaction with reservoir engineers, *in* M. Shepherd, Oil field production geology: AAPG Memoir 91, p. 201–207.

Interaction with Reservoir Engineers

INTRODUCTION

Production geologists work closely with reservoir engineers in managing reservoirs. One of the main tools to help do this is the reservoir engineer's simulation model. This is coupled with the geological model. Sometimes, it is found that the changes that a reservoir engineer needs to make to a simulation model provide feedback to the geologist to help understand the detailed flow geology behavior of the reservoir.

WHAT RESERVOIR ENGINEERS DO

Reservoir engineers are responsible for estimating and auditing the reserves for the fields they work on. Additionally, they respond to demands from the senior management to produce a considerable number and variety of reserves forecasts and production profiles for the coming year. These tie in to the monetary estimates that the management requires for financial planning and company budgets.

They will also take the lead in reservoir management, keeping a watchful eye on day to day production and pressures. If production problems arise, they will take action to remedy the situation. Reservoir engineers will use a variety of analytical tools to understand the reservoir performance. These include material balance, decline curve analysis, and reservoir simulation.

Material balance techniques are based on calculating the changes in pressure and relative volumes of oil, gas, and water within a reservoir as the hydrocarbons are produced. The technique is used for predicting the pressure response that is likely to result from the production of a given volume of hydrocarbons. In turn, material balance analysis can also be used to get a better handle on reservoir volumes once a series of pressure measurements has become available (Jahn et al., 1998).

Production profiles in long-lived oil wells commonly show a long lengthy decline phase later on in their producing history. Reservoir engineers can fit curves to these trends and from these, make an estimate of the ultimate recoverable volume for the well. *Decline curve analysis* thus gives a prediction of how much more oil a well is likely to produce.

RESERVOIR SIMULATION MODELS

The main analytical tool used by reservoir engineers is a *reservoir simulation.* This is a computer model containing a simplified version of the geological and rock property model. However, a reservoir simulation is anything but simple as it involves a large number of complex analytical procedures that are very computer intensive. The reservoir is represented by cells, which are assigned values for properties such as initial pressure, permeability, porosity, relative permeability, capillary pressure, and oil saturation (Figure 148).

Transmissibility multipliers may be defined to restrict or increase flow from one cell to another, both vertically and horizontally, thus representing baffles or barriers to fluid flow. The geologist should assist the reservoir engineer in defining the transmissibility barriers for the simulation model. These may be required for layer boundaries and for the interfaces between individual macroforms. In the latter case, annotated lithofacies maps showing features likely to effect transmissibility will be of help to the reservoir engineer at this stage (Figure 149).

DOI:10.1306/13161208M913372

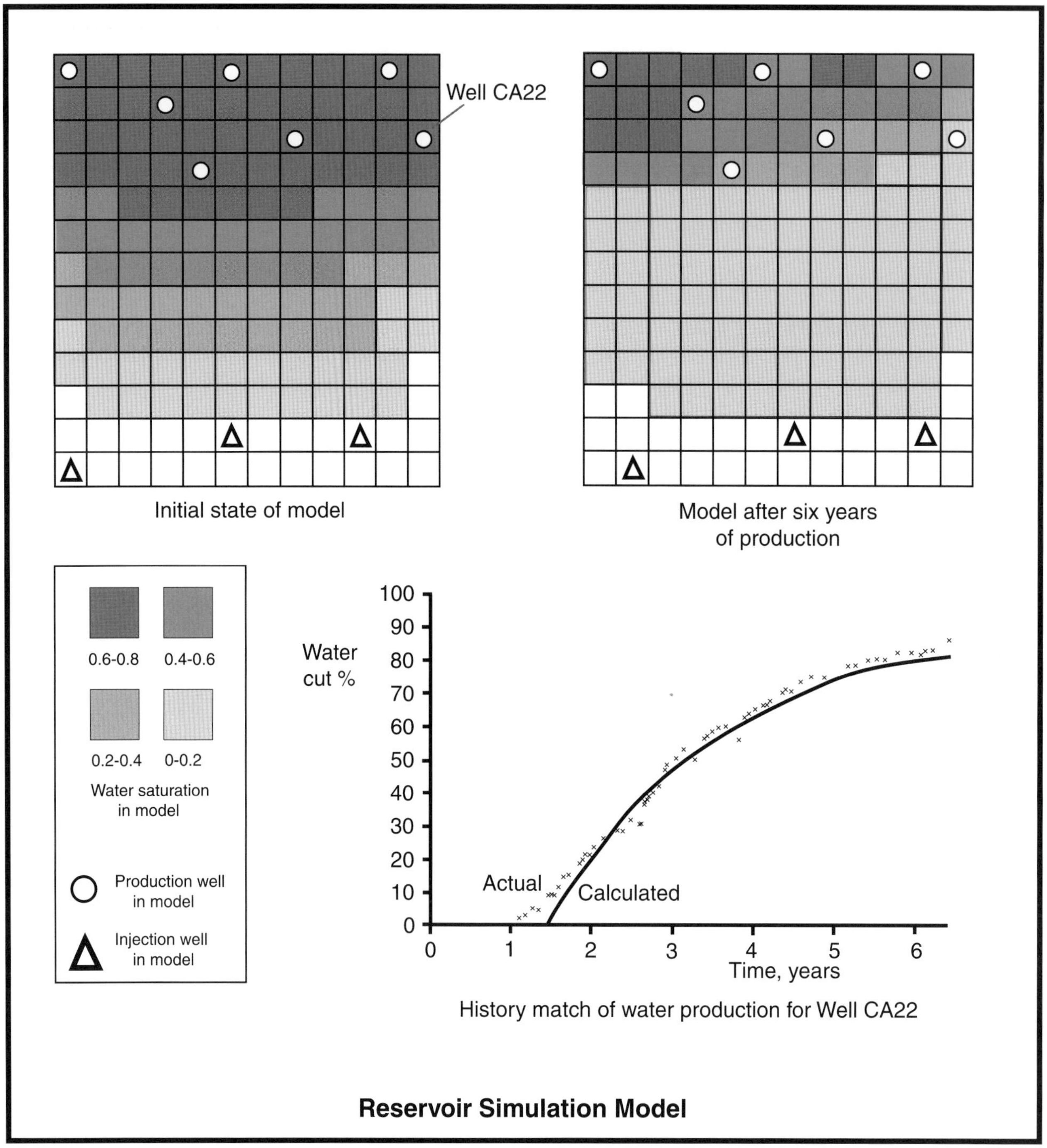

FIGURE 148. A reservoir simulation model is the main tool used by reservoir engineers. It is a computer model with a simplified version of the geological and rock property model represented as a 3-D grid. It is used for estimating reserves and long-term production profiles for a field.

Transmissibility multipliers can also be applied to faults in the model so as to represent variation in sealing properties along their length, in particular, accounting for the fault rock permeability and thickness (Walsh et al., 1998; Fisher, 2005). A common method of estimating fault transmissibility multipliers on this basis is that of Manzocchi et al. (1999, 2002).

INITIALIZATION OF THE SIMULATION MODEL

Once the reservoir engineer has constructed his model, there is a need to *initialize* it. The initial state of the simulation is defined in parameters, including

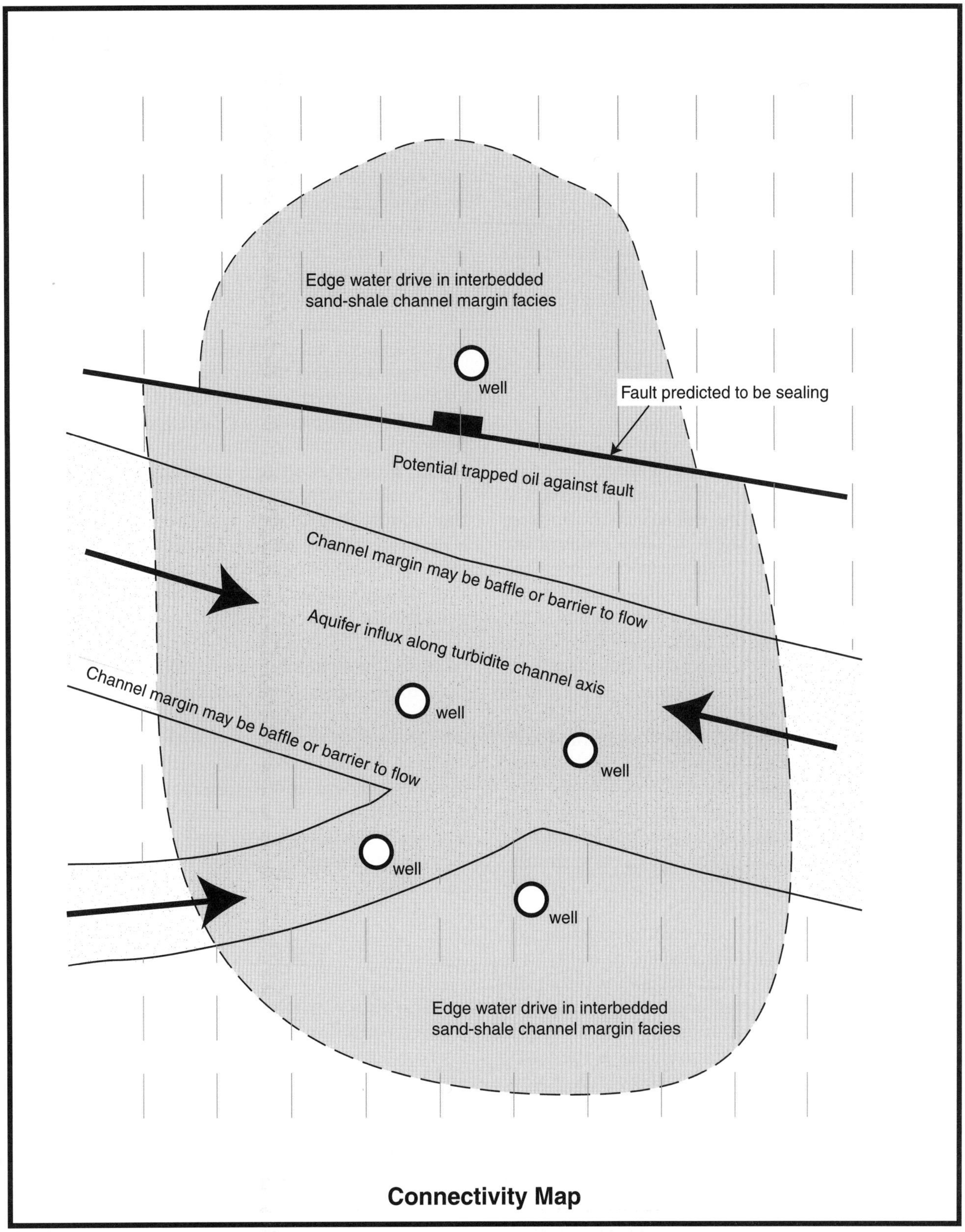

FIGURE 149. Connectivity maps provide the reservoir engineer with information on the flow geology of a reservoir interval. They can be used to help modify transmissibility within the simulation model.

pressure, fluid saturations, relative permeability, and a simulation-derived HIIP volume. The latter will be cross checked against the geological volumes, in case errors have crept in at the model construction stage. Reasons for a mismatch may be the result of a simulation grid that is too coarse or because of inaccurate rock property upscaling. Another cause for discrepancies may be because different methods may be used to characterize fluid saturations in the geological model and in the simulation model. The engineer should modify his model to better match the geological net pore volumes. This can be done by changing the distribution of the porosity within the model using *pore volume multipliers*. To further cut down on the model size, the reservoir engineer may deactivate certain cells that are outside the reservoir or that are non-net. The aquifer may be simplified by merging the grid cells into a smaller number of larger cells.

HISTORY MATCHING

The model is analyzed at discrete *time steps*. For each time step in the simulation of a waterflooded oil field, a certain volume of oil will be produced from the wells. The loss of fluid volume will reduce the pressure in the cells penetrated by the wells. As a result of the pressure drop, water will be drawn in from both the injector wells and the aquifer, moving incrementally from one cell to the next. At every time step, the simulation program will solve reservoir engineering equations related to material balance, fluid flow, changing fluid properties, and fluid phase changes. At the end of each step, the pressure and fluid saturations in each cell are recalculated. The model will ultimately be run to the model time equivalent to the present day. At this stage, the engineer will be looking to see how closely the pressures and fluid production rates at each well match the actual historical field data. This is called *history matching* (Cosentino, 2001). It is unlikely that the first simulation run will be close or sometimes anywhere near close to replicating the field performance. For instance, the model may be indicating that a specific production well should be producing 100% water when, in reality, it is producing no water at all. What this means is that the geological and/or the reservoir engineering input used for the simulation model is not totally valid. In fact, it has been said that the simulation model is the major tool for finding the effects of everybody's uncertainty (Haldorsen and Damsleth, 1993).

What the reservoir engineer will do at this stage is to modify the model to try and get a better history match. The technique is to preserve those features that seem to work in the model while at the same time altering individual elements in an attempt to improve the match. The geological model used for the reservoir simulation is unlikely to be perfect despite the geologist's best efforts. In fact, the way in which the reservoir engineer needs to modify the model to get a good history match can provide useful feedback to the geologist. For instance, if the water is not breaking through to the wells fast enough, then it may be that there is a larger volume of movable oil between the injectors and the producers than is present in the model. There can then ensue what has been called a volume hunt. The reservoir engineer should contact the production geologist to inquire whether the geological model can accept more pore volume in this part of the model without invalidating it. If so, then the reservoir engineer can add the volume using a pore volume multiplier. There may not be a unique solution to this problem. The same problem of water not breaking through fast enough to a production well may also be explained by modeled permeabilities that are too low or because the model has not captured the wide range of permeability values that are present in reality.

The reservoir engineer will modify any number of attributes within the model. These include the relative permeability characteristics, fault transmissibilities, permeability distribution, hydrocarbon volumes, and the aquifer properties. History matching typically involves a series of iterations between the reservoir engineer and the production geologist in an attempt to achieve an acceptable history match. This is one of the reasons why the model is simplified in order to cut down the run time for the various iterations that may be required to get a match.

It should be recognized that a simulation model can be history matched by any combination of modifications to the attributes used to build it. There will be no unique solution to achieving a match, in much the same way as 3 + 3 = 6 is as valid an arithmetic statement as 4 + 2 = 6. However, close cooperation between the production geologist and the reservoir engineer will help to ensure that the simulation model is reasonably constrained to the flow-geology model and to make the history match a realistic solution to the known reservoir performance.

PREDICTION

The history match phase is deemed complete once the well performance is reasonably well matched or the reservoir engineer decides that the match is good enough for practical purposes. At this stage, the model will be put into *prediction mode*. That is where the reservoir model is simulated in the period between the present day and the anticipated end of field life at 1–3-month time steps. The production estimates made from running a simulation model in prediction mode provides the basis for reserves and production forecasting.

The simulation model can also be used to make an assessment of sweep in the reservoir. Where there are large volumes of unswept oil indicated in the simulation at the end of field life, the engineer may put in a dummy well to see what volumes are produced by the well in the model. If the volumes are large enough to be economic, then this can be the basis for proposing a new infill well to the subsurface team members.

ITERATION BETWEEN THE GEOLOGIST AND THE RESERVOIR ENGINEER

The reservoir simulation model is of great importance in the modern oil company. It is the tool for estimating the field reserves and hence much of the value of the company. It is also the main instrument on which many economic evaluations are based.

The reservoir simulation has the geological model at the core, and it depends on this being valid. Nevertheless, the way oil companies presently perform the geological model to simulation model workflow can be less than ideal. Current practice in many oil companies is to construct a single geological model that is then used for the simulation model. This single best guess approach often disappoints (Anderton, 1995). There are many reasons for this (Smith et al., 2005):

1) The geological model may not address the problem it was intended to solve.
2) The conceptual geological model may be poorly thought out or far too complex ('if you cannot draw it then do not model it').
3) The model may be a totally bad representation of the geology and will have no predictive capability.
4) There may be no plan for handling reservoir uncertainty.

Awareness of this problem is growing, and remedies are being found to improve matters.

A powerful case can be made for making a flow geology analysis as part of any geology to simulation workflow. If the geological framework is based on an understanding on the controls on fluid flow, then this gives a better chance that the simulation model will be matched without too much difficulty.

Another approach is to consider producing more than one geological model for simulation. This is the scenario approach (Taylor, 1996). There are two common methods used here. The first is to construct a small number of geological models using different geological schemes. For example, at the appraisal stage, the environment of deposition may not be obvious from the limited well data available. Several different models could then be constructed to investigate how different scenarios for the reservoir depositional scheme affect the reservoir performance and project economics.

The second method is to produce multiple realizations of a 3-D geological model using stochastic methods. A hundred or more realizations may be made. The advantage of this is that there is a very quick method of testing the performance of each of these models by using *streamline simulation*. The streamline simulation method approximates to 3-D fluid flow calculations by summing 1-D analytical solutions along streamlines (Figure 150). The advantage of streamline simulation over conventional reservoir simulation methods is that it is very quick to run (Blunt et al., 1996). Additionally, the simulation can be run on geological models without having to upscale them. A drawback is that they work best with two-phase flow, oil and water for instance. They perform less well with three-phase flow; oil, water, and gas flowing together.

The best-fit geological model from a hundred or more realizations can be found relatively quickly with streamline simulation (Datta-Gupta, 2000; Wang and Kovscek, 2003). Nevertheless, the results from a streamline simulation modeling exercise should ideally be fed back to the production geologist as a means of constraining the static model for the field. The message here from the reservoir engineer should be along the lines that "It is important to look at this parameter a bit more closely and to get a better understanding of how it affects connectivity. If you do this, it looks as if we might get a better history match." This will result in a simulation model and a 3-D geological model that are compatible with each other.

This type of analysis can also be formalized as an extra step in the transition of a geological model into a reservoir simulation. Simple reservoir engineering models can be built with the idea of assessing specific aspects of connectivity within the reservoir. For instance, take the situation of a hypothetical field with a single production well. The field has many small faults and if they are sealing, the well will only be producing from one small compartment. If the faults are non sealing, the production well will potentially access oil from most of the reservoir. A simple reservoir engineering material balance model can assess whether the amount of oil produced by the well and the degree of pressure depletion seen is consistent with production from a small contactable volume or a large one.

A similar approach is used by BP with their proprietary top-down reservoir modeling technology (Durham, 2006). The key uncertainties affecting the performance of the reservoir are identified. Many simple geological models representing various scenarios are built and are simulated. An example of this approach has been published for the Teak field, offshore Trinidad (Kromah et al., 2005). In that example, BP wanted to establish the feasibility of drilling infill wells in the field and to assure that the

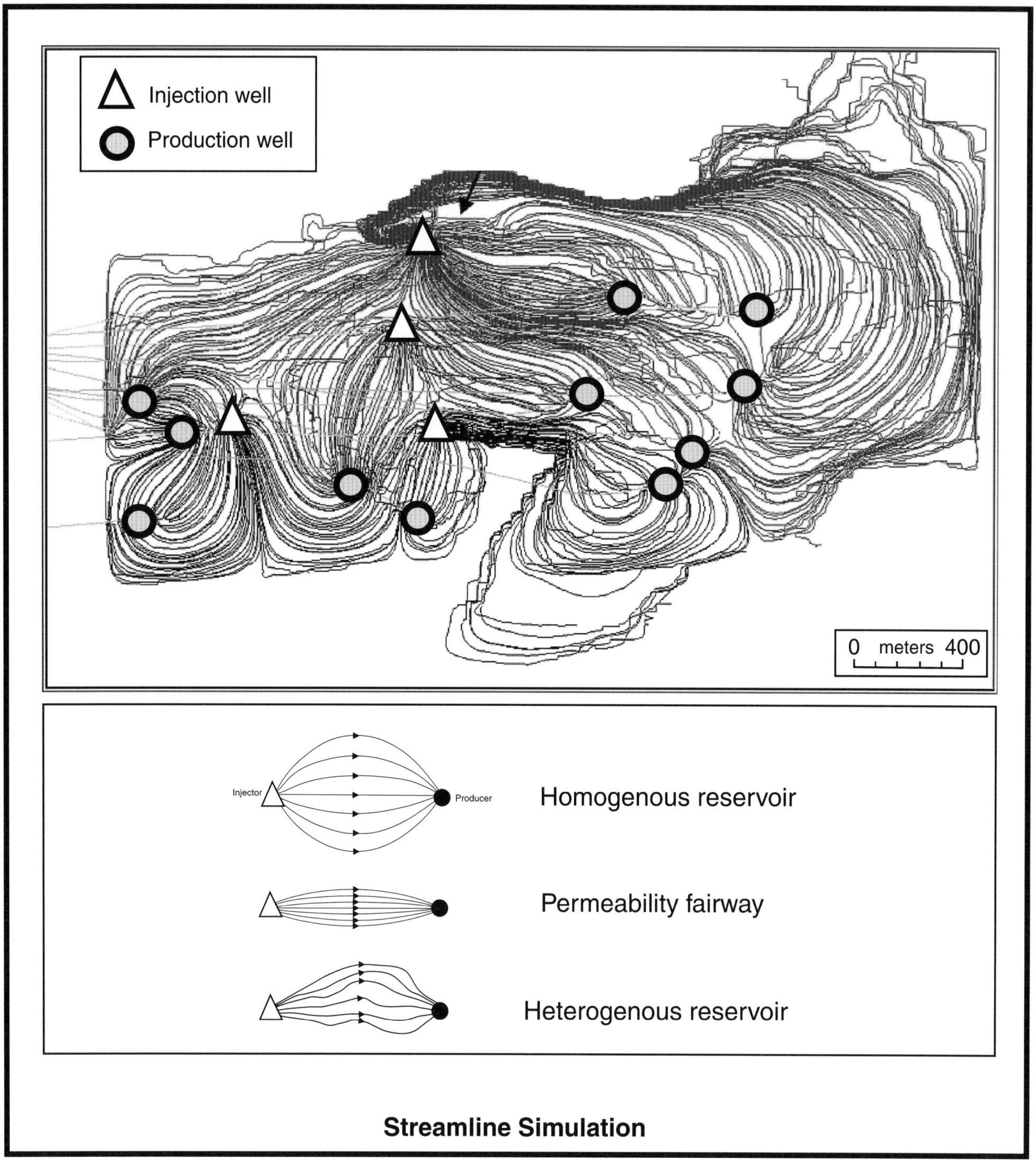

Figure 150. Streamline simulation is a method of assessing numerous stochastic realizations very quickly. The example shown is from the Alba field in the UK North Sea (from Fretwell et al., 2007). Reprinted with permission from the AAPG.

predicted incremental volumes would be sufficient to justify the operation. The evaluation involved taking a structural geological framework and making 500 realizations by varying the pore volume, fault transmissibility, transmissibility across an extensive shale, and the aquifer strength. Each realization was automatically history matched and this resulted in sixteen best matches to production performance. The main critical features influencing field performance were found to be the structural model and the fault transmissibility. This information was fed back to the rest of the subsurface team with the result that a new seismic interpretation was made and a detailed fault seal analysis conducted. The previous geological model had the field fully compartmentalized by faults. The updated model indicates that the reservoir is better connected, with the faults more open rather than sealing. The effort made to build a reliable model meant that it was possible to justify a new infill well location with reasonable confidence.

COMMENTS ON THE RESERVOIR SIMULATION METHODOLOGY

The reservoir simulation methodology works best in layer-cake reservoirs. If the geology is reasonably predictable, then the dynamic behavior of the reservoir is also likely to be predictable. However, with sedimentologically or structurally complex reservoirs, there can be a large departure from the 'geological reality' and the predictive capability of the model will be less reliable. A threshold degree of complexity in reservoirs probably exists such that modern techniques of reservoir analysis, both geological and engineering, become impractical (a test for this situation is that the geologist is unable to draw a representation of the reservoir geology [Dubrule, 2003]). In these high-complexity reservoirs, simpler data integration techniques may prove a better way of understanding the field for reservoir management purposes.

Reservoir engineers can be skeptical about the detailed predictive value of a full-field simulation model once a reservoir starts showing complex sweep patterns. The simulation will replicate the sweep patterns in the simple large-scale production fairways such as sheet sands or large channels. In those areas with more complex geology, the model may be too coarse to match the level of detail required to pick out the smaller unswept oil volumes (Wetzelaer et al., 1996). It may instead show a diffuse patchwork of remaining oil saturations, that area with slightly higher oil saturations, another area with slightly lower saturations, but nothing that corresponds to an obvious oil target. The geologist may be in a better position to locate the remaining oil opportunities in a mature oil field by using the techniques described in the next section of this publication. Alternatively, small-scale sector simulation models may have more scope for predicting sweep performance in these fields.

Shepherd, M., 2009, Locating the remaining hydrocarbons, *in* M. Shepherd, Oil field production geology: AAPG Memoir 91, p. 209.

Section 5

Locating the Remaining Hydrocarbons

INTRODUCTION

The most important task for the production geologist is to help get more hydrocarbons out of the reservoirs they work on. The geologist has the most intimate knowledge of the reservoir architecture and is the best placed to find out where the unproduced volumes of oil or gas are to be found.

A problem area for the modern production geologist is that modern methods of production geology are so biased toward computer analysis that it is easy to overlook that certain aspects of the job still involve old fashioned pencil, paper, and thinking power. This is the effort of data integration and the related activity of locating the remaining oil. These tasks are so essential to the success of the subsurface operation, it is important that the production geologist does not become occupied in only building geological models. Once a systematic search for the remaining oil is made, it can be surprising how much hitherto unsuspected stranded volumes can be found. It is considered that this section is the most important in this book. It will describe the various patterns in which oil (and gas) can be stranded in reservoirs. A workflow will then be followed through giving a methodology for locating the remaining oil using both qualitative and quantitative methods. The key method involves identifying and if possible validating the number and location of drainage cells in a reservoir. Maturity tables can then be compiled to determine which drainage cells have enough remaining oil volumes to warrant further investigation for infill wells or recompletion of existing wells. An opportunity inventory is then compiled, listing the potential locations of undrained oil pockets in a reservoir along with suggestions for the type of well operation that may be required to produce the oil.

DOI:10.1306/13161209M913372

24

Shepherd, M., 2009, Where hydrocarbons can be left behind, *in* M. Shepherd, Oil field production geology: AAPG Memoir 91, p. 211–215.

Where Hydrocarbons Can be Left Behind

INTRODUCTION

Geological features control how oil flows through a reservoir. Structure, sedimentology, and diagenesis all combine to create pathways, baffles, and barriers that enhance or retard the movement of oil toward the production wells. Moving oil has to negotiate a complex 3-D maze to get produced. Not all of the oil will be recovered; some of it will be trapped in dead ends or abandoned in slow moving volumes. Various patterns of remaining oil can be found depending on the geological elements that make up the reservoir architecture.

STRUCTURAL DEAD ENDS

The structure will have a major influence on the remaining oil patterns especially in reservoirs that have a moderate to high density of sealing faults. Some of the *isolated fault blocks* may have no production wells and will remain undrained as a result. The larger volumes are obvious drilling opportunities (Figure 151a).

Attic oil is where oil is trapped by a structural culmination generally above the highest producing interval in all of the wells in a field or a specific field compartment. Attic oil volumes can be common remaining oil targets in well-swept reservoirs (Figure 151b).

Cellar oil occurs where oil is trapped at a structural level lower than a nearby production well. It can occur for instance where deeper hydraulic units subcrop an unconformity updip or where a lower hydraulic unit onlaps onto a basement high (Figure 151c, d).

SEDIMENTOLOGICAL DEAD ENDS

Sedimentological and stratigraphic configurations can create dead ends. Experience with specific depositional environments shows that certain types of macroforms are better swept than others. An example of *facies controlled bypassed volumes* occurs in channelized turbidite systems, where it is often found that levee overbank deposits contain stranded oil (see Chapter 37, this publication).

Updip pinch-out traps are a feature where sandstones pinch out into shale, for instance barrier bar sandstones pinching out into lagoonal shales (see Figure 33). Bypassed oil can be common in reservoirs with jigsaw-puzzle geometries.

SHINGLES

Many depositional environments show a shingled geometry whereby inclined shales, for instance, separate individual shingled compartments (see Chapter 10, this publication). Experience has shown that bypassed oil may go unrecognized in these systems because a layer-cake geological geometry and good lateral reservoir connectivity had been assumed. However, compartmentalization by shingled barriers will result in poor to no lateral communication with the potential for uncontacted oil in the undrilled shingles. This problem has been recognized in oolite shoal bodies and fluvial deltaic sediments. Evidence for the presence of shingles may come from correlation, biostratigraphy, and unexplained poor recovery.

SLOW HYDRAULIC UNITS

Where a reservoir consists of numerous stacked hydraulic units, some of them will be thicker and more permeable than the others. These fast hydraulic units will deliver a significant part of the production to a well. The slow hydraulic units will show sluggish production by comparison. The oil can lag behind so much that there is

DOI:10.1306/13161210M913372

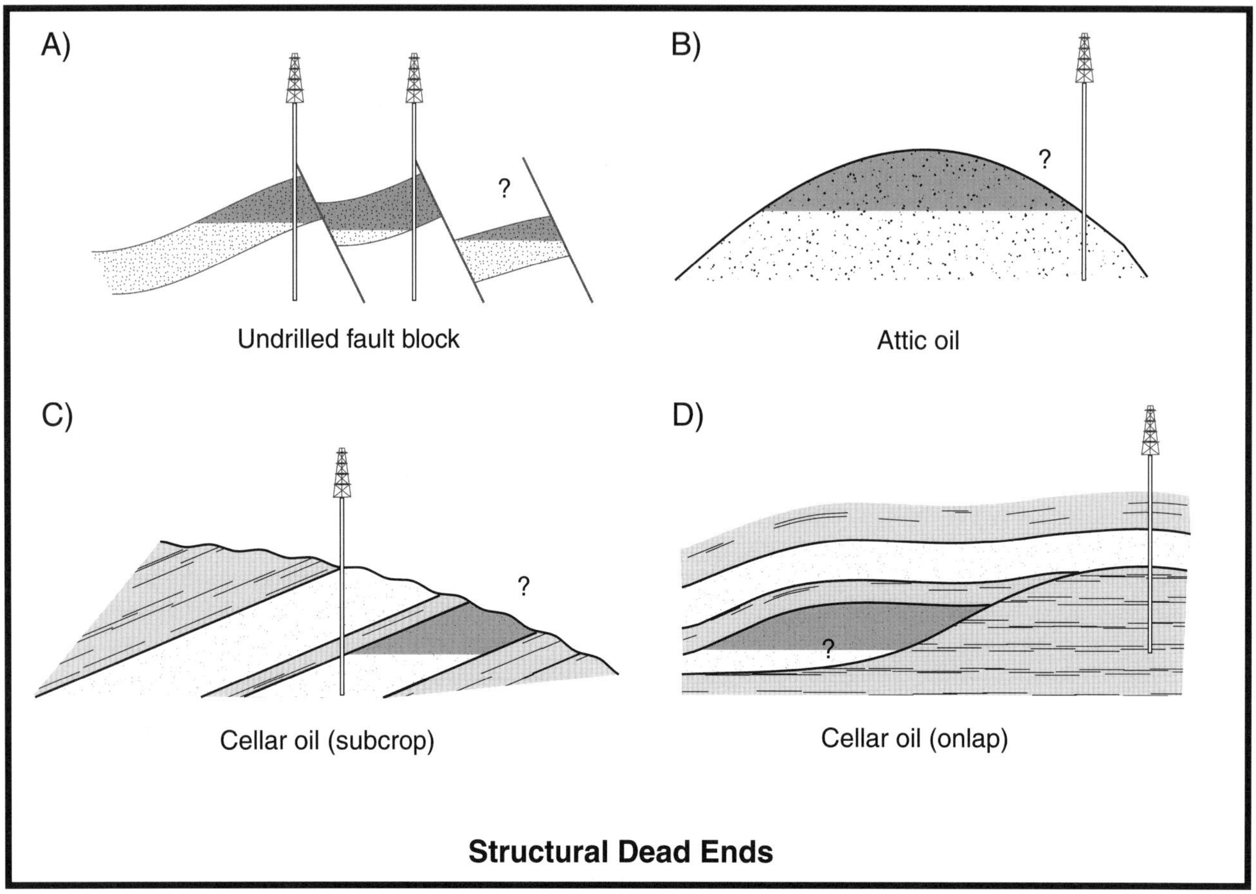

FIGURE 151. Structural dead ends can occur in several structural configurations, including undrilled isolated fault blocks, attic oil, and cellar oil.

a chance that it will not all be produced by the end of field life. If ways and means can be found to accelerate the flow contribution of slow hydraulic units, then it may be possible to recover this oil in a timely manner. A common technique is to drill horizontal wells into slow hydraulic units (see Chapter 28, this publication). Although the permeabilities are low, a sufficient length of well may allow reasonable production rates to be achieved.

BANKED OIL

U.S. geologists often describe what they call *banked oil* in onshore fields. This is where a waterflood has pushed oil up against a fault or a sand pinch-out edge, beyond the location of existing producers. Continuing sweep can subsequently strand these volumes and leave them isolated. Pranter et al. (2004) described banked oil against a sealing fault in the Vacuum field in New Mexico. These have large enough oil volumes to justify targeting with horizontal wells (Figure 152). Clark et al. (1997) proposed that water injection patterns in the Yowlumne field, California, have swept oil toward the pinch-out margins of turbidite channels. The banked oil is considered to represent a significant volume of the remaining reserves.

LOW-RESISTIVITY PAY

A shaly sand interval may not look as if it contains oil pay from wireline log analysis but appearances can be deceptive. The high water saturations may be the result of large amounts of irreducible bound water within the clays (Hurst and Nadeau, 1995). Such intervals, termed *low-resistivity pay*, can be more productive than expected (Figure 153). Formation resistivities can be lower than 3 ohm m, and yet the reservoir may still produce commercial volumes of oil (Worthington, 2000).

Numerous fields produce from low-resistivity pays, offshore in the Gulf of Mexico (Moore, 1993). An example of a low-resistivity pay field is the Little Creek field in Mississippi (Werren et al., 1990). In this fluvial reservoir, as much as 50% of the measured porosity is microporosity associated with grain-coating chlorite. The

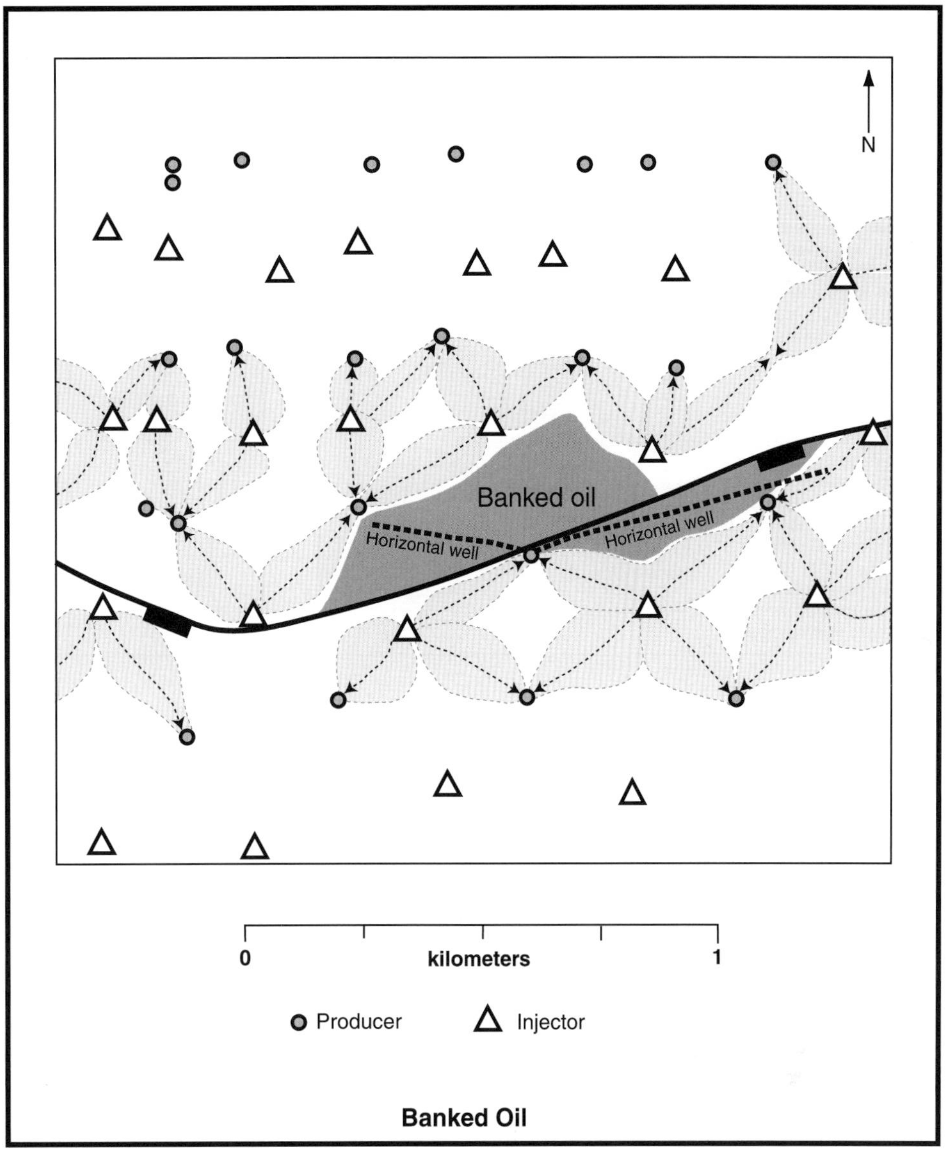

FIGURE 152. Banked oil in the Vacuum field, New Mexico. The map shows a streamline simulation of the fluid-flow regions between wells based on pressure data, production, and injection rates. Regions that are interpreted to be unaffected by the waterflood with potential banked oil are shown on both sides of the sealing fault. These regions were targets for horizontal wells (from Pranter et al., 2004). Reprinted with permission from the AAPG.

resistivity of the oil zone is less than 1 ohm m with the average water saturation in excess of 55%. Nevertheless, dry oil production occurs despite the very high water saturations; the bound water within the chlorite is tightly held by adhesive forces and is immobile.

Sneider (2003) listed several causes for low-resistivity pay. These include

- thin clean sandstones interbedded with shales, siltstones, or shaly sandstones
- sandstones with clay coated grains
- glauconitic sandstones
- sandstones with interstitial dispersed clay
- pyritic sandstones
- sandstones with clay-lined burrows
- sandstones with abundant clay clasts
- very fine-grained sandstones with highly saline connate water

High-resolution logging tools and appropriate techniques of petrophysical analysis can help identify low-resistivity pay where they are suspected (Boyd et al., 1995; Worthington, 2000).

UNPERFORATED INTERVALS

Sometimes the simplest method of finding unproduced hydrocarbons is to look for them in the existing production wells. It may be found that only part of the reservoir interval in a production well is perforated. There may be intervals of unswept oil or gas opposite the unperforated sections. Often, there is a good explanation for this. For instance, zones with friable sandstones may not be perforated as there could be a risk of sand production. Yet, it often happens that the historical reason for

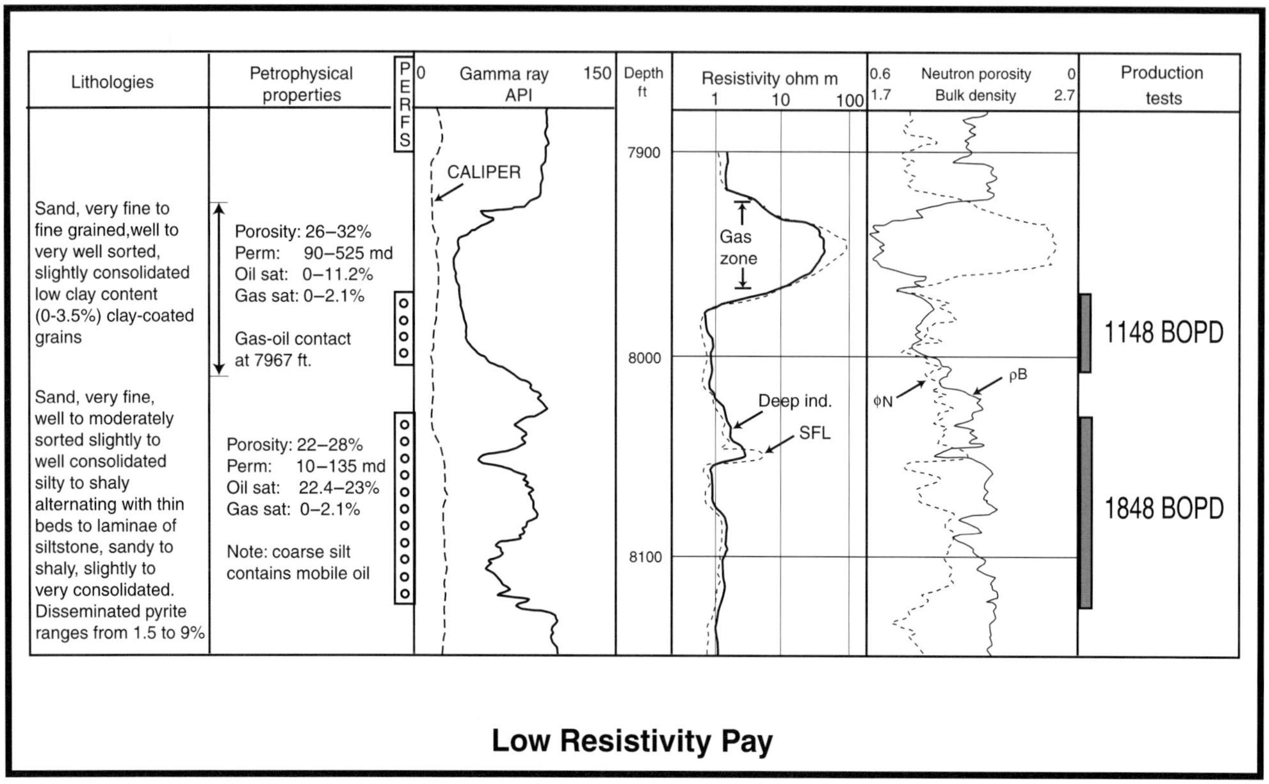

FIGURE 153. Low-resistivity pay in a discovery well, offshore west Africa (from Sneider, 2003). Reprinted with permission from the Houston Geological Society. Perfs = perforations; Deep ind = deep induction log; SFL = spherically focused laterolog.

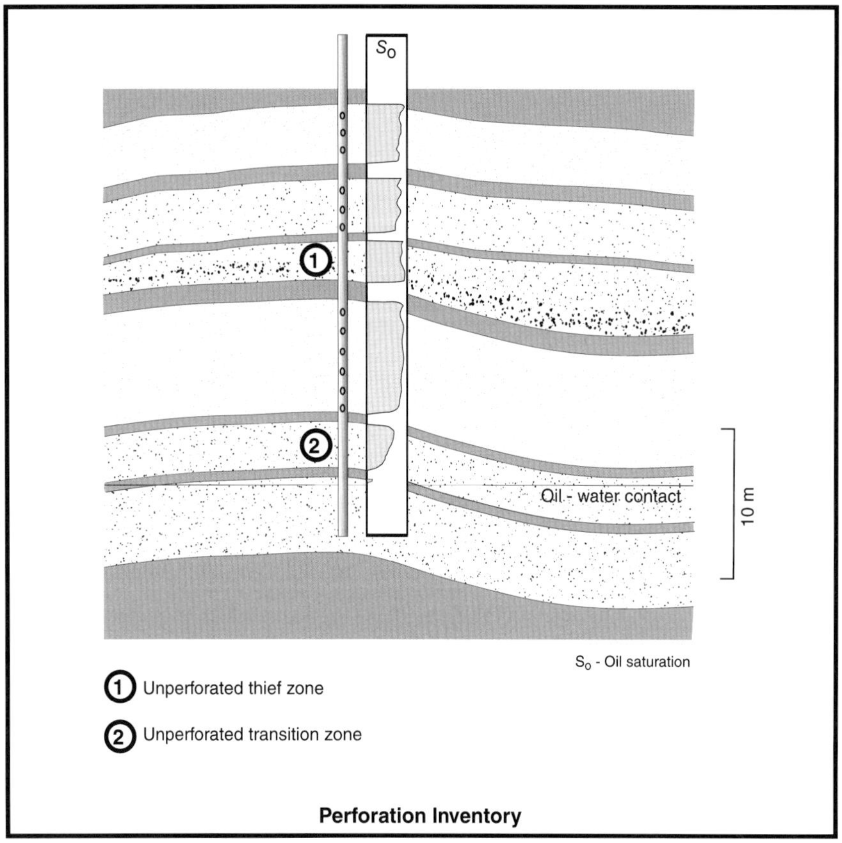

FIGURE 154. A perforated inventory for a production well shows the presence of oil-saturated zones that are unperforated. Adding perforations in these potentially unswept zones can improve recovery and increase production rates. Reprinted with permission from the Houston Geological Society. S_o = oil saturation.

not perforating an interval of net pay becomes forgotten with time.

A good example of stranded oil behind casing is an unperforated thief zone. This zone will not have been perforated originally as it would have induced early water breakthrough and killed the well. Once the well has reached a very high water cut and is close to being shut in, then this may be the right time to add perforations opposite the thief zone. Adding perforations at this stage will provide a welcome boost to production. These opportunities can come to light by compiling a *perforated interval inventory*, a list of unperforated intervals with potential net pay. This can be illustrated for each well by a diagram showing the perforated intervals for the well cross plotted with the interpreted petrophysical curves (Figure 154).

25

Shepherd, M., 2009, Qualitative methods for locating the remaining hydrocarbons, *in* M. Shepherd, Oil field production geology: AAPG Memoir 91, p. 217–220.

Qualitative Methods for Locating the Remaining Hydrocarbons

INTRODUCTION

The main qualitative method of locating the remaining hydrocarbons involves overlaying maps showing fluid flow patterns onto geological and volumetric maps. The main geological maps to use for this are fault maps and lithofacies maps at the level of individual hydraulic units.

BUBBLE PLOTS

The overlay of bubble plots on geological and volumetric maps is a typical data integration technique (see Chapter 17, this publication). Bubble plots of cumulative production can be plotted onto faulted structure maps (Figure 155a). This method may allow undepleted or underperforming fault blocks to be picked out that are worth targeting with an infill well.

Bubble plots of cumulative production can be overlain onto lithofacies maps (Figure 155b). These will give the geologist an idea of where the sweet spots are in the reservoir and where zones of bypassed oil may be located. The patterns seen on these plots can be compared to maps showing water influx fairways as inferred from water-cut maps and oil-water contact rise domains (see Chapter 17, this publication).

Bubble plots of cumulative production can also be plotted onto hydrocarbon pore thickness maps, both at the reservoir and hydraulic unit scale (Figure 155c). Hydrocarbon pore thickness maps indicate where the largest volumes of oil are to be found in a reservoir or reservoir interval (see Chapter 21, this publication). The overlays should be studied carefully. Wells with high cumulative production volumes should correspond to the thickest hydrocarbon columns. If there are any areas of thick hydrocarbon columns with no wells, then these may be worth investigating for potential infill well locations. A related method is to plot production bubbles onto *isocapacity maps*. These are made by mapping out the product of permeability and net thickness for a reservoir interval. These maps will give an indication of where the most productive parts of the reservoir are likely to be found.

Where 4-D seismic data are available, maps showing the relative change in amplitude over time can be cross-checked against production bubble plots (Figure 155d).

The features seen from data integration displays can be summarized as an areal sweep map. This is the production geologist's interpretation of the areal sweep patterns (Figure 156).

VERTICAL SWEEP PLOTS

Vertical sweep patterns are illustrated on *vertical sweep plots*. These are a series of cross sections across the field showing a representation of the vertical sweep. Various data can be shown on these plots, including perforations, isolated perforations, production logs, formation tester, and oil-water contact data. They can also be drawn sequentially at different time intervals to show the progression of sweep over time. For example, Hamilton et al. (1998) constructed a vertical sweep plot for the Jackson field in Australia. This indicated two potential zones where unswept oil may be targeted (Figure 157).

SURVEILLANCE ATLASES

Areal sweep maps and vertical sweep cross sections can be compiled and bound within a *surveillance atlas*. This is a document that shows a graphical representation of the sweep performance of the field and the location of the fluid contacts at the current time. Areal sweep maps are made for each individual hydraulic unit. The vertical sweep patterns are illustrated on a regular grid of structural cross sections.

The atlas is updated on a yearly basis, incorporating the latest production data. This is a good method of

DOI:10.1306/13161211M913372

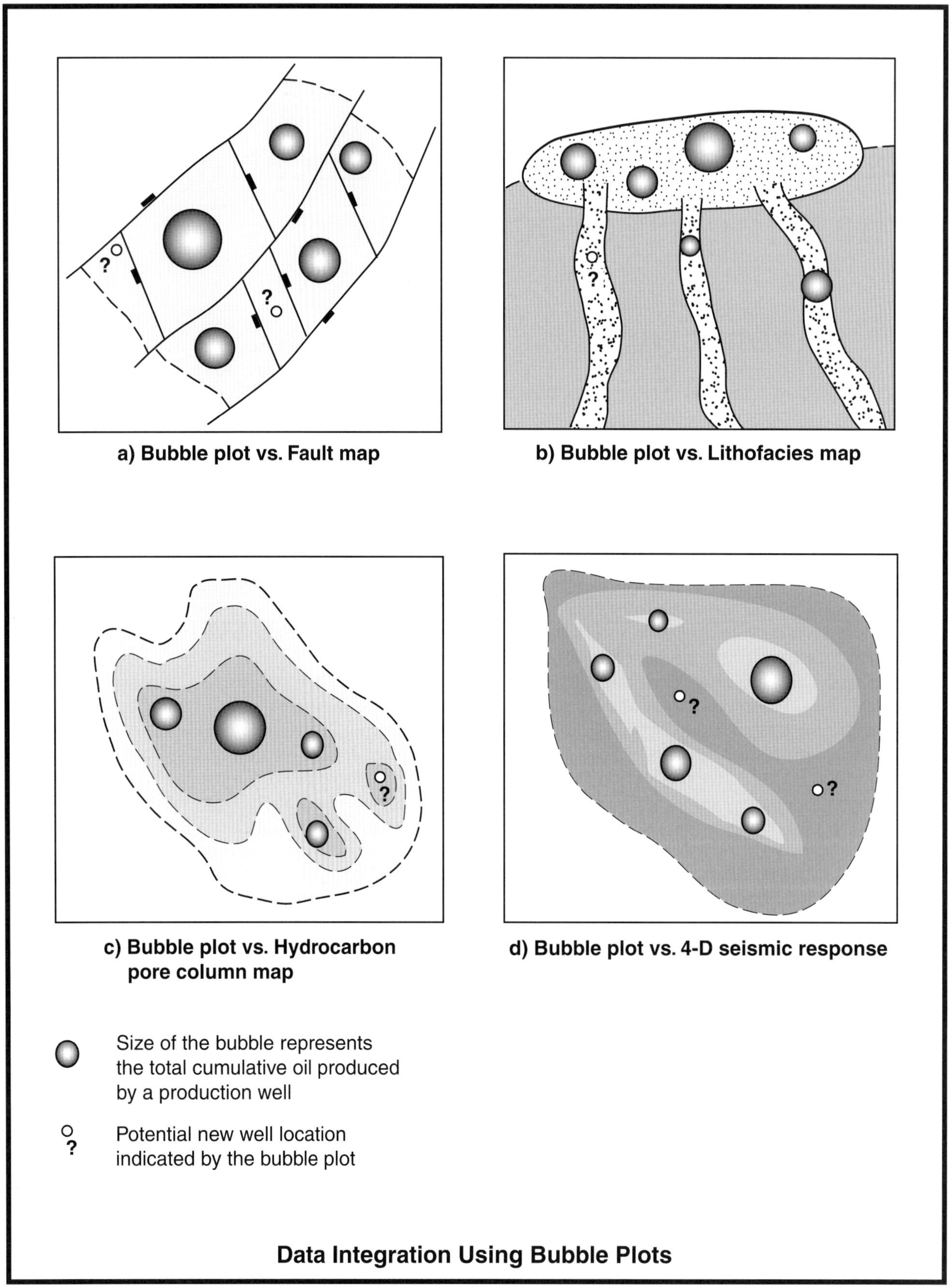

FIGURE 155. Bubble plots showing the total cumulative hydrocarbon production for wells overlain on a variety of maps. These pick out the main production fairways and also give an idea of where there may also be stranded hydrocarbons.

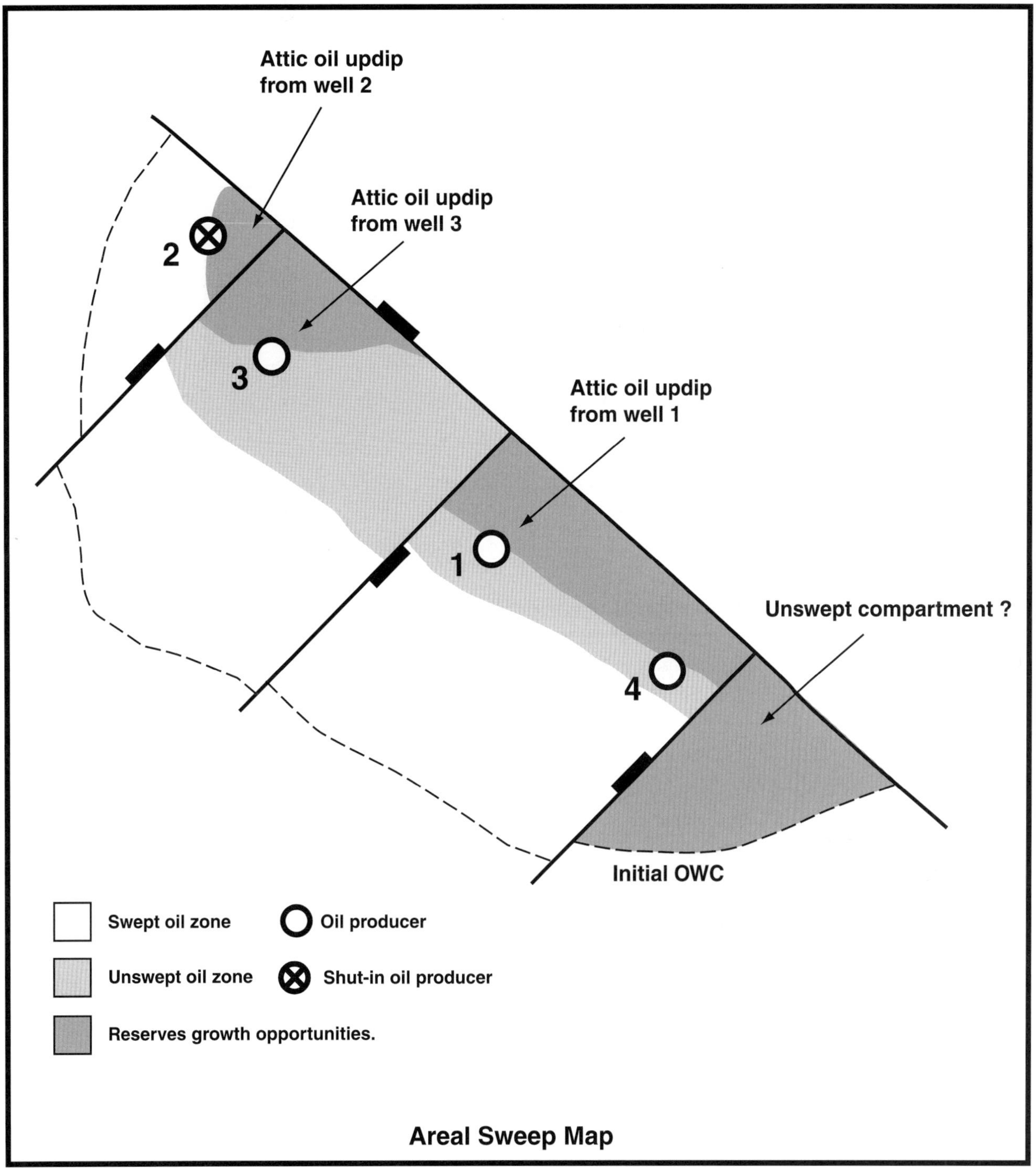

FIGURE 156. Areal sweep map. This summarizes the areal distribution of the remaining oil for a reservoir interval. Data integration methods are used to construct these maps. OWC = oil-water contact.

picking up on potential zones of bypassed oil. The technique has also been used as a method of evaluating old fields for bypassed oil potential in order to assess rehabilitation.

A surveillance atlas can be made for large fields with more than a hundred wells drilled. With this amount of production data, reservoir simulation methods can become unwieldy and too crude to use for detailed reservoir management purposes. The simulation will still be required for economics and forecasting, but it will usually be supplanted by the reservoir atlas technique to help with reservoir management.

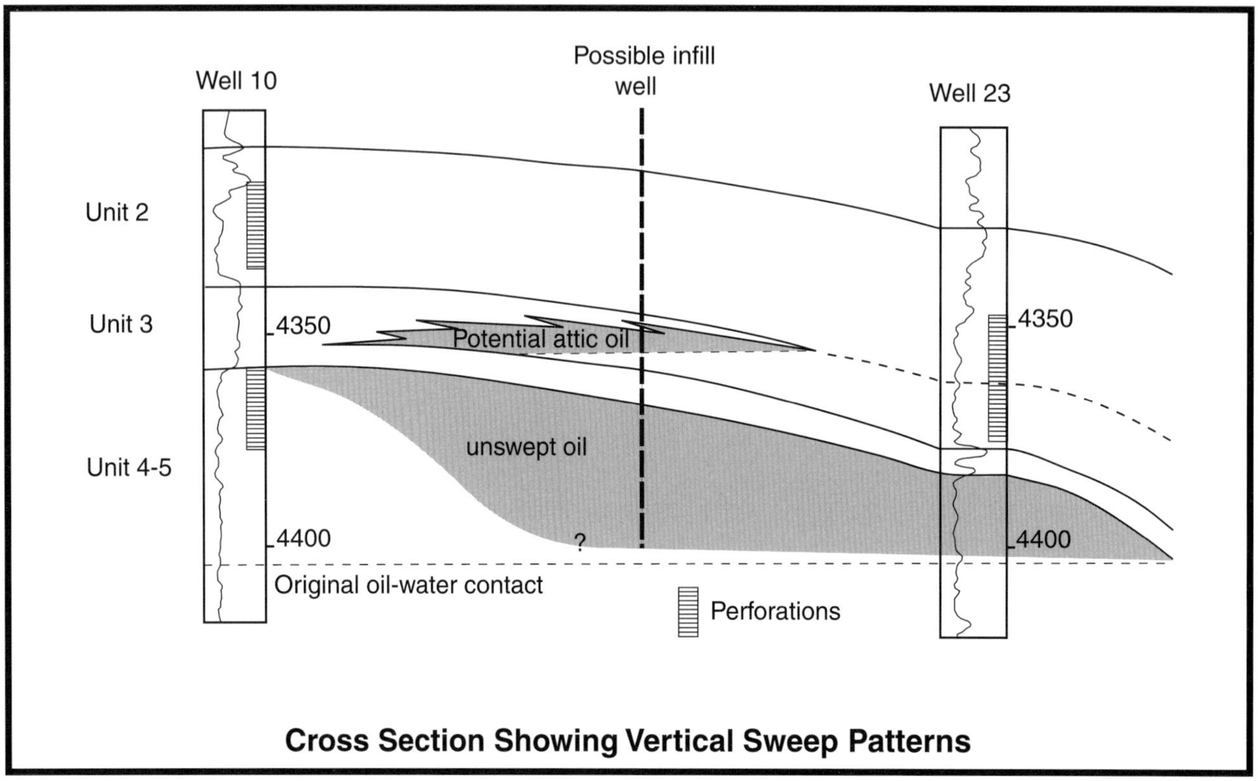

FIGURE 157. Potential infill well opportunity indicated by a vertical sweep cross section, Jackson field, Australia (from Hamilton et al., 1998). Reprinted with permission from the AAPG.

26

Shepherd, M., 2009, Quantitative methods for locating the remaining hydrocarbons, *in* M. Shepherd, Oil field production geology: AAPG Memoir 91, p. 221–226.

Quantitative Methods for Locating the Remaining Hydrocarbons

INTRODUCTION

The qualitative methods mentioned in the previous chapter can be used to work out where the remaining hydrocarbons are likely to be found. The current chapter summarizes quantitative methods of doing this (Figure 158). This involves defining drainage cells and validating their size by making drainage charts. Maturity tables are compiled for each drainage cell, and these can be screened for significant volumes of remaining oil. The underperforming drainage cells are likely to be those with potential for target oil volumes.

DETERMINING THE VOLUME OF UNRECOVERED MOBILE OIL

The first task is to estimate how much unrecovered mobile oil (UMO) is present in the field under evaluation. As mentioned in Chapter 5, UMO is the remaining movable oil volume predicted to be present after the field has been abandoned. The best way of illustrating the UMO is on a maturity pie chart (Figures 31, 158). This can show management and partners the value of detailed production geological work on a mature asset by pointing out that what was once considered a played-out old field will still have many millions of barrels of mobile oil left behind at the anticipated end of field life.

VALIDATING THE VOLUME OF DRAINAGE CELLS

As mentioned previously in Chapter 18, drainage cells are self-contained volumes that act as minireservoirs within a field. Drainage cells are therefore the important volumetric elements in reservoirs. Quantitative methods of determining the remaining oil apply to volumes within drainage cells. One method of validating the oil volume of a drainage cell is to draw a drainage chart. This method applies to bottom-water drive reservoirs.

DRAINAGE CHARTS

Drainage charts show how the oil-water contact is rising with time within a drainage cell. The simplest method of showing this is to plot the depth of oil-water contacts found in the wells from production logs and postproduction infill wells against the year of measurement. These values define a *drainage path* for the drainage cell. The oil-water contact will rise with continuing production over a period of time.

A more detailed method of drawing drainage charts has been used by the author as a way of checking that a drainage cell has been defined correctly (Shepherd, 2007). This involves constructing a chart showing the movable oil volume plotted against height above the initial oil-water contact for a specific drainage cell (Figure 159). The chart shows, for a given height above the oil-water contact, the volume of movable oil in the drainage cell between that depth and the original oil-water contact. This represents the volume of oil that needs to be drained to get an oil-water contact rise to that height (under ideal conditions).

The cumulative oil production volume on a year-by-year basis is tabulated for wells accessing the drainage cell. A point marking the total cumulative oil production figure from the drainage cell for each year is put onto the theoretical drainage path. This calibrates the ideal drainage path according to time. If the size of the drainage cell tank is correct, then the curve represents

DOI:10.1306/13161212M913372

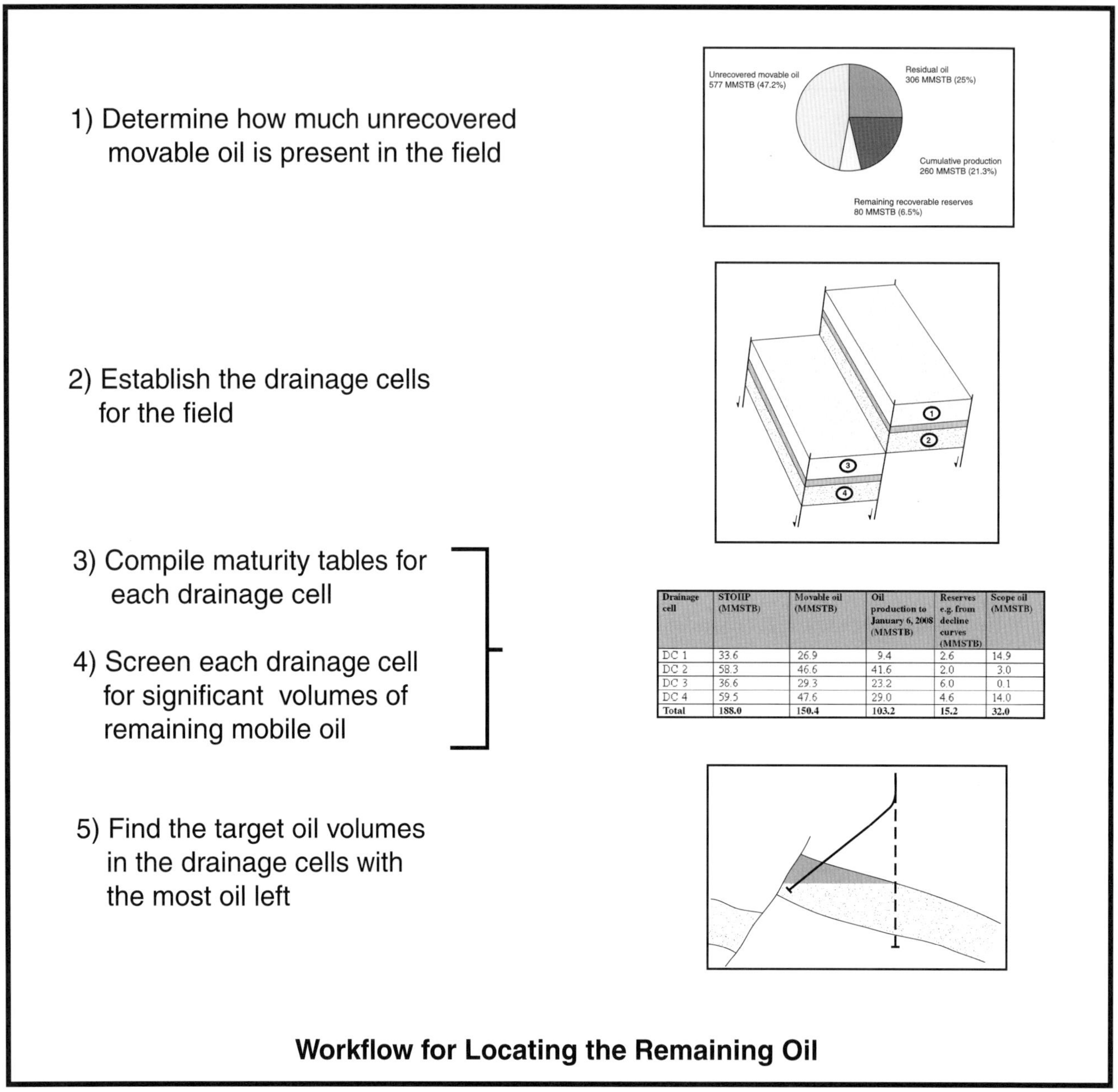

Drainage cell	STOIIP (MMSTB)	Movable oil (MMSTB)	Oil production to January 6, 2008 (MMSTB)	Reserves e.g. from decline curves (MMSTB)	Scope oil (MMSTB)
DC 1	33.6	26.9	9.4	2.6	14.9
DC 2	58.3	46.6	41.6	2.0	3.0
DC 3	36.6	29.3	23.2	6.0	0.1
DC 4	59.5	47.6	29.0	4.6	14.0
Total	**188.0**	**150.4**	**103.2**	**15.2**	**32.0**

FIGURE 158. Workflow for locating the remaining oil.

the amount that a cleanly rising oil-water contact will rise with a given amount of production from the cell.

Having done this, the height of the actual oil-water contact rise data derived from production logs is plotted at the appropriate point vertically above (or below) the time-calibrated volume-depth curve, corresponding to the year the logs were run.

The actual and ideal oil-water drainage paths will not be expected to coincide exactly. For the values to coincide, the following ideal conditions need to be met:

- The volume of the drainage cell has been identified correctly.
- The correct value of residual oil saturation has been subtracted from the stock tank oil initially in place (STOIIP) to give the movable oil volume.
- The oil-water contact has been rising uniformly as a level surface throughout the drainage cell.
- The sweep efficiency is 100%.
- The drainage cell is undergoing predominantly bottom-water drive.
- No oil is leaking into or out of the drainage cell.

It is unlikely that all of these conditions will be satisfied, and it is normal for the actual oil-water contact values to lie above the theoretical drainage path. For a

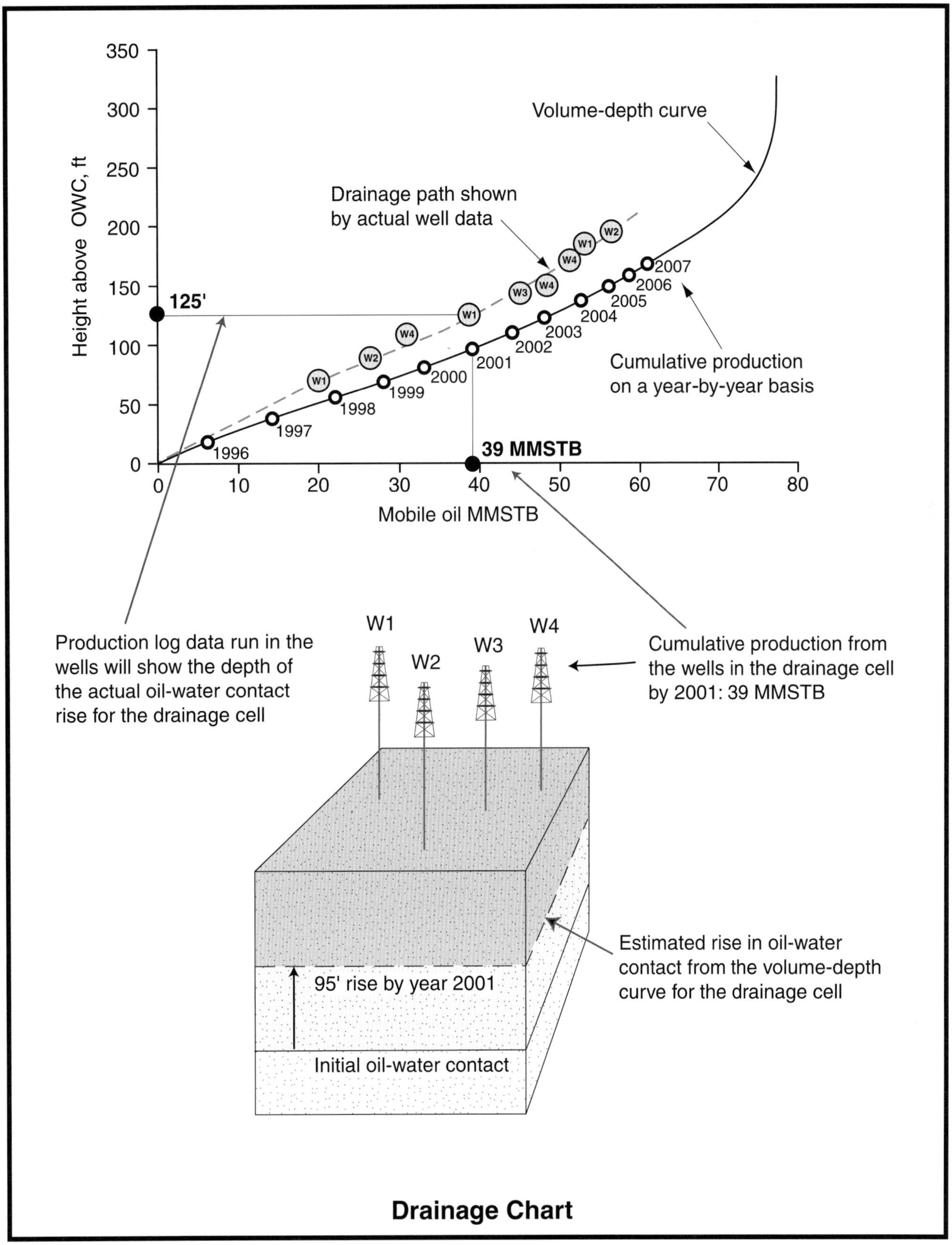

FIGURE 159. As a drainage cell is produced, the oil-water contact will rise. The drainage chart compares the actual measurements of the oil-water contact rise with the theoretical rise, assuming a given drainage volume in the geological model. If the latter is more or less right, the two paths should lie close to each other. OWC = oil water contact.

Table 20. Example of a maturity table. STOIIP = stock tank oil initially in place.

Drainage Cell	*STOIIP (MMSTB)*	*Movable Oil (MMSTB)*	*Oil Production to January 6, 2008 (MMSTB)*	*Reserves, e.g., From Decline Curves (MMSTB)*	*Unrecovered Mobile Oil (MMSTB)*
DC 1	58.6	46.9	18.3	2.6	26.0
DC 2	58.3	46.6	34.6	2.0	10.0
DC 3	36.6	29.3	23.2	6.0	0.1
DC 4	59.5	47.6	29.0	4.6	14.0
Total	213.0	170.4	105.1	15.2	50.1

clean, low-heterogeneity drainage cell that has been correctly defined, the actual and theoretical values should not be too different. If there is a big difference between the two, then this is useful information, and the reason for the difference can lead to modifications of the flow-geology model.

The reason for a large difference may be that the original definition of the drainage cell is wrong. Where, for instance, the actual oil-water contact depths lie below the theoretical drainage path, the real drainage volume is probably larger than it has been inferred to be. Sometimes the actual oil-water contact data can fall on more than one drainage path. This pattern may be showing that what was initially thought to be a single drainage cell is more likely to comprise two separate drainage cells with two different drainage paths. A reiteration may be required to split the drainage cell further.

Even where there is no or little direct oil-water contact movement data, a drainage chart can still prove useful. The perforated interval of a well producing dry oil should not lie below the calculated oil-water contact from a drainage chart. This type of observation can give an indication that there are larger connected volumes to a well than expected.

The drainage chart method can be effective in relatively simple, low-heterogeneity reservoirs but may be impractical in complex reservoirs, edge-water drive systems, or in oil fields with active gas caps. Here, reservoir engineering mass balance methods may give a better indication of the manner in which the drainage cells are depleting.

MATURITY TABLES

Maturity tables are used to screen remaining oil volumes in drainage cells (Vining, 1997; Bush et al., 2001). Table 20 shows the oil in place, mobile oil, cumulative produced oil, estimated reserves, and the unrecovered mobile oil (UMO) on a cell-by-cell basis. The unrecovered mobile oil is what is left over after both the cumulative production and the estimated reserves have been subtracted from the movable oil volume. The maturity table in Table 20 shows that there is a significant volume of UMO in drainage cells one and four. This is illustrated by the fuel tank display shown in Figure 160. In this way, drainage volumes can be screened to localize those areas of the field where there are likely to be significant volumes of remaining oil.

The cumulative oil volume produced from a drainage cell is easy enough to calculate if the wells only produce from that drainage cell. However, if some wells produce from more than one drainage cell, then an allocation split of production will need to be made according to each drainage cell penetrated. The allocation from the reservoir simulation model can be used or alternatively the cumulative production data for a well can be allocated to individual drainage cells according to historical flowmeter data. The reservoir engineer can provide the volumes of expected reserves to the end of field life for each well within the drainage cell. This can be calculated from decline curve analysis.

LOCATING THE REMAINING OIL

Maturity tables will provide the geologist with an estimate of the remaining unrecovered mobile oil (UMO) for each drainage cell. The task is now to determine where the target oil volumes are to be found within the drainage cell. It may be obvious on a preliminary analysis as to where the remaining oil can be found: in an attic oil volume or an undrilled fault block for instance.

At other times, it may be far from clear where the remaining oil is to be found within a drainage cell. The reason for this may be that the remaining oil is associated with numerous small-scale heterogeneities, and is found as dispersed patches of uneconomic oil volumes. For a drainage cell to be worth targeting, there should be just enough reservoir complexity to create large volumes of trapped oil but not so much complexity that the volumes are patchy and uneconomic to drill (see Figure 32). Analog field exposures can be examined with the idea of understanding the nature of *macroscopic sweep*. A small-scale reservoir simulation model of an outcrop analog

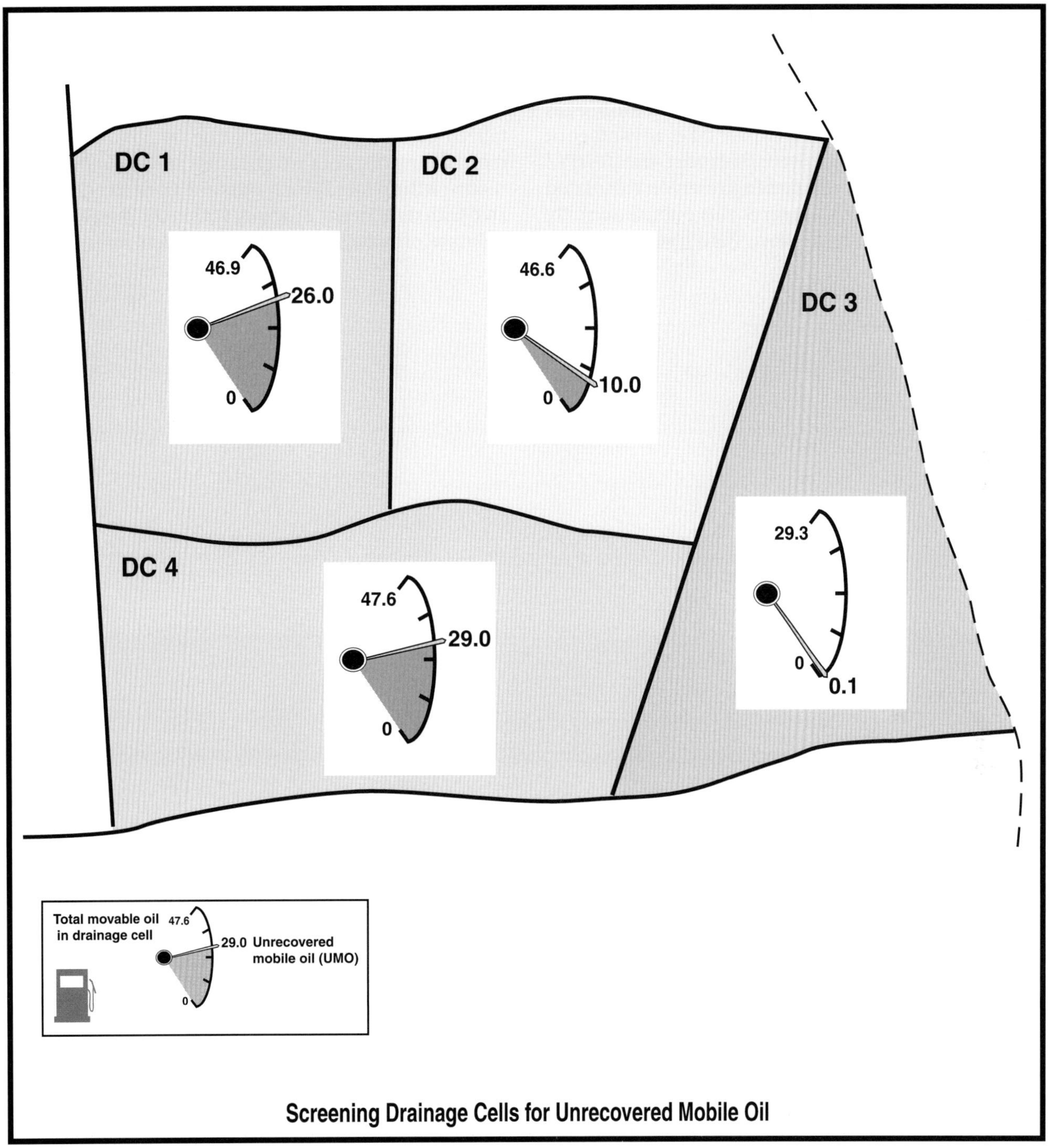

FIGURE 160. The calculation of unrecovered mobile oil volumes on a drainage cell basis is the main method of screening a field for remaining oil. These are illustrated on a fuel tank display. Drainage cells one and four are worth investigating for target oil volumes and possibly drainage cell two. DC = drainage cell.

may help to quantify the macroscopic sweep efficiency at this level of detail (e.g., Willis and White, 2000).

LOCATING THE REMAINING GAS

Similar techniques to those described for locating the remaining oil can be used to find the remaining gas reserves. An additional resource here takes advantage of the reservoir engineering material balance estimates (*P*/*Z* plots) of contacted gas volumes in a reservoir calculated from pressure data (see Chapter 5, this publication). By comparing gas initially in place (GIIP) volumes with material balance volumes, the difference can be attributed to bypassed gas. These volumes will not influence the P/Z plots as they are not in pressure communication with the existing wells.

Jackson and Ambrose (1989) estimated the volumes of the remaining gas in the I-92 reservoir in the Julian North field of South Texas using this method. The difference in volumes between the GIIP and material balance estimates was 4.8 Bcf. It was inferred that compartmentalized gas was present along the boundaries between crosscutting distributary channels and delta front sediments, within local pinch-outs and in dead ends associated with diagenetic variability within the reservoir. These volumes provide small potential targets for infill wells.

27

Shepherd, M., 2009, The opportunity inventory, *in* M. Shepherd, Oil field production geology: AAPG Memoir 91, p. 227–228.

The Opportunity Inventory

OPPORTUNITY INVENTORY

A key objective for the production geologist is to produce an *opportunity inventory*. Opportunities are pockets of remaining hydrocarbons within a reservoir that will remain unproduced unless targeted by drilling a new production well or recompleting an existing well. They are also called *reserves growth opportunities*. Once these have been identified, they should be cataloged in an opportunity inventory. An entry in an opportunity inventory may look like the following example shown on the next page.

The opportunities can be ranked in the inventory according to specific criteria. For example, Ambrose et al. (1997) proposed the following criteria for selecting opportunities for new wells or workovers for a reservoir interval in Venezuela:

1) Significant volumes of remaining oil
2) Optimal distance to current and previous producing wells
3) Relatively low water cut
4) Favorable structural locations

The ranked opportunities can then be screened to establish their economic value. A major part of the evaluation here is to determine the most appropriate well operation required to recover these hydrocarbons, whether this is a new conventional well, a coiled tubing sidetrack, or the workover of an existing well for example.

It is essential to develop an optimistic and imaginative mind set when it comes to evaluating opportunities in the subsurface. Every potential target volume should be itemized no matter how small or poor the reservoir quality. Technology, oil price, and economic environments all change with time and today's marginal opportunity can become tomorrow's prize.

DOI:10.1306/13161213M913372

6) <u>Attic oil near well B-28</u> P_{50} STOIIP: 6.6 MMSTB

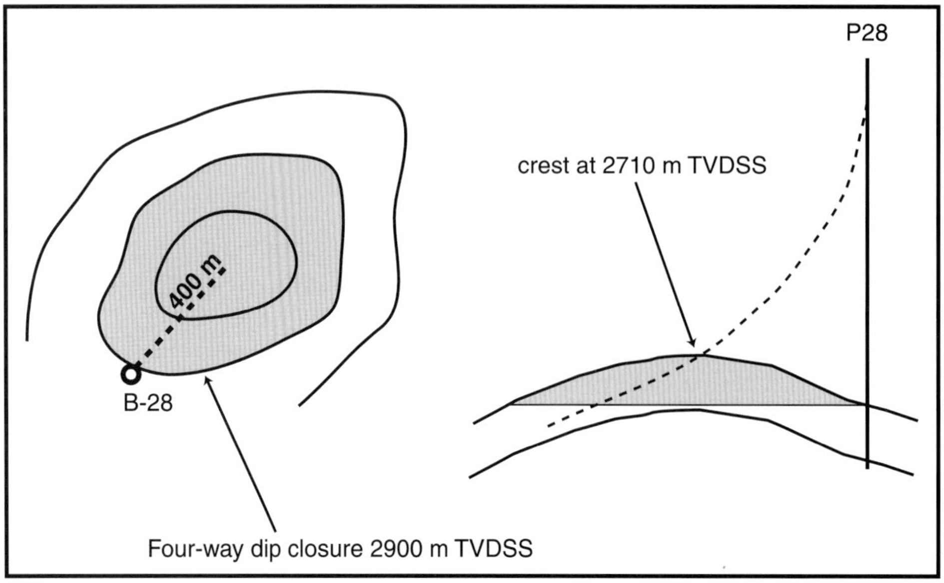

Description: A four-way dip closure is located northeast of the existing oil producer B-28. The effective oil-water contact is at 2900 m (9514 ft) TVDSS, equivalent to the top perforated interval in the well. The crest of the attic oil accumulation is at 2710 m (8891 ft) TVDSS.

Oil in place: Estimated oil in place is 6.6 MMSTB based on the 2007 3-D geological model (Company report no. 1283-2007). The probabilistic range is P_{90}; 4.2 MMSTB; P_{50}; 6.6 MMSTB; P_{10}; 9.3 MMSTB.

Uncertainties: The structure is low lying with shallow relief. It is sensitive to the depth conversion uncertainty.

What could be done to access the oil: The existing B-28 producer can be sidetracked updip to the crest of the attic oil location 400 m (1312 ft) northeast of the well. B-28 is currently producing 1200 BOPD at a 68% water cut. Given the current production rate, it is not considered to be available for sidetracking until at least 2010.

Comments: Recheck volumes once the new 3-D seismic data set is interpreted next year.

Example of an entry in an opportunity inventory.

STOIIP = stock tank oil initially in place; TVDSS = true vertical depth subsea.

Shepherd, M., 2009, Well planning, *in* M. Shepherd, Oil field production geology: AAPG Memoir 91, p. 229.

Section 6

Well Planning

Production geologists take an active part in well planning within an oil company. Not only are they involved in working out where the unswept petroleum is to be found in a producing field, they will also take a leading role in proposing and planning well locations to recover these volumes. Wells are the key to unlocking extra reserves from a reservoir.

There are a variety of well types that can be drilled. These can be tailored to optimize the drainage of oil and gas fields according to thier specific reservoir geometries, rock properties, hydrocarbon fluid properties, and remaining hydrocarbon configurations. Wells can be aligned in various drilling patterns to ensure optimal recovery from a field.

Guidelines are given as to how to plan a well. This takes into account the well objectives, the justification for drilling the well, the well targets, and any drilling hazards likely to be encountered.

DOI:10.1306/13161214M913372

28

Shepherd, M., 2009, Types of wells, *in* M. Shepherd, Oil field production geology: AAPG Memoir 91, p. 231–237.

Types of Wells

INTRODUCTION

There are a number of different types of wells that can be drilled, and these are described in the following text. A particular well type may be best suited or most economic in the efforts to drain a specific configuration of hydrocarbons. Various drilling strategies can be adopted to place wells in specific patterns with the aim of optimizing production from a field.

CONVENTIONAL WELLS

In the early days of the oil industry, drilling wells was a simple operation. A well location was picked at top reservoir, and the well was drilled directly down to the target as a *vertical well*. Then drilling became more sophisticated when the art of *deviating* wells was perfected. Here, the drill bit is deflected at an angle from the vertical toward a specific target. Deviated wells are commonly drilled from fixed drilling locations such as an offshore platform (Cheatham, 1992). One method of *directional drilling* uses an assembly with a *mud turbine* and a bit. The flow of mud through the turbine causes the attached bit to rotate while the drill string remains stationary. Drillers refer to this type of drilling as being in *sliding mode* because the drill pipe slides along the hole behind the turbine. To deflect the bit in the appropriate direction, a *bent sub* is used; this is a piece of drill pipe bent to about 1–2° angle, which is inserted behind the mud turbine and oriented from the surface along the planned direction for the well (Inglis, 1987).

A more recent technique for deviating a well involves using a *rotary steerable assembly*. Signals from the surface can be sent to the tool to deflect the bit in the appropriate direction while it is still drilling ahead in rotary mode. Drilling can be more efficient this way because there is less risk that the drill pipe will get stuck, it turns instead of slides, and the rate of penetration is faster (Downton et al., 2000).

Vertical and moderately deviated wells are called *conventional wells*. They are the most common well configurations because they are relatively cheap to drill.

SIDETRACK WELLS

A typical operation is to sidetrack a well. This is where a well has already been drilled or partly drilled and there is a need to exit out of one side of the well to a different target. A sidetrack may be required if there is an object stuck in the original hole, which cannot be fished out. In producing fields, an existing well may be sidetracked if there is no further use for that well, e.g., the oil well has watered out. A *window* will be cut in the casing of the original well by a special *milling assembly*, and drilling will then proceed out of the window toward a new target.

HORIZONTAL WELLS

Horizontal wells are wells where the reservoir section is drilled at a high angle, typically with a trajectory to keep the well within a specific reservoir interval or hydrocarbon zone. In a strict sense, these wells are rarely perfectly horizontal, but they tend to be near horizontal mostly, generally at an angle greater than 80° from vertical.

Horizontal wells are drilled in a specific configuration. The *tangent section* of the well is drilled along a deviated well path to just above the reservoir section, to what is known as the *kick off point*. From the kick off point, the well is drilled at an increasingly higher angle, arcing around toward an angle close to horizontal. The point at which the well enters (or lands on) the reservoir is called the *entry point*. From there on, the well continues at a near-horizontal orientation with the intention of keeping it substantially within the reservoir target until

DOI:10.1306/13161215M913372

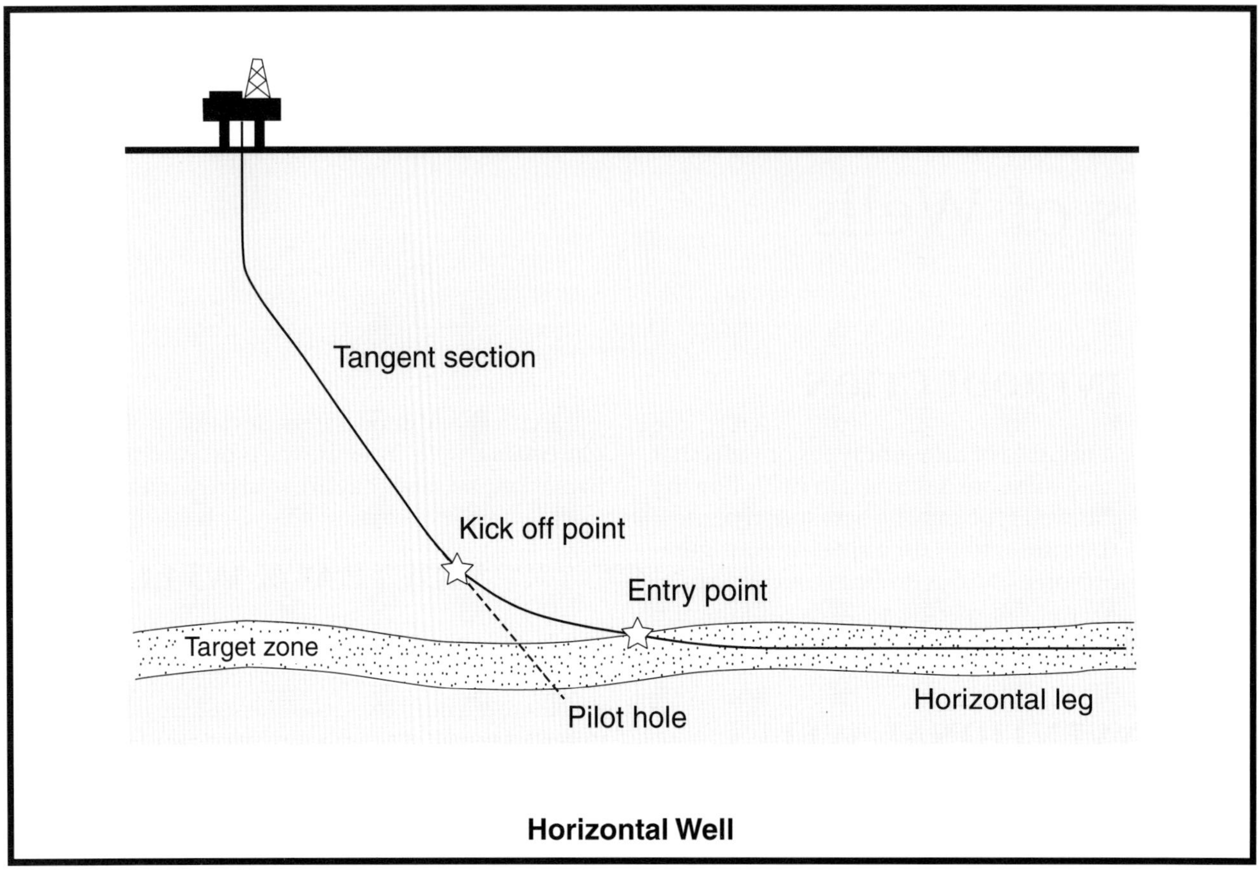

FIGURE 161. Horizontal wells are drilled at a high angle, generally greater than 80°, with the intent of keeping the well within a specific reservoir interval or hydrocarbon zone.

the desired length of horizontal penetration is reached (Figure 161).

One problem in drilling a horizontal well is in locating the kick off point at about the right distance above the reservoir (Figure 162). The kick off point will be planned for a specific depth above the prognosed target zone depth, such that there will be enough room to turn the well around, so as to enter the target at a near horizontal angle. If the target zone comes in high on prediction, the chances are that the well will be drilled all the way through the reservoir before being able to turn round quickly enough to establish a horizontal trajectory. If the target zone is deeper than expected, then quite a long distance of well can be drilled at a very high angle before the reservoir is entered. Given the normal uncertainty on establishing the depth of a target zone from the seismic method, it is common for a *pilot hole* to be drilled first to get this information directly. Pilot holes may be vertical, although it is better to deviate the pilot hole in the direction of the horizontal well path, and closer to the planned entry point for the horizontal section. If a horizontal well is planned near an appraisal well, then this can be used as a proxy for a pilot hole.

A horizontal well can be drilled *geometrically* where there is a reasonable confidence in the expected reservoir geometry. The targets are defined at the entry point and at total depth, and the well is drilled according to a set geometrical plan between them.

The alternative is to *geosteer* a horizontal well, particularly where there is less confidence in predicting the reservoir geology. Geosteering involves using geological information obtained as the well is being drilled to try and keep the well path within the target. This can involve the use of real-time log data but may also include input while drilling from well-site biostratigraphy or from examination of drill cuttings if the lithologies at the top and base of the reservoir are distinctive.

The main technique in geosteering involves the use of a *real-time log data display* while the horizontal well is being drilled. The downhole log data can be directly transmitted to a computer screen in the geologist's office from the well site. This allows the geologist to establish which part of the reservoir is being drilled through and then decide where the well should be steered to next. This is done by comparing the real-time logs with data from nearby wells. Log responses in horizontal wells can look different from that in conventional wells (Meehan, 1994). A catalog of expected log responses, as they would appear in a horizontal trajectory, can be created by computer modeling. If the geologist thinks that the well is

Figure 162. Problems can be encountered with landing a horizontal well if the target zone is too high or too low compared to what is predicted.

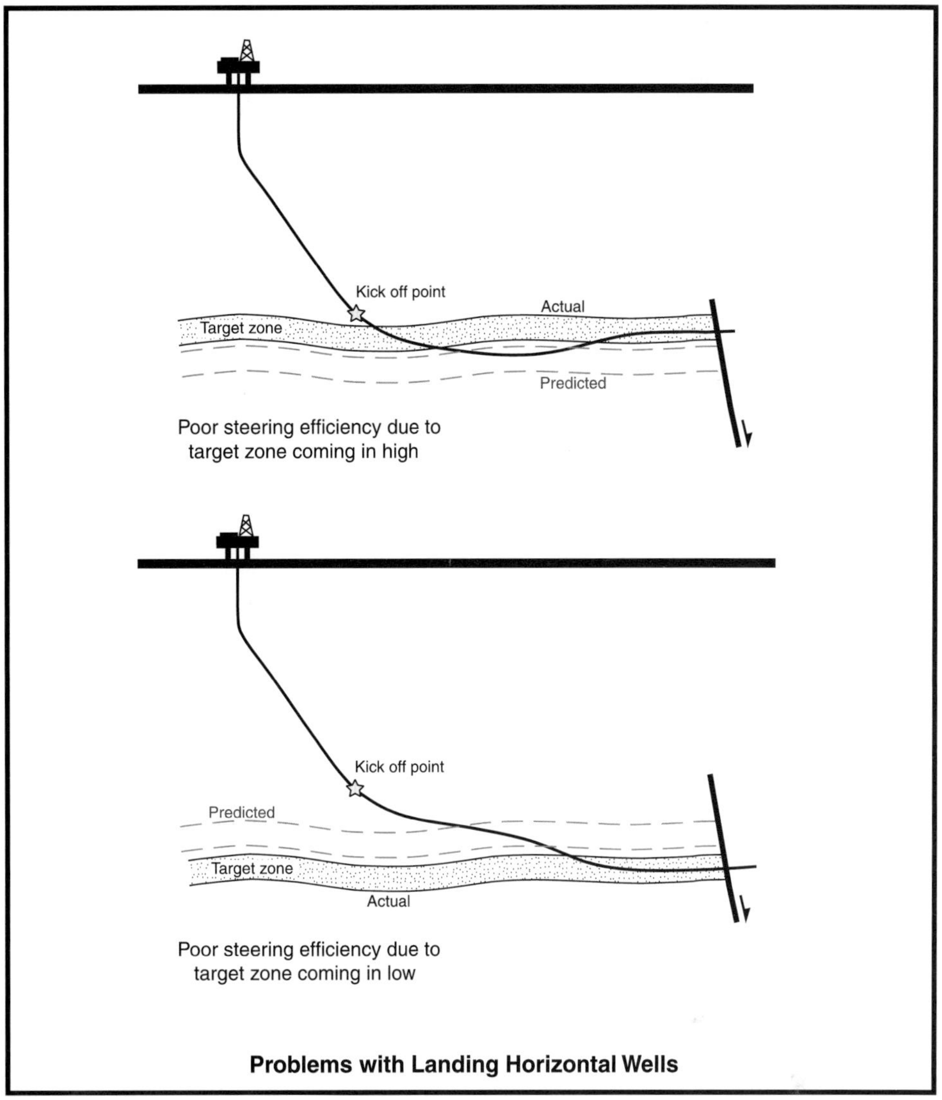

above the target zone, they will ask the directional driller at the rig site to steer down; if the geologist believes they are below the target, they will ask the driller to steer up.

Geosteering is at times a high-risk operation, and it can be stressful. In the early days of drilling horizontal wells, it was found that just under half of all the horizontal wells that were drilled ended up as failures or underperformed compared to expectation (Beliveau, 1995). The record may have improved since then; nevertheless outright failures still occur today.

Frequently, when drilling new well locations, the geology will turn out quite different from what was expected and this reflects the nature of reservoir uncertainty. Even so, the outcome from the vertical penetration of a reservoir interval is a lot more predictable than when a horizontal well is drilled. Random geological uncertainties that will have a relatively trivial effect on the drilling outcome of a vertical well can cause serious problems with a horizontal well operation.

At very high angles, if the top reservoir is 15 m (49 ft) deeper than predicted, the target will be penetrated much later than planned, or maybe missed altogether (Figure 162). Sometimes, after tracking the target interval, the well may then cross an unexpected subseismic fault and exit out of the target zone. It may not be clear which stratigraphic interval has been found on the other side of the fault. The geologist monitoring the well may not know if the target is above or below the well path. Another problem that can occur is that the predicted formation dip angle is wrong by a few degrees. In this instance, the well will quickly exit out of the top or base of a thin target. It can take a long section of the drilled interval before it can be steered back into the target horizon again (Figure 163).

Some geologists refer to the *steering efficiency* of a horizontal well; the percentage of the total well length within the target zone beyond the entry point. Modern LWD resistivity logs used in geosteering assemblies have

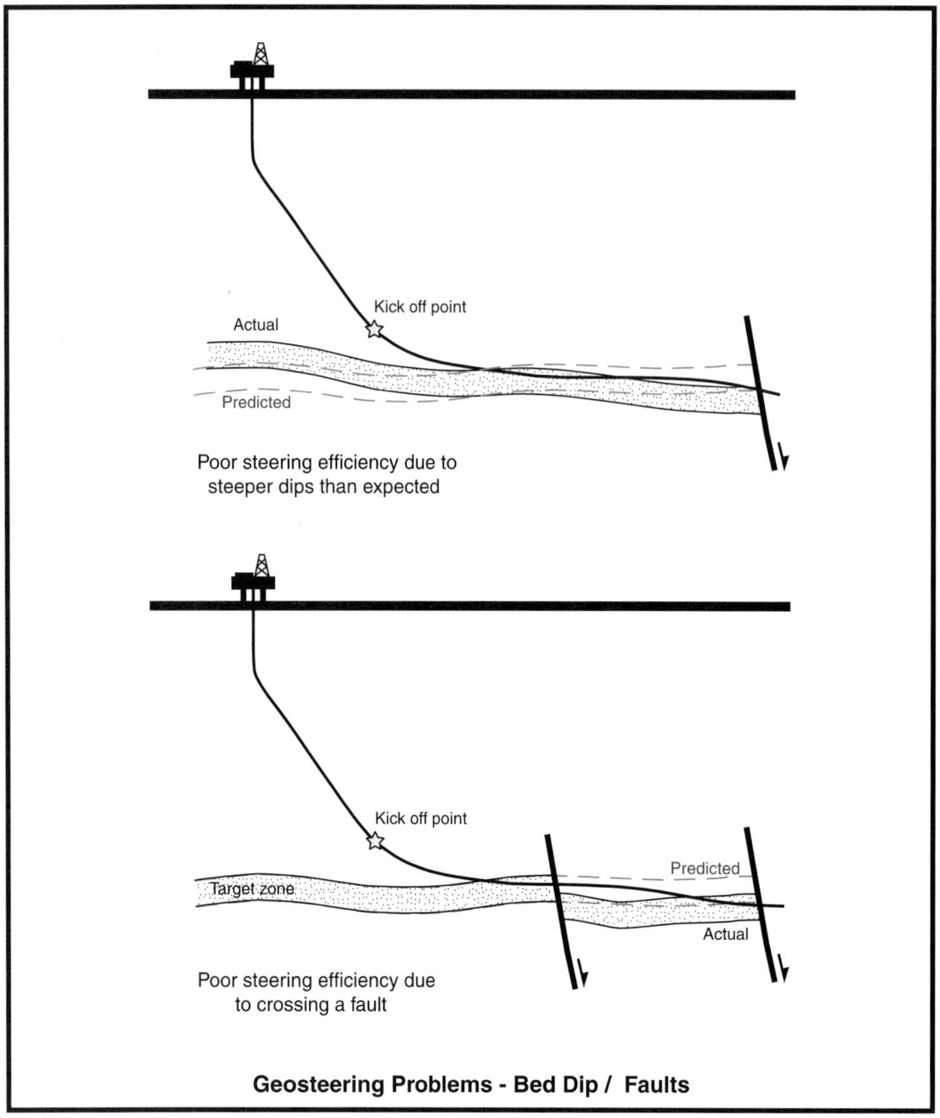

FIGURE 163. A horizontal well will be geosteered through a target zone by assuming the bed dip. If the assumed dip is wrong, the well may exit the target zone. Problems also occur if the well crosses an unexpected fault.

some degree of look-ahead capability to try and maximize the steering efficiency. The current created by the tool can have a sufficient depth of penetration to detect if the drilling assembly is converging on a bed boundary. This can give enough warning to allow the well to be steered away from the bed boundary.

Despite these problems, horizontal wells often end up as the best producers in a field. There are many reasons for drilling a horizontal well as opposed to a conventional well. They can produce considerable volumes of incremental reserves from what would otherwise be an underperforming area of the reservoir. Although they are more expensive to drill and are more prone to failure, horizontal wells often produce at several times the rate of an equivalent conventional well in the same reservoir. For example, experience in the Heavy Oil Belt of Venezuela has shown that flow rates are increased significantly by producing from horizontal wells, yet they cost only 1.5 times more than vertical wells (Hamilton et al., 2003). In the Widuri and adjacent fields, offshore Sumatra, 15% of the producers are horizontal wells, yet these provide 30% of the oil production volume (Carter et al., 1998).

Reservoirs tend to be much longer and wider laterally compared to their thickness, so a horizontal well is more likely to be in significantly greater contact with a given length of reservoir than a vertical well. Another feature of a horizontal well is that, for a given flow rate, a longer well needs less pressure drawdown to produce at that rate.

All this can create the outcome by which horizontal wells are much more productive or economic than conventional wells. This tends to be true of the following situations:

- Thin reservoirs. A conventional well will intersect a relatively thin section of the reservoir, whereas a horizontal well can run the length of the reservoir

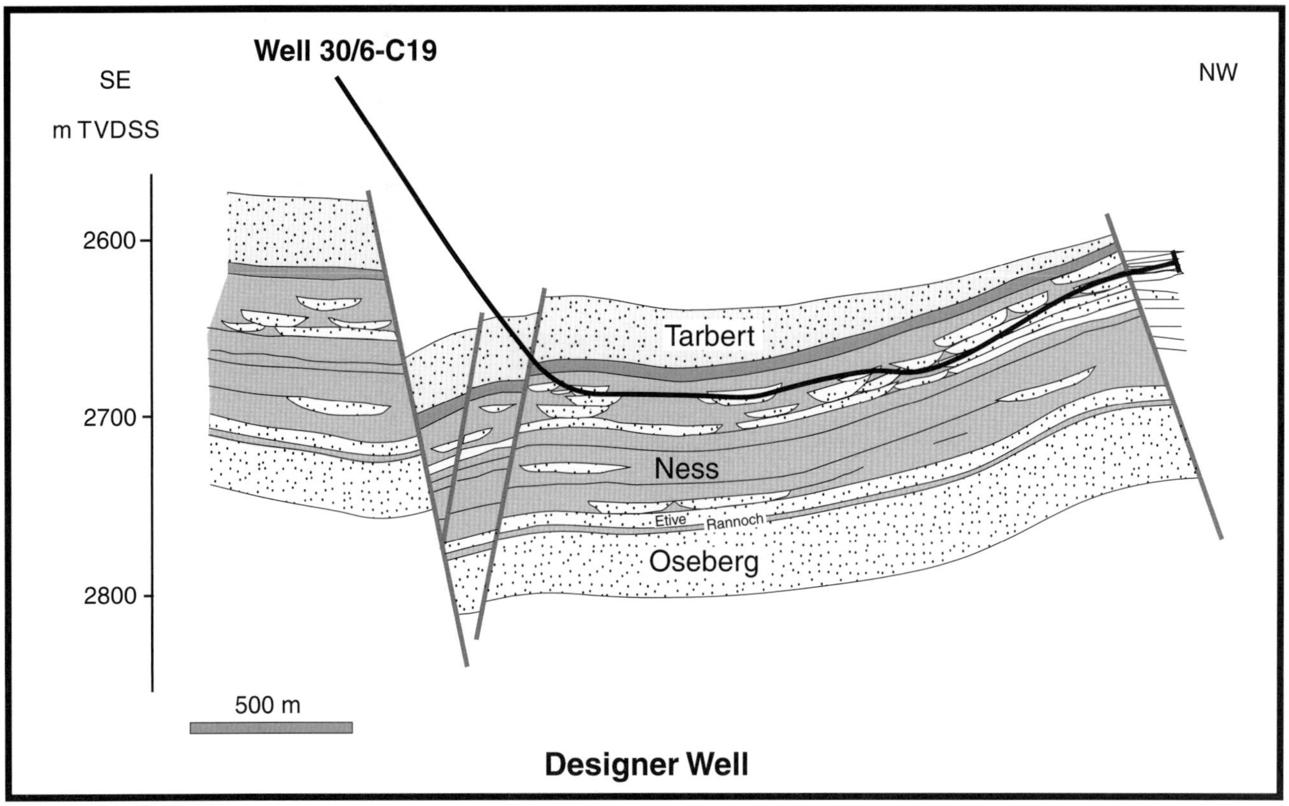

FIGURE 164. A designer well in the Oseberg field, Norwegian North Sea. The horizontal well section was planned to target several seismically defined, fluvial channel bodies within the Ness Formation (from Ryseth et al., 1998). Reprinted with permission from the AAPG.

and produce much more hydrocarbons (Fayers et al., 1995).

- Horizontal wells can target long, narrow macroforms such as channel fill sandstones.
- Fractured reservoirs. A horizontal well has a much greater chance of intersecting vertical or steeply dipping natural fractures compared to conventional wells. This can be a particularly effective way of producing fractured reservoirs with very low matrix permeabilities (Major and Holtz, 1997).
- Low-permeability reservoirs. Where an interval shows low permeabilities, horizontal wells can make up for this by maximizing the contact length with the reservoir. This means that low-permeability rocks such as chalk can produce at economic rates that would be marginal to uneconomic with conventional wells.
- Reservoirs prone to coning. Because of the lower drawdown, horizontal wells may be less prone to water or gas coning behavior. For example, horizontal wells have been drilled in the Widuri field, offshore Sumatra, so as to minimize water coning. High vertical permeabilities and viscous oil are factors likely to promote coning behavior in the vertical wells in the field (Carter et al., 1998).
- Similarly, individual horizontal wells produce more oil in heavy oil reservoirs because the lower pressure drawdown tends to keep water and gas away from the well longer. For example, a total of 110 horizontal wells had been drilled prior to 2002 in the Hamaca field in Venezuela's Orinoco Heavy Oil Belt. The development plan is to ultimately drill over 1000 horizontal laterals to produce the 8–10° API gravity oil (Tankersley and Waite, 2002).
- *Oil rims*, thin oil columns typically lying below a gas cap, can be targeted with horizontal wells. The reduced drawdown minimizes the chances of coning water up from the water leg or drawing gas down from the gas cap.

In certain parts of the world, horizontal wells are the well type of preference, whereas conventional wells are much less common. This is true of the Danish North Sea, where chalk is the main reservoir interval, and also in parts of the Middle East such as Qatar, Abu Dhabi, and Oman (Nurmi, 1996).

There are situations where it is not advantageous to drill horizontal wells. In reservoirs where there is a very low K_v/K_h because of small-scale bedding-parallel baffles, bedding-parallel horizontal wells are not effective

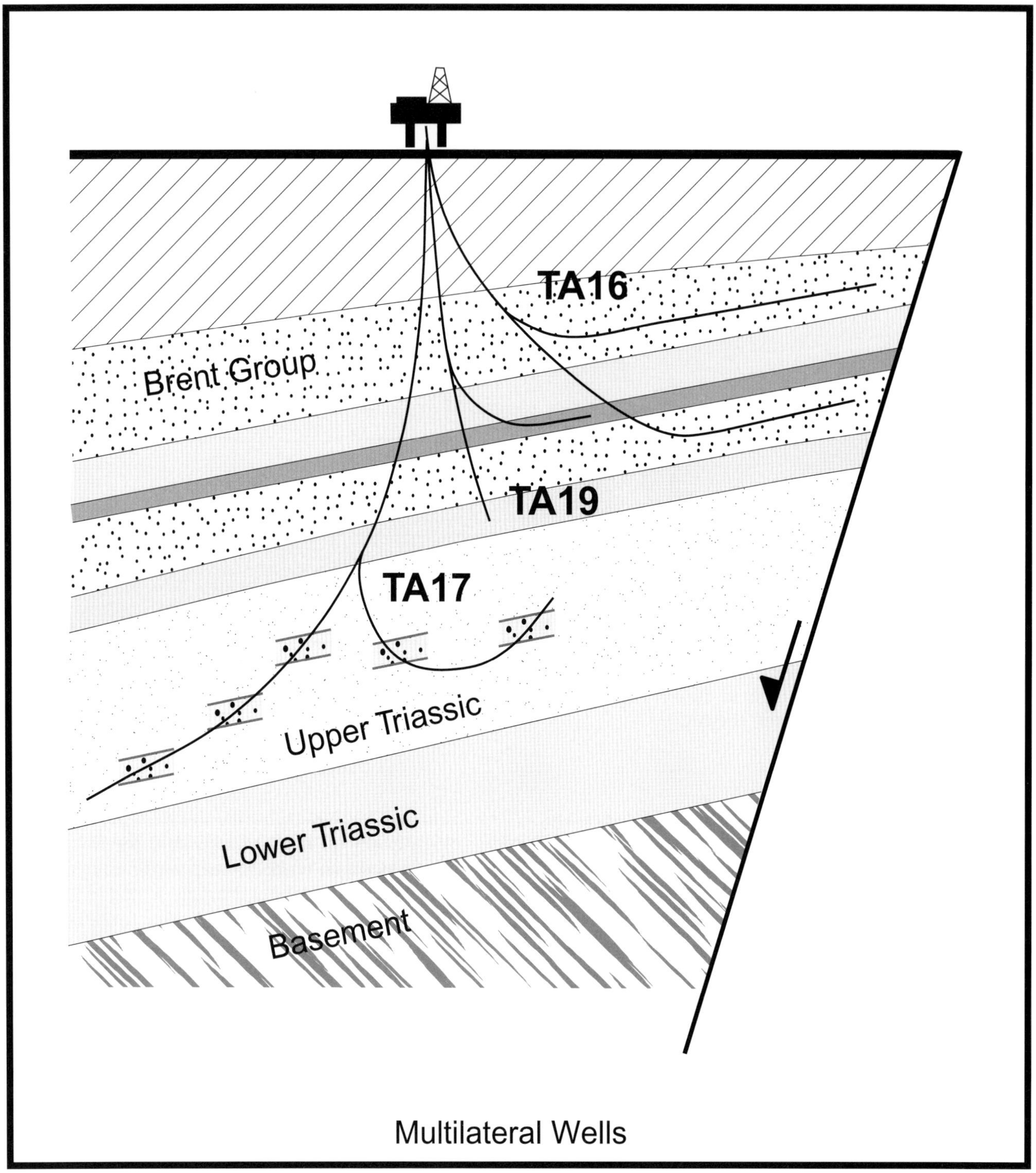

FIGURE 165. Multilateral wells in the Tern field, UK North Sea (from Black et al., 1999). Reprinted with permission from the Geological Society.

(Haldorsen et al., 1987). Numerous baffles parallel to the wellbore will severely restrict the contactable drainage volume. It is better to drill strongly layered reservoirs like these with slanted instead of horizontal wells. Some subsurface professionals will advise against drilling horizontal wells if it can be more practical to drill a slant well. These are less risky to drill, and there is a better chance of establishing which part of the reservoir stratigraphy has been penetrated by the well. Slant wells may be a better option for drilling injection wells where it is important to ensure waterflood support to a specific reservoir interval.

DESIGNER WELLS

Designer wells are types of high-angle or horizontal wells that have more than one intended target. This makes them more cost effective because the individual targets would have otherwise required several conventional wells to drain them effectively. One aim of a designer well could be to penetrate and drain more than one fault block. In mature fields, multitarget infill wells can increase the chances of finding an economic volume of oil. For example, in the Oseberg field, Norwegian North Sea, a designer well successfully targeted and lined up several fluvial channel sandstone bodies (Figure 164) (Ryseth et al., 1998).

MULTILATERAL WELLS

Multilateral wells are wells that have more than one branch radiating from the main borehole (Figure 165). Each branch can drain a separate part of the reservoir and produce into a common single wellbore. The advantage of multilateral wells is that, for the same number of drainage points, they can be somewhat cheaper than if separate wells had been drilled.

COILED TUBING DRILLING

Coiled tubing is continuous, small-diameter steel pipe stored on a reel at the surface in lengths of up to 6000 m (19,685 ft) long. Coiled tubing can be used in place of drill pipe for new wells and short-length to medium-length horizontal sidetracks (typically with a step-out of less than 800 m [2625 ft]). A mud turbine and drill bit combination is used for coiled tubing drilling. The turbine is powered by the mud moving through it; the tubing itself does not rotate. The advantage of coiled tubing drilling is that the drilling operation is quicker than normal drilling in that the connection time involved with a jointed drill pipe is eliminated. The tubing is simply rolled in and out of the well.

THROUGH TUBING ROTARY DRILLING

Through tubing rotary drilling is a relatively inexpensive method of creating a short-length to moderate-length sidetrack of an existing well (with a step-out of up to 1000 m [3381 ft], sometimes longer). Slim-bore drill pipe is used to drill the well, and the benefit of this is that the drill pipe is narrow enough to be run through the existing production tubing (Reynolds and Watson, 2003). This eliminates the time and cost involved with pulling the completion in an existing well to start drilling and then rerunning it after the well has reached total depth. Through tubing rotary drilling has been used in the Gullfaks field in the Norwegian North Sea. 4-D seismic data is used to identify remaining oil targets. Many of these targets are small but can be drilled cheaply by the use of through tubing rotary drilling. This has contributed to a reversal of the oil production decline at the late mature stage of field life (Todnem et al., 2005).

WELLS, THE PRODUCTION GEOLOGIST'S TOOL KIT

As can be seen, numerous types of wells can be drilled. These are the production geologist's tool kit. If a reservoir target is uneconomic or unfeasible with one type of well, try another type. If the opportunity does not produce enough oil with conventional wells because the permeability in the reservoir is too low, try a horizontal well. If there are numerous unswept oil targets in a series of fluvial point bars, each individually with low volumes and rather risky to define, drill several of them with a designer well. If there is attic oil updip from a high water-cut producer, sidetrack it. If the opportunity looks attractive but does not quite make enough money to justify drilling it with a conventional well, try a cheap drilling method such as coiled tubing drilling.

29

Shepherd, M., 2009, Well patterns, *in* M. Shepherd, Oil field production geology: AAPG Memoir 91, p. 239–240.

Well Patterns

INTRODUCTION

The optimal drainage that can be achieved for an oil field will depend on the number and configuration of production and injection wells. Specific well patterns can be adapted to various reservoir geometries and locations.

HOW MANY WELLS?

How many wells does it take to produce oil or gas from a field in the most effective way? For an oil field, this very much depends on how many drainage cells there are. For low-complexity fields, such as a high-permeability shoreface reservoir, not many wells will be needed. A single well may produce oil from a very large volume in this instance.

Good reservoir management involves finding an appropriate well spacing such that a minimum number of wells will produce the largest volume of oil without interfering with each other. For complex reservoirs that have poor connectivity or are heavily faulted, many wells may be required. Wells are expensive, and if too many have to be drilled to optimally drain an oil field, then the operation will be marginal to unprofitable.

WELL SPACING

A typical *well spacing* offshore is about 0.5 to 1 km (0.3 to 0.6 mi) between wells. Onshore, the well spacing can vary. A typical practice onshore, particularly in the United States, is to start at a large well spacing and to progressively infill to a closer well spacing. The idea is that a shorter well spacing will result in wells contacting more reservoir dead ends, particularly in the more complex reservoirs.

WATER INJECTION WELLS

If water injection wells are to be drilled, set patterns can be chosen according to the structure of the field. If the field has a simple domal anticline structure, then the injection wells can be configured in a *peripheral pattern* around the margins of the field at or below the oil-water contact. Where the structure is a simple dipping fault block, then the producers can be located at the crest of the fault block with a *line drive* of injectors downdip at the base of the hydrocarbon leg (Figure 18). Sometimes in this type of structure, there will be a middle row of producers half way between the producers and the injectors. Once the middle row of producers have watered out, then these can be turned around to provide injection support to the crestal wells.

In onshore fields, wells are drilled in regular grids with injection wells located according to well-established patterns, such as five-spot, seven-spot, nine-spot, or line-drive patterns (Figure 166).

There is evidence that flood fronts are influenced by the in-situ horizontal stress directions. Water breakthrough is more rapid between injectors and producers aligned close to or along the maximum horizontal stress orientation (Heffer and Dowokpor, 1990). For this reason, it has been suggested that the alignment of producers and injectors along the maximum horizontal stress orientation should be avoided (Bell and Babcock, 1986; Hillis and Nelson, 2005).

In layered reservoirs, some intervals may get better injection support than others. At the mature stage of field life, this can result in a combination of swept fast hydraulic units and several poorly swept slow hydraulic units. Injection and production wells can be selectively recompleted. The swept zones will then be isolated to ensure that the slow hydraulic units can be produced in a timely fashion.

DOI:10.1306/13161216M913372

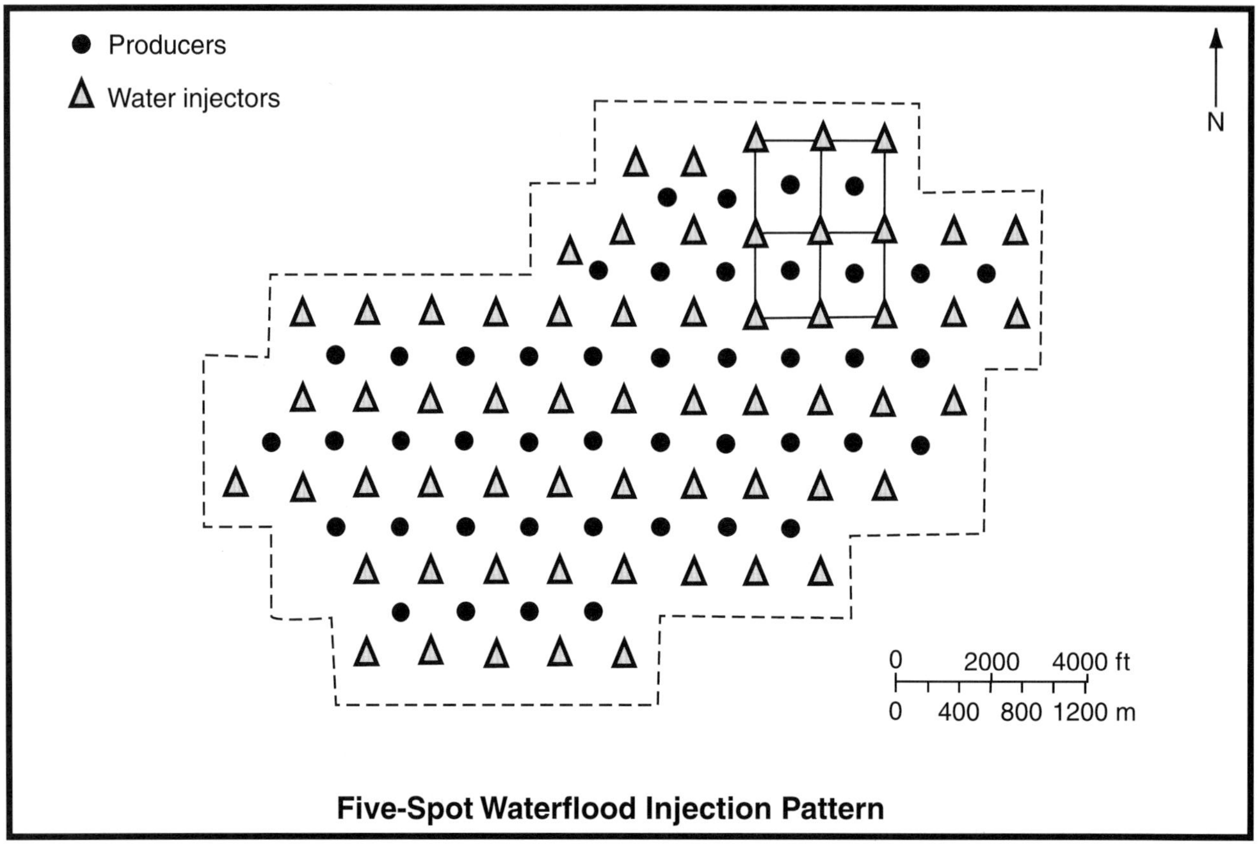

FIGURE 166. Five-spot waterflood injection pattern in the San Andres C waterflood unit, Permian Basin, Texas (from Sneider and Sneider, 2001). Reprinted with permission from AAPG.

WHEN DO YOU STOP DRILLING IN A FIELD?

Field operations can involve extended phases of drilling, with well after well going down. By drilling wells, more hydrocarbons are recovered from a field and the production decline of the reservoir will be arrested. The wells will be drilled providing there are opportunities of sufficient size left to drill; the task of the production geologist is to find the remaining hydrocarbons.

There will eventually come a time when the targets become more and more marginal as the better ones will have all been drilled. Experience has shown that there will always be failures when drilling targets in a reservoir. The geological description is often too uncertain to guarantee success every time. However, drill two failures in a row and confidence will start to decline. Three failures in a row may indicate that there is a need to take a time out on drilling. Perhaps the reservoir is played out or it is just possible that the geologist is concentrating too much on the wrong type of target, e.g., structural targets as opposed to layer targets. This is a good time for a rethink.

A word of caution though; general experience shows that the spread of high-, medium-, and low-rate production wells in a field follows a lognormal distribution (Beliveau, 1995). In any reservoir, a small number of star producers will be supported by a large number of low-rate performing wells. This may also be the case for a drilling sequence of new wells. Out of 10 new wells, possibly two or three will end up as high-rate producers, six or seven will show lower rates, and there may be one or two failures. This is a normal pattern, but if the drilling has been based on the premise of every well being successful, then the project as a whole may be deemed to have failed. The result may be to call a halt to any further drilling because of this. It is best not to overplay expectations to the management and partners at the start of a new drilling campaign.

Shepherd, M., 2009, Well planning, *in* M. Shepherd, Oil field production geology: AAPG Memoir 91, p. 241–248.

30

Well Planning

INTRODUCTION

The production geologist takes a leading role in proposing and planning well locations in oil and gas fields. This chapter provides an outline of the procedure involved in well planning. This includes establishing the well objectives, the justification for drilling the well, the location of well targets, and the description of any drilling hazards likely to be encountered.

In practice, procedures for well planning vary considerably between the oil companies. Some companies insist on a rigorous process-driven approach, whereby each stage of the planning procedure follows preset guidelines. The logic here is that drilling wells is the most expensive operation in the subsurface. Thus, if all the guidelines are followed through to the rule, then the chances of making expensive mistakes should be reduced.

HOW NEW WELL LOCATIONS ARE PICKED

It is normal practice to keep drilling wells for much of the field life, including both producer and injector wells. New wells will arrest the natural tendency of a mature field to decline in production every year. A new production well will be warranted if it can be shown that it is likely to produce enough oil or gas not only to pay for the expenses involved in drilling it, but also to give the oil company and its partners a suitable rate of return for the money spent. A production well will be justified if it can be demonstrated that it will produce a significant volume of oil and gas that would not otherwise be produced by the existing wells (Shirzadi and Lawai, 1993). The extra hydrocarbons produced like this are called *incremental reserves*.

WELL PLANNING

The following is a general guideline as to how well planning is conducted. It is written with a bias toward the way wells are planned for offshore locations. The whole procedure tends to be more elaborate for offshore wells compared to onshore operations because of the much greater expense and financial risk involved.

PLANNING PROCEDURES

The first step in well planning is to justify the new well in terms of extra production and economic value. If this can be demonstrated, then a proposal to drill the well will be presented to the management and partners. Many companies kick off this phase with an *internal peer review*. The well concept will be presented to a subsurface team from another asset so as to get feedback as to whether the logic for drilling the well is reasonable. Any overlooked problems can be picked up at this stage before the planning process is completely closed out.

A *well proposal document* will then be circulated around to the relevant parties. It is at about this point that the budget for drilling the well is approved. The document summarizing the latter is called an *AFE*, an acronym for approval for expenditure. The arrangements for approving a budget by partners will vary from field to field, but usually there will be a pass mark specified in the joint operating agreement. This can be any number between 50 and 100% of the partners involved.

Once the budget is approved, then the drilling engineers will start working on the detailed *drilling program*. This will cover technical details such as casing selection, directional drilling design, and drilling mud type. Much of this will involve geological input. The first point of contact here is the *operations geologist* if there is one. The

DOI:10.1306/13161217M913372

operations geologist is a specialist responsible for organizing the geological operations at the well site, and liasing with the well-site geologist and with the drilling department in-house. The larger companies will employ an operations geologist, whereas the smaller companies will expect the production geologist to look after the geological aspects of the operation by themselves.

Normally, there will be regular meetings where everybody involved in the drilling operation will come together to discuss the progress of the well planning. If a directional well is being drilled, it can sometimes take several iterations before the ideal well path can be determined. Some modern 3-D modeling packages include a *well planning module*. When these are used in sessions involving the geologist and the drilling engineer, the time involved in the iterative search for the optimum well trajectory can be significantly reduced.

It is also good practice for the drilling engineer and the geologist to review the experience of previous wells drilled in the area to find out if any previous problems were encountered. This simple procedure can save a lot of money as the drilling operation can be modified to mitigate against any reoccurrence of these problems.

Close to spudding the well, a *prespud meeting* will be held. This involves presentations by the well planning personnel to the contractors and drilling crew representatives to make sure everyone is aware of the well objectives, the plan to be followed, and the safety issues. It is normal for the production geologist to give a short presentation at the meeting on well objectives, the expected lithologies, data acquisition, and any drilling hazards that could be expected.

WELL PLANNING AND THE WELL PROPOSAL DOCUMENT

The *well proposal document* gives details of the well location and its justification. The document is sent to the internal management and then to partners with the specific aim of receiving budget approval for drilling the well.

Well planning is an iterative procedure, with ideas being passed around within the subsurface team and with the drilling engineers in order to produce a sound drilling plan that will be economically successful. The well proposal document itemizes the general procedure of the well planning in a more structured fashion than tends to happen in practice. So the text here describes the structure of a well proposal document as a way of explaining the well planning process with some degree of logical progression.

The primary aim of the well proposal document is to obtain the approval for the well. As such, it should explain the logic for drilling the well, simply and clearly.

A well proposal document will typically include information on the well objectives, target data, total depth (TD) criteria, the justification for the well, and any potential drilling hazards.

OBJECTIVES

The objectives are the basic reasons for drilling a well. For example, an objective could be to drill and complete an oil producer to access the attic oil updip from a watered out well. Sometimes there may be more than one objective for a well, particularly if there are multiple reservoir horizons to be accessed by the well path.

Some companies will use an *objective matrix summary*. An example is shown in Table 21.

The objective matrix summary is an excellent way of presenting the basic reasons behind the well in a simple format that is comprehensible to anyone responsible for approving the well plan. It outlines the key objectives, explains the logic behind these objectives, and gives an idea of what might go wrong.

TARGET DATA

The target for the well is specified by giving the target horizon, the target depth, and the geographical coordinates for the target (latitude and longitude or universal transverse mercator [UTM] coordinates). For example, the following could be a typical definition for a target: Top Painter Member 2550 m (8366 ft) TVDSS, UTMs 5,459,515 m (17,911,795 ft) north, 437,626 m (1,435,780 ft) east. The target location is usually given at the top of the reservoir.

Details will also be provided for the *target area* (Figure 167). Although the subsurface team will specify the well target as a single point, it is nearly impossible for the drillers to hit this point exactly. Nevertheless, the drilling engineers can guarantee to hit a target area around the target point with a high degree of confidence, providing it is large enough. A target area comprising a circle of 30-m (98-ft) diameter around the drilling target is achievable under normal conditions. It can be expensive to drill smaller targets than this. It is sensible to make the target area as large as possible without compromising the original objectives of the well.

The logic used to define the target area will also need to be discussed. There may be a good reason for not wanting the well to drift beyond the specified target edges. These may include such factors as the proximity to faults, the chance of collision with nearby wells, erosion edges, the loss of structural height, and hence thickness of oil column. Some edges are more problematic to go beyond than others, and the reason for this should be specifically mentioned.

Table 21. Example of an objective matrix summary.

What Are the Key Objectives and How Do We Measure Success?	*What Are the Key Drivers for These Objectives?*	*What Are the Main Risks or Hazards to These Drivers?*
1. To drill and complete a Ferret Formation oil producer to access the attic oil in the southeast fault panel of Block 2. The well should be capable of producing at an initial rate of 6500 BOPD.	To recover the attic oil in the southeast fault panel. To improve the recovery from Block 2.	The well is close to a major fault zone, and if the damage zone or the fault itself is drilled, this could seriously impair the ultimate productivity of the well.
2. To drill and complete the well safely within budget timing and the cost estimate.	Maximize the well value. Meet the production target.	Mudstones at the top of the Ferret Formation may be potentially mechanically unstable if drilled at a high angle.
3. To recover incremental reserves of 5.4 MMbbls.	Maximize the field Net Present Value.	If the seal integrity of the southeast-panel-bounding fault is poor, there may be less attic oil than predicted.

There are occasions where it may be necessary to specify more than one target. For a geometrically defined horizontal well, a target will be defined at both top reservoir and the TD location.

TD CRITERIA

The TD for the well must be specified. A sufficient hole length will be drilled below the base of the sand to ensure that the entire reservoir interval can be logged and to allow various completion operations to be conducted. For horizontal wells, TD may be called after a set length of the reservoir target has been drilled. The TD criteria should be discussed with the drilling or completion engineer to establish that there is enough hole length to accommodate any operational requirements.

WELL JUSTIFICATION

It is normal practice for the geophysicist, geologist, and reservoir engineer to write a section in the well proposal justifying the well location from their different perspectives. The details should be kept to a minimum so as to concentrate succinctly on the main features defining the well concept.

The geophysicist will describe the structural configuration of the well target on the basis of the seismic interpretation. This section will include a depth map of the target horizon, the error bar on the target depth, and the presence or absence of faults in the area. The error bar on the target depth can be large; it will depend on how good the well control is in the area. The precision of the seismic method is such that a well coming in at top reservoir within this envelope can be defined as being on target.

The geologist will justify the well in terms of the reason for drilling. Inherent in this is the idea that the well will produce oil that will not be produced by the existing set of wells. One or two structural cross sections will be used to illustrate the geometry around the well. The style of faulting and the potential for fault sealing should also be described. Key production elements should be mentioned such as vertical permeability barriers and permeability profiles. It may also be necessary to show how the geological model and well control have been used to predict the expected lithofacies and rock properties at the well location. This can be illustrated with facies and rock property maps. Following on from this, there should be a discussion of the hydrocarbon volumes in the target area and the uncertainties involved in estimating these volumes. Finally, a well prognosis will be included showing the depths to the key formation tops that are expected in the well path (Figure 168). It is also good practice to list the uncertainties in the prognosed well tops relative to the estimated depth.

The reservoir engineer will describe the target in terms of the expected fluid distribution, pressures, and the production history of nearby wells. There will also be a description of the analytical methods used to justify the well location, including details of any analysis based on running the reservoir simulation.

A prediction will be given for the expected *production profile* from the new well and the value of the expected production. The production profile is typically calculated using the reservoir simulation, and, from this, the total incremental oil for the well is estimated. The reservoir engineer will also quote the drilling and completion costs. The main issue concerns whether the well project will make enough money to cover the drilling and completion costs and then go on to make enough profit to satisfy the economic criteria of both the operator and the project partners. Typically, three production profiles will be provided. These will use risking factors to derive a downside, base, and upside case. A *risk matrix table* is presented as a basis for this (Table 22).

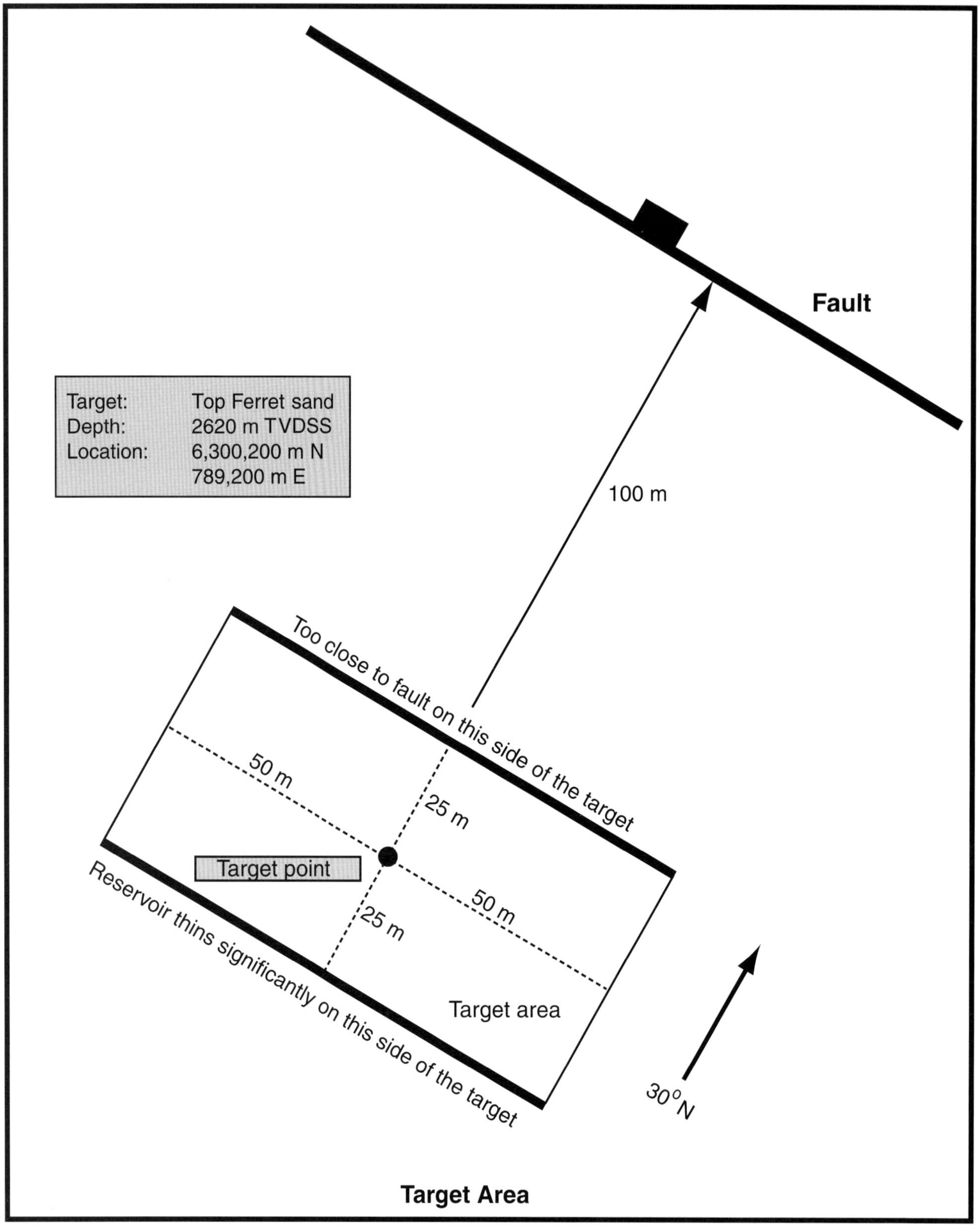

Figure 167. Geological target area for a new well. TVDSS = true vertical depth subsea.

The incremental reserves and the *net present value* (NPV) of expected production from the well after costs will be estimated for all three cases. The NPV is the present day value of future production discounted by the expected inflation and a factor that considers the cost of money to the company.

A section will be included in the well proposal on the well planning and operational requirements. This

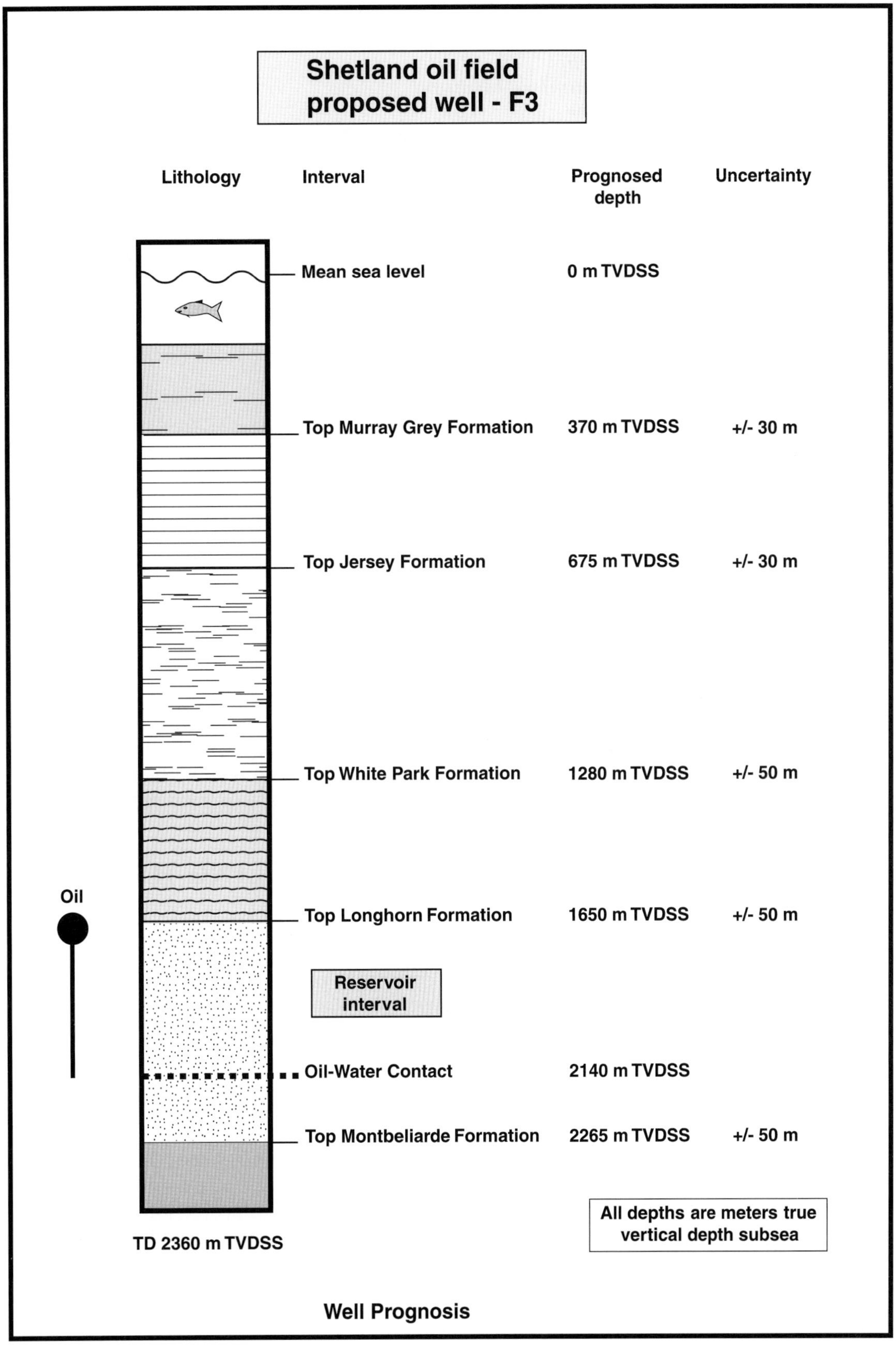

FIGURE 168. A well prognosis shows the expected depths for the tops of each formation likely to be penetrated along the proposed well path.

Table 22. Risk matrix table. STOIIP = stock tank oil initially in place.

Factor	*Risk*	*Impact*	*Impact*	*Mitigated by*
	High/ Medium/ Low	*High/ Medium/ Low*		
Well comes in low to prognosis	M	L	Less STOIIP	Sufficient volumes above the nearest producer to justify the well
Poor reservoir quality (because of fault proximity)	M	M–H	Poorer productivity	Similar well, P_{15} produces at a high rate, although near the fault
Inefficient sweep resulting from faults or heterogeneity	L	M	Faster water breakthrough	Completion strategy is flexible enough to cope with early water production
Missing upper sand (erosion or facies change)	L	M	Less STOIIP and production	
Missing reservoir interval caused by minor faults	L	L	Possibly less production	Two reservoirs targeted
Poor injection support	L	L–M	Poorer productivity, costly workovers	Two reservoirs targeted
Reservoir damage, e.g., scale	L–M	L–M	Drop in production	The well can be worked over

provides details as to how the well will be drilled, the type of well that has been planned, and the casing setting criteria.

There should also be a section on the range of expected reservoir and overburden pressures; this will be used to decide on the mud weights for each section of the well. It is advisable to recalculate the reservoir pressures and advise the drilling department shortly before the well is expected to spud, as there may be a several-month gap between the writing of the subsurface well proposal and the spud date. The reservoir pressures may change significantly in that period in an actively producing field. Sometimes nearby injection wells will be switched off for several months before drilling a new well so as to avoid increasing pressures too much in the target intervals.

The final section will state the data gathering requirements. The petrophysicist will coordinate this and will discuss the program with everyone in the subsurface team. This section will give details of the coring operation, if any, and the logs to be run in the well. Guidelines are given for the intervals where drill cuttings are to be collected for further analysis (e.g., biostratigraphy).

Coring and wireline logging operations are expensive; they add to the overall length of time the well takes and introduces the risk of equipment getting stuck downhole. Because of this, it is necessary to justify a business case for the benefits of the information gained from data gathering. The justification may follow on from the need to make an operational decision; for example, selecting the perforated intervals in the reservoir. Data may also be required to help with reservoir characterization and future reservoir management. For instance, formation tester data may be needed to define hydraulic units within the reservoir, or borehole image log data may be required to help with the understanding of a structurally complex or fractured reservoir.

DRILLING HAZARDS

A section must be written to outline any potential drilling hazards. This cannot be missed out; oil industry operations can end up killing people if they go very wrong. It should include anything that might possibly constitute a hazard while drilling, (Table 23) (Haskell et al., 1999; Campbell, 1999).

Communication is of critical importance with drilling hazards. Everyone must be aware of any potential risks involved in the operation. Beware of finding yourself thinking "I'm not sure if this is important, but..." and saying nothing. You may think the problem is trivial or even silly, but people's lives are more important than personal dignity, so speak up. There are no stupid questions or suggestions.

An excellent method of itemizing drilling hazards and problems is the *well problem matrix chart* (Table 24). This is compiled from examination of mud logs and drilling reports from offset wells near the proposed well location. Each problem found in the wells is itemized.

Table 23. Some possible drilling hazards.

Hazard	*Comments*
Shallow gas	If shallow gas is encountered unexpectedly during drilling operations, a blowout can occur. Gas trapped in shallow sediments can be sourced from deeper hydrocarbon reservoirs but may also result from biogenic activity.
Overpressure	Rapid burial of sediments under a low-permeability seal can create higher than normal pressures and undercompacted sediments. If these are not expected while drilling, there is the potential for taking a kick or the borehole may collapse.
Differential depletion in multilayered producing reservoirs	Differential sticking occurs where the drill pipe comes into stationary contact with the borehole wall. A high mud weight and a low formation pressure can create a differential force acting to immobilize the drill string.
Lost circulation zones	Loss of the drilling mud into fractures or cave systems can result in well control problems, potentially leading to blowouts if not dealt with.
Weak formations	When the weight of the drilling mud exceeds the fracture strength of a weak formation, fracturing will occur leading to lost circulation problems.
Hydrogen sulfide (H_2S)	Common in carbonate reservoirs but also occasionally found in sandstones, hydrogen sulfide is a highly toxic gas that is a known killer when it leaks at the surface.
Gas hydrates	Gas hydrate is a frozen form of gas (mostly methane) and water that occurs in sea floor oceanic sediments and permafrost regions at moderate pressures, low temperatures, and high gas saturations. Drilling problems are generally caused by the dissociation of the gas hydrate as a result of heating by warm drilling fluids. These include uncontrolled gas releases during drilling, collapse of wellbore casing, and slope failure of the seabed sediments.

HIGH-PRESSURE/ HIGH-TEMPERATURE RESERVOIRS

High-pressure/high-temperature (HP/HT) reservoirs are particularly difficult to drill. The United Kingdom oil and gas regulatory body defines HP/HT reservoirs as those with pressures more than 10,000 psi and/or with temperatures in excess of 300°F (Loth, 1998). Some definitions will also stipulate a pore-pressure gradient of at least 0.8 psi per foot. This type of reservoir is common in the deeper parts of the UK North Sea, offshore west Africa, and the Gulf of Mexico for instance.

Wells in HP/HT reservoirs require special rigs designed for high pressures. There is commonly a low drilling margin between the fracture strength of the rock and pore pressures. Strong casing is required for these wells to contain the high pressures. They are expensive wells to drill, and planning can take many months, sometimes years. If things go wrong, it can be very expensive to remedy the situation. Conventional logging tools may not work at the high temperatures found in these reservoirs. Even wireline logs specially designed for high temperatures can be operating at the upper end of their reliable working range in HP/HT wells. Mud coolers may be required to give the logs a chance of working.

Drilling can be very difficult in depleted HP/HT reservoirs. In layered reservoirs, some hydraulic units may become much more depleted than others. The fracture strength may have decreased in the more depleted layers to less than the pore pressure in the higher pressured units. It is not possible to drill safely in this situation because a mud weight sufficient to counter the higher pore pressures will fracture the depleted zones. Mud will be lost into the fractures, and the loss of a sufficient weight of mud to balance the formation pressures will cause serious well control problems.

Table 24. Well problem matrix chart.

Interval	*Drilling Hazard Encountered in the Offset Wells*		
	Well 1	*Well 2*	*Well 3*
Tertiary	Shallow gas		Swelling shales
Cretaceous		Gas shows in the Chalk	
Jurassic	Lost circulation into fractured dolomites		Variable pressure depletion within the reservoir caused differential sticking of the wireline logs

GEOLOGICAL WELL PLANNING DATA DOCUMENT

The well proposal document is typically issued months even years before the well is actually spudded. By the time it comes around to spud the well, much of it may already be out of date. Well planning specifics are likely to have changed, not the least because the detailed drilling engineering planning will only usually start once the budget has been approved for the well. Details of the logging, formation pressures, etc., may also have changed in the interim. Thus, it is good practice to issue a *geological well planning document* before the detailed drilling engineering analysis has been finalized. This should be a short document restating the target parameters, formation pressures, well prognosis, data acquisition program, and the possible drilling hazards.

WELL OPERATIONS

The operations geologist will normally handle the day-to-day well operations. In some of the smaller offices, there may not be an operations geologist and the production geologist will be expected to perform this role. The production geologist will have a watching brief on the details of well operations, and they will be copied on the daily drilling and geology reports from the rig. The operation will start to get critical as the top reservoir is approached and the production geologist will liase closely with the operations geologist at this stage. Sometimes, expedient changes are required to the well planning program as a result of unforeseen problems or if the well is significantly off prognosis. The operations geologist will coordinate with both the production geologist and the drilling engineer to come up with any alternative plans for continuing the well.

After the well has reached TD, the final well results will be analyzed to see how they have come in relative to prognosis. The petrophysicist will be active at this stage, gathering all the log data to make a petrophysical interpretation (see Figure 49). In parallel, the geologist will make a log correlation to see how the new well results fit in with the existing geological scheme.

A well will rarely turn out exactly as predicted. When it comes to drilling new wells, it is wise to prepare for the unexpected. Sometimes the well results are better than predicted, sometimes worse.

There is often a rush to evaluate any new well results, and many weekends or a late night can be spent in an office working on an analysis of the new well results. The reason for the rush is that the operation to complete the well will be waiting on this evaluation.

If the well comes in with good results then a decision will be made as to what intervals are to be produced. An e-mail will be sent to the rig and copied to the completion engineer with details as to which reservoir intervals are to be perforated. If the well has come in a lot worse than expected, then there may either be a decision to abandon the well, to make the best of a poor well and produce what is there, or else sidetrack the well hopefully to a better location.

END OF WELL REPORT

Once the well has been drilled and all the data received and analyzed, then an *end of well report* should be compiled. This should include information on the basic data, the original objectives, when the well was drilled and completed, formation tops, a list of wireline logs run, core retrieved, the producing interval of the well, and the petrophysical analysis log. It is worth putting in a brief evaluation of what the well shows that is new or different from the existing geological scheme and some speculation as to why this is the case.

31

Shepherd, M., 2009, Problem wells, *in* M. Shepherd, Oil field production geology: AAPG Memoir 91, p. 249–251.

Problem Wells

INTRODUCTION

Sometimes a well is drilled and although the initial logging results have looked impressive, the well has never produced a large volume of hydrocarbons. On occasion, a well can start to produce at high initial rates only to be dead a matter of weeks later. This happens every now and again with new wells in most fields. These are problem wells. The various problems that can arise include formation damage, poor reservoir permeability, fault damage, poor reservoir pressure, water production, and mechanical problems.

FORMATION DAMAGE

A well can suffer *formation damage*. This results from the introduction of incompatible fluids and solid particles into the formation as a result of drilling a well, workover operations, or the use of inappropriate stimulation fluids (Figure 169a). The well fluids may chemically or physically react with minerals within the rock to produce fines that clog up the pore throats, reducing or destroying the permeability (Krueger, 1986).

One of the parameters calculated from well test analysis is the *skin factor*. *Skin* is a zone of altered permeability near the wellbore, which results either from formation damage (positive skin) or well stimulation (negative skin).

Sandstones with significant amounts of clay minerals are prone to formation damage (Eslinger and Pevear, 1988). Where clays with a high swelling potential such as smectite occur, changing the composition of the associated pore fluids (e.g., drilling mud invasion) will cause damage. Even where nonswelling clays are present, clay cements are generally weakly attached to the grain framework. The clay minerals can be disaggregated by a combination of changes in water chemistry and the force of fluids passing through the pores. Loosened clay minerals such as kaolinite will be carried through the pores with a tendency to mat across pore throats like leaves in a drain, reducing the permeability in the near wellbore area.

Formation damage can also be caused by the inappropriate use of acid during well stimulation operations. If iron minerals such as chlorite are present in significant quantities within the reservoir sandstone, the mineral can react with the acid to form iron hydroxide gels.

The introduction of fluids, including water, into the near wellbore area can reduce the relative permeability to hydrocarbons. Precipitated solids can form as a result of the reaction of the well fluid with saline formation water.

Formation damage can be prevented. The production geologist will often be asked to provide core material for laboratory testing with potential drilling, workover, or stimulation fluids. If a well suffers formation damage, there are remedies that can be taken. The well can be acidized, fractured, or reperforated. It can be cleaned up by nitrogen gas lifting or by other methods of artificial lift. In an extreme case, the well can be sidetracked using more appropriate drilling fluids.

POOR RESERVOIR PERMEABILITY

If a new well is producing less than predicted, and the well test permeabilities are lower than they should have been, chances are that the well has encountered an unpredicted localized reduction in permeability that may be facies or diagenetically related. The remedy may be to sidetrack the well or to fracture stimulate the reservoir.

DOI:10.1306/13161218M913372

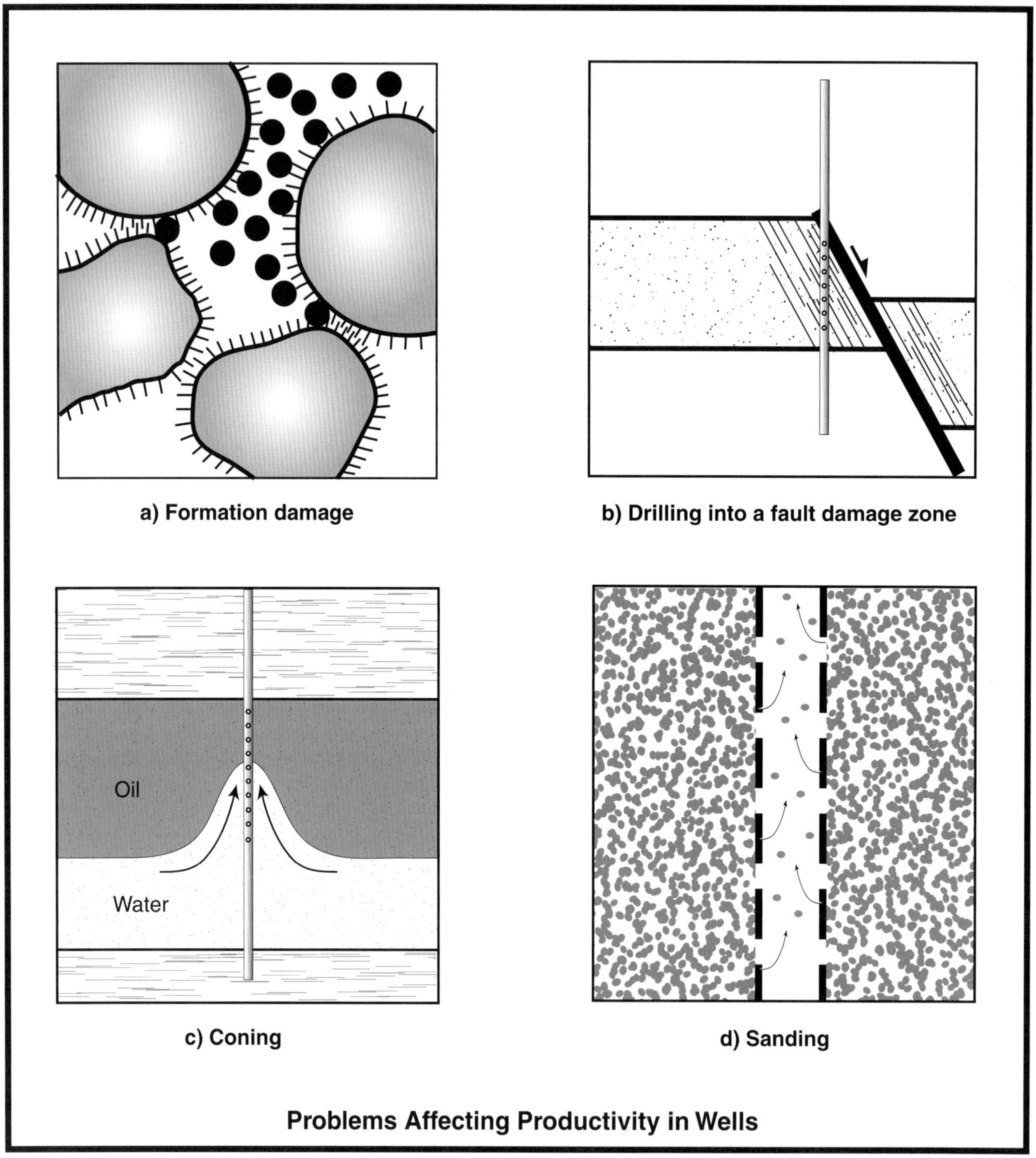

FIGURE 169. Various problems can cause the productivity to be reduced in wells.

FAULT DAMAGE

There may be occasions where a well has been drilled close to a known fault. The reservoir interval looks reasonable on the petrophysical interpretation and the well produces at a high rate initially. Then the pressure falls very rapidly, and the well is dead within a matter of days. If formation damage is not the problem, then this may be an indication that the well has been drilled into a *fault damage zone* (see also Chapter 13, this publication) (Figure 169b). It may be worth hydraulically fracturing the well (Fossen and Bale, 2007), and if this does not work, then drilling a sidetrack farther away from the fault should be considered.

POOR RESERVOIR PRESSURE

Every now and again, a well will come in with the reservoir interval severely pressure depleted. If the poor reservoir pressures were not expected, the chances are that the well has been inadvertently drilled into a depleted fault compartment. An injector may be needed to support the well or a sidetrack may be more practical if the volumes contacted are too small to keep the well going for any period of time.

EXCESSIVE WATER CUT

The well has been drilled into a very clean, excellent section of the reservoir with high permeabilities. The celebrations were perhaps premature; the well starts cutting a lot of water very quickly. The well is choked back and the water cut falls back a small amount but not by too much. The well may be coning (Muskat and Wyckoff, 1934). This is where a more mobile fluid rapidly rises through clean, very permeable rocks into the lower perforations (Figure 169c). Reducing the effects of a cone means cutting back on the production rate.

Rapid water breakthrough can also occur if a thief zone is perforated. Some faults can be effective fluid transmitters along their length. Water influx can typically be reduced by the mechanical isolation of the problem zone.

MECHANICAL PROBLEMS

Excessive *sand production* may be experienced from compacted, friable sandstone reservoirs (Figure 169d). The sand can be highly erosive to any metalwork in the wellbore. In the worst cases, the wellbore can fill up with sand. If the problem has been anticipated, then *sand screens* can be installed in the well. A sand screen is a perforated tube with close wrapped steel wire that is set across the perforations. The mesh size is selected to match the grain size of the reservoir sandstone. The space between the screens and the liner can be filled with sand or gravel, and this is termed a *gravel pack* (Conway, 2007). These filter out the sand but will allow hydrocarbons to be produced. Alternatively, selective perforation can be made so as to avoid any potentially weak zones that could produce sand.

Scaling can occur in production wells. Scale forms from the crystallization of solid material onto metal surfaces as a result of the precipitation of various compounds from aqueous solutions. A common problem in North Sea wells is the formation of a *barium sulfate scale* in the wellbore. This results from the mixing of sulfate-rich injected seawater with barium-rich formation waters. In the worst cases, the well can end up totally blocked by scale. The scale is hazardous to remove because of the presence of small amounts of radioactive material. If injected sea water has broken through to a producer, *scale inhibitor* chemical treatments can be placed in the well to prevent scale from forming in significant amounts.

Shepherd, M., 2009, Depositional environments and their flow characteristics, *in* M. Shepherd, Oil field production geology: AAPG Memoir 91, p. 253.

Section 7

Depositional Environments and their Flow Characteristics

The theme of this section is that specific depositional environments, deltas for instance, show common patterns of production behavior. Many of these themes repeat again and again from field to field in different basins and in various geographical areas around the world. The knowledge of the production patterns from one field in a specific environment may be applicable to reservoirs with the same depositional environment elsewhere.

Each of the depositional environments has a specific character in terms of the flow geology. The macroforms making up a reservoir will have a typical size range and geometry for a particular depositional environment. These combine to give one of the distinctive geometric types; shoreline deposits are commonly layer-cake in geometry whereas meander belts are labyrinthine in form. Certain lithofacies will act as flow barriers; for example, marine shales in deep-water marine reservoirs or coals in fluvial reservoirs. Permeability profiles can be particular to a specific type of depositional environment; channel fills commonly show upward-decreasing permeabilities whereas sediments prograding into standing water will typically exhibit an upward-increasing permeability profile. Production characteristics, the sweep efficiency, and the location of bypassed hydrocarbon volumes can to a large degree be predicted according to the particular depositional environment.

The flow character of a reservoir is strongly influenced by how the various macroforms within a reservoir connect to each other (Larue and Friedmann, 2005). Although individual sand-prone packages tend to link up as connected volumes, on occasion, the boundary between two sand prone macroforms can form a flow restriction. This latter behavior can be characteristic of certain depositional settings.

DOI:10.1306/13161219M913372

32

Shepherd, M., 2009, Eolian reservoirs, *in* M. Shepherd, Oil field production geology: AAPG Memoir 91, p. 255–259.

Eolian Reservoirs

Eolian sediments comprise desert sediments and coastal sand dunes, much of which are wind deposited. They can form thick and laterally extensive sandstone reservoirs. Eolian sandstone reservoirs are commonly found in two time intervals in the geological record, the early Paleozoic and between the Permian and the Early Jurassic (North, 1985).

EOLIAN LITHOFACIES ASSOCIATIONS

Eolian lithofacies associations include dune, interdune, fluvial, and sabkha environments (Figure 170). *Dunes* form where large volumes of dry sand are blown across the landscape. Lying between the dunes are the *interdune* areas, which are flat-lying belts or depressions. These areas may be subjected to either erosion or deposition. In wetter conditions, alluvial fans may extend outwards from upland areas, and fluvial sediments can be deposited by ephemeral streams. Large damp to wet areas between the dunes may dry out to form flat-lying evaporitic crusts called *sabkhas*. *Playa lakes* are desert lakes that episodically dry out.

Several different dune types are found in deserts. They can form as *crescentic dunes* (including *barchan dunes*), as long *linear dunes*, or as *star dunes* with crest lines radiating from one or two central peaks. Different dune types may be superimposed to form *complex dunes*, whereas the same type of dune is superimposed to form *compound dunes*. The term *draa* has been used to refer to large compound or complex dunes (Kocurek, 1996).

In deserts, dune sediments aggrade laterally and vertically as large-scale sand blankets. These may be very thick (more than 300 m [1000 ft]), and cover hundreds of square kilometers (Richardson et al., 1988a). Internally, dunes comprise thick, cross-bedded intervals of well-rounded, well-sorted sandstones. They are normally the most productive lithofacies in eolian reservoir systems. Flatter-lying *eolian sand sheets* may be found along the margins of dune systems (Kocurek and Nielson, 1986).

Interdune and fluvial sediments generally show poorer reservoir characteristics by comparison to dune lithofacies. They are poorly sorted and are more likely to contain evaporite cements. Intercalated fine-grained sand and silt laminations together with diagenetic cementation tend to produce reservoir intervals with very poor vertical permeability. Nagtegaal (1979) used multivariate analysis to determine the factors controlling the porosity of eolian sediments from the Permian of the southern North Sea. He found that the main control on porosity is grain sorting, which varies from well sorted in dune sandstones to less well sorted in the other associated sediments. The relationship between original sedimentary texture and porosity has survived even after extensive diagenesis. The very poor permeability characteristics of interdune sediments are commonly reported. Lindquist (1983) found a contrast in permeability of four to five orders of magnitude between interdune and dune deposits in the Nugget Sandstone of southwestern Wyoming.

GEOMETRY

Eolian sediments form layer-cake to jigsaw-puzzle geometries. Dune sand bodies may intercalate with or pinch out into poorer quality sabkha and fluvial facies associations. Eolian environments tend to occur on a large scale and dune sandstones can be greater in length than typical well spacings. Weber (1987) described outcrops of eolian sandstone in the Permian

DOI:10.1306/13161220M913372

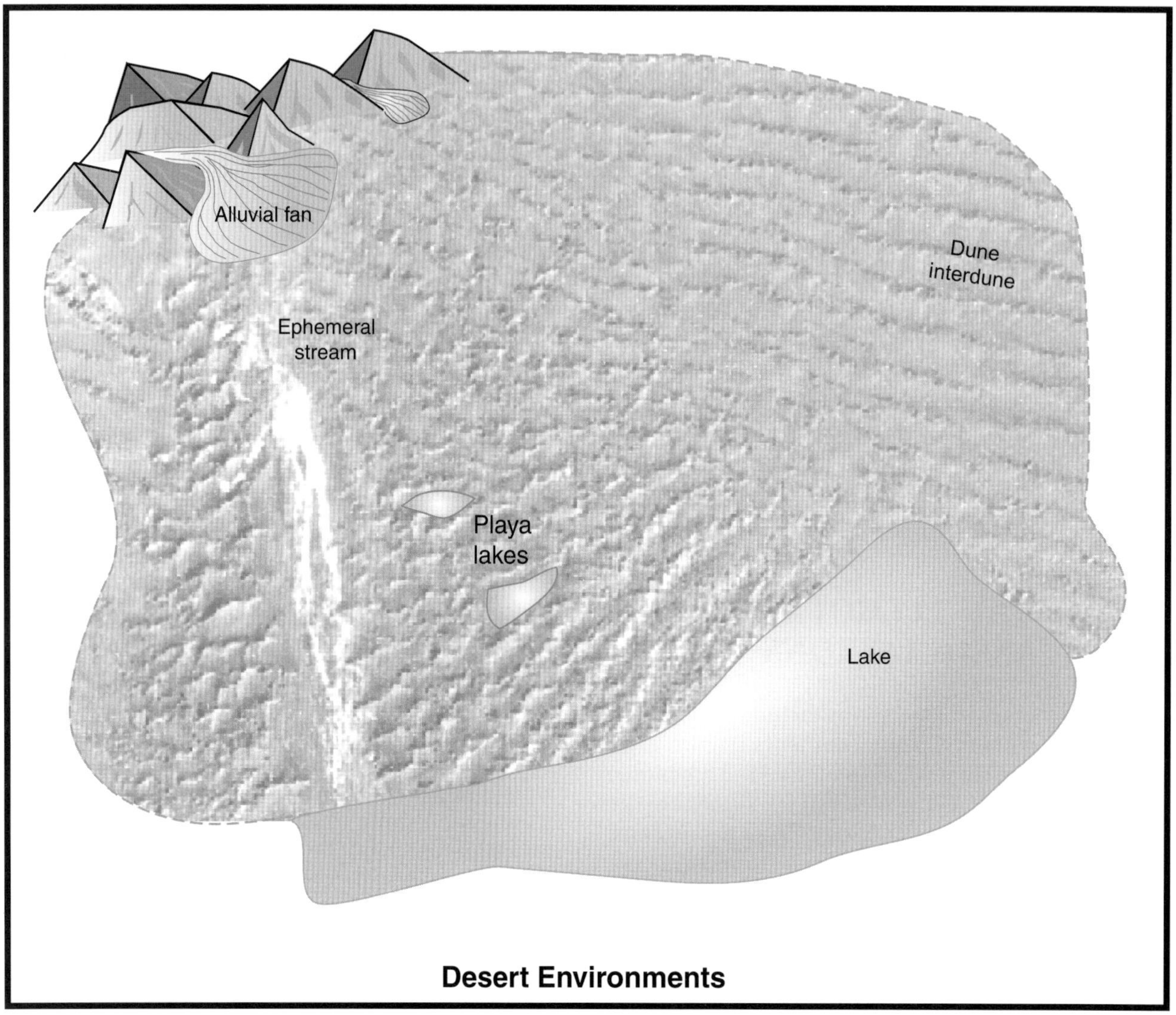

FIGURE 170. Eolian lithofacies associations include dune, interdune, fluvial, and sabkha environments.

De Chelly Sandstone in northern Arizona. The average thickness of cross-bed sets is about 6 m (20 ft). The width-to-thickness ratio is estimated as between 50:1 and 100:1. The length-to-thickness ratio is estimated as 200 to 1.

HETEROGENEITY OF EOLIAN SANDSTONES

It is difficult to make generalizations about the effect of heterogeneity on hydrocarbon recovery from eolian sandstones as recovery factors seem to vary enormously worldwide. It is possible that eolian sandstones are much better suited as gas than oil reservoirs. Some of the Western European Permian desert sandstone gas reservoirs show very high recoveries; a 91% recovery factor is quoted for the Leman gas field for instance (Hillier, 2003). By comparison, the viscous oil reservoirs of the Pennsylvanian to Permian Tensleep Sandstone of Wyoming and Montana show low recoveries by primary production. In the Little Buffalo Basin field of Wyoming, well spacing has been reduced successively from 40- to 20- to 10- to 5-ac spacing, in some cases without interwell interference (McCaleb, 1979; Ahlbrandt and Fryberger, 1982).

Ciftci et al. (2004) attributed the poor recovery in the Tensleep Sandstone to low permeability baffles and barriers along *bounding surfaces* within the eolian dune sets. Bounding surfaces are subhorizontal to inclined discontinuities that divide eolian cross-beds into subsets, sets, and cosets (Figure 171a). These form as a result of dune migration at the smaller scale and from regional discontinuities at the larger scale. Bounding surfaces have a tendency to act as baffles or barriers as a result of facies and grain size contrasts across them (Shebi, 1995). Perhaps the considerable difference in recoveries between the eolian reservoirs of Northwestern Europe and the

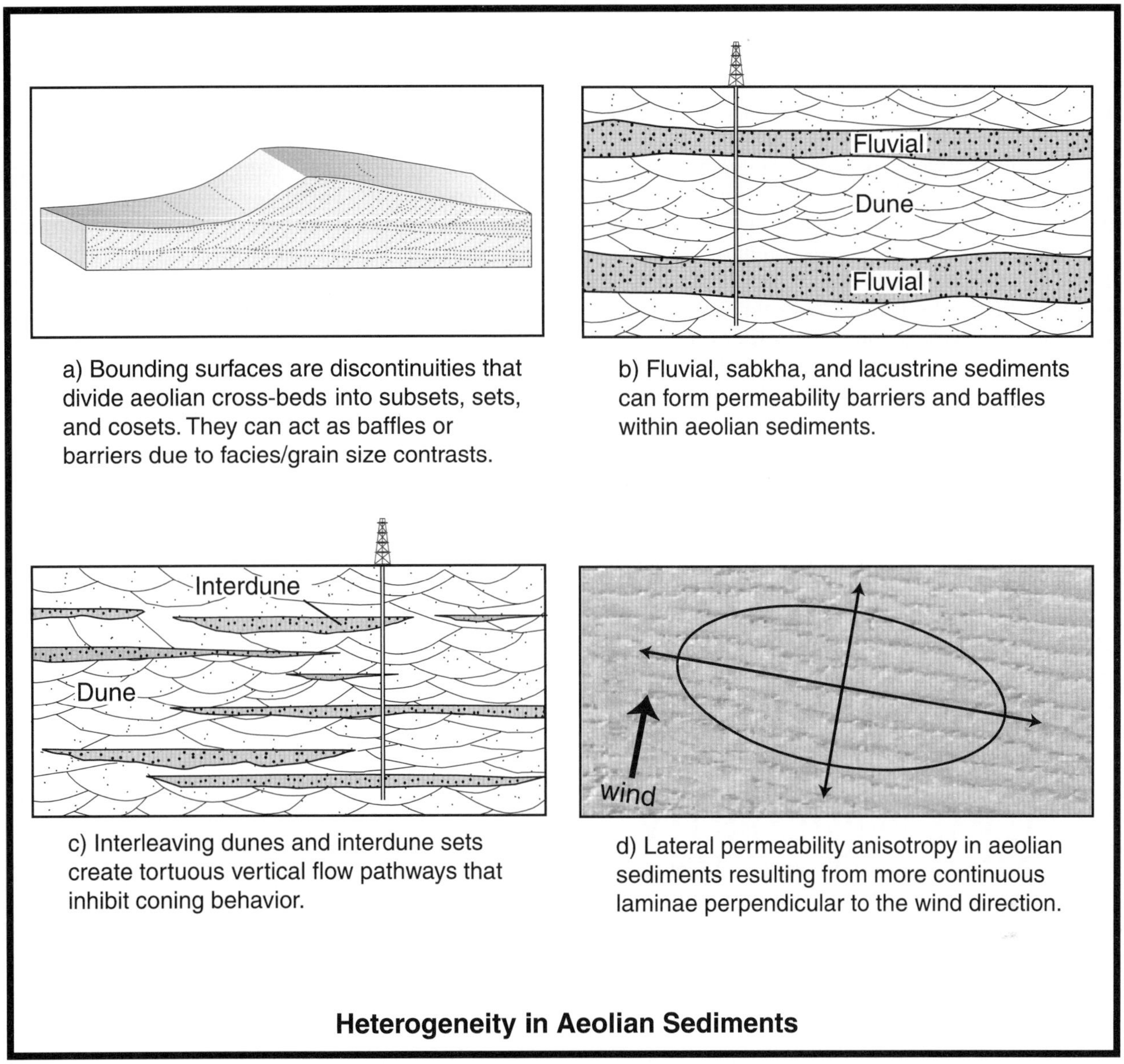

FIGURE 171. The influence on fluid flow by heterogeneity in eolian sediments.

United States is a function of how bounding surfaces influence fluid flow in each area. These features may provide less of an impedance to the flow of highly mobile gas in the Northwestern Europe gas fields than they do for the viscous oil found in the Tensleep Sandstone of Wyoming and Montana.

One other factor may contribute to poorer oil than gas recoveries in eolian sediments. In oil fields, a significant volume of capillary-trapped oil can result from the waterflooding of dune-bedded sandstones. Huang et al. (1995) showed that between 30 and 55% of the oil was trapped in a coreflood experiment on cross-laminated eolian sandstone under conditions of low-rate flooding.

VERTICAL PERMEABILITY BARRIERS IN EOLIAN SANDSTONES

Fluvial and sabkha sediments deposited in interdune areas can be permeability barriers and baffles within eolian sediments (Figure 171b). These may either be confined to interdune areas and of limited extent or they can be laterally extensive on a basin scale.

Baffles of limited areal extent in interdune areas are described by Shebi (1995) from the Tensleep Sandstone of the Bighorn Basin in northwestern Wyoming and southwestern Montana. These are thin, discrete intervals of dolomite and anhydrite, about 0.15–0.7 m (0.5–2 ft) thick and with lateral dimensions on the scale of a few meters to tens of meters. The dolomite and anhydrite intervals are interpreted as sabkha deposits, which formed in wet interdune areas and playa lakes.

Studies in the western United States have shown that some sabkha units can be traced for several kilometers within the Mesozoic eolian sediments (Crabaugh and Kocurek, 1993). Cyclic climatic conditions resulted in alternating dune sandstone and widespread sheet-like fluvial deposits in the Jurassic Kayenta-Navajo Formations of northeastern Arizona (Herries, 1993).

The degree of layering within an eolian reservoir can therefore range from moderate to intense (Krystinik, 1990). Probably a major control on this is as to whether dry desert or wet desert conditions prevail, the latter associated with extensive fluvial, sabkha, and lacustrine interbeds.

Vertical permeability barriers can also be formed by diagenetic cements. Chandler et al. (1989) noted that meteoric water can seep along bounding surfaces with the preferential formation of carbonate and silicate cements. Where the lower part of a dune is below a water table, early cementation may form a permeability barrier (North and Prosser, 1993).

LATERAL PERMEABILITY ANISOTROPY WITHIN DUNE SANDSTONES

Horizontal and vertical permeability can be highly variable at the laminar scale in dune sandstones (e.g., Prosser and Maskall, 1993). This results from the configuration of the three basic strata types in dune sandstones: *wind-ripple*, *grain-flow*, and *grain-fall deposits* (Hunter, 1977).

Sand grains migrate over dunes, forming rippled surfaces. The grains pack together relatively closely in wind-rippled strata, and the porosity is lower in these units. Inverse grading is common with low-permeability pin-stripe laminae reducing the vertical permeability. Chandler et al. (1989) found in the Page Sandstone of Arizona that the grain size ratio from the coarse-grained to the fine-grained parts of each wind-rippled strata averages 3:1 and can be as much as 7:1. The permeability ratio between the coarse and fine laminae gives an average value of 11:1 with a maximum value of 75:1.

When the wind-blown sediments reach the brinkline of the dune, the wind speed drops and the grains fall onto the leeward dune slip face, coming to rest as *grain-fall deposits*. These form parallel-laminated, slightly tapering strata. Grain size can vary between the individual strata.

On steep surfaces, avalanches may occur, forming *grain-flow deposits*. These develop as cone- or tongue-shaped geometries at the base of the slipface. Grain-flow cross strata are thicker than other eolian strata, up to a maximum of 2–5 cm (0.7–1.9 in.) thick. They are internally structureless or show subtle grading. The grain packing of grain-flow deposits is relatively loose as a result of the very rapid deposition of the grains. Thus, grain-flow deposits tend to be more porous and permeable compared to the other strata types.

The contrast in grain size and sorting between the individual sand streaks in dune beds results in large variations in permeability. Permeability differences can be exacerbated by early diagenesis. Fine-grained laminae can potentially draw in cementing solutes by capillary action.

All three strata types may be found on dune foresets. By contrast, wind-ripple lamination dominates the interdune sediments. These are typically poorly sorted and finely laminated. Interdune sediments probably create an interleaving network of permeability baffles, which serve to create tortuous flow pathways upward through stacked dune reservoirs (Figures 171c, 172). They can act to inhibit coning in thick dune sandstone reservoirs (Weber, 1987).

Eolian dune sets show strong *lateral permeability anisotropy* within the reservoir. Reservoir fluids flowing across the wind-flow direction are impeded by pin-stripe lamination of fine-grained material along the dune cross sets. By contrast, the individual layers and laminae are much more continuous along the depositional strike trend of the dune system, perpendicular to the wind flow direction (Figure 171d) (Weber, 1987). Krystinik (1990) stated that, in most eolian reservoirs, the *anisotropy permeability ratio* is between 4:1 and 25:1, although the overall range may be approximately 1: 1 and up to 200:1 locally. He recommends that horizontal core plugs should be taken both along and perpendicular to the wind-flow direction in order to assess the lateral permeability anisotropy in dune sandstones. Follows (1997) described how a horizontal well was planned to be drilled along depositional dip in the Auk oil field in the UK North Sea. The intention was to connect up the highest number of grain-flow sets between bounding laminae so as to maximize production.

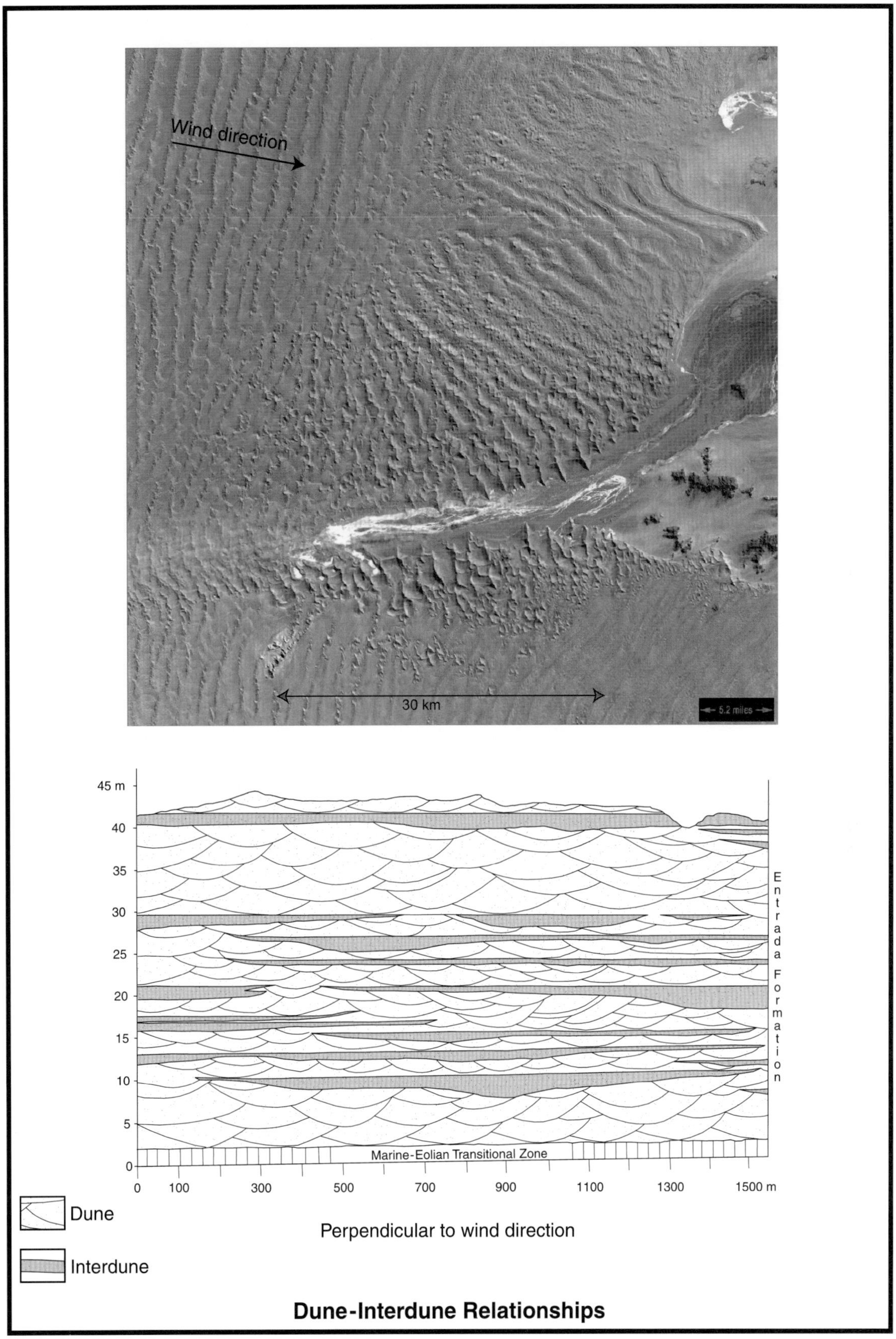

FIGURE 172. Dune-interdune relationships in the Entrada Sandstone, northern Utah and Colorado. Interdune sediments act as discontinuous baffles in eolian sediments (from Kocurek, 1981). Reprinted with permission from Elsevier. Satellite photograph of Namib Desert, Namibia, Courtesy of NASA. Web site www.earthasart.gsfc.nasa.gov.

33

Shepherd, M., 2009, Meandering fluvial reservoirs, *in* M. Shepherd, Oil field production geology: AAPG Memoir 91, p. 261–272.

Meandering Fluvial Reservoirs

INTRODUCTION

Fluvial reservoirs are difficult for the production geologist to understand, characterize, and model. One major problem involves trying to classify fluvial reservoirs in the subsurface. The system used in this book broadly categorizes fluvial systems into meandering and braided fluvial reservoirs. Although this is a classification used by many production geologists, not all experts are happy with this approach; some believe the classification to be too prescriptive. They consider that only limited inferences can be made from core and log data as to the overall geometry of a fluvial reservoir in the subsurface (e.g., Bridge, 2003). Because of this, some geologists prefer to use a simple nongeneric description by classifying subsurface fluvial geometries as either sheets or ribbons (Friend et al., 1979).

Despite the above difficulties, the production geologist will nevertheless try and find some basis for providing a predictive model for the subsurface geology of a fluvial reservoir. Seismic data can help to determine the planform geometry where it is of sufficient resolution (see Figure 70). Fluvial geometries can sometimes be well differentiated on horizon slice amplitude displays (e.g., Brown et al., 1981; Rijks and Jauffred, 1991; Noah et al., 1992; Carter, 2003).

Meander belt reservoirs show different production behavior characteristics from braided river reservoirs; in the absence of seismic geomorphology evidence, the production geologist should intuitively pick the fluvial geometry type most likely to fit the available data and the reservoir performance. Perhaps because of the uncertainty involved in determining the planform geometry in fluvial reservoirs, the scenario approach mentioned in Chapter 22 of this publication may be an appropriate tool to help evaluate fluvial reservoirs.

GEOMETRY OF MEANDER BELTS

Meandering rivers deposit sand and mud within well-defined *meander belts*. The appearance of a meander belt in plan and cross section is of a complex labyrinth of interlocking sand bodies on the scale of hundreds of meters, embedded within varying volumes of mud (Figure 173). The mud can make up 50% or more of the volume. Channel features, where they survive, tend to be plugged with clay (Figures 173, 174).

Gibling (2006) provided data on width and thickness relationships for fluvial systems in various settings from Quaternary and older outcrops (Table 25). He found that meandering rivers do not generally create thick sedimentary packages. The maximum thickness for meandering river deposits in his database is only 38 m (124 ft), with 4–20 m (13–65 ft) as a common thickness range. Gibling makes the comment that despite their familiarity in the modern landscape, meandering river deposits probably constitute only a minor portion of the fluvial rock record by comparison to braided systems. This may be because the organized flow patterns associated with meandering rivers rarely persist for long periods.

MEANDERING FLUVIAL MACROFORMS

Macroforms found in meander belts include point bars, crevasse splays, and mud-rich channel plugs within a background of floodplain muds. Coals are found in fluvial systems with high water tables. Levees sometimes border rivers but they do not appear to be a major feature preserved in the subsurface (Gibling, 2006).

DOI:10.1306/13161221M913372

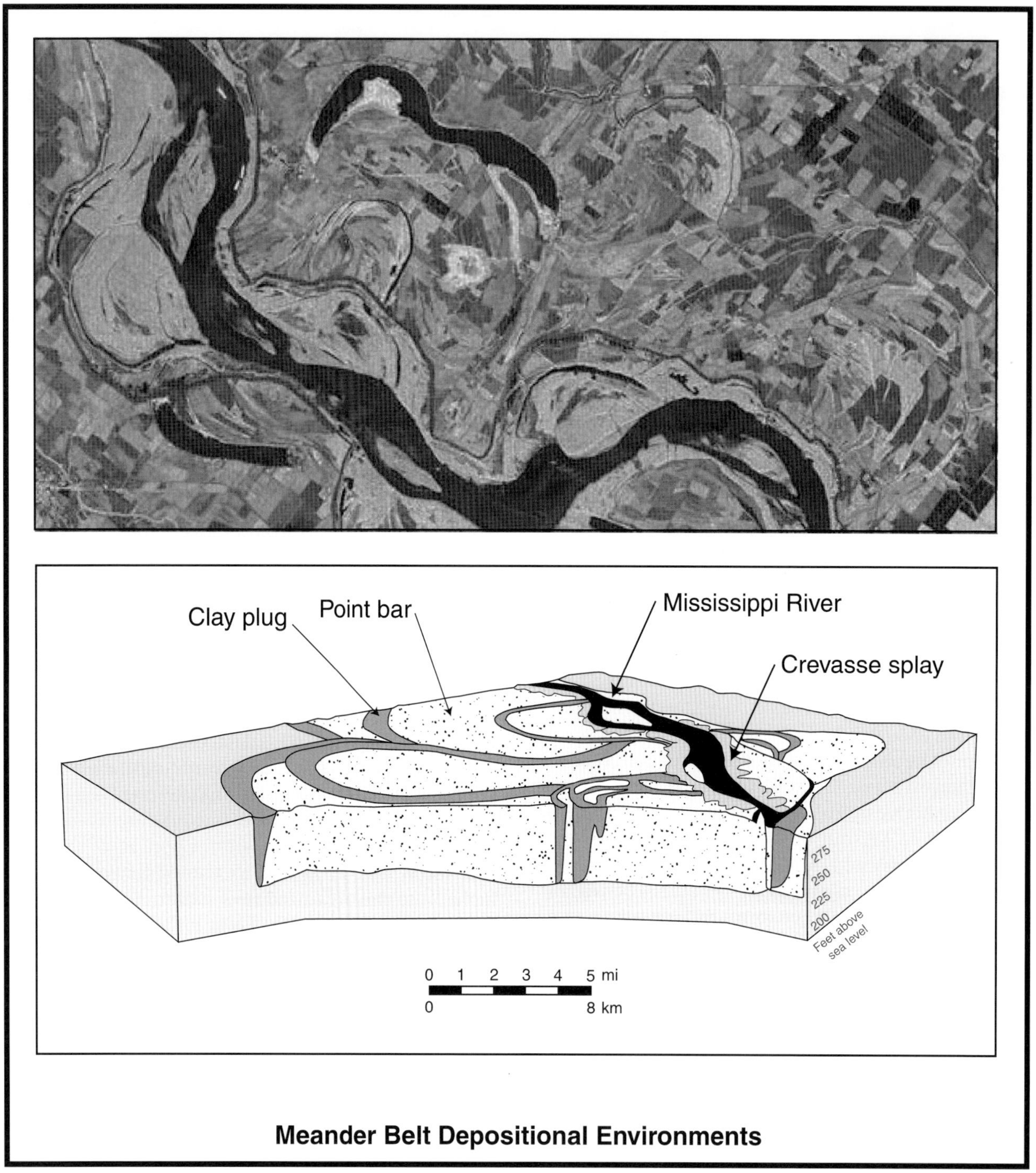

FIGURE 173. Satellite photo of a fluvial meander belt, United States. Courtesy of the NASA Web site (www.earthasart.gsfc.nasa.gov). The lower diagram shows the internal geometry of the present-day Mississippi meander belt (from Jordan and Pryor, 1992). Reprinted with permission from the AAPG.

POINT BARS

The main sand-prone macroforms found in meandering river sediments are *point bars* (Figure 175). These form by lateral accretion of sediment on the inside of meander bends, and they occur as discrete sand bodies with a lenticular or half-moon shape in plan view. A multitude of point bar sandstone bodies may be found studded within a meander belt.

Jordan and Pryor (1992) made detailed measurements on sediment body dimensions along a 10-mi

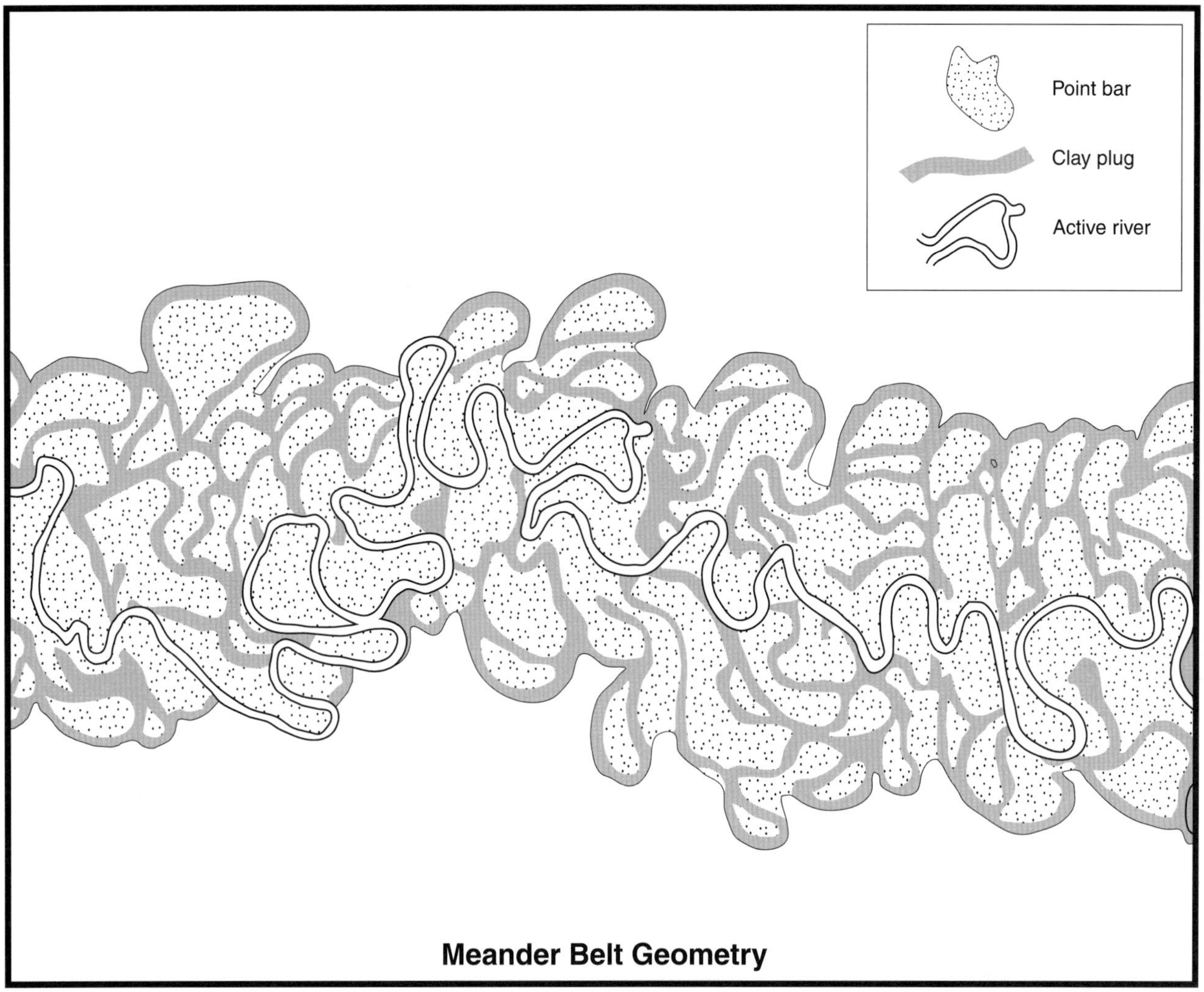

FIGURE 174. Computer simulation of a meander belt geometry from Sun et al. (1996). The meander belt comprises a complex labyrinth of point bars and clay plugs. The meander belt width is on the scale of a few hundreds of meters. Reprinted with permission from the American Geophysical Union.

(16-km) stretch of the Mississippi River meander belt system in southeastern Missouri. The point bars here are 15–45 m (49–147 ft) thick, a few kilometers long (typically 3 km; 2 mi), and between 600 and 1800 m (1968 and 5905 ft) wide. The Mississippi is a continental-size river stretching the length of the United States. As a general rule, big rivers like the Mississippi will tend to deposit large point bar sand bodies; lesser rivers will tend to produce smaller point bars (Figure 176a). For example, the typical width of individual point bars in the 35-1 sand of the Widuri field in the Java Sea is 1200-1500 m (Carter, 2003). A well located in one of the point bars has produced 3.2MM bbls of oil. By contrast the fluvial sands in the Jonah Gas field of Wyoming are estimated to have a P_{50} width ranging from only 60 to 210m (Shanley, 2004).

Several technical papers give cross plots of fluvial sand-body widths versus the maximum bankful depth of rivers (e.g., Bridge and Mackey, 1993). These plots have been used to model thickness-to-width ratios for 3-D geological models.

Miall (2006) criticized the use of empirical relationships for fluvial geometries in too prescriptive a manner. He suggests that they should only be used as approximate guidelines for developing alternative scenarios of fluvial reservoirs for modeling purposes. Shanley (2004), characterizing the Jonah field in Wyoming, preferred to estimate a range of possible dimensions for fluvial bodies instead of using a single unique value for the width-to-thickness ratio.

Werren et al. (1990) described a vertical profile for a point bar deposit in the Cretaceous reservoir of the Little Creek field in Mississippi. An erosional base is overlain by channel lags with intraformational shale rip-up clasts. Above this are large-scale cross-bedded sandstones, which pass upward to beds showing horizontal

Table 25. Width and thickness relationships of fluvial sediments in various settings.*

Depositional Environment	*Thickness*	*Width*	*Width/Thickness Ratio*
Braided and low sinuosity rivers	1–1200 m (3–3937 ft), most < 60 m (197 ft); common range 5–60 m (16–197 ft)	50 m–1300+ km (164 ft–808+ mi); many > 1 km (0.62 mi); common range 0.5–10 km (0.3–6 mi)	15–15,000+; some > 1000; common range 50–1000
Meandering rivers	1–38 m (3–125 ft); common range 4–20 m (13–65 ft)	30 m–15 km (98 ft–9 mi); most < 3 km (1.8 mi); common range 0.3–3 km (0.1–1.8 mi)	7–940; most < 250; many < 100; common range 30–250
Delta distributaries	1–35 m (3–115 ft); most < 20 m (65 ft); common range 3–20 m (10–65 ft)	3 m–1 km (10 m–0.6 mi); most < 500 m (1640 ft), common range 10–300 m (33–984 ft)	2–245; most < 50; many < 15; common range 5–30
Channels in eolian settings	1–19 m (3–62 ft)	2.5–1500 m (8.2–4921 ft); most < 150 m (492 ft)	1–80; most < 15
Valley fills on bedrock unconformities	12–1400 m (39–4593 ft); most < 500 m (1640 ft)	75 m–52 km (246 ft–32 mi); most < 10 km (6 mi)	2–870; highly variable; mainly 2–100
Valley fills within alluvial and marine strata	2–210 m (6–689 ft); most < 60 m (197 ft)	0.1–105 km (0.06–65 mi); common range 0.2–25 km (0.1–15 mi)	4.6–3640; highly variable; common range 10–1000; many from 100 to 1000

**From Gibling (2006), Journal of Sedimentary Research. Reprinted with permission from the SEPM (Society for Sedimentary Geology).*

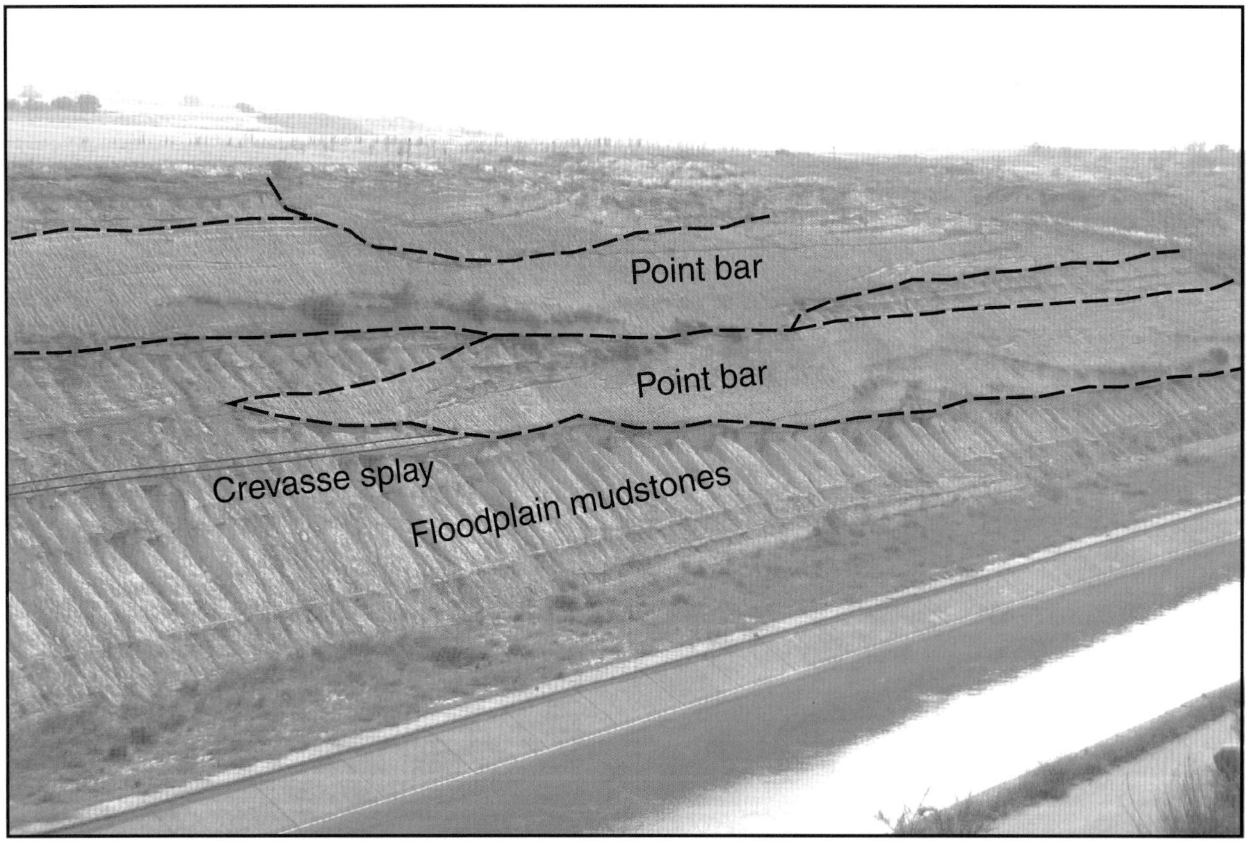

FIGURE 175. Point bars within background floodplain shales and crevasse splays, Ebro basin, Spain.

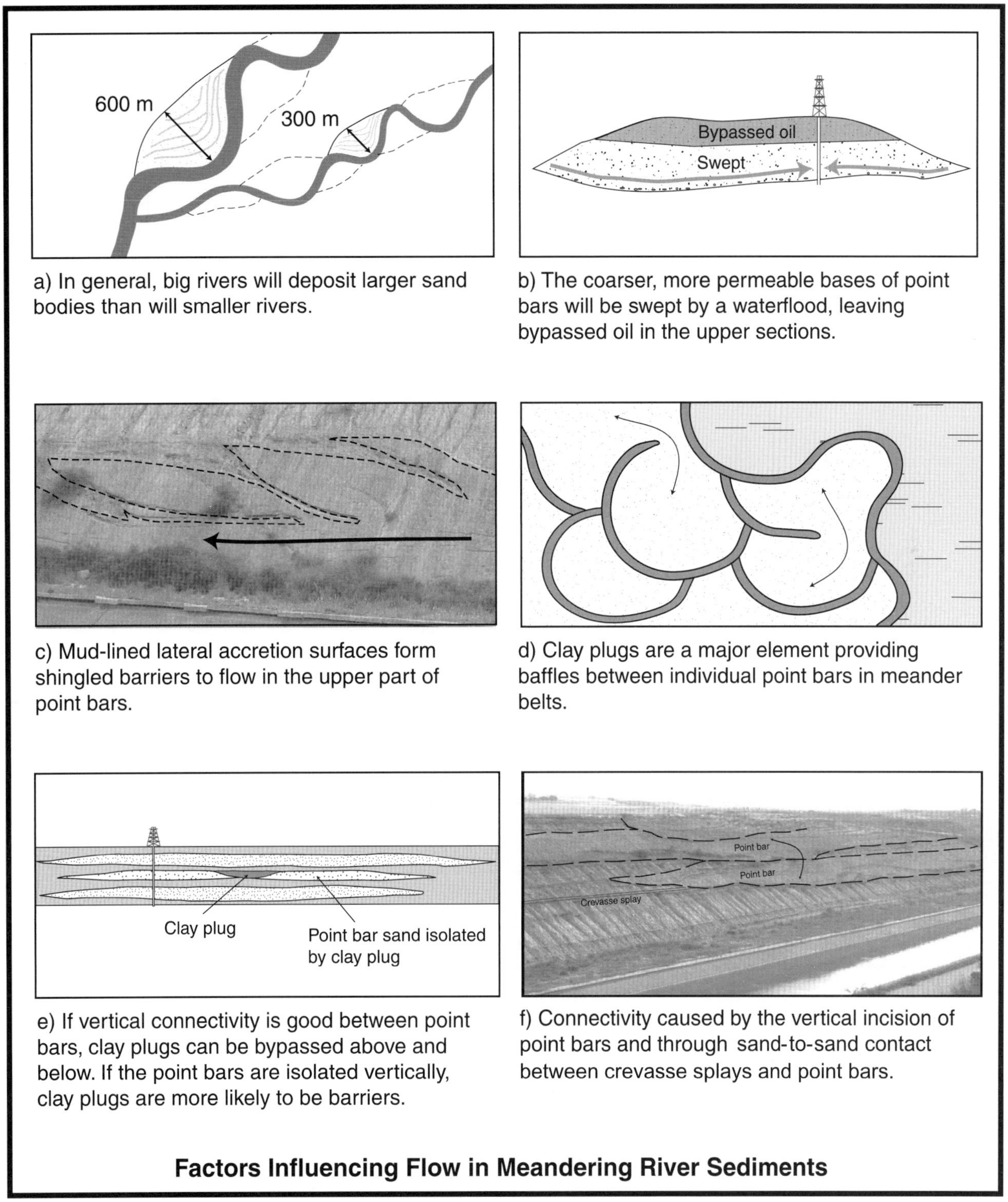

FIGURE 176. Flow geology influences in meandering river sediments.

and small-scale ripple cross laminae, clay drapes, micaceous and carbonaceous streaks, local mud balls, and intraclasts. The overall pattern is fining upward.

Fining-upward profiles are typical for point bars; the permeability declines upward with decreasing grain size (Figure 177). The decrease in permeability commonly occurs in a step-like fashion rather than showing a gradual upward decrease.

Upward-decreasing permeability profiles are unfavorable to efficient sweep. Water will flood through the

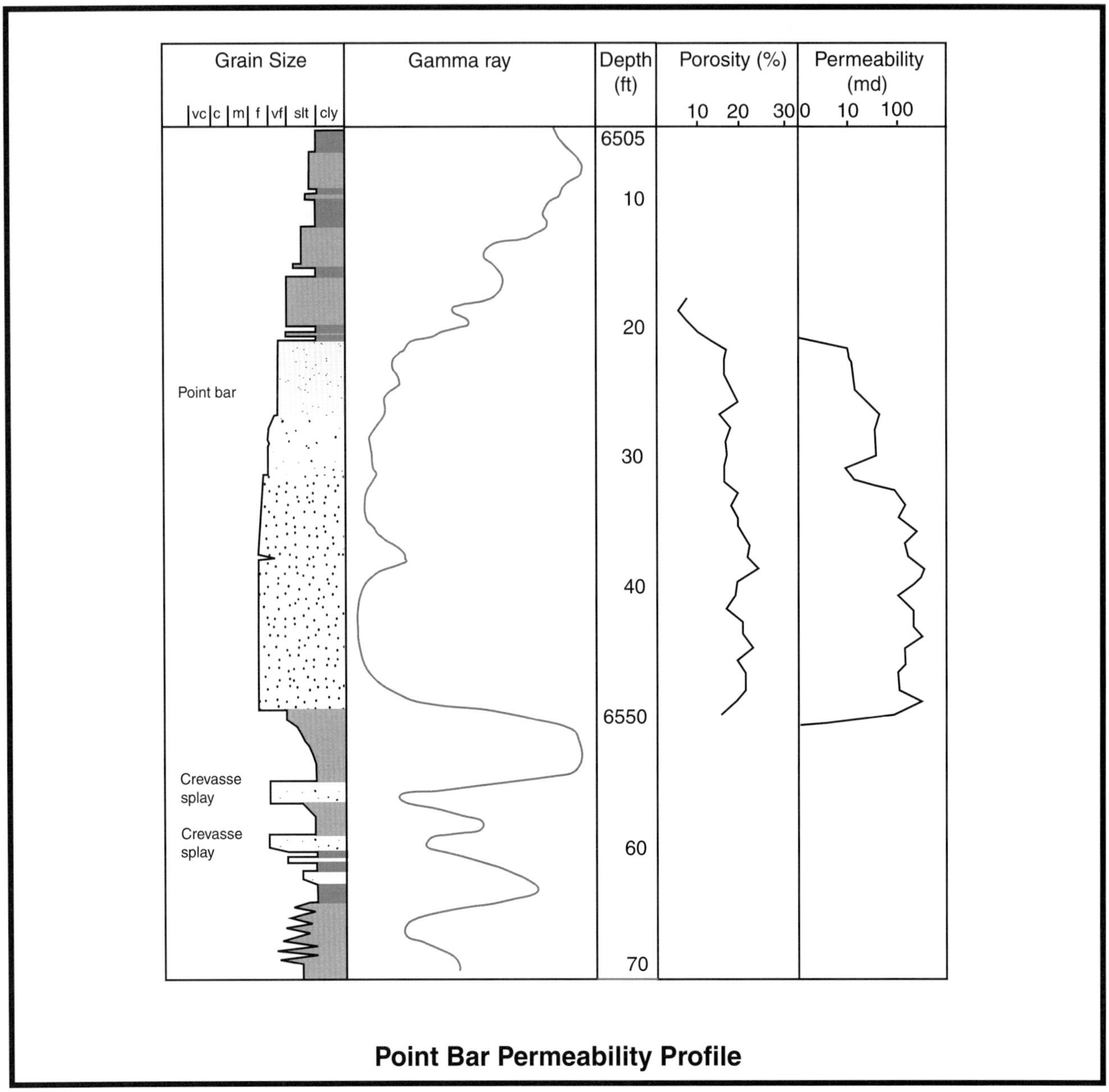

FIGURE 177. Upward-decreasing permeability profile in a point bar sandstone in the Peoria field, Colorado (from Chapin and Mayer, 1991). Reprinted with permission from the SEPM (Society for Sedimentary Geology).

high-permeability basal part of the point bar leaving the uppermost section unswept (Figure 176b).

Sweep will be retarded where the upper sections of point bars form mud sheets along *lateral accretion surfaces* (Figures 176c, 178). These are inclined surfaces formed by the lateral growth of the point bar as the meander loop migrates. Mud-lined lateral accretion surfaces develop by mud deposition during ponding at low river stages (Jordan and Pryor, 1992). The inclined mud drapes form a series of shingled barriers to both vertical and horizontal flow (Ma et al., 1999; Pranter et al., 2007).

Detailed data integration analysis has been made for oil recovery from meander belt sandstones in the Daqing field in China (Xue, 1986; Fu Zhiguo et al., 1998). After 30 yr of production from one of the reservoir units, the sweep efficiency is only 29.8%. It was found that although the basal intervals of the fluvial sandstones are well swept, there is much oil left behind in the upper part. Water cuts in the production wells can reach 90% with only the basal section of the fluvial sandstones contacted by the waterflood. A horizontal well was drilled as a pilot trial to determine whether this could recover the oil in the upper sections of the point bar sandstones. The

FIGURE 178. Mud-lined lateral accretion surfaces in the upper section of a point bar sandstone, Ebro basin, Spain. The point bar grew by accretion from left to right.

well found an estimated net pay of 2–4 m (7–13 ft) but produced less than expected as a result of a combination of formation damage and poorer than predicted vertical permeability (Fu Zhiguo et al., 1998).

Meander belt sediments may be better suited as gas reservoirs than oil reservoirs. The labyrinth of numerous dead ends in these systems will not tend to trap nearly so much gas as oil. The expansion of gas on the reduction of pressure with depletion will cause much of the gas to spill out of the dead ends in fluvial reservoirs. Gas can also flow through the low-permeability connections that exist in fluvial systems, which would otherwise not allow oil to pass.

CREVASSE SPLAYS

Crevasse splays are sandy overbank deposits, which are found intercalated with background floodplain muds. They form fan-shaped sheets, tens to hundreds of meters wide and typically 0.3–2 m (1–6.5 ft) thick. Mjøs et al. (1993) gave a width-to-thickness ratio for crevasse splays of 150–1000. The splays thin laterally toward the margins of the floodplain (Miall, 1996). Individual flows may amalgamate into thicker composite intervals.

Crevasse splays do not normally contain large volumes of hydrocarbons, although they may provide a target for infill wells onshore. Ambrose et al. (1991) noted that crevasse splays in the meander belt reservoir of the La Gloria gas field in Texas were only partially depleted or undepleted. These have limited lateral extent and pinch out less than 460 m (1509 ft) along depositional strike from the channel fill deposits. The crevasse splays are often found with much higher pressures than the main producing intervals in the field. Nevertheless, they deplete rapidly on production, indicating that they contain only small, isolated volumes of gas. However, the overall production potential is thought to be significant as these splay compartments are numerous.

MUD PLUGS

Fluvial reservoirs may be partially or totally compartmentalized by *abandoned channel mud plugs* (Ambrose et al., 1991). A meander loop can be cut off by the river breaking through into a new course during a flood. The abandoned meander channel is quickly isolated from the river flow, typically forming lakes and ponds for a period of time. The channels slowly fill up with clay, silt, and peaty organic material.

Mud plugs are crescentic in plan view and lenticular in cross section. They are narrow with widths of tens to hundreds of meters. In the Widuri field, offshore Java, mud plugs are described as 50–150 m (164–492 ft) wide and up to 5 m (16 ft) thick (Carter, 2003). A meander belt may contain many individual mud plugs, which can collectively make up a large volume of the total system. In a study area comprising 16 billion m^3 of a meander belt in the modern day Mississippi River, Jordan and Pryor (1992) estimated that 11.1 billion m^3 of this volume comprises point bar and splay sandstones and the remaining 4.9 billion m^3 consists of clay plugs within numerous abandoned channels.

Clay plugs can be difficult to recognize in the subsurface. It is sometimes possible to identify them on correlation panels if the wells are closely spaced. They can also be picked out from amplitude displays on good quality seismic data.

CONNECTIVITY IN MEANDER BELTS

Determining the connectivity of the sand bodies in a meander belt system is critical to evaluating the commerciality of types of reservoirs. Individual point bars are relatively small reservoir bodies likely to contain only a few million barrels of recoverable oil at best. They may be successfully drilled onshore where wells are relatively cheap, but they are less likely to make much profit as a primary target offshore. However, if several of these sand bodies overlap with each other, then they can combine to form a larger connected sand volume.

Technical papers indicate that connectivity in meander belt sediments can be highly variable and prone to chance factors. An example of this is the Little Creek field in Mississippi (Werren et al., 1990). The lower reservoir unit comprises three connected point bar sandstones (Figure 179). The Sweetwater field immediately to the north is believed to form part of the same fluvial system and produces from a fourth point bar sand body along the same trend. Nevertheless, the Sweetwater field is isolated from the Little Creek field on the evidence of a 24-m (79-ft) higher oil-water contact. The two fields are thought to be separated by a shale plug or an area with relatively high capillary displacement pressure. A similar observation was made by Carter (2003) for a meander belt reservoir in the Widuri field in the Java Sea. Following the depletion of a well on the updip side of a 100-m (328-ft)-wide abandoned channel, a second well was drilled on the opposite site of the clay plug. A full oil column was found in the new well, unaffected by production from the previous well. In the Saddle Lake area of Alberta, Canada, oil and gas pools are restricted to point bars completely surrounded by clay plugs (Edie and Andrichuk, 2003).

It seems from these case examples that clay plugs can be an important element limiting horizontal connectivity in meander belt sediments (Figure 176d). Some point bars show flow connectivity with each other, others

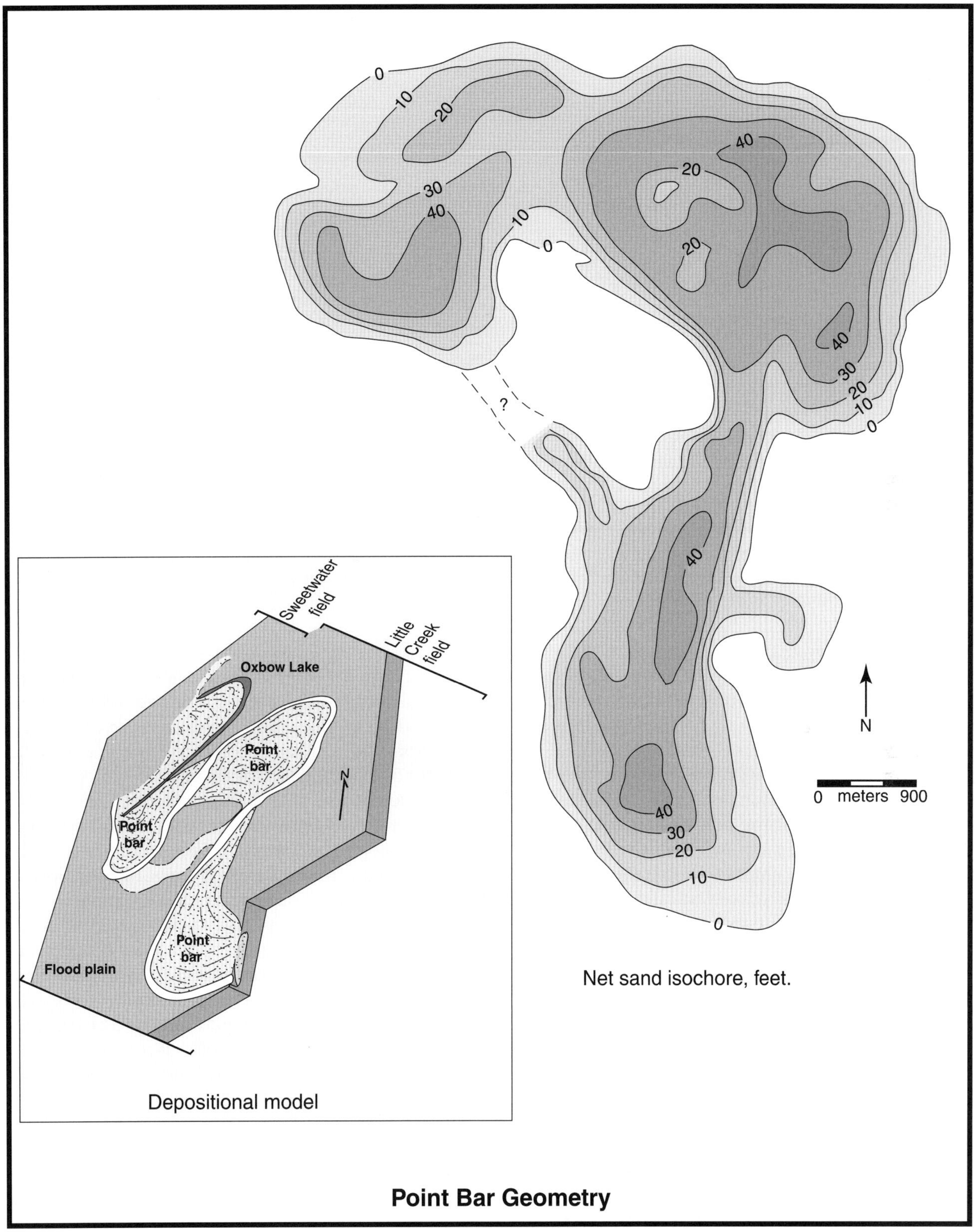

FIGURE 179. Net sand isochore map of the Q reservoir in the Little Creek field in Mississippi. The reservoir comprises three connected point bar sandstones in a background of floodplain mudstones and siltstones. Just to the north is the Sweetwater field, which produces from a depositionally isolated point bar in the same meander belt system (from Werren et al., 1990). Reprinted with permission from Springer Ltd.

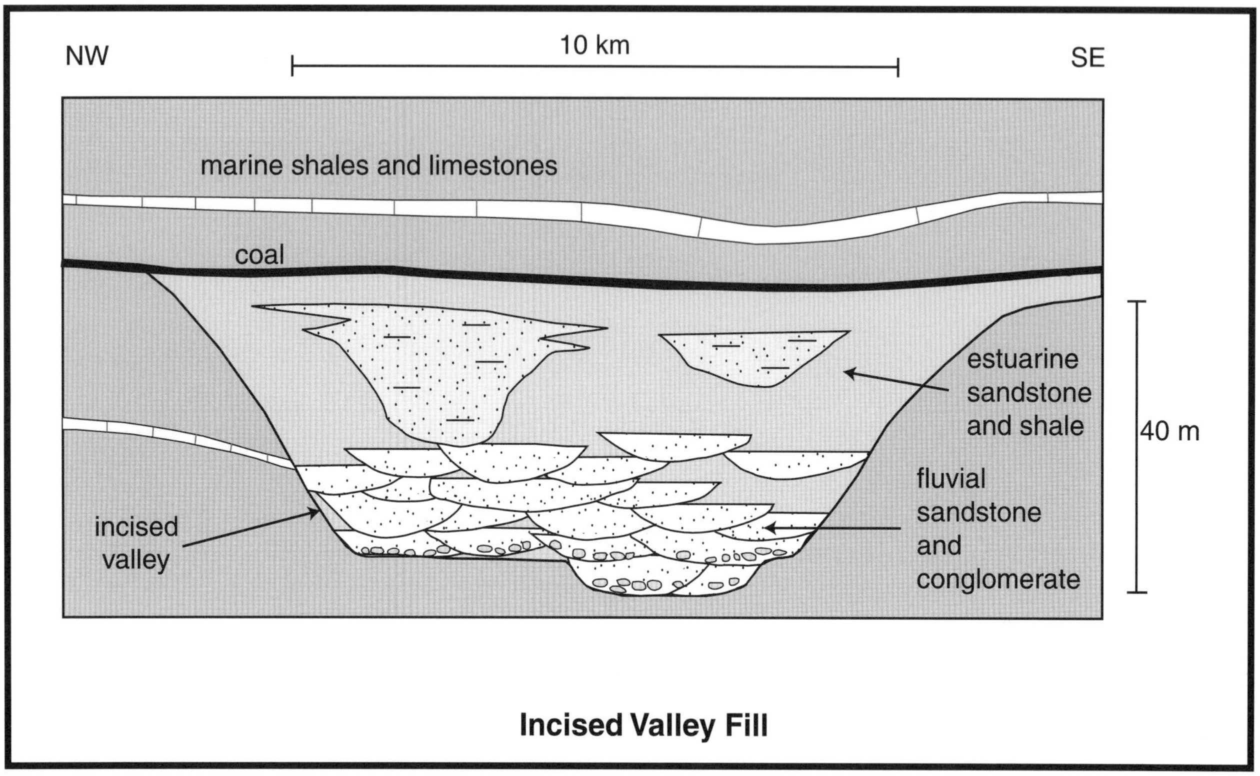

FIGURE 180. Cross section through an incised valley fill, Tonganoxie, Northeastern Kansas. Amalgamated fluvial channel fill sandstones are overlain by estuarine deposits of sandstone, sandy mudstones, and coal (from Feldman et al., 1995). Reprinted with permission from the AAPG.

do not. Connectivity may be effective where the clay plug does not totally separate two point bars areally. Richardson et al. (1987) noted that there is commonly some sand or gravel underneath clay plugs that can allow communication. Similar observations have been made by Donselaar and Overeem (2008). Clay plugs occur at the same level of the point bar that it partially encloses. If vertical connectivity is effective between incised point bars, the clay plug obstruction can be bypassed above and below. If no effective vertical communication occurs, then clay plugs are more likely to act as lateral flow barriers (Figure 176e).

WHAT OUTCROPS OF MEANDER BELT SEDIMENTS INDICATE ABOUT CONNECTIVITY

Well-exposed outcrops of meander belt sediments can be used to get an understanding of the reservoir connectivity in three dimensions. Connectivity can result from the vertical incision of one point bar into an older underlying point bar, creating *multistory sand bodies* (Figure 176f). Point bars commonly connect with each other across the shallow, sandy course of the river where the tips of point bars overlap on opposite banks. Crossovers like this can create a connected system of point bars which Donselaar and Overeem (2008) describe as a string-of-beads sandstone body. Crevasse splays may also link up one point bar with another.

Nevertheless, it probably does not take much for the connectivity between the various sand bodies in a meander belt to be disrupted. The connections between the various macroforms are likely to be through apertures of limited cross-sectional area such as erosional windows, crossovers, and crevasse splay-point bar intersections. Carbonate cementation of the basal lag by circulating groundwater can create permeability barriers at the base of individual point bars. Precipitation of carbonate cement may be accentuated where calcrete fragments form part of the basal lag (Mckie and Audretsch, 2005).

Abundant mud chips caked along the base of point bars have the potential to attenuate communication. For example, Chapin and Mayer (1991) found that the vertical connectivity between stacked point bars was severely impeded by mudstone-rich lags at the base of individual point bars in the reservoir of the Peoria field in Colorado. Doyle and Sweet (1995) found that mudclast lags at the base of point bar sandstones have a patchy distribution in the Gypsy Sandstone of Northern Oklahoma. They consider them more likely to form baffles to flow instead of continuous barriers. Shanley

Table 26. Factors influencing connectivity and reservoir development in meander belt reservoirs.

Characteristic	*Favorable for Reservoir Development*	*Unfavorable for Reservoir Development*
Large rivers	Larger point bars with larger in-place volumes	
Small rivers		Smaller point bars with smaller in-place volumes
Upward-decreasing permeability profiles		Poor sweep caused by water contacting the basal section only
Mud-lined lateral accretion surfaces present		They form shingled barriers to flow in the upper part of point bars
Gas reservoir	Heterogeneity and low-permeability connectivity less of a factor with gas	
Mud plugs		Can act as lateral barriers to isolate individual point bars and create low volume compartments
Multistory sand bodies	Composite sand bodies with larger volumes	
Single-story sand bodies with poor to no vertical permeability		Single sand bodies with small in-place volumes
Basal lags have good vertical permeability	Allows connectivity between superimposed macroforms	
Basal lags are cemented or are caked with clay chips		Poor to no connectivity between superimposed macroforms
Coals present	Can act as permeability barriers to vertical flow	
High density of faulting	Small faults can create connectivity in layered sandy fluvial systems.	Likely to create numerous, small, marginal to uncommercial reservoir compartments in low net-to-gross systems

(2004) noted that where basal lags contain abundant mudclasts, they can be mistaken for shales on the gamma-ray log. Caution should be taken where a shale-like wireline log response is seen within thick multistory fluvial sandstones.

Coals commonly act as significant flow barriers where they occur in more humid fluvial systems. The precursor peat deposits to coal occur as thick mats of flexible and intertwined plant material and these can withstand strong erosive forces to stay substantially intact.

HOW BASINAL CONTROLS AND NET-TO-GROSS VARIATION AFFECT CONNECTIVITY

Several computer modeling projects have provided general rules as to whether fluvial sandstones will stay isolated in a given rock interval or whether they can connect up to form larger bodies. One of the first studies involved an analysis of how net-to-gross values control the chances that individual sand bodies will overlap with each other. This is on the premise that if there is more sand in the system relative to the background shale, there will be more connectivity. In the early computer models, the sands were represented as a series of straight ribbons within a background of floodplain mud. With only a few of these sands, the fluvial ribbons pass through the mass of shale with only one or two ribbons touching, if any at all. Increasing the number of sand ribbons increases the chance of contact either vertically or laterally. It was found that, with a net to gross of less than 25%, there was almost no connectivity in the system (e.g., Clemetson et al., 1990; King, 1990). A more recent analysis indicates that connectivity is widespread (more than 90% connected) in channel systems that have a net to gross greater than 30%. With net-to-gross values less than 30%, the connectivity rapidly diminishes as a function of the net to gross (Larue and Hovadik, 2006).

The fluvial style within a thick rock interval can switch episodically from aggradational to erosional as a result of episodic changes in base level, tectonics, or climate. A river will then erode into the underlying sediments and rock to form an *incised valley*. The incised valley can subsequently backfill with sediments once the depositional system switches back over from net erosion to net aggradation. *Incised valley fills* can be tens of meters thick, several kilometers wide, and sand rich (Galloway and Hobday, 1996). A commonly described pattern is for a valley fill to pass upward from sand-rich fluvial sandstones and conglomerates to estuarine sandstones and shales (Figure 180) (Gibling, 2006).

Another factor that can localize river flow and cause the vertical stacking of sands is the presence of fault-defined topography. Synsedimentary faults can act to focus and stack the deposition of channel sand bodies parallel to the fault planes. Fielding (1984) described the vertical stacking of fluvial sandstones in an extensional fault setting in the Durham Coal Measures onshore in the United Kingdom. The fluvial sandstone forms elongate belts in topographic lows bounded by syndepositional faults. Tectonic uplift can cause localized river incision and vertical stacking of sediment when a river crosses the fault at a high angle (Kosters and Donselaar, 2003).

The volume of sand supply in a fluvial system can therefore vary significantly with time and place. Intervals of fluvial sediments typically show a wide variation in net to gross both vertically and areally (Miall, 2006). Sandy intervals with higher net-to-gross values are more likely to be better connected than the less sandy intervals. Careful vertical zonation can commonly pick out volumes of fluvial reservoir that are more likely to be producible than others. Low net-to-gross intervals have the potential to act as vertical baffles between the higher net-to-gross intervals.

THE EFFECT OF FAULTING ON CONNECTIVITY IN MEANDER BELT SEDIMENTS

Small faults can create vertical connections across the fault plane between sand bodies in high net-to-gross intervals. Where low net-to-gross fluvial sediments are dissected by faults, this can significantly reduce connectivity. The reduced degree of sand-to-sand juxtaposition across faults is the controlling factor (Bailey et al., 2002). The potential for clay smear and sealing faults will also be high particularly in low net-to-gross fluvial systems (Table 26).

34

Shepherd, M., 2009, Braided fluvial reservoirs, *in* M. Shepherd, Oil field production geology: AAPG Memoir 91, p. 273–277.

Braided Fluvial Reservoirs

BRAIDED FLUVIAL SYSTEMS

Sandy braided river systems show an intricate geometry of small bars, sand flats, and vegetated islands (Figure 181). The river flows across and between the various bars, splitting and joining continuously in a braided pattern (Walker and Cant, 1984). Repeated avulsion generates a complex of amalgamated channel segments.

BRAIDED FLUVIAL SEDIMENTS MAKE GOOD RESERVOIRS

Braided river reservoirs typically make excellent, very productive reservoirs. The net to gross can be much higher than in meander-belt reservoirs, and there is normally much less in the way of interbedded shales. Braided river deposits can occur on a continental scale with individual systems that are very thick and laterally very extensive. Gibling (2006) quoted widths in excess of 40 km (25 mi) and thicknesses up to 1200 m (3937 ft) for the very large braided river systems (Table 25). These form in response to periods of active tectonism, rapid subsidence, and a large volume of coarse sediment influx. These conditions are typical for foreland basins where braided fluvial sediments are commonly found.

Some very big oil fields are known from braided river reservoirs, including the Prudhoe Bay field in Alaska and several giant oil fields in the Sirte basin of Libya.

LATERAL CONTINUITY

Lateral continuity is typically excellent in braided fluvial reservoirs. The net to gross of these systems is normally very high (>80%), and, as such, these types of reservoirs are usually well connected laterally. In detail, they can be internally complex with intervals of upward-decreasing permeability profiles, but the lack of organized stratification or laterally continuous shales results in braided fluvial reservoirs showing effectively layer-cake geometry and acting as a single integrated reservoir at the larger scale (Galloway and Hobday, 1996).

Braided river systems normally comprise medium to coarse-grained sands and gravels, and the rock properties can be excellent. Oil recovery factors can be very high in braided river reservoirs, commonly more than 50% (Martin, 1993). Laterally extensive braided river reservoirs tend to be in communication with strong aquifers.

BARRIERS TO VERTICAL FLOW

Not much mud is preserved in braided river systems under normal conditions. What mud there is, collects as bar tops or represents fragments of floodplain silts and muds. The shales tend not to be very extensive and are randomly distributed. Formation tester data in braided river reservoirs typically show good vertical pressure communication with only localized evidence for flow baffling (Martin, 1993).

In some braided river sediments, shales are laterally more extensive and may be correlatable (Geehan et al., 1986; Davies et al., 1993). Where floodplain shales cover a large area, they can act as permeability barriers bounding hydraulic units. Widespread lacustrine shale beds act as barriers within the Jurassic braided river reservoir of the Jackson field of Australia (Hamilton et al., 1998) and in the Tertiary Merecure unit B interval of the Budare field in Venezuela (Hamilton et al., 2002).

Robinson and McCabe (1997) studied braided river sediments from the Morrison Formation of Utah. Individual single-story sand bodies have a mean width of 271 m (889 ft) and a mean thickness of 5 m (16 ft) in the lower interval. In the upper interval, the mean width

DOI:10.1306/13161222M913372

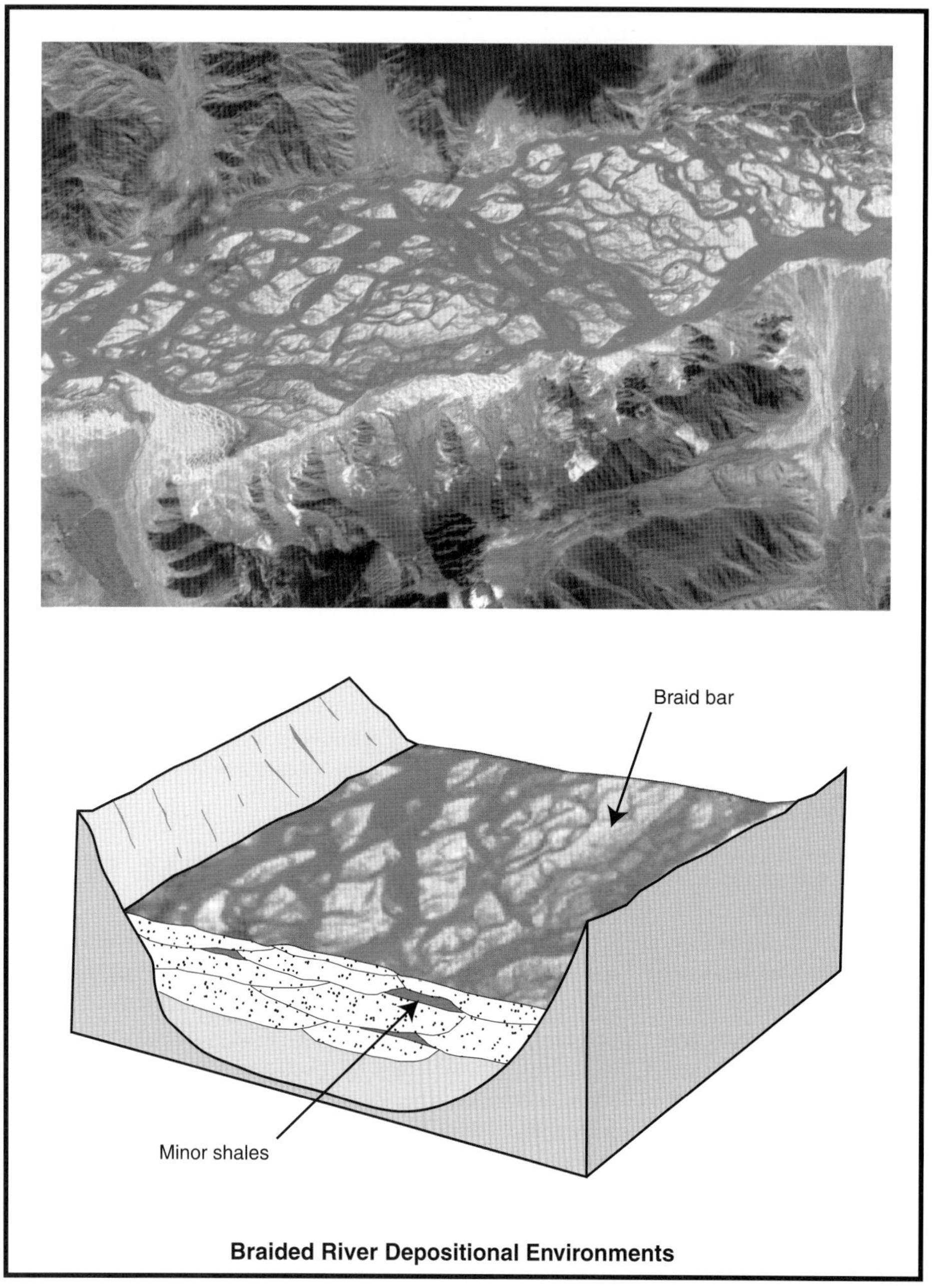

Figure 181. Satellite photo of a braided river in Tibet. The braided river belt is about 7 km (4 mi) wide. Courtesy of the NASA Web site (www.earthobservatory.nasa.gov). The lower figure was modified from Cant (1982). Reprinted with permission from the AAPG.

is 530 m (1739 ft) with a mean thickness of 10 m (33 ft). Two types of shale-prone facies have the potential to form baffles to flow. Overbank or floodplain deposits are laterally extensive (mean width 305 m; 1000 ft) and have the potential to form barriers or baffles to vertical flow communication. Abandoned channel-fill deposits are shorter ranging with mean widths of 99 m (325 ft) (Figure 182).

Local shale barriers in the braided river deposits of the Prudhoe Bay field in Alaska are advantageous to production where they occur because they inhibit both gas coning and water influx to wells. Wells with shale barriers can be produced at higher than normal rates as a result (Atkinson et al., 1990). However, in certain situations, they can be detrimental to production. Shales can act as baffles to pressure support from gas cap expansion to the producing intervals underneath them. Gas underrun can also be a problem.

In the Lower Cretaceous Cutbank Sandstone of Southern Alberta, channel lag deposits with clay-rich rip-up clasts have undergone later diagenetic modification. These form flow barriers within the braided river sediments (Farshori, 1989). Likewise, intraformational mud-clast conglomerates were considered potential permeability barriers in ephemeral braided fluvial deposits of the Lower Jurassic Kayenta Formation of Utah (North and Taylor, 1996).

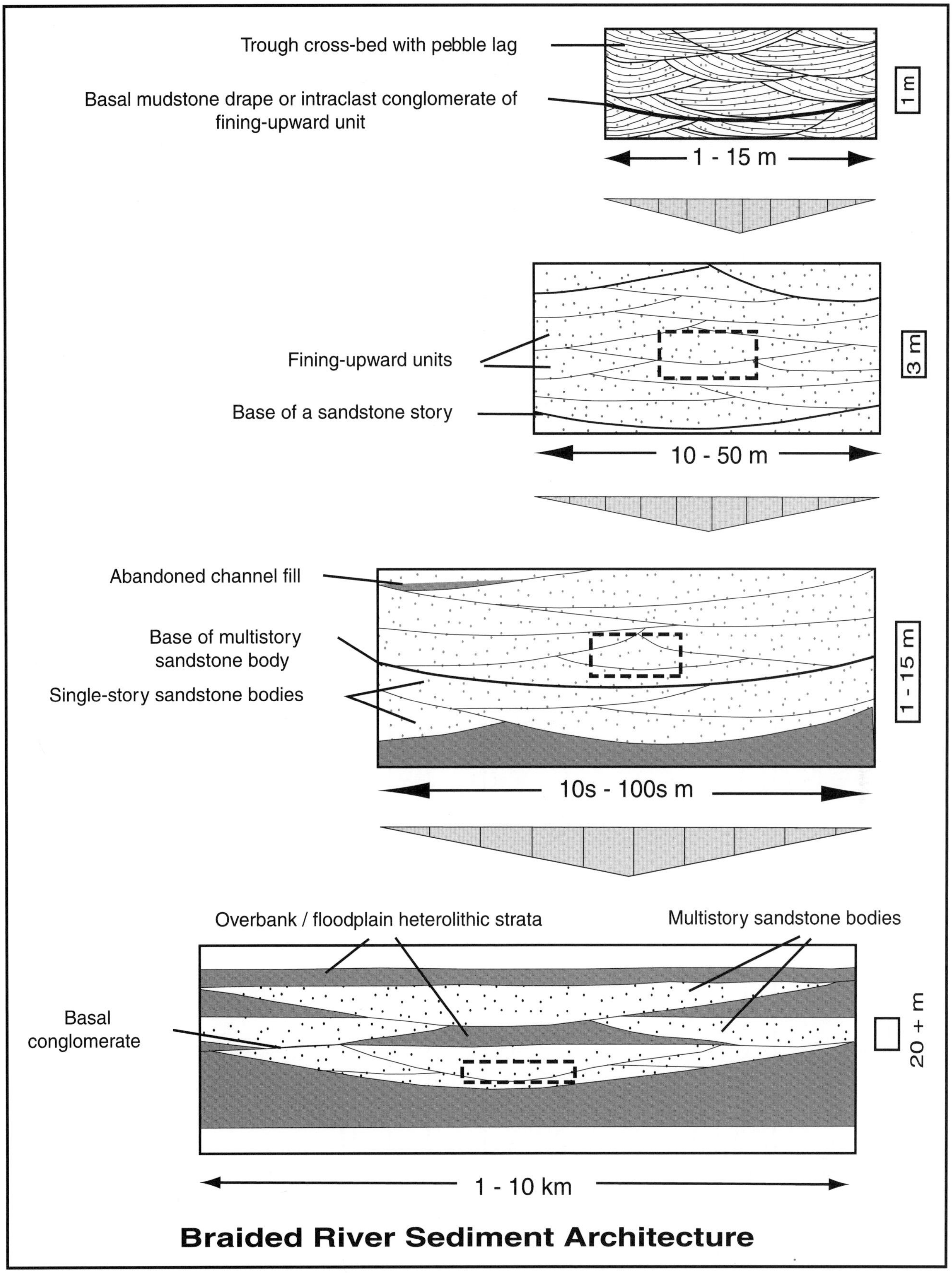

FIGURE 182. Hierarchy of sandstone and shale bodies in braided river sediments from the Morrison Formation, Garfield County, Utah (from Robinson and McCabe, 1997). Reprinted with permission from the AAPG.

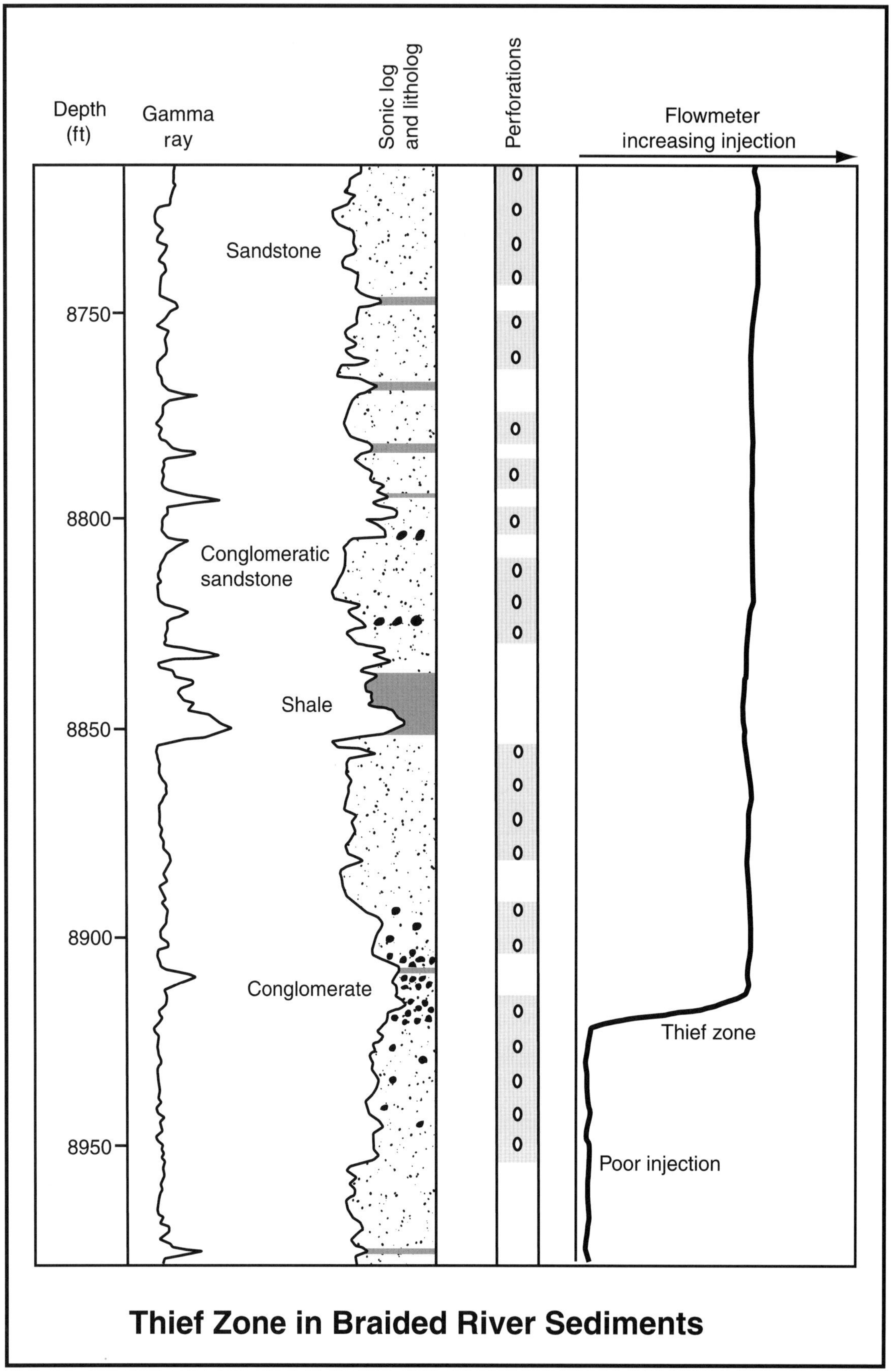

Figure 183. A high-permeability conglomeratic thief zone from the braided river reservoir of the Prudhoe Bay field, Alaska (from Atkinson, 1990). Reprinted with permission from Springer-Verlag Ltd.

PERMEABILITY VARIATION

Vertical permeability variation can be a significant feature in braided river reservoirs (Table 27). High permeabilities in the coarser pebbly layers are known to act as thief zones. An example of this is given by Atkinson et al. (1990) from the Prudhoe Bay field, Alaska. One injection well shows 95% of the total injected water entering a 3-m (10-ft)-thick conglomeratic zone with a permeability of 4 darcies (Figure 183).

Table 27. Factors influencing connectivity and reservoir development in braided river reservoirs.

Characteristic	*Favorable for Reservoir Development*	*Unfavorable for Reservoir Development*
High net to gross, thick sand units	High connectivity, commonly acting as a single hydraulic unit	
Thick sand bodies, areally extensive, commonly on a continental scale	Typically have big aquifers with efficient water drive	
Shales not commonly laterally extensive	Good vertical pressure communication; where present shales can inhibit coning behavior	
Floodplain and lacustrine shales can act as laterally extensive permeability barriers		Less effective vertical pressure communication
Cemented channel lag deposits can form permeability barriers		Less effective vertical pressure communication
Thief zones present in coarse-grained intervals		Poor sweep

35

Shepherd, M., 2009, Deltaic reservoirs, *in* M. Shepherd, Oil field production geology: AAPG Memoir 91, p. 279–288.

Deltaic Reservoirs

INTRODUCTION

Deltas are major sites of sand and mud deposition. They contain significant volumes of hydrocarbons worldwide (Figure 184). Major petroleum provinces include the Niger Delta in west Africa, the Mahakam Delta in Borneo, the Caspian Sea, and the Maracaibo Basin in Venezuela.

DELTAIC SEDIMENTS OFTEN FORM COMPLEX RESERVOIRS

Delta systems make heterogeneous reservoirs, typically with a jigsaw-puzzle to labyrinthine sediment geometry (Table 28). There may also be considerable structural complexity. Many rivers, particularly the larger ones, dump very large quantities of sediment into deltas on top of an unstable substrate containing mobile salt and/or shale. Salt deformation features and growth faulting are common in deltaic sediments, and these can result in numerous segmented reservoirs, such as in the Tertiary Niger Delta in west Africa (Evamy et al., 1978; Tuttle et al., 1999).

DELTAS ARE OFTEN GAS RESERVOIRS

Many deltaic reservoirs, particularly long-lived Tertiary to present-day delta areas, contain more gas than oil. This is because they can be particularly rich in coals and woody kerogen, which form gas-prone humic source material. Gas fields are found in the Mackenzie, Nile, and Irrawady deltas, for instance. Deltas can contain oil or mixed oil and gas where sandstones interfinger with a marine source rock (Galloway and Hobday, 1996).

TYPES OF DELTA

Deltas have been categorized into three classes in terms of sedimentary process: *wave dominated, tidal dominated,* and *fluvial dominated* (Figure 185) (Galloway, 1975). Coarse-grained deltas include fan deltas and braid deltas. Each specific environment has its own geometries and typical reservoir characteristics. The geometrical patterns shown by the various types of delta can often be recognized on isochore, net-sand, and log-facies maps (Coleman and Wright, 1975). For example, a wave-dominated delta will show a T motif on these maps as a result of fluvial lineaments converging at a high angle to a shoreline trend (see Figure 71). The lobate shape of the delta front may also be recognized.

DEPOSITIONAL ENVIRONMENTS

The influence of river, wave, and tide on deltaic sediments produces a complex mix of macroforms. Fluvial processes dominate in the upper delta plain, although this area can also be swampy with marshes and lakes present. The lower delta plain is subjected to marine influence, acting to modify the fluvial-derived sediments. Delta fronts comprise nested complexes of distributary channels, mouth bars, tidal bars, and reworked delta-front sediments (Figure 184).

DISTRIBUTARY CHANNELS

Distributary channels are so called because of the way in which they branch off from the main feeder river and distribute water and sediment across the delta. Where the distributary channels split off from the main feeder river, the volume of water carried by individual channels will be a fraction of that in the main river. By comparison

DOI:10.1306/13161223M913372

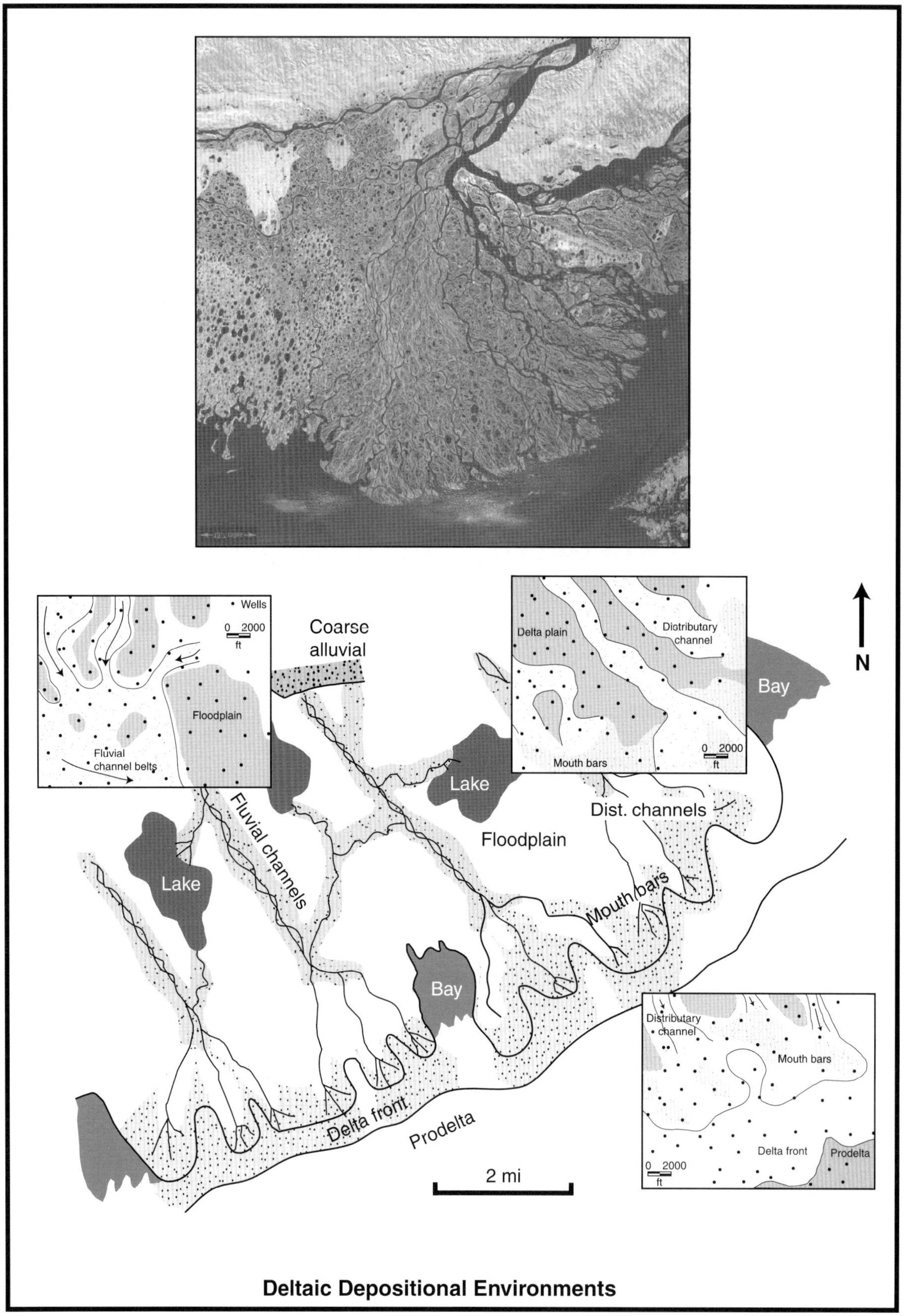

FIGURE 184. Photograph of the Lena delta, Russia. Courtesy of the NASA Web site (www.earthasart.gsfc.nasa.gov). The delta is about 200 km (124 mi) across in this view. The photograph has been rotated such that north faces down the page. The lower diagram is a lithofacies map of the basal Ivishak Formation, Prudhoe Bay field, Alaska (from Tye et al., 1999). Reprinted with permission from AAPG.

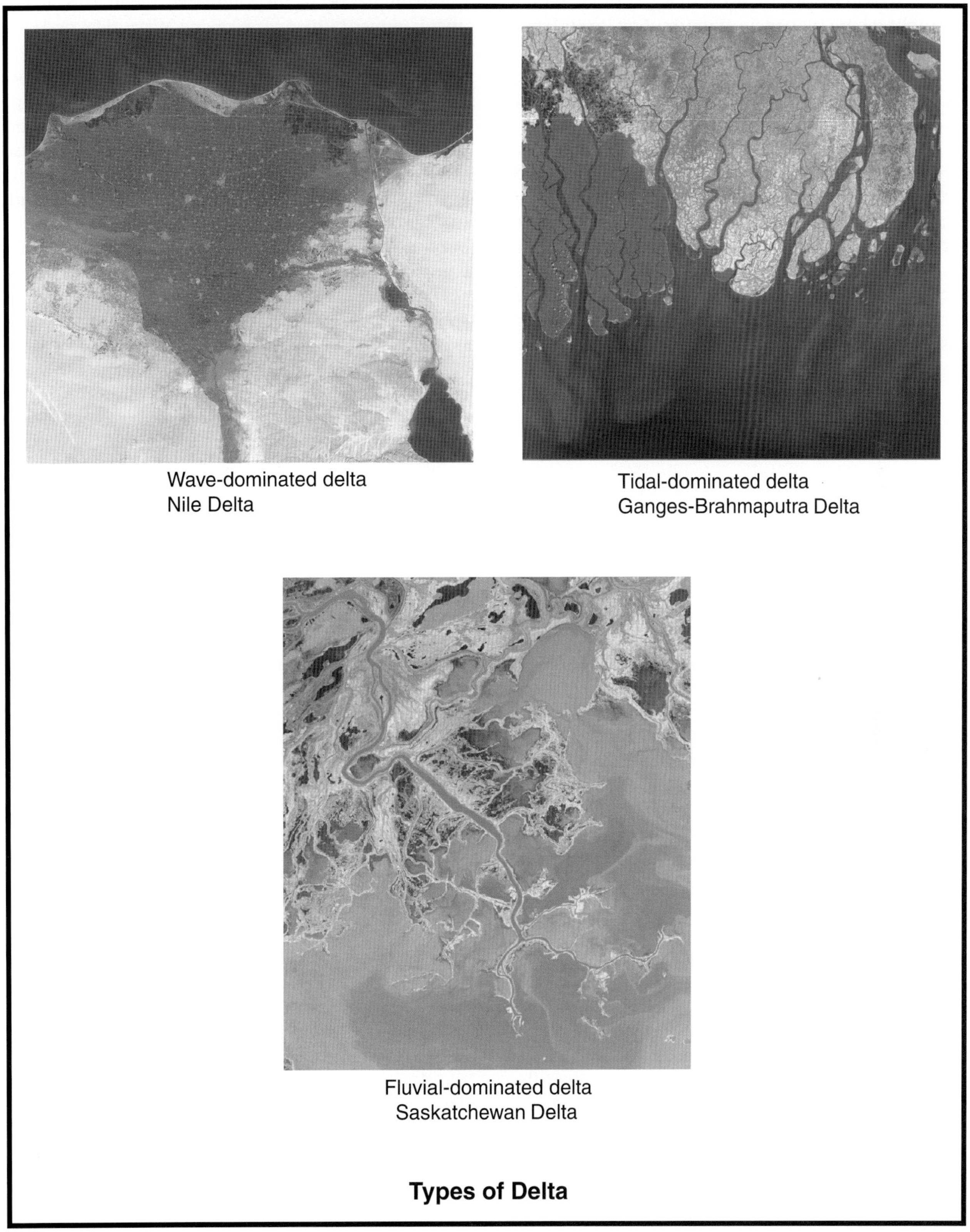

FIGURE 185. Three categories of delta can be defined according to the dominant sedimentary process. These are wave-dominated, tide-dominated, and fluvial-dominated deltas. Courtesy of the NASA Web site (www.earthobservatory.com).

to fluvial channels, distributary channels tend to be narrower and shallower. Gibling (2006) noted that distributary channels show a common width range of 10–300 m (33–984 ft) (see Table 25). Distributary channels tend to be straight where they incise a mud substrate and more sinuous within a sand substrate (Sneider et al., 1978).

Sand is deposited within linear distributary channels as side bars. In the modern-day Mahakam Delta, Borneo, side bars alternate on both sides of the distributary channels. These form elliptical sand pods, 5–8 km (3–5 mi) or more long and up to 1 km (0.6 mi) wide (Allen and Chambers, 1998). Channel fills typically show an upward-fining sediment profile and an upward-decreasing permeability profile. From the base upward, a distributary channel comprises the active channel fill, showing decimeter-scale trough cross-bedded sets; a partial abandonment fill with mainly centimeter-scale cross-beds; and sometimes an abandonment channel fill of thinly interbedded fine sand, silt, and shale.

MOUTH BARS

Mouth bars form where a distributary channel enters a standing body of water and sediment drops out. A shoaling to emergent sand body grows at the channel mouth. The resulting obstruction can cause the channel to bifurcate at the upstream head of the mouth bar. Mouth bars show an arcuate fan shape in plan view and a wedge-shaped profile in cross section. Reynolds (1999) gave average dimensions for mouth bars of about 3 km (1.8 mi) wide and about 6.5 km (4 mi) long (see Figure 123; Table 18). Relatively straight distributary channels building out into deep water will form more linear deposits known as *bar fingers* (Figure 186) (Fisk, 1961).

An upward-increasing grain size profile is characteristic for mouth bars. The lower parts are finer grained, more poorly sorted, and with common shale intercalations. Upward, the texture is coarser although there may be many laminations of clays and organic material. Permeability typically increases upward (Figure 187).

Mouth bars usually show lower overall permeabilities than distributary channel fills (Richardson et al., 1989). For example, Tye et al. (1999) gave average rock property values for the various lithofacies associations within the Ivishak Formation of the Prudhoe Bay field in Alaska. The mouth bars have a mean permeability of 151 md compared to 315 md for the distributary channel fills.

The coarsest and best sorted sediments in the mouth bars form near the stream mouth and along the bar margins adjacent to the distributary channels. Tye and Hickey (2001) found an order of magnitude higher permeability in this part of the point bars in Prudhoe Bay field, Alaska. Outward and down slope, the sediment becomes finer grained. Downstream, along the outer edge of the mouth bar, fine sand and silts interfinger with prodelta muds.

DELTA FRONT

The *delta front* area comprises a jigsaw-puzzle to labyrinthine complex of channel sandstones, mouth bars, and sediments formed by marine reworking. Wave, tidal, and fluvial processes act to rework the sediments on the delta front.

Wave reworking tends to produce relatively smooth lobate delta fronts. As the degree of wave reworking of the isolated mouth bars increases, the sediments become more continuous, coalescing to form laterally extensive beach-ridge and coastal-barrier sand bodies. The outlines of individual mouth bar forms start to become indistinct as they are reworked. Sometimes the sites of mouth bar deposition may only be recognizable by local thickening of the delta front (Galloway and Hobday, 1996).

Delta-front sandstones tend to be finer grained, although better sorted, than distributary channel-fill sandstones. Later stage channels may cut into or overlie mouth bars and delta-front sandstones, creating jigsaw-puzzle geometrical complexity. Laterally and offshore, the sandstones become interbedded with background *prodelta mudstones* and will pinch out into them. The sandstone quality deteriorates laterally toward the margins.

TRANSGRESSIVE SANDSTONES

Delta lobes will persist as areas of active sedimentation for a period of time, eventually becoming abandoned once the locus of sediment input switches elsewhere. The lobe eventually founders as a result of subsidence and a marine transgression follows. The transgressive sediments are thin, forming a distinctive facies association consisting of a series of coarse-grained, shelly beach ridges and barrier-bar inlet complexes (Galloway and Hobday, 1996). These units are laterally extensive and can be important marker beds in delta systems where the reservoir comprises stacked delta lobes. The generally heterogeneous nature of delta sediments can make correlation difficult otherwise. Other important marker horizons are thin marine shales, impure limestones, and coal beds.

TIDAL PROCESSES

Tidal deltas are deltas where tidal flow is important in the reworking of the delta front. Tidal currents ebb and flow out of the channel mouths creating tidal sand flats, sand ridges, and shoals that may be isolated within prodelta muds. Elongate sandstone bodies, broadly perpendicular to the shoreline, form at the seaward end of many tide-dominated deltas. These sand bodies can be tens of meters thick, several kilometers wide, and several tens of kilometers long (Willis et al., 1999).

Tidal channels are commonly branched, with the branches tending to converge down stream. The channel

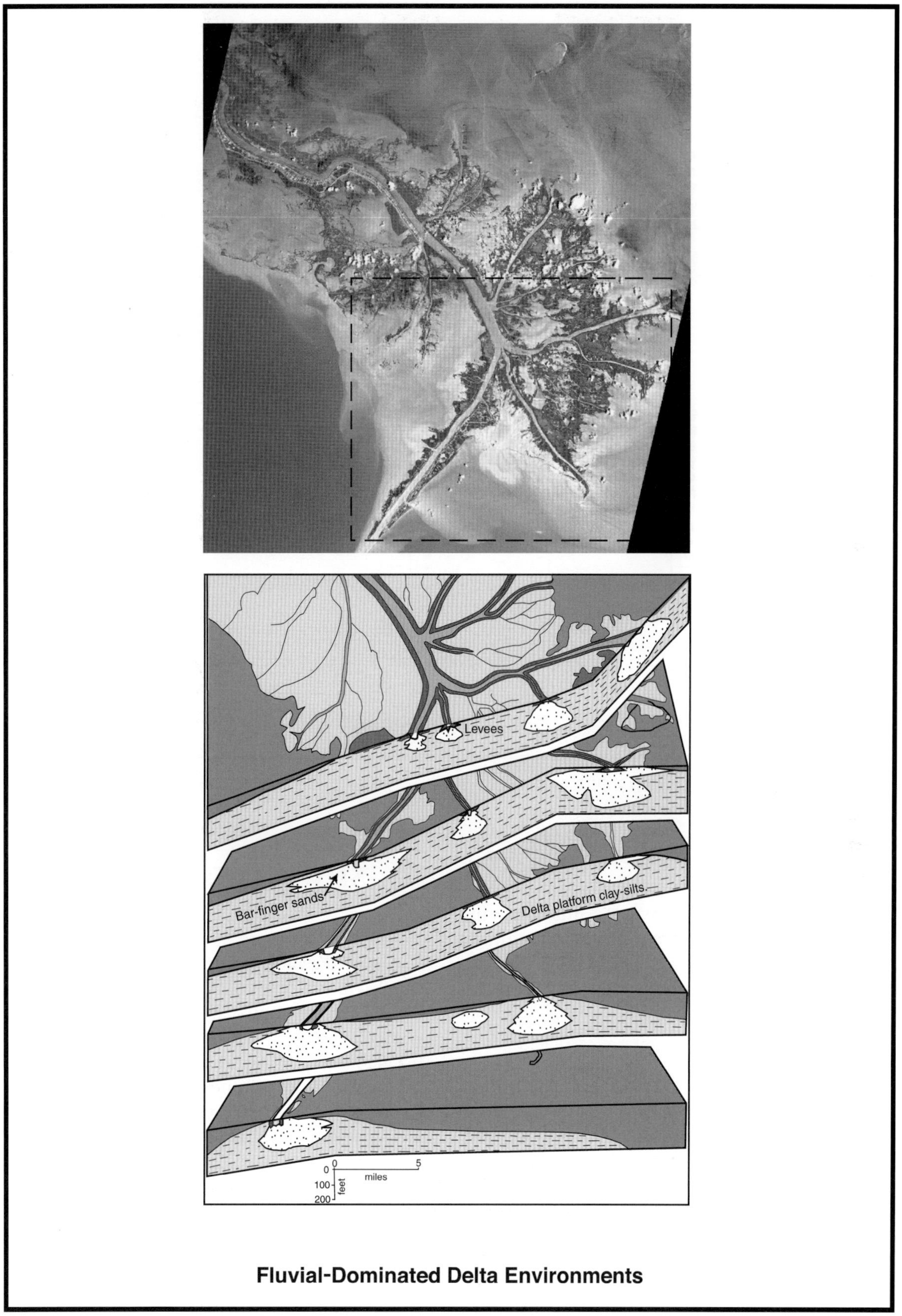

Figure 186. Fluvial-dominated delta environment, Mississippi Delta. Photograph courtesy of the NASA Web site (www.earthasart.gsfc.nasa.gov). The inset box on the photograph measures 34 × 42 km (21 × 26 mi). The lower diagram is a box diagram showing the sedimentological relationships within the inset box (after Fisk, 1960, courtesy of the AAPG).

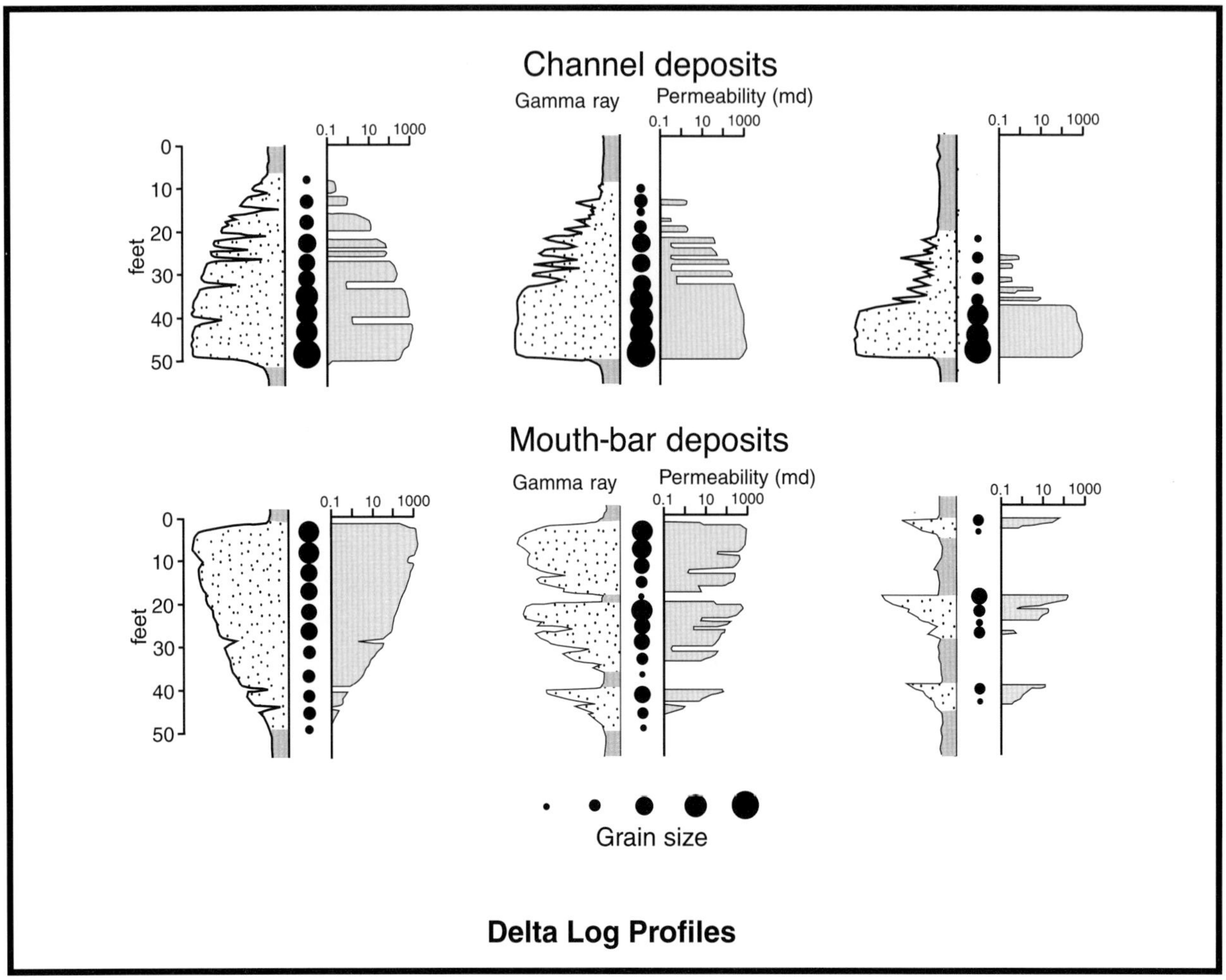

FIGURE 187. Idealized log and permeability profiles for deltaic sand bodies (from Sneider et al., 1978). Reprinted with permission from the Society of Petroleum Engineers.

fills comprise multiple stacked, fining-upward depositional units with abundant intercalated mud and silt drapes. Individual sand bodies are complex in architecture with numerous mud baffles and barriers. Given the heterogeneity of these systems, oil recover can be poor from tidal delta sediments.

MUDSTONES IN DELTAS

Weber (1982) stated that shales are not so laterally extensive in deltas by comparison to shoreface systems as fluvial and tidal channels commonly erode them. In distributary channels, the shale breaks can be short, commonly less than 10 m (33 ft) laterally.

Mudstones may be more extensive along the delta front. Tye et al. (1999) found that mudstones deposited following delta lobe abandonment formed locally significant flow barriers between delta lobes within the Ivishak Formation, the basal reservoir interval of the Prudhoe Bay field in Alaska.

SHINGLED GEOMETRY OF DELTAS

Shingled geometries are common in deltas (Figure 188). Shingled motifs may be seen on regional seismic lines and large-scale correlation diagrams (Bhattacharya and Walker, 1992). There is evidence from some fields that the shingled patterns can result in compartmentalization and bypassed oil volumes; for example, the Teal field, offshore Louisiana (Sibley and Mastoris, 1994; Hart et al., 1997), the Sirikit field, onshore Thailand (Ainsworth et al., 1999), and the Ivishak Formation of the Prudhoe Bay field, Alaska (Tye et al., 1999).

COARSE-GRAINED DELTAS

Fan Deltas

Fan deltas are coarse-grained deltas that form where alluvial fans deliver sediments into a lake or the sea.

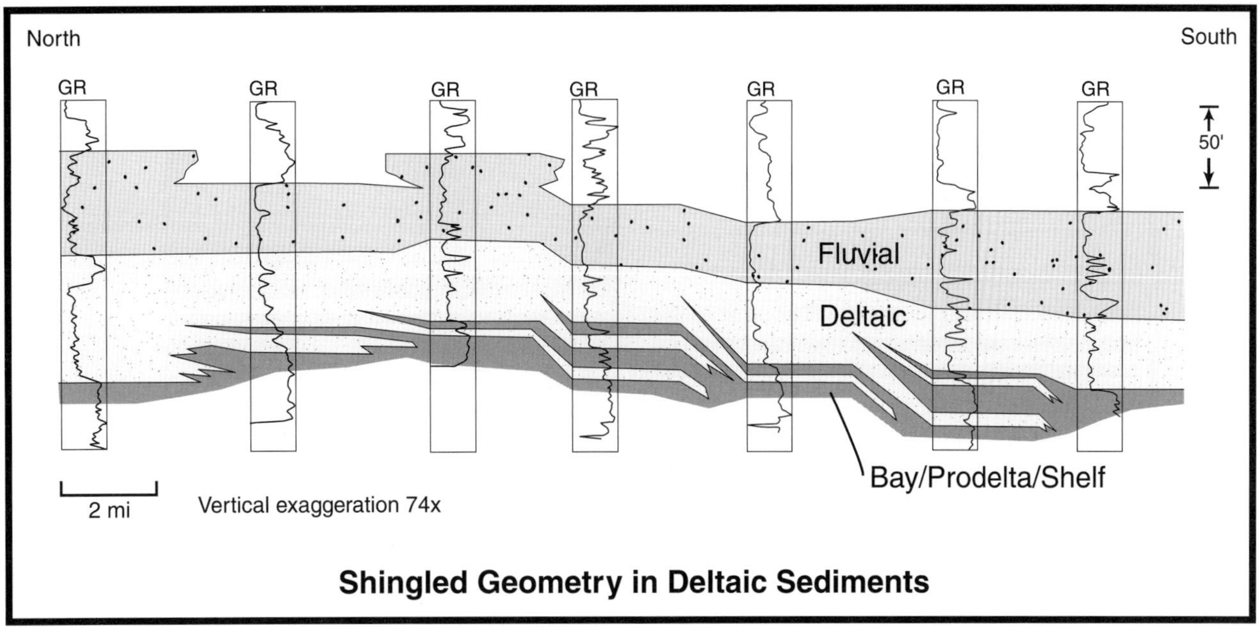

FIGURE 188. Seaward-dipping shingles in the Ivishak Formation, Prudhoe Bay field, Alaska (from Tye et al., 1999). Reprinted with permission from the AAPG. GR = gamma ray.

Geometries vary from wedging to tabular to sheet-like. Dreyer (1993) described exposures of fan-delta sediments in the Miocene Ridge Route Formation of California as an analog for the Tilje Formation of the mid-Norwegian shelf. The main permeability barriers occur as transgressive prodelta mudstones separating individual regressive fan-delta sedimentary episodes. Minor heterogeneity within individual fan-delta mouth bars is provided by intramouth bar shales and carbonate-cemented sandstones.

Braid Deltas

A *braid delta* is a coarse-grained delta fed by a braided river (McPherson et al., 1988). Braid-delta sediments from the Tertiary of the Apsheron Peninsula, Azerbaijan, have been described by Reynolds et al. (1998) as the onshore outcrop analog for the reservoir interval for offshore oil fields in the South Caspian Basin. This technical paper is an excellent example of reservoir characterization in practice. The sedimentology has been analyzed with a focus on making predictions of the flow-geology behavior for the offshore fields.

Four facies associations were defined:

1) Alluvial braided river, comprising stacked units of fine to coarse-grained, poorly sorted, cross-bedded sandstone. There are no shales present to act as permeability barriers to vertical flow. Excellent sweep could result from a wide well spacing in these rocks. However, there may be preferential flow through the coarser base of the sandstones, and the flood front could be unstable at the laminar scale as a result of grain size variation in the cross-beds.
2) Delta plain characterized by multistory sandstone channels and laterally extensive delta-plain siltstones. The bases of the channels commonly show silty conglomerate lags that are likely flow barriers or baffles. The persistent, thick siltstone intervals impart a strong degree of flow layering and create several stacked hydraulic units within this facies association.
3) Proximal delta front (Figure 189). This comprises a series of thick mouth-bar sandstones and channel systems separated by extensive delta-front siltstones and mudstones. The latter can potentially form permeability barriers and create stacked hydraulic units.
4) Distal delta front. This facies association comprises interbedded thin fine-grained sandstone, siltstone, and mudstone beds. The thin sandstones are extensive, highly layered, and have low permeability.

SWEEP PATTERNS IN DELTAIC SANDSTONE RESERVOIRS

Deltas comprise stratigraphically complex jigsaw-puzzle and labyrinth reservoirs with significant potential for bypassed oil. Tyler et al. (1987) commented on the recovery efficiencies of the various types of delta systems in Texas. Fluvial-dominated reservoirs show low to average recoveries because of the predominance of fluvial channels and an overall labyrinthine geometry.

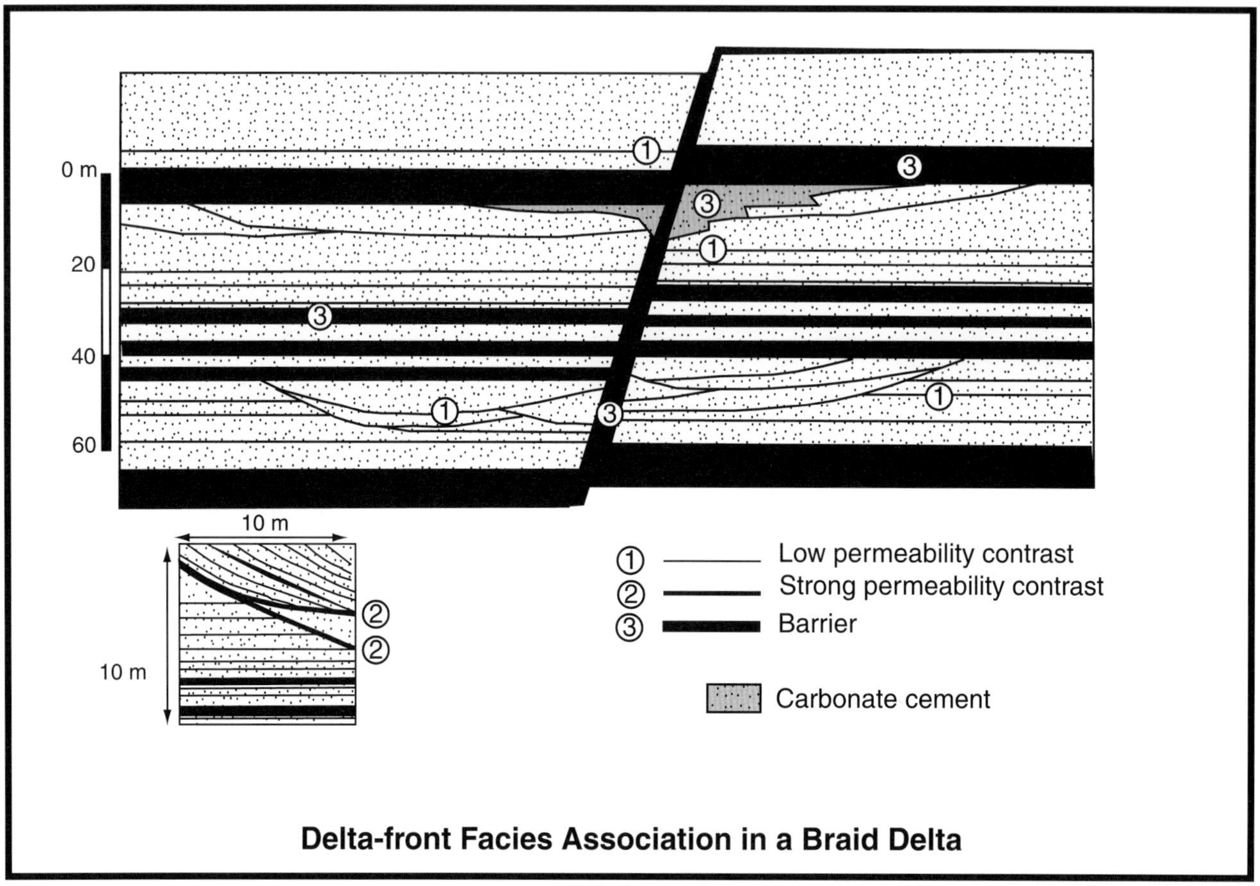

FIGURE 189. Delta-front facies association from the Tertiary of the Apsheron Peninsula, Azerbaijan. A stacked succession of mouth-bar and channel sandstones is vertically sealed by delta-front siltstones. Internally, the sandstones display both coarsening-upward and fining-upward trends with low internal heterogeneity (from Reynolds et al., 1998). Reprinted with permission from the AAPG.

Wave-dominated deltas show more lateral continuity and have significantly higher recoveries. They commonly have large aquifers, and as a result these reservoirs will be supported by strong water drives.

Tyler and Ambrose (1986) described the sweep patterns in a wave-dominated delta system from the Cayce reservoir of the North Markham-North Bay City field in Texas. Two distinct patterns occur. Water flows preferentially along fluvial channels. The beach ridge, delta front, and shoreface sandstones are more laterally continuous and show broader zones of edge-water influx.

SWEEP PATTERNS IN DISTRIBUTARY CHANNELS

Distributary channels tend to be narrow with a common width range of 10–300 m (33–984 ft) (Gibling, 2006). It is possible that hydrocarbons in isolated distributary channels may be missed in fields with larger well spacings. Richardson et al. (1989) stated that it could be impractical to try and locate both injection and production wells to sweep individual distributary channels. The sweep efficiency in the distributary channels will be low without direct injection support, particularly if the sand bodies are isolated. There is a better chance of improving recovery by waterflooding the delta-front sandstones.

SWEEP PATTERNS IN DELTA-FRONT SEDIMENTS

Sneider et al. (1978) showed how considerable flow complexity can be present in the delta-front environment. This results from the incision of fining-upward channels into coarsening-up mouth bars. The coarser basal sections of the channels act as permeability fairways for water ingress to production wells with the potential to leave oil bypassed within the finer grained parts of the mouth bars. Similar sweep patterns were noted by Hartman and Paynter (1979) in Tertiary fluviodeltaic sandstones in fields offshore of Louisiana (Figure 190a).

Swanson (1979) noted that distributary channels tended to be more productive than mouth bars in the

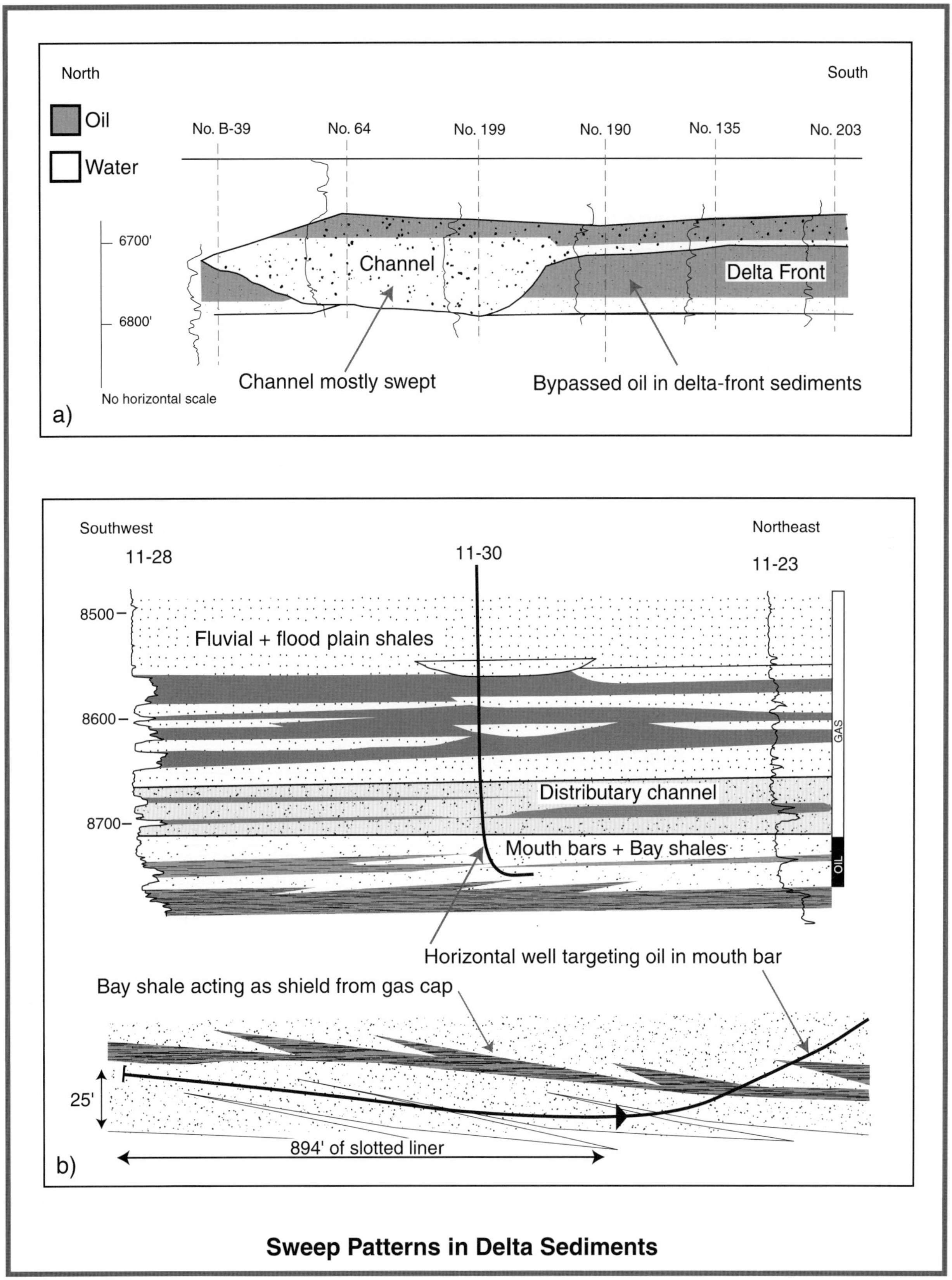

FIGURE 190. Sweep patterns in delta sediments. (a) Preferential water ingress along channel sediments can result in bypassed oil in the surrounding sediments. From the South Pass Block 27 field, offshore Louisiana (from Hartman and Paynter, 1979). Reprinted by permission from the Society of Petroleum Engineers. (b) Horizontal well drilled to target oil within a mouth bar in the Ivishak Formation, Prudhoe Bay field. Bay shales above the mouth bar act to prevent gas ingress from a gas cap immediately above (from Tye et al., 1999). Reprinted with permission from the AAPG.

Mississippian upper Morrow Sandstone of the Anadarko Basin, United States. The basal sections of the distributary channels appear to be well connected, whereas the extremely laminated lower parts of the upper mouth bar facies tend to restrict flow. The degree to which mudstone laminations are present in mouth bars may control the vertical sweep of these bodies.

Repeated delta lobe switching can result in a series of stacked mouth-bar intervals within a thick interval of deltaic sediments. Vertical amalgamation can produce thick multistory bodies as a result, Elsewhere, individual mouth bars can be separated vertically by bay mudstones and marine flooding events.

Hamilton et al. (2002) described a wave-dominated delta reservoir in the Budare field, Venezuela. Mouth-bar facies amalgamate to produce vertical continuous sand bodies up to 27 m (90 ft) thick. These are in vertical communication. Distributary channel complexes also comprise stacked sandstone units up to 17 m (55 ft) thick and likewise show good vertical communication. By contrast, the strand-plain complexes flanking the mouth bars received only peripheral sediments separated by widespread numerous marine shales. Vertical sweep is poor in the strand-plain sandstones, and these are targets for recovering bypassed oil.

Eight delta-front cycles comprise the Miocene reservoir interval of the Peciko gas field in the Mahakam Delta, Indonesia. The cycles are predominately mouth-bar sand bodies, 30–50 m (98–164 ft) thick. They are separated by marine-shale flooding events, which act as permeability barriers. Each delta cycle represents a distinct hydraulic unit or drainage cell within the field (Lambert et al., 2003).

Tye et al. (1999) found examples of mouth bars vertically separated and isolated by bay mudstones in the Ivishak Formation of the Prudhoe Bay field, Alaska. They successfully targeted bypassed oil in individual mouth bars by drilling horizontal wells (Figure 190b).

Table 28. Factors influencing connectivity and reservoir development in deltaic reservoirs.

Characteristic	*Favorable for Reservoir Development*	*Unfavorable for Reservoir Development*
Growth faults common		Numerous sealing fault compartments
Shingled geometry		Results in bypassed oil in individual shingles
Increasing marine reworking of delta front	Creates increasing lateral connectivity in the delta-front sediments	
Wave-dominated delta	More continuous, may have an aquifer	
Fluvial-dominated delta		Can show low recoveries caused by labyrinthine geometry
Tidal-dominated delta		Low recoveries caused by complex geometry and numerous mud and silt baffles
Distributary channels form narrow sand bodies		May be missed by wells in fields with a large well spacing; difficult to locate injection wells
Distributary channel sands commonly the highest permeability facies association in deltas	Can be the most productive intervals in a delta	
Stacked distributary channels	Larger sand bodies with good vertical connectivity and sweep	
Mouth bars contain extensive mudstone laminae		Mouth bars may have poor vertical connectivity and sweep
Stacked mouth bars	Larger sand bodies with good vertical connectivity and sweep	
Mouth bars separated vertically by shales	Individual mouth bars can be targeted by horizontal wells; shale prevents water influx from swept units above and below	Poor to no vertical connectivity between mouth bars
Coarser grained distributary channels cutting finer-grained delta-front sandstones		Can result in preferential water ingress into the delta-front area, bypassing oil in the delta-front sediments
Peripheral strand-plain complexes with high frequency marine shales		Poor vertical sweep with potential for bypassed oil

Shepherd, M., 2009, Siliciclastic shorelines and barrier-island reservoirs, *in* M. Shepherd, Oil field production geology: AAPG Memoir 91, p. 289–293.

36

Siliciclastic Shorelines and Barrier Island Reservoirs

SHORELINES AND BARRIER ISLANDS

Shoreface sands are deposited along shorelines, and they generally form extensive, high-quality reservoir systems (Figure 191). Wave action and occasional storms act to deposit sand along the shoreface. The *lower shoreface* lies below fair-weather wave base but can be affected by storms; the sands tend to be siltier and more poorly sorted by comparison to the *upper shoreface*, where the sands have been subjected to wave winnowing. A shoreface deposit separated by a lagoon from the land is known as a *barrier island*.

SHOREFACE SANDS FORM LAYER-CAKE GEOMETRIES

Shoreface sands prograde by lateral accretion with a tendency to produce layer-cake tabular geometries. Depositional dead ends are rare within individual shoreface sandstones, and sweep efficiencies are generally high as a result; for example, Tyler and Ambrose (1986) described excellent continuity and efficient simple sweep in the Carlson shoreface reservoir of the North Markham-North City Bay field of Texas. The large size and excellent lateral continuity of shoreface reservoirs give a reasonable chance that these systems will be in contact with an aquifer (Table 29).

PARASEQUENCES AND PARASEQUENCE SETS

Shoreface sandstones commonly form parasequence sets (see Chapter 10, this publication). An individual parasequence can comprise a series of facies belts showing a progression from coastal to offshore sediments. For example, in the Scott field in the UK North Sea, back barrier, foreshore, upper shoreface, and lower shoreface facies belts can be mapped out in the Upper Piper Sandstone Member (see Figure 73). (Guscott et al., 2003). An analysis of parasequence stacking patterns can help the geologist to predict and map facies belts in the areas beyond the well control. For example, Spaak et al. (1999) used stacking analysis on the Jurassic shoreface sediments of the Fulmar field, UK North Sea, to help construct the depositional scheme for the reservoir.

The basal section of individual parasequences is defined by a flooding surface that is commonly a marine shale. Shales can isolate individual parasequence shoreface cycles vertically, and they can be laterally extensive for several hundreds of meters or more. Fluid flow communication may occur between parasequences where the shales are absent as a result of erosion or nondeposition. It can be useful to produce vertical flow barrier maps for parasequence boundaries (see Figure 102). The localized presence or absence of bounding shales can be a critical feature in the flow geology characterization of a shoreface reservoir (Larue and Legarre, 2004).

VERTICAL PERMEABILITY PROFILES IN SHOREFACE SANDSTONES

Shoreface sandstones characteristically show upward-increasing permeability profiles. This in turn reflects increasing grain size and better sorting higher up the shoreface profile (see also Figures 42, 108). A contrast in rock properties is characteristically seen between the lower and upper shoreface intervals. Upper shoreface beach facies associations generally show higher permeabilities

DOI:10.1306/13161224M913372

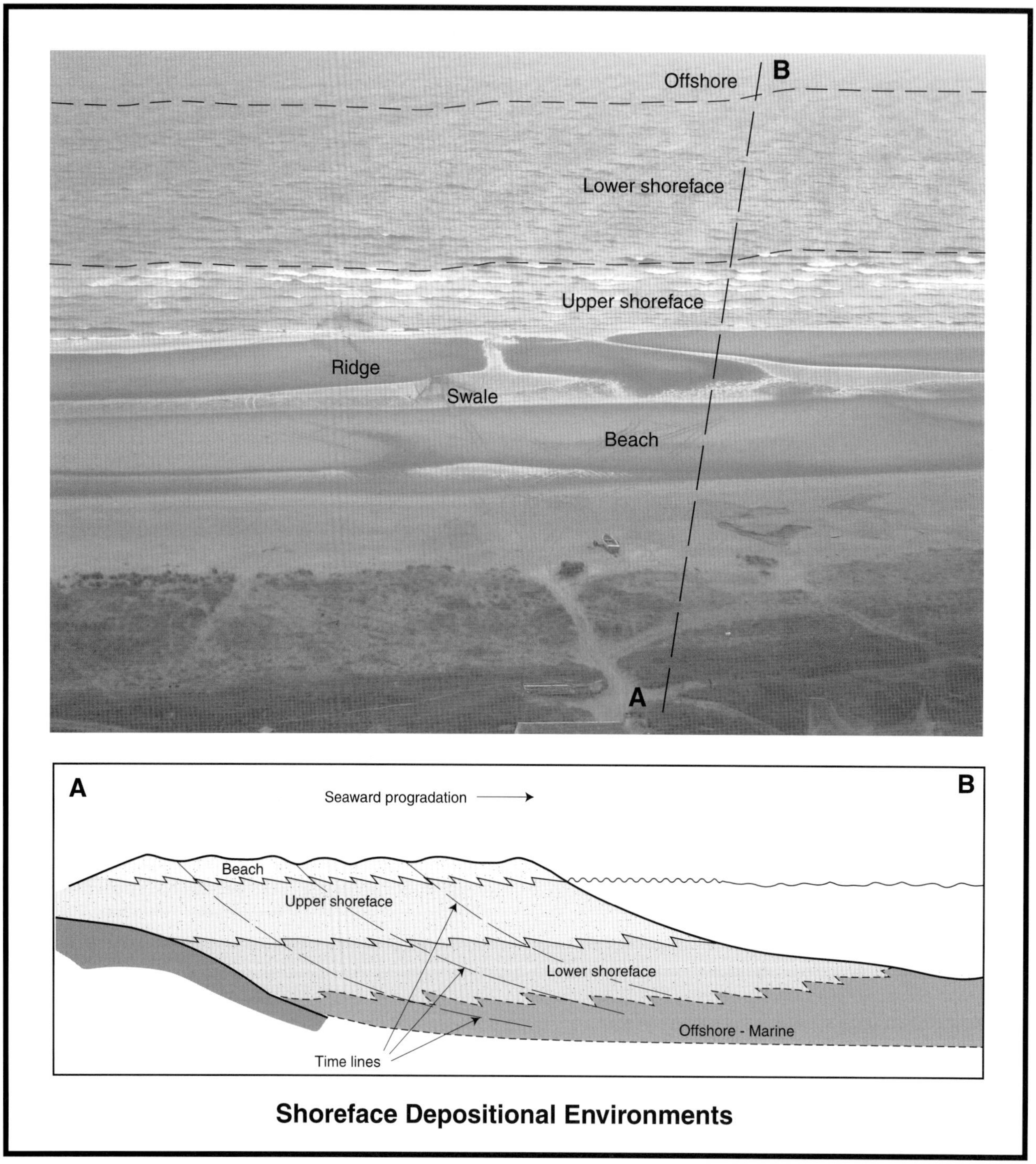

FIGURE 191. The photograph shows a shoreface profile on St. Cyrus Beach, Scotland. The beach is just over a hundred meters wide. Reservoir properties are influenced by the degree of wave reworking up the shoreface profile. Lower figure from McCubbin (1982). Reprinted with permission from the AAPG.

than lower shoreface sediments. When a shoreface sand is subjected to a waterflood, the water tends to edge ahead through the high-permeability tops of these cycles by viscous forces. Gravity and capillary action will then draw the water down through the shoreface cycle into the lower units, displacing oil upward. Sweep efficiencies can be high as a result.

The degree to which the lower part of the shoreface is swept by water will depend on the magnitude of the vertical permeability within the lower shoreface. In the Middle Jurassic Brent Province of the UK North Sea, bypassed oil is often found within the lower shoreface facies association (Rannoch Formation). The overlying upper shoreface (Etive Formation) is typically an interval of water overrun (Thomas and Bibby, 1991). This behavior can be reinforced by a zone of mica concentration at the top of the lower shoreface, which acts as a baffle to vertical flow (Wetzelaer et al., 1996). Horizontal wells have been drilled in several Brent Province fields to target bypassed oil in the lower shoreface (Braithwaite et al., 1989; Black et al., 1999). The success of these wells depends on the presence of low vertical permeabilities at the top of the lower shoreface interval in order to prevent water coning down from the swept upper shoreface interval.

Shallow marine sandstones can contain extensive stratiform carbonate-cemented bands and nodules. Taylor et al. (1995) found widespread cemented intervals below major flooding surfaces in the Upper Cretaceous Blackhawk Formation in Utah. The reduced sedimentation rates associated with these surfaces are thought to have allowed more time for cements to form. The cemented horizons have the potential to form baffles.

BEACH SANDSTONES

Beach sandstones form as single belts or accrete laterally to form strand plains many kilometers long and several kilometers wide. Modern strand plains such as the Nayarit strand plain of western Mexico show a ridge and swale topography on their surfaces (McCubbin, 1982). Mud fills in the interridge swales can act as permeability barriers to lateral flow in the subsurface. One example in the Frio Formation of south Texas is known to have caused the compartmentalization of a beach ridge interval containing several million barrels of recoverable oil (Reistroffer and Tyler, 1991). Water ingress may preferentially occur along the low-lying swales (Tyler and Ambrose, 1986) (see also Figure 117).

TIDAL CHANNELS

Although shoreface sandstones typically show a simple layer-cake geometry with minor internal heterogeneity, some reservoirs may contain an element of jigsaw-puzzle geometry provided by tidal and fluvial channel fills. Tidal channels may migrate laterally with sand accreting to form long, thin, sheet-like bodies parallel to the shoreface. Bypassed oil volumes may be found within tidal channel macroforms in shoreface sandstone intervals.

BARRIER ISLANDS

Barrier islands form thick, well-sorted sand bodies with a tabular geometry (Figure 192). They typically comprise a composite of beach, dune, and upper shoreface sandstones (Galloway, 1986). Barrier islands can be continuous for tens of kilometers along strike but may only be a few kilometers wide. Local heterogeneity can be provided by tidal channel inlet deposits. These form crosscutting lenticular pods, disrupting the layer-cake continuity of the barrier island body. Recent barrier island sediments on the South Carolina coast provide a modern analog and are described in detail by Sexton and Hayes (1996).

Ambrose et al. (1997) gave an example from an oil field in Venezuela where sweep has resulted from preferential water encroachment along the sandstone-rich core of the barrier island depositional axis with bypassed oil remaining along the landward pinch-out edge.

Richardson et al. (1988b) noted that severe problems can arise where barrier islands show a jigsaw-puzzle arrangement with poorer quality and discontinuous macroforms. Displacing fluids are channeled through the barrier island sandstones and bypass the other units. Efforts should be made to target the poorer quality intervals for this reason. Ambrose et al. (1991) described a relatively low-permeability tidal inlet channel within the high-permeability barrier island facies association of the West Ranch reservoir of the Frio barrier-strand-plain play of Texas. Unswept oil is commonly found within the tidal inlet sand bodies or along the permeability contrast marking the boundary between the tidal inlet sediments and the enclosing barrier island sandstones.

BACK-BARRIER ENVIRONMENTS

Background deposition within the lagoon behind the barrier island is generally mud, but with some sand bodies present. These include *washover fans* and *flood-tidal deltas* (Figure 192). Washover fans form when storms pitch sand over the barrier bar into the lagoonal area. Flood-tidal deltas develop as a result of the tidal movement of sand through an inlet into the lagoon. These lagoonal sand bodies are characterized by a pinch-out geometry into the lagoonal shales. The shales interfinger with the sandstones, commonly enveloping the sandstones and sometimes isolating them as discrete hydraulic units.

Washover sandstone complexes may be rather patchy and laterally heterogeneous. Production from washover sandstones in the Glasscock reservoir of the West Ranch field in Texas is described by Galloway (1986). Waterflooding has proceeded irregularly with injected water preferentially flowing along the washover channels. As a result, the Glasscock reservoir has the lowest projected recovery factor (38%) of all the major intervals in the West Ranch field.

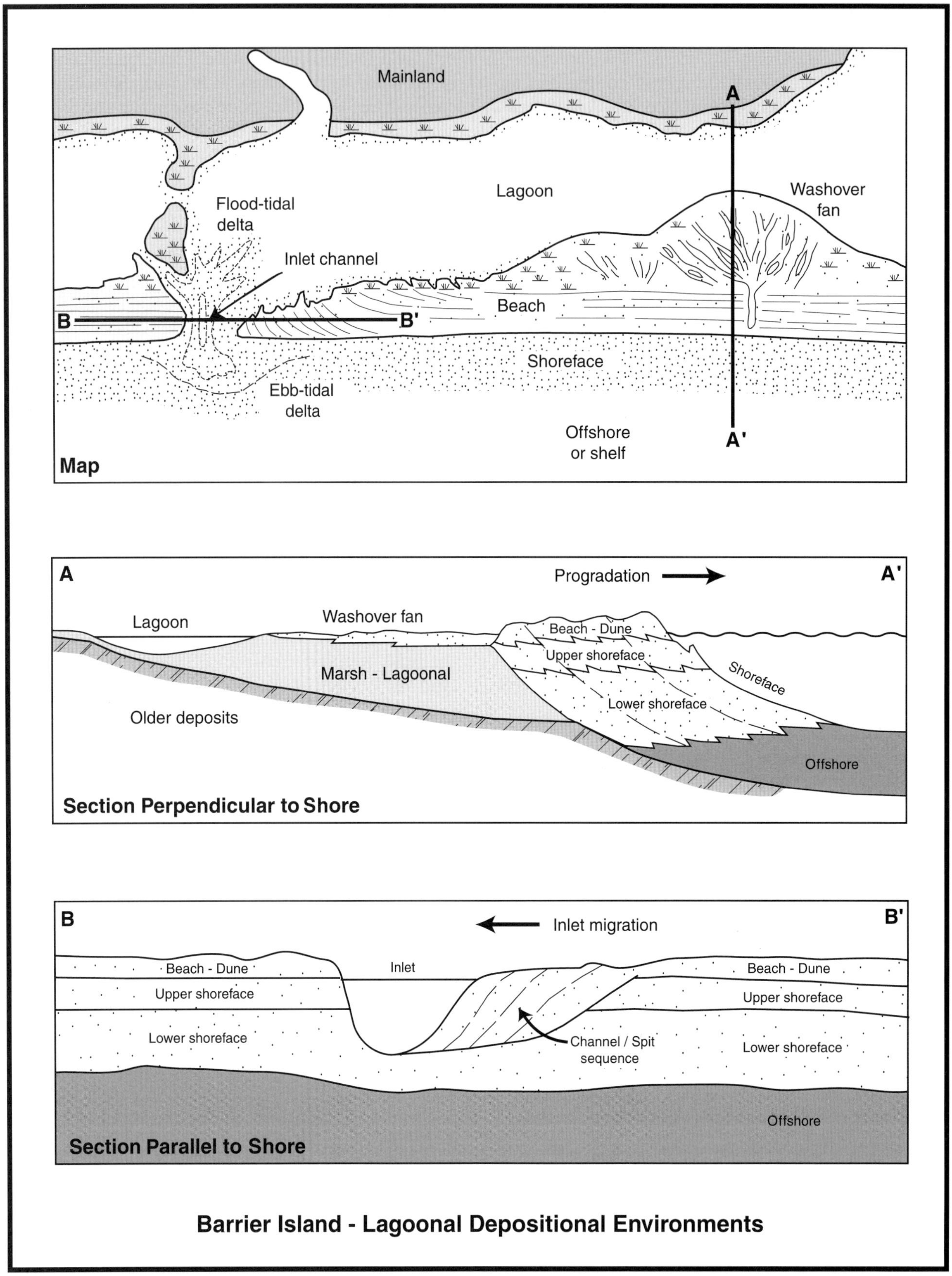

FIGURE 192. Generalized map and cross sections showing major environments and facies associations of a barrier island-lagoonal system (from McCubbin, 1982). Reprinted with permission from the AAPG.

Table 29. Factors influencing connectivity and reservoir development in siliciclastic shorelines and barrier island reservoirs.

Characteristic	*Favorable for Reservoir Development*	*Unfavorable for Reservoir Development*
Shoreface sandstones often show layer-cake tabular geometries	Excellent lateral continuity, high sweep efficiency, aquifers common	
Shoreface sandstones commonly stack in parasequence sets		Marine-flooding shales may create poor to no vertical connectivity between individual parasequences
Shoreface sandstones show upward-increasing permeability profiles	Favorable to high vertical sweep efficiencies	
Poor vertical permeability at upper to lower shoreface boundaries (e.g., mica plating)		Lower shoreface sediments may be poorly swept; horizontal wells may improve sweep in favorable circumstances
Extensive layer-parallel carbonate cements below flooding surfaces		Poor vertical connectivity
Mud-filled interridge swales in strand-plain sandstones		Can create barriers to vertical flow with compartmentalization
Common tidal and fluvial channel fills		Permeability contrast with shoreface or barrier island sandstones can result in bypassed oil
Lagoonal sandstone bodies (washover fans and flood-tidal deltas) can be wholly or partially enclosed in mudstone		Can form discrete hydraulic units or compartments with bypassed oil

37

Shepherd, M., 2009, Deep-water marine reservoirs, *in* M. Shepherd, Oil field production geology: AAPG Memoir 91, p. 295–300.

Deep-water Marine Reservoirs

INTRODUCTION

Deep-water marine reservoirs have been increasingly found since the 1970s, particularly as a result of an increase in offshore drilling activity. Many of these are Tertiary in age, although large reservoirs of Jurassic and Cretaceous age have also been found, particularly in the North Sea.

The term deep water has been used in two different ways. It applies in a geological context to deep-water systems that have been transported by gravity flow processes in a marine setting (Weimer and Slatt, 2004). Deep water is also defined as present-day sea depths in excess of 500 m (1640 ft) deep.

Since 1984 there has been an intensive effort in exploring for reservoirs located in present-day deep water with numerous prolific discoveries (Pettingill and Weimer, 2001). Deep-water exploration in the Gulf of Mexico, Brazil, and west Africa is targeting and finding a large number of hydrocarbon pools in deep-water marine-sand systems. Only about 20% of these reservoirs had been developed to 2004 (Weimer and Slatt, 2004).

DEEP-WATER MARINE RESERVOIRS CAN BE PROLIFIC RESERVOIRS

Deep-water marine sandstones can be prolific reservoirs where they occur. They commonly contain oil fields as a result of the interfingering of gravity-flow sandstones with marine oil-prone source rocks. An example is the interfingering of the Upper Jurassic submarine fans of the UK North Sea with the Kimmeridge Clay Formation source rock for the province, e.g., the Magnus and Claymore fields (Shepherd et al., 1990; Harker et al., 1991).

The reservoir quality of deep-marine sandstones is among the best of the various sedimentary environments that comprise reservoirs. Porosities, permeabilities, and net-to-gross ratios are typically high. Under favorable conditions, deep-water sandstones may be ponded and stacked vertically into very thick, sand-rich intervals (Table 30). These reservoirs are very profitable as they can be produced by a small number of wells at very high rates (Weimer and Slatt, 2004).

TYPICAL SETTINGS FOR DEEP-WATER MARINE RESERVOIRS

The main settings include

1) channelized systems;
2) channel-levee complexes; and
3) sheet complexes (including lobes).

Other settings include mass transport complexes, which are generally too heterogenous to form major reservoir intervals, and remobilized sandstone reservoirs. The latter are reservoirs that have undergone significant postdepositional mobilization with the formation of deformed sand pods, dykes, and sills (injectites). They are common in the Tertiary of the North Sea (e.g., Hurst and Cartwright, 2007).

PROBLEMS IN CHARACTERIZING DEEP-WATER MARINE SANDSTONES

Deep-water marine sandstone systems can be difficult to characterize in the subsurface. The basic problem is in trying to differentiate sheet sandstones from

DOI:10.1306/13161225M913372

Table 30. Factors influencing connectivity and reservoir development in deep-water marine reservoirs.

Characteristic	*Favorable for Reservoir Development*	*Unfavorable for Reservoir Development*
Under favorable conditions, deep-water marine sandstones can form very thick, high net-to-gross reservoirs	Produce at high rates with high ultimate recovery; can provide reservoirs for very profitable fields	
Depositionally isolated channel-fill sandstones in channelized systems		Create compartmentalized reservoirs requiring a multiwell development scheme
Widespread amalgamation of channel-fill sandstones in channelized systems	Creates laterally and vertically connected high-volume reservoirs	
Shale drapes or late-stage channel-fill shales common in channel-fill sandstones		Reduces vertical and lateral connectivity between individual channel-fill sandstones
Preferential water ingress along channel axes		Banked oil may form along channel margins
Levee sediments are commonly in poor communication with the channel-fill sandstones in channel-levee complexes		Bypassed oil in levee sediments
Levee sediments in channel-levee complexes are thin bedded but can show reservoir connectivity across a large area	Levee sediments can be a production target in their own right	
Laterally extensive mudstones commonly form permeability barriers to vertical flow	Encourages edge-water drive and can suppress early water production	Creates hydraulic units; water overrun is common
Fill and spill geometries		Potential to create bypassed oil volumes in cellar oil accumulations

channel complexes. *Sheet sandstones* are typically well connected laterally and can be very productive. By contrast, the large-scale reservoir connectivity of channel complexes can vary between good, where they have coalesced into connected bodies, and none at all, where they are depositionally isolated.

Channel complex connectivity will depend upon whether the individual channels show extensive sand-sand contacts with each other or not. It is not always easy to assess this with limited well control. This is a particular problem when it comes to the costly appraisal of deep-marine sandstone reservoirs offshore (Steffens, 2004).

Weimer and Slatt (2004) gave guidelines on how to differentiate between sheet and channelized deep-marine sandstones. If the log patterns and net-to-gross ratios vary considerably over short distances between wells, then the chances are that this is a channelized system (Chapin et al., 1994). A typical pattern is for channels to have high net-to-gross values and a blocky log response in the channel axis, with lower net-to-gross values and a serrated log response toward the margins (Figure 193). In core, channel-fill sandstones can show erosional features such as erosional bases, shale clasts, and an abundance of chaotic looking sediments.

Sheet sandstones and their associated shale interbeds are more layered and massive both at the core and interwell scale. Erosive features are relatively rare with few or no shale rip-up clasts. Post-production formation tester data can help to recognize the high degree of lateral continuity likely to be present here.

PRODUCTION FROM CHANNELIZED SYSTEMS

The channel axes may act as pathways for the preferential ingress of water. This can result in the stranding of banked oil along the channel margin pinch-out edges (Figure 194a) (Clark et al., 1997).

Channel fills can show differing degrees of amalgamation laterally and vertically (Figure 193). The degree of sandstone continuity is influenced by the stacking patterns. In channelized systems, this can be a critical parameter in assessing economic feasibility for reservoir appraisal, e.g., as in the appraisal of the Schiehallion field, offshore United Kingdom (see Chapter 3, this publication) (Leach et al., 1999). Eubanks (1987) described channelized turbidites from the Oligocene lower Hackberry

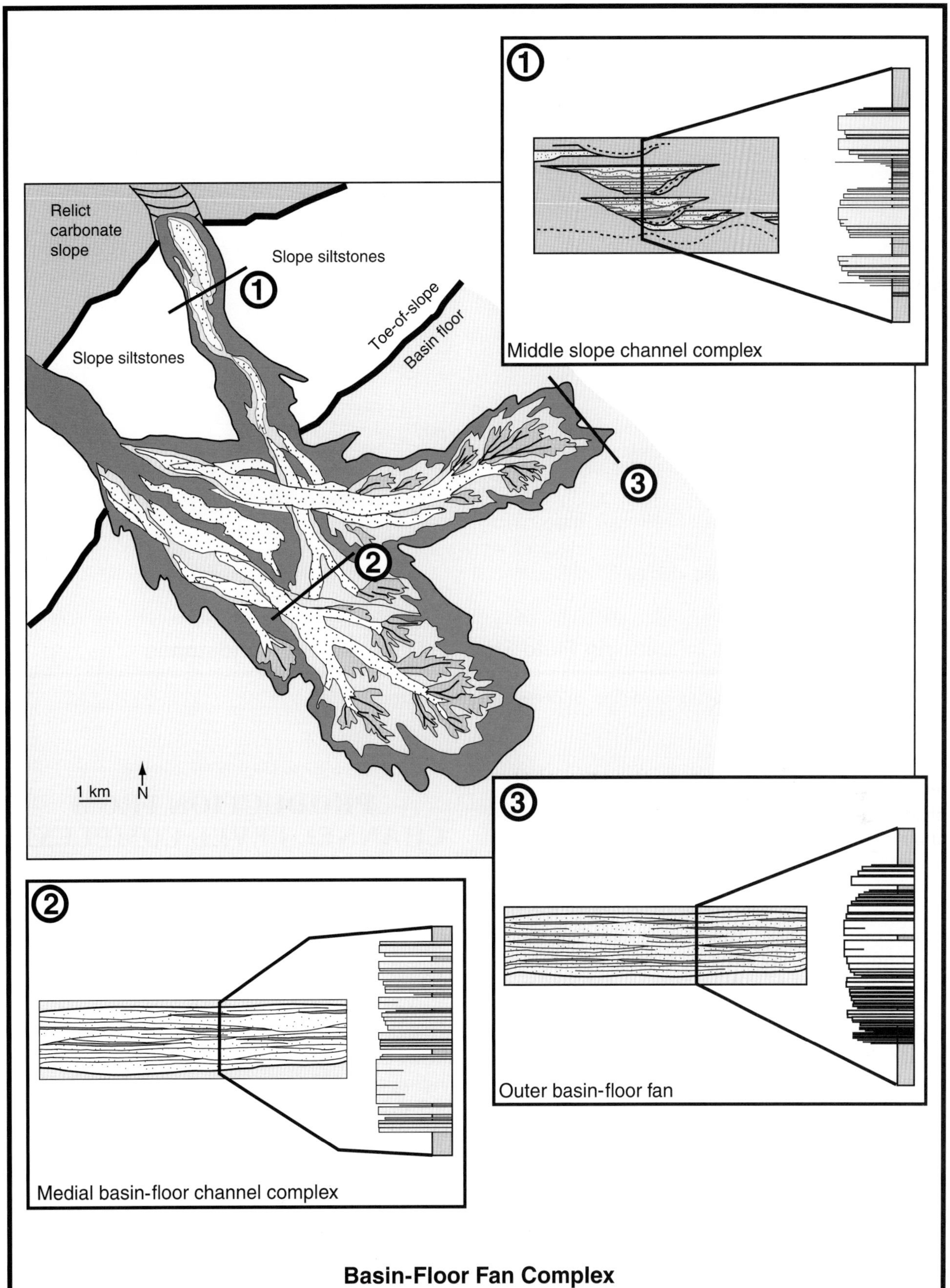

FIGURE 193. Depositional model for channelized turbidites and a basin-floor fan complex, Brushy Canyon, Texas. From Beaubouef (1999). Reprinted with permission from the AAPG.

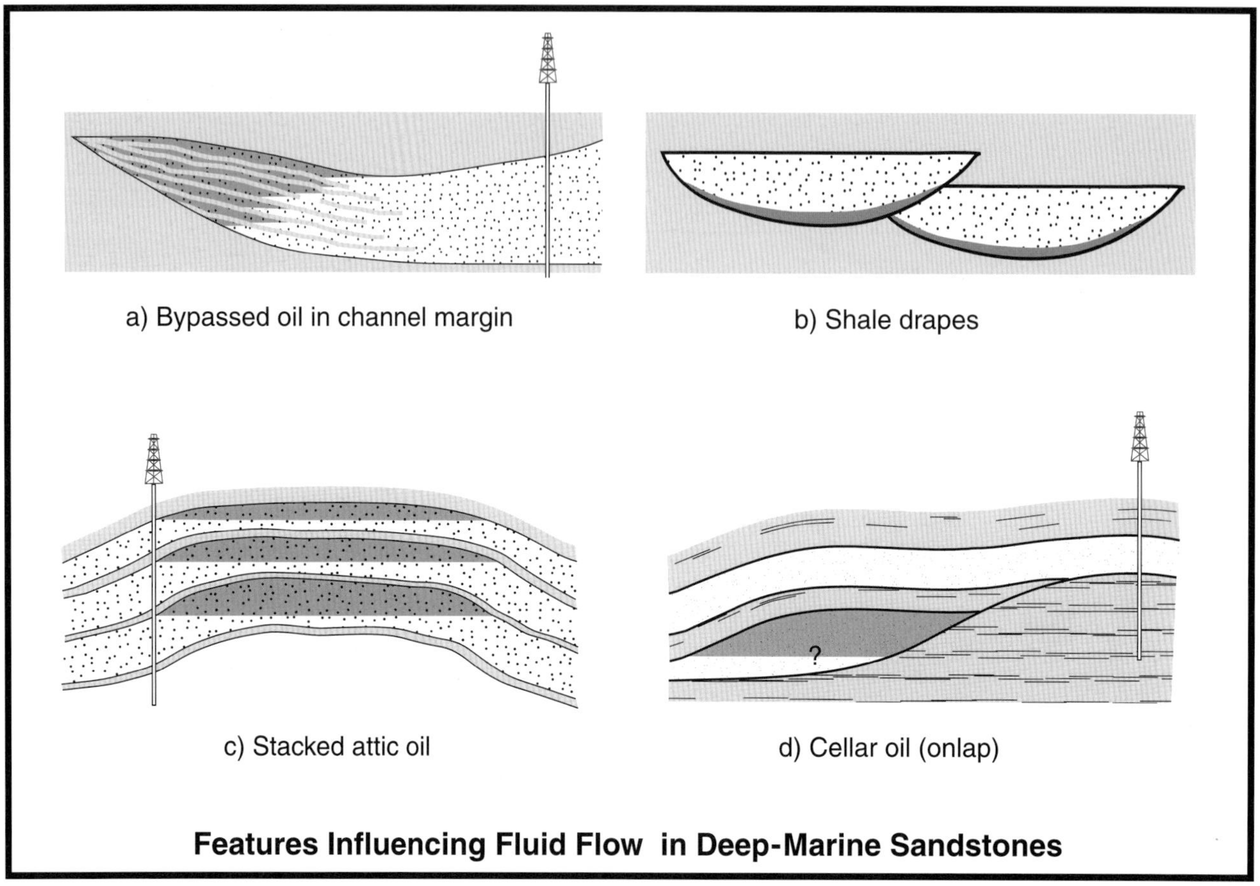

FIGURE 194. Features influencing fluid flow in deep-marine sandstones.

sands of the North Sabine Lake field, Louisiana. Stacking of individual channels resulted in a large amalgamated reservoir interval with pressure communication throughout. Isolated channels also occur, but these have shown rapid depletion within 2 months following production start-up.

Connectivity between individual channels will depend on how much shale is present. The effect of increasing shale content is to reduce vertical connectivity through sand-on-sand contacts and also to increase lateral variability (Weimer and Slatt, 2004).

Shales may be present either as extensive late-stage channel-fill shales or as *shale drapes* at the base of the channels. They may be composed of mudstone, siltstone, or heterolithic sediments (Figure 194b) (Beaubouef et al., 1999). Simulation modeling indicates that shale drapes may be a significant feature reducing connectivity between channel complexes and impairing the recovery efficiency (Larue, 2004). This mechanism has been invoked to explain why channel margins in the Forties field, UK North Sea, appear to act as baffles to fluid flow (Vaughan et al., 2007).

According to Weimer and Slatt (2004), the width to thickness ratio of channels typically ranges from 10:1 to 300:1.

PRODUCTION FROM CHANNEL-LEVEE COMPLEXES

Channel-levee complexes can show highly variable continuity between the channels and levees (Figure 195). It is common for hydrocarbons in the channel-fill sandstones to be poorly connected with the levee sediments. The channel fills may be younger than the levees themselves, and the beds in the proximal levee deposits may be discontinuous (Cronin et al., 2000; Beaubouef, 2004). Kneller et al. (2007) noted that collapse structures, including rotated blocks, slide sheets, slump folds, and thick debris flows, are common on levee margins and may contribute to poor reservoir continuity.

Injection wells located in overbank splays have poor communication with wells in the channel fill of the East Ford field in Texas (Dutton et al., 2003). The levee deposits can be a production target in their own right. In the Ram-Powell field in the Gulf of Mexico, production data indicate that levee deposits, although thin bedded, can show reservoir connectivity over a large area and give impressive production rates (Shew et al., 1995; Weimer and Slatt, 2004).

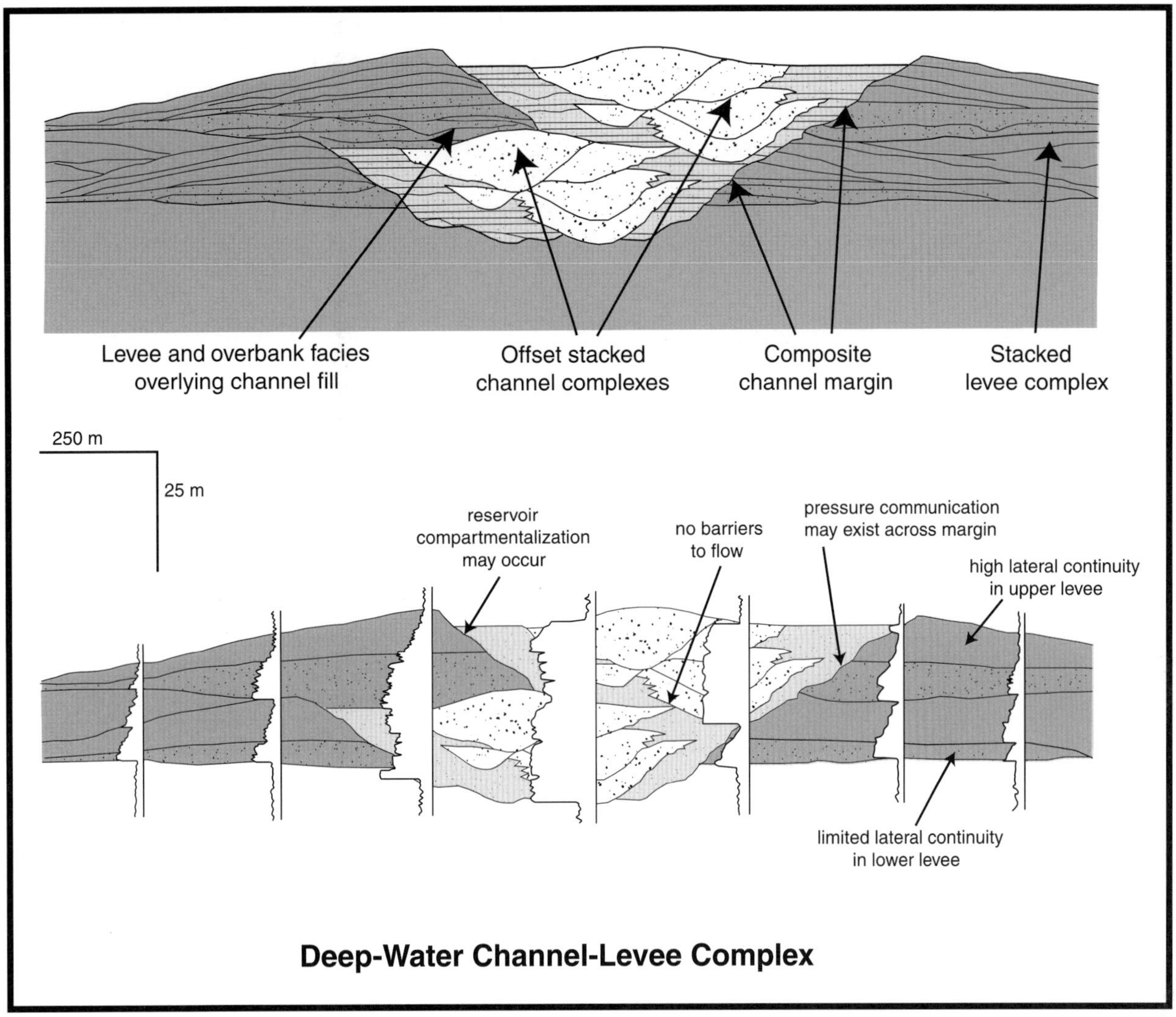

FIGURE 195. Schematic section across a deep-water channel-levee complex based on outcrops from the Cerro Toro Formation, Upper Cretaceous, southern Chile (from Beaubouef, 2004). The lower section shows a series of idealized gamma ray logs. Reprinted with permission from the AAPG.

PRODUCTION FROM SHEET COMPLEXES

Sheet sandstones form excellent reservoirs. Their characteristics include simple tabular geometries, good lateral continuity, and few erosional features (Weimer and Slatt, 2004). A large volume of deep-water sheet sandstones can be produced by a single production well. Width-to-thickness ratios are large, more than 500:1 for sheet complexes compared to a range of 10:1 to 300:1 for channels (Weimer and Slatt, 2004). Vertical connectivity can be variable depending on the amount of interbedded shales or the degree of sand-on-sand amalgamation.

Thin but laterally extensive mud blankets, either deposited from hemipelagic settling or the muddy tails of turbidity flows, can form permeability barriers to vertical flow in these systems (e.g., Hempton et al., 2005). Shales representing maximum flooding surfaces are commonly permeability barriers. Large-scale debris flows also have the potential to form baffles and barriers. Shales can subdivide the reservoir into several hydraulic units (Lowry et al., 1993). In certain favorable circumstances, these blanket shales can lead to a more efficient recovery by encouraging edge-water drive and suppressing bottom-water influx into the basal perforations of production wells. This type of flow behavior can be recognized on the basis of formation tester pressure discontinuities (see Figure 112) and slow-rising oil-water contacts on pulsed neutron logs. A typical management strategy in deep-water reservoirs with extensive shales is to isolate water-producing perforations in production wells by setting a plug in the well opposite one of these shale barriers.

Water overrun above laterally extensive shales is a common feature in sheet complexes. Stranded oil can be found under these shales. Blanket shales also have the potential to form multiple attic oil targets under local structural culminations in deep-water sediments (Figure 194c).

Cellar oil targets can also occur (Figure 194d). Early sand input into a receiving basin tends to pond into the bathymetric lows. The seabed will eventually become smoother as the lows are filled in. Later sand flows will spill beyond the extent of the previous flows. These show a more tabular geometry and will be spread across a larger area than the underlying sediments, creating a fill and spill geometry. The ponded sand bodies can potentially hold isolated oil volumes by comparison to the more extensive later flows, which may be better swept.

38

Shepherd, M., 2009, Carbonate reservoirs, *in* M. Shepherd, Oil field production geology: AAPG Memoir 91, p. 301–309.

Carbonate Reservoirs

CARBONATE RESERVOIRS

A significant proportion of the world's oil reserves are found in carbonate reservoirs. Many of these are located in the Middle East, Libya, Russia, Kazakhstan, and North America. Some very large oil fields have carbonate reservoirs, including the largest conventional oil field in the world, the Ghawar field of Saudi Arabia.

The reason for the very large size of some carbonate reservoirs is not surprising when one considers the sheer scale of even modern-day carbonate settings. The shallow submerged platform area of the Bahamas extends more than 400 km (248 mi) north–south and covers an area of about 125,000 km^2 (48,263 mi^2). The size of individual sediment bodies on the Bahama Banks can be impressive too (Figure 196). The Joulters Cay ooid shoal is a single carbonate sand body with a mobile border 25 km (15 mi) long and between 0.5 and 2 km (0.3 and 1.2 mi) wide (Major et al., 1996).

CARBONATES ARE DIFFERENT FROM SANDSTONES

Carbonate sediments have several features that set them apart by comparison with siliciclastics. Carbonate sediments tend to form and be deposited in situ, with enormous volumes of calcareous material provided by the death, disintegration, or digestion of plant and animal matter (Ginsburg and James, 1974). The coarser material tends not to be widely spread or abraded by waves and currents. Consequently, uniform grain sorting is not a major characteristic of carbonates. There can be a great diversity of grain sizes and shapes in most carbonate sediments compared to sandstones.

There are some similarities to siliciclastic environments. Various sedimentary bodies such as beaches, barrier islands, shelf sediments, gravity flows, and dune sands are also found in carbonate settings.

MANY CARBONATE RESERVOIRS OFFER A CHALLENGE TO THE PRODUCTION GEOLOGIST

Carbonate reservoirs can be difficult to develop for a variety of reasons. They generally have poorer recoveries than siliciclastic sediments (e.g., Sun and Sloan, 2003). A combination of depositional geometry and diagenesis creates highly heterogeneous reservoirs (Table 31). They can have lower primary recoveries as connected volumes may be areally limited with no contact to a large aquifer. The lower energy drive mechanisms such as solution gas drive are common. Heterogeneity at all the reservoir scales can make them a challenge to model, and it is not an easy task to make reliable predictions about their production performance. Reservoir management is difficult because the accurate targeting of production and injection wells is problematic, and sweep may be inefficient as a result of this.

Pore sizes in carbonates vary from micron scale to cave systems. Carbonates with vuggy porosity can store significant volumes of oil, yet sometimes the vugs are largely unconnected, yielding low flow rates. Tiny pores on a micron scale can form a high component of the porosity. The porosity may look impressive on logs, yet much of this may be microporosity and unproducible (Pittman, 1971; Cantrell and Hagerty, 1999). The petrophysical analysis of carbonate reservoirs is difficult and prone to greater uncertainty than with sandstone reservoirs. The uncertainty in the determination of water saturation, effective porosity, net pay, and permeability will impact the estimation of in-place volumes and reserves. Carbonates have a tendency to oil-wet characteristics or show mixed wettability. Typical behavior in oil-wet systems includes early water breakthrough and high water production rates (see Chapter 4, this publication). Carbonates can have thick transition zones in reservoirs with low matrix permeability (Masalmeh et al.,

DOI:10.1306/13161226M913372

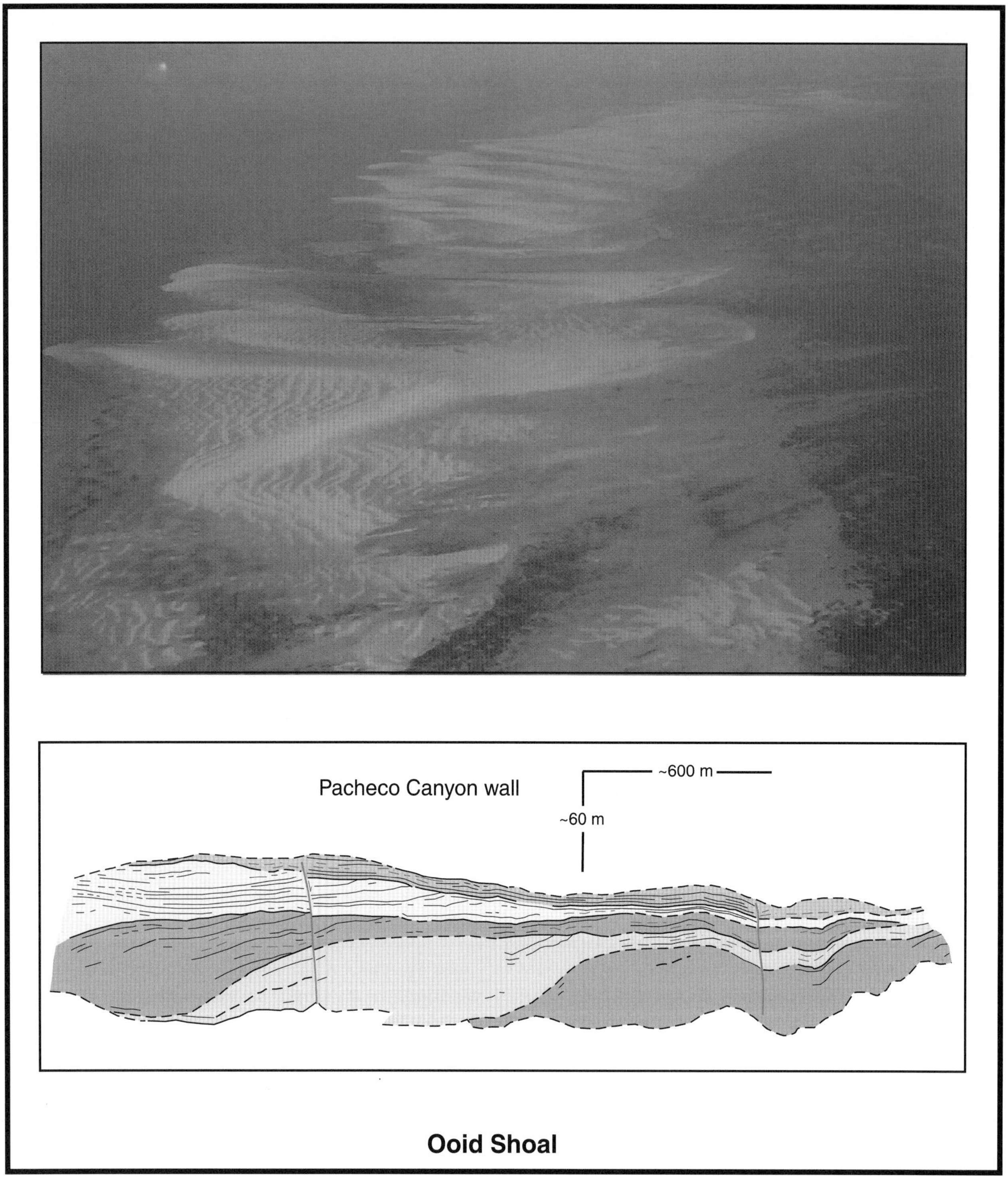

FIGURE 196. Ooid shoal, Bahamas; the bottom edge of the photograph represents a 4.5-km (2.7 mi)-wide transect. The lower inset is an illustration of a cliff exposure of laterally accreting (shingled) oolites from the Lower Cretaceous of Northern Mexico (from Osleger, 2004). Reprinted with permission from the AAPG.

2005). Residual oil saturations can also be high (Holtz et al., 1992; Kamath et al., 2001).

Carbonates are typically brittle rocks and are commonly fractured. The fractures can be a major component of the field performance, enhancing effective permeability and creating connectivity within otherwise heterogeneous reservoirs. Fractures will influence sweep patterns and will cause considerable variability in well-flow rates. Thief zones in fractures and high-permeability intervals can cause early water breakthrough.

GEOMETRY

Carbonate sediments tend to show a ribbon-like geometry and are less commonly developed as widespread sheets. Examples of both geometries are shown by two of the major carbonate reservoir intervals in the Middle East (Ehrenberg et al., 2007). Sediments of the Permian–Triassic Khuff Formation were deposited on a very low relief shelf, sheltered from the open ocean by a barrier reef. These show a layer-cake geometry consisting of interbedded mudstones and fine-grained grainstones (Alsharhan, 2006). By contrast, sedimentation in the Jurassic Arab Formation occurred on a shelf differentiated into shallow shoals and intrashelf basins. These exhibit a progradational geometry (Meyer and Price, 1992).

Carbonate sediments with ribbon geometries show a complex lateral facies progression in map view. A tendency for lateral accretion in successive cycles creates a subtle shingled geometry, which can make accurate correlation difficult (see Chapter 10, this publication, and Figure 67). For example, laterally accreting grainstones show a shingled geometry on a kilometer scale in Albian carbonates in northern Mexico (Figure 196) (Osleger et al., 2004). It can be a mistake to fit a layer-cake geometry to these systems because this results in reservoir models where lateral connectivity is predicted to be more extensive than is the case (Tinker, 1996). Facies belts may be difficult to define as lithofacies variation in carbonates is frequently transitional rather than sharp.

Carbonate sedimentation is very rapid and the buildup of carbonate sediment can exceed sea-level rise in a short period of time. For example, Neumann and Land (1975) estimated that the carbonate sediment accumulation rate in the Bight of Abaco in the Bahamas is 120 mm (5 in.) per thousand years. This is about three times the estimated subsidence rate of 38 mm (1.4 in.) per thousand years. The phrase carbonate factory is commonly used to describe the manner in which large volumes of sediment are produced on tropical shelfs.

Vertically, carbonates can be characterized by *high-frequency stacking*, with shoaling-upward cycles a few meters thick. Westphal et al. (2004) described high-frequency depositional cycles from the Mississippian Madison Formation in the Wind River Basin of Wyoming. The cycles occur over a meter-scale thickness and consist of a lower transgressive and an upper regressive hemicycle. The transgressive hemicycle is dominated by tidal flat sediments (laminated mudstone and wackestone) and subtidal deposits (e.g., stromatilites). The regressive hemicycle comprises high-energy carbonate sand-shoal facies (Figure 197).

High-frequency upward-shoaling cycles commonly comprise individual hydraulic or flow units within carbonate reservoirs (Kerans et al., 1994). Porosity variation in carbonate reservoirs occurs at the scale of high-frequency cycles (Ehrenberg, 2004). Larger scale trends in porosity variation can also occur at the systems tract or sequence level (Ehrenberg et al., 2006).

The measurement of carbonate body dimensions is a topic that gets less attention than is the case for siliciclastic reservoirs. A recent exception is Qi et al. (2007), where geometric data for ooid shoal, tidal flat, and eolian carbonate macroforms were used for constructing a 3-D reservoir model for the Big Bow and Sand Arroyo Creek fields in Kansas. The model has four zones with the ooid grainstone lithofacies showing the highest porosities and permeabilities. These form large linear shoals associated with structural highs. The model can be used to make predictions as a result of the simple zonation, large macroforms, and a reasonable correspondence between facies and rock properties. Many carbonate reservoirs are more complex than this and rather more difficult to model.

There may be several reasons why there is not so much measured data available for carbonate body dimensions compared to siliciclastics. Many carbonate reservoirs are characterized by rock types instead of lithofacies. Here, a combination of lithofacies and diagenesis acts as a control on rock properties. Thus, it is not always possible to make a predictive rock property model of a carbonate reservoir that is allied to the lithofacies model as in the example above.

The choice of reservoir analogs can be problematic. Carbonate environments have changed substantially over geological time by comparison to siliciclastic environments; for example, the type of organism responsible for building reefs has varied throughout the geological record. For this reason, it is advisable to select an outcrop analog that was deposited at roughly the same time as the carbonate reservoir interval under investigation (Markello et al., 2006).

IMPORTANCE OF DIAGENESIS

Diagenetic alteration tends to be the rule rather than the exception in carbonates and will act to modify or obscure the original depositional porosity (Jardine and Wilshart, 1982). It can happen that the best potential reservoir intervals such as reefs have their porosity totally occluded by diagenetic cement. However, poor or nonreservoir facies such as tidal mud flats can be modified into reservoir rock by dolomitization. Diagenetic reactions can reorganize the pore system significantly, commonly crosscutting stratigraphic boundaries. The process of rock property characterization in carbonates has to take into account both the stratigraphic and diagenetic model as a result (Lucia, 1995, 1999).

The diagenetic history of a carbonate reservoir can be complex, involving various phases of cementation, dissolution, compaction, and mineral transformation

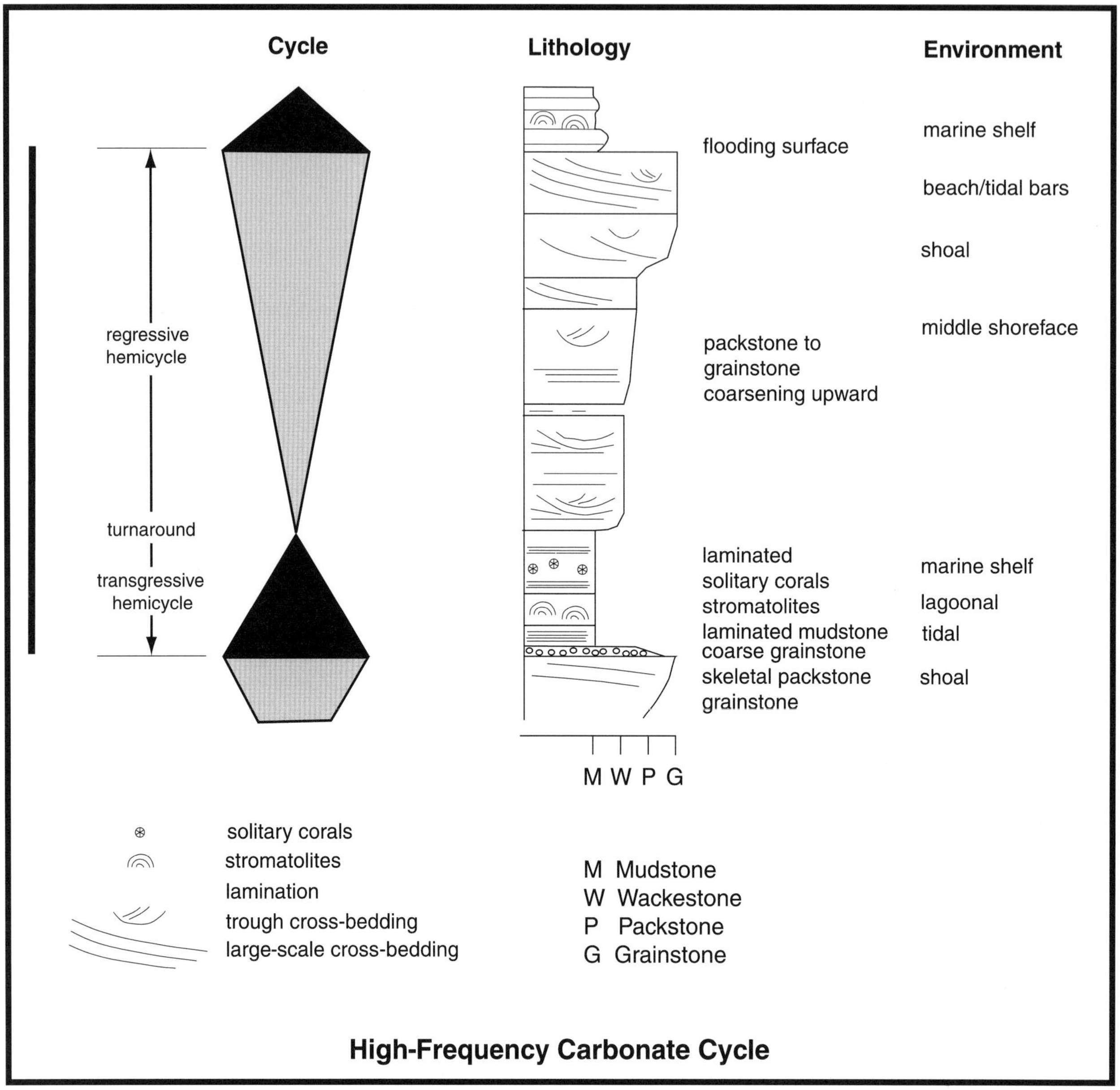

FIGURE 197. High-frequency carbonate cycle on a meter scale from the Mississippian Madison Formation in the Wind River Basin of Wyoming (after Westphal et al., 2004). Reprinted with permission from the AAPG.

(Tucker and Wright, 1990). Early oil migration can inhibit further diagenesis and preserve porosity in carbonate reservoirs (Neilson et al., 1998).

Dolomitization is the process by which calcium carbonate is altered to the magnesium-rich carbonate mineral dolomite. It has been estimated that about 80% of the reserves in the carbonates of the United States are in dolomite with 20% in limestone (North, 1985). Dolomitization materially affects the pore distribution of carbonate sediments. Dolomitization can act to eliminate heterogeneities in minor lithofacies that would otherwise form barriers or extensive baffles. Muddy carbonates can be transformed into porous dolomites with good intercrystalline connectivity. Dolomites tend to show higher porosities at increased depths of burial by comparison to limestones (Ehrenberg et al., 2006).

The process of dolomitization requires a large source of magnesium ions and a fluid transport path for the magnesium to move through the pore space. Several mechanisms have been proposed to explain dolomitization (Machel, 2004). For instance, in the *reflux model* of dolomitization, dolomite can form where hypersaline conditions exist in peritidal, lagoonal, and restricted basinal environments. Intense evaporation in the tropical heat will result in brine concentrations. The precipitation of gypsum and anhydrite removes calcium from the saline fluids, leaving a magnesium-rich residual brine. The dense, concentrated brine solution will subsequently filter down, reacting with the underlying sediments to form dolomite (Adams and Rhodes, 1960).

ROCK TYPES IN CARBONATES

It is an established procedure to characterize the rock properties of carbonates by rock types instead of lithofacies (Lucia, 1995, 1999). These are textural classes that are related to both depositional and diagenetic processes. Sandstone rock properties are dominated by intergranular pore systems, which exhibit a strong lithofacies control on grain size, shape, and sorting. By contrast, carbonate pore systems are much more complex (Choquette and Pray, 1970). The primary intergranular porosity is more variable because of the greater range in grain sizes and shapes. In addition, skeletal materials common in carbonates will show intraparticle porosity. The primary rock texture will often then be overprinted by postdepositional leaching, replacement, and cementation to form an even more complex pore network (Jardine and Wilshart, 1982).

TYPICAL SETTINGS FOR CARBONATE RESERVOIRS

Typical settings for carbonate reservoirs include

1) organic build-ups including reefs;
2) grainstone shoals on shelves;
3) subtidal and intertidal complexes;
4) leached zones below unconformities;
5) karst; and
6) chalk.

ORGANIC BUILD-UPS INCLUDING REEFS

Organic build-ups and reefs can be excellent reservoirs where the primary porosity has been preserved and is not occluded by internal sediments and secondary cements. They have the highest recovery factors among carbonate sediments according to Sun and Sloan (1993). Vertical permeability is typically good, and large pore systems are common in the reef core and in the near reef facies.

Major reef-forming organisms at various periods in geological time have included, amongst others, corals, algae, stromatoporoids, and rudist bivalves. Four main periods of reef reservoir formation have been described by Kiessling et al. (1999). These are the Silurian to Late Permian, the Late Jurassic, the middle Cretaceous, and the Miocene. Late Middle–Late Devonian reef reservoirs are particularly common worldwide.

Barrier reefs form thick massive sheets or ribbons parallel to the shoreline (Figure 198). Some of these can be very long, up to many tens of kilometers in length. The reef is the result of the growth of the calcareous framework created by the reef-forming organisms. This framework is interspersed with sands, silts, and muds that have formed from the erosion of the reef by biological activity and the occasional storm. The reefs themselves can act as a source of sediment, which may either be transported landward or seaward. The *back reef* can show impressive areas of skeletal sand deposition up to several kilometers wide. Localized patch reefs are also found here. *Reef aprons* form seaward from the reef and are composed of silt to boulder-size debris, derived from the reef front. The reef apron sediments can be stabilized or encrusted by in-situ fore reef biota such as foraminifera, sponges, or algae.

Barrier reef reservoirs are found in major oil fields such as the Oligocene to upper Eocene Kirkuk field of Iraq or the Lower Cretaceous fields found in the Golden Lane of Mexico (Viniegra-O and Castillo-Tejero, 1970).

Organic build-ups tend to be found encased in marine shales and/or evaporites. Massive reservoirs of this type are observed in relatively small *dome-shaped reefs*. The more complex *pinnacle reef* systems display a layered and lenticular distribution of zones with better reservoir properties. Where fractures occur, these can connect isolated porous and permeable zones into a dynamically unified system. Low-energy drive mechanisms tend to operate in these isolated systems. Pressure maintenance is often required. Secondary recovery operations can be efficient because the organic build-ups are typically thick and well connected (Sun and Sloan, 1993).

GRAINSTONE SHOALS ON SHELVES

Grainstone shoals form large elongate sheets that can extend for tens of kilometers in length (Figure 196). They are commonly found on the seaward edges of banks, platforms, and shelves (Halley et al., 1983). The grainstone shoals are composed of sand-size grains, which can be skeletal or non-skeletal in origin. The latter includes ooids. *Ooids* are coated grains with a calcareous outer cortex and nuclei that are variable in composition (Tucker and Wright, 1990). *Oolites* are rocks formed from ooids. Where oolites are relatively uncemented and not too deeply buried, they can form world-class productive intervals such as in the Jurassic Arab-D reservoirs of the Middle East. However, oolites can undergo cementation such that the interparticle volume is pervasively cemented, whereas the ooids dissolve out to form *oomoldic porosity*.

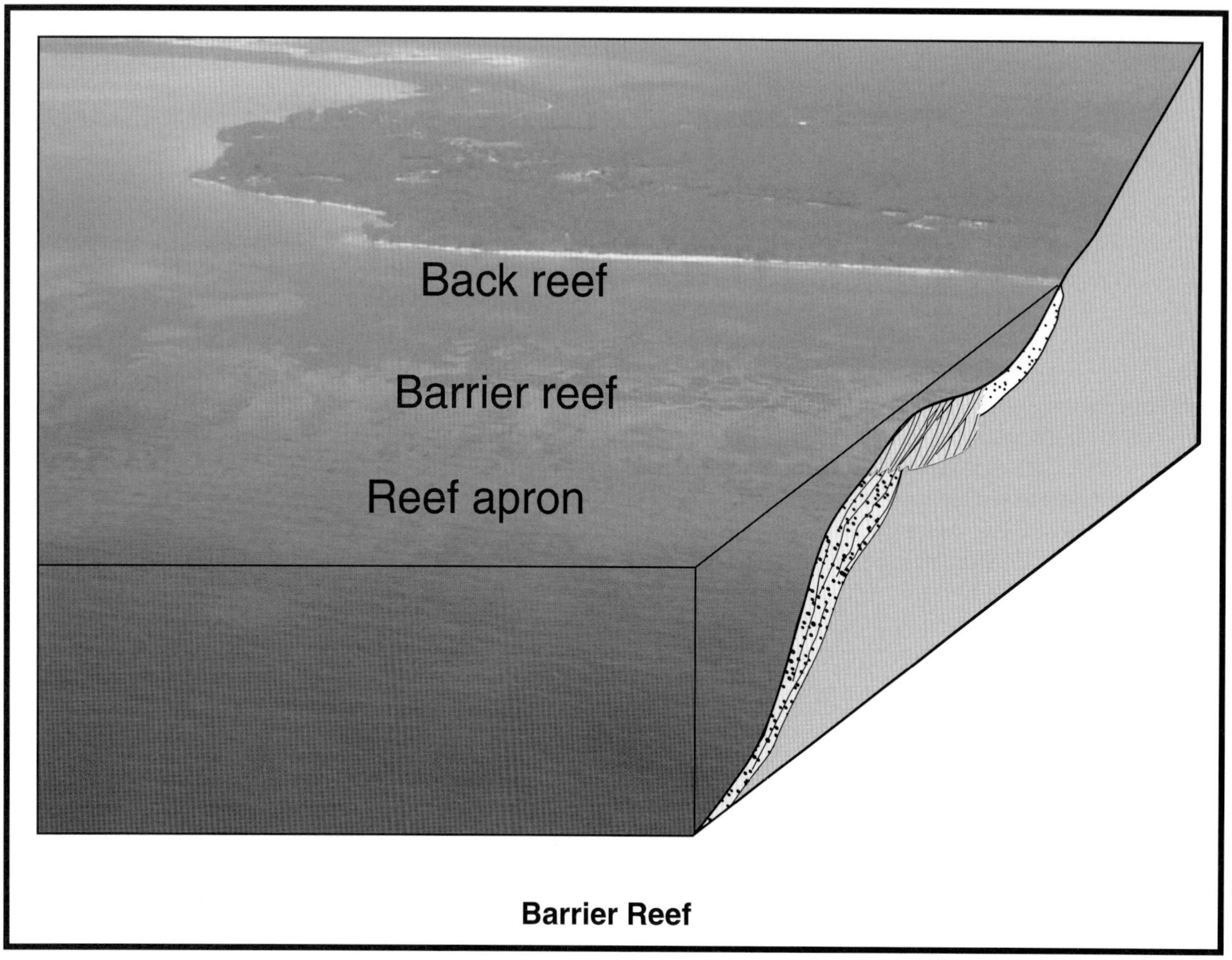

Figure 198. Barrier reef, Bahamas. The back reef between the barrier reef and the shoreline is 700 m (2296 ft) wide.

The ooids are typically poorly connected. One example, described from the Upper Jurassic Smackover Formation in Arkansas and Louisiana, shows 30% porosity but only one millidarcy or less permeability (Halley et al., 1983). Grainstone shoals are known to accrete laterally as a series of shingled units that may be compartmentalized by muddy barriers (Sneider and Sneider, 2000). Minor lateral heterogeneity occurs where tidal channels cut the ooid shoals.

SUBTIDAL AND INTERTIDAL COMPLEXES

The shelf interior in carbonate systems commonly shoals to a *tidal flat* environment that may be extensive in area (Figure 199). The highest porosities and permeabilities are found in the subtidal to intertidal facies with the best reservoir quality in tidal channel sediments. Supratidal sediments show the poorest reservoir quality and are typically barriers to vertical flow (Shinn, 1983). In arid environments, *supratidal sabkha* may be found. The evaporites can act as internal seals (Wilson, 1980).

Tidal flat mudstones can be extensively dolomitized to form significant reservoir intervals. Examples of this are found in reservoirs of the Ordovician Ellenburger Formation in the United States, the Ordovician Red River Formation of the Williston basin, the Permian Basin carbonates of Texas, and the Cretaceous offshore of west Africa.

KARSTIFICATION AND PALEOCAVE SYSTEMS

Karstified landscapes and *paleocave systems* form an important class of carbonate reservoirs. Caves present within a limestone bedrock are liable to collapse on compaction, creating a collapse breccia and with associated fracturing of the roof rock. Not all caves fall in with increasing burial; some can survive. When these are penetrated during drilling, the bit can suddenly drop

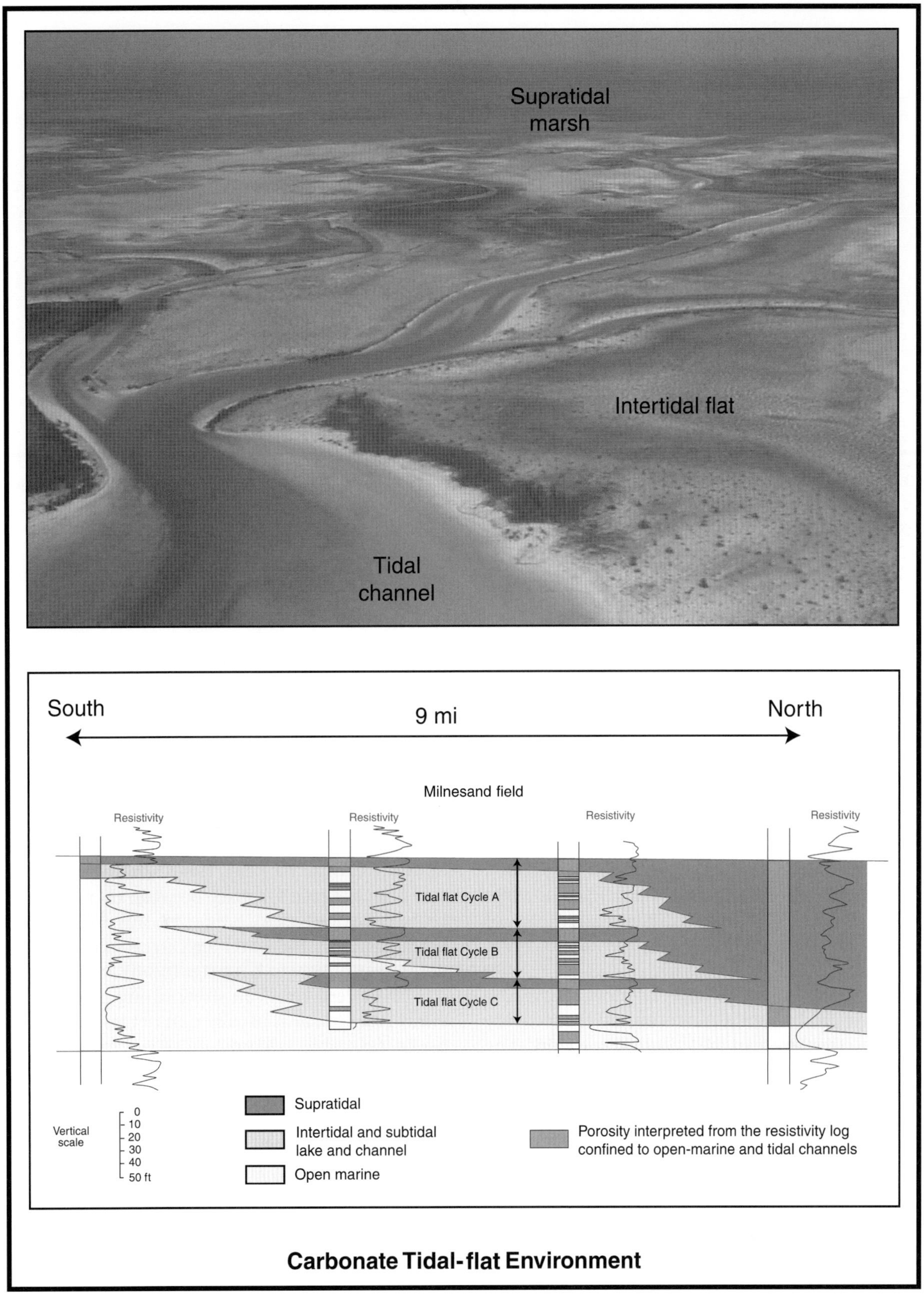

Figure 199. The upper photograph shows a Carbonate tidal flat on Andros Island, Bahamas. The tidal channel is about 150 m (492 ft) wide at the bottom of the photograph. The lower diagram shows three tidal flat reservoir cycles in the Permian San Andres dolomite of the northern Delaware basin in New Mexico and Texas (after Shinn, 1983). Repeated transgression and regression create cycles of tidal flat reservoirs, each sealed by impermeable anhydritic supratidal facies toward the north. Reprinted with permission from the AAPG.

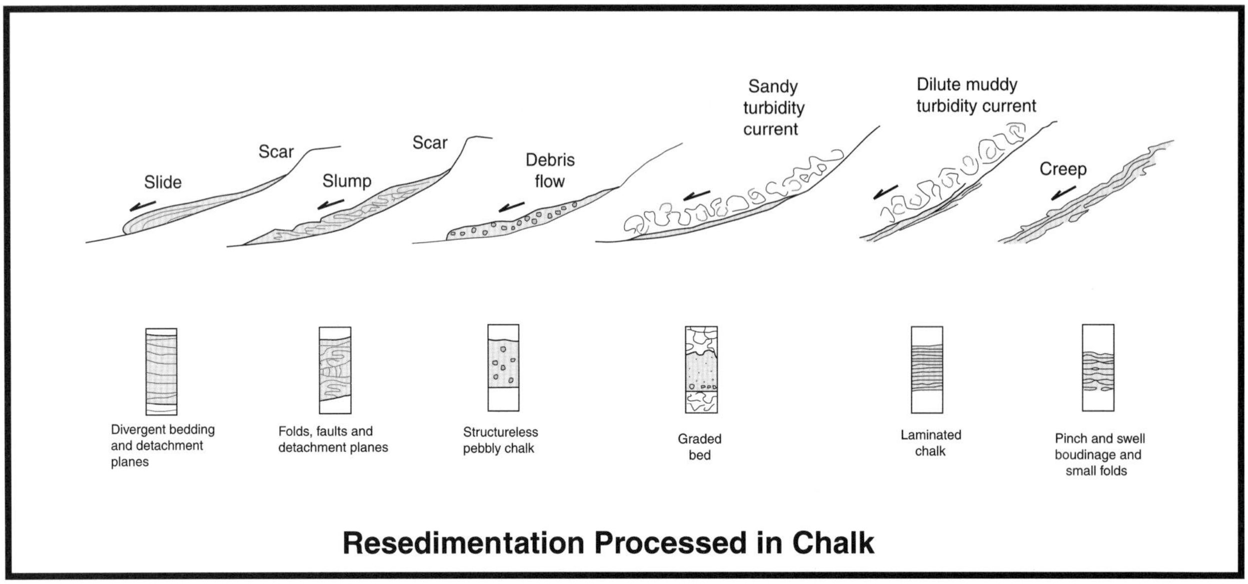

FIGURE 200. Redeposited chalk provides the main reservoir intervals in chalk fields. Resedimentation processes include sliding, slumping, debris flows, turbidity currents, and creep (from Surlyk et al., 2003). Reprinted with permission from the Geological Society.

several meters and large losses of drilling mud into the cave system can ensue.

Numerous cycles of cave formation and subsequent collapse can result in coalescing collapsed cave systems of considerable size, typically hundreds to several thousands of meters across. These systems may be mappable on 3-D seismic data. Collapse and sag structures form circular karst features that may be discernable from amplitude displays (Loucks, 1999).

Paleocave systems contain some very large hydrocarbon accumulations, such as the Lower Ordovician Puckett field in west Texas (Loucks and Anderson, 1980), in the Permian Yates field in west Texas (Craig, 1988), and in the Lower Cretaceous Golden Lane fields of eastern Mexico (Viniegra-O and Casstillo-Tejero, 1970; Coogan et al., 1972).

Karst and paleocave reservoirs can show poor recoveries. Fracture production is common, and the recovery is sensitive to the nature of the fracture framework. The better reservoirs have a fracture system that connects to an aquifer with a water drive operating. However, overproduction of these systems is detrimental to recovery because this will result in rapid water breakthrough and an early production decline (Sun and Sloan, 1993).

CHALK

Chalk is very fine-grained carbonate sediment, comprising skeletal calcitic debris of algae platelets. Porosity in chalk can be high, sometimes as high as 40–50%. Nevertheless, given the very fine-grained nature of the rock, permeabilities are low; 1–7 md is typical of the productive intervals. Factors influencing porosity preservation in chalk are overpressure, early oil migration, burial depth, chalk lithofacies, mud content, and grain size (Scholle, 1977; Nygaard et al., 1983; D'Heur, 1986; Brasher and Vagle, 1996). A correlation is found between the clay content of the chalk and the degradation of reservoir quality; clay hinders early lithification. As a result, clay-rich chalks are less rigid and will tend to undergo more compaction (Kennedy, 1987). It is a common pattern in chalk oil fields to find the highest porosity in the crest of the field, decreasing incrementally toward the oil-water contact (D'Heur, 1986). This character may result from the race for space between oil migration and cementing fluids (see Chapter 12, this publication). The permeability in the water leg can be so poor that chalk fields are unlikely to have significant aquifers.

Chalk reservoirs can show strong permeability layering. *Pelagic chalk* is usually non-net reservoir although under favorable circumstances it can be productive (Megson and Tygesen, 2005). Pelagic or *autochthonous* chalk results from the slow settling of sediment on the sea floor. Pervasive early cementation and extensive bioturbation significantly reduce the porosity and permeability from an early stage.

Pelagic chalk on the seabed is easily disturbed and remobilized. Clean chalk lacks any significant sediment cohesion as it has no unbalanced interparticle electric charges or platy interlocking grains to hold it together (Bramwell et al., 1999). Processes tending to redeposit chalk include debris flows, turbidity currents, slumps, and slides (Figure 200) (Kennedy, 1987).

Redeposited, *allochthonous* chalk typically shows much better porosities and permeabilities compared to autochthonous chalk in the same interval. The rock properties are thought to have been enhanced by several processes (Kennedy, 1987; Taylor and Lapre, 1987):

1) The chalk is loosened up as it is remobilized, with the break up of any early diagenetic cements that may already have formed.
2) Porosity is preserved as a consequence of minimal dewatering on burial.
3) The redeposited chalk tends to form as thicker masses and this results in the bulk of the sediment escaping bioturbation and early cementation at the sediment-sea water interface.

Given the low permeability of chalks, the presence of fractures can significantly enhance the productivity of chalk fields (see Chapter 14, this publication). Sorenson et al. (1986) differentiated between two classes of producing chalk fields in the North Sea: low-porosity chalk (15–30%) and permeabilities in the range of 0.2–1 md, which need an extensive natural fracture system to be productive, and high porosity chalk with 30–50% porosity and permeabilities between 1–10 md.

Horizontal wells are used to develop chalk fields (Megson and Hardman, 2001). Permeabilities are too low for conventional wells to be effective. Long horizontal wells, commonly 2 km or more in length, maximize the permeability-thickness and productivity of chalk fields. Fracture stimulation is used to enhance productivity (e.g., Cook and Brekke, 2004). Waterfloods can be highly effective in chalk because the fine capillary structure will draw in water very efficiently, displacing much of the oil (Surlyk et al., 2003). The injection wells should be drilled to avoid any open fractures that are likely to connect up with production wells, as rapid water breakthrough will ensue.

Table 31. Factors influencing connectivity and reservoir development in carbonate reservoirs.

Characteristic	*Favorable for Reservoir Development*	*Unfavorable for Reservoir Development*
Carbonates form highly heterogenous reservoirs		Generally have lower recoveries than sandstone reservoirs; difficult to locate wells
Generally do not have large aquifers		Poor primary recoveries
Tend toward oil-wet behavior		Early water breakthrough and high water production rates
Brittle rocks and commonly fractured	Fractures can create widespread connectivity in an otherwise heterogenous matrix rock	Can form thief zones with rapid water breakthrough
Common high-frequency cycles on a meter cycle		Numerous hydraulic units and highly layered reservoirs
Shingled geometries can be present		Potential to create bypassed oil volumes, particularly in shingled oolites
Diagenesis can significantly modify the original depositional connectivity in carbonate sediments	Dolomitization can potentially create good connectivity by modifying fine-grained sediments	Pervasive cements can significantly reduce rock properties and connected volumes

39

Shepherd, M., 2009, Less common reservoir types, *in* M. Shepherd, Oil field production geology: AAPG Memoir 91, p. 311–312.

Less Common Reservoir Types

ALLUVIAL FAN RESERVOIRS

Alluvial fans are found where mountain streams disperse sediments to form a fan-shaped body at the base of a mountain front or upland area (Nilsen, 1982) (Figure 69). Alluvial fans are not common reservoir intervals because they are seldom in direct contact with source rocks and rarely have proper seals. Examples of alluvial fan reservoirs include the Quiriquire field of Venezuela (Salvador and Leon, 1992) and the Chaunoy field in France (Eschard et al., 1998). Where they do occur as reservoirs, they tend to be not very productive. They show a disorganized aggregation of zones of porous and permeable streamflow deposits along with nonpermeable debris flow and mudflow deposits. Connectivity can be very poor. Correlation is difficult because of the absence of fossils and short-ranging abrupt facies changes. Cementation is common, particularly carbonate cements although siliceous and iron oxide cements may also occur (Nilsen, 1982).

The more distal fan environments have a greater chance of showing reasonable reservoir quality. Although the sediments here are finer grained, there is less interbedding of the impermeable mudflow and debris flow deposits that are more prevalent in the proximal and medial part of the fan.

GLACIAL SEDIMENTS

Oil and gas are produced from glacial sediments in Oman, Australia, Algeria, Argentina, and Bolivia. Reservoirs of glacial origin are prolific producers in Oman where more than 3.5 billion barrels of oil have been discovered in reservoirs of the Al Khlata Formation of the Permian–Carboniferous lower Haushi Group (Levell et al., 1988). One of these, the Marmul field, has a stock tank oil initially in place of more than 2 billion barrels of heavy oil (de la Grandville, 1982). The reservoir geology is highly complex. The main producing intervals are glaciofluvial and outwash fan sandstones.

ESTUARIES

Sediments are deposited in estuaries by varying elements of fluvial, tidal, and wave energy. A *bayhead delta* may form where a river discharges sediment into the head of the estuary. Where tidal processes dominate, reservoir sandstones are found in tidal bars, narrow linear bodies oriented parallel to the estuary margins. Wave energy can on occasion be strong enough to create a barrier bar partially enclosing the mouth of the estuary. Fluvial-estuarine channel complexes are significant producing reservoirs both onshore and offshore western Trinidad (Wach et al., 2004). The individual sandstone bodies are difficult to correlate with numerous permeability baffles and barriers present.

SHELF SANDSTONES

Shelf sandstones can contain significant volumes of oil and gas, for example, in the Cretaceous shelf sandstones of the United States. Galloway and Hobday (1996) distinguished between progradational and transgressive shelf systems. The former consists of typically low energy muddy sediments in layer-cake sheets or aprons, locally thickening into broad shoals or elongate bars. These comprise amalgamated storm beds and cross-laminated sands. Transgressive systems consist of isolated lensoid sandbars and ridges within an extensive sheet deposit of irregular interlaminated poorly sorted sand and mud.

Krause et al. (1987) described the Upper Cretaceous shelf sandstone reservoir of the Pembina field of Central

DOI:10.1306/13161227M913372

Alberta, Canada. This comprises irregularly interbedded, lensed, and shingled sand units. These have a jigsaw-puzzle geometry with poor reservoir continuity. The beds show a ridge and swale topography typical of a storm-dominated marine shelf. Thick conglomerates act as thief zones.

RESERVOIRS IN FRACTURED IGNEOUS AND BASEMENT ROCKS

Any rock can be a reservoir if it has fracture porosity and permeability, including igneous and metamorphic rocks of most types (Landes et al., 1960; Petford and McCaffrey, 2003). A common trap type, particularly in China and Southeast Asia, is the *buried hill structure*. These are fractured basement highs that are flanked by source-rock sediments. In north China, several oil fields have been discovered in Archean metamorphic rocks located in buried hill topographies (Guangming and Quanheng, 1982; Xiao Guang and Zuan, 1991). Five major oil fields produce from fractured granite in the Cuu Long Basin, offshore Vietnam. Individual field reserves are between 100 and 1400 MMbbls (Nguyen and Hung, 2003). The fractured granite basement high has been fed laterally by upper Oligocene source rocks. Oil columns of about 1000 to 1500 m (3281–4921 ft) are recorded.

METEORITE IMPACT STRUCTURES

Nine producing fields have been found in *meteorite impact structures* in North America (Donofrio, 1998). Production comes from impact-affected granites, carbonates, and sandstones. Reservoirs are found in central uplifts, rims, slump terraces, and ejecta deposits. One example is the Newporte field in North Dakota, which produces from an impact crater involving both Precambrian basement and lower Paleozoic sediments (Forsman et al., 1996). Oil and gas production started in 1977 from the brecciated basement on the rim of the 3.2-km (1.9-mi)-diameter crater.

The largest field in Mexico, the Cantarell field, produces about 60% of its total daily production from a 300-m (984-ft)-thick carbonate breccia. It is thought that this formed as a debris flow apron after the continental margin collapsed on the impact of the Chicxulub meteorite at the end of the Cretaceous (Grajales-Nishumura et al., 2000, 2002). The producing interval is sealed by a 30-m (98-ft)-thick impermeable dolomitized bentonitic bed, believed to be an ejecta layer formed from material thrown up by the impact.

40

Shepherd, M., 2009, Concluding comments, *in* M. Shepherd, Oil field production geology: AAPG Memoir 91, p. 313–314.

Concluding Comments

SUMMARY

A very large resource is available from our existing oil fields. If the recovery could be substantially increased from them, then this could go some way towards maintaining global energy supplies. The production geologist has a major role here.

The task is not easy. Very little data are available from our oil fields; indeed, a reservoir scheme comprises less than 0.1% information and more than 99.9% prediction. Nevertheless, it is possible to understand and make predictions about reservoirs such that multimillion-dollar decisions can be made to drill wells in them. Sedimentary environments typically show a continuity of process that allows predictions to be made. The sediments tend to be organized into discrete packages with a specific range of rock properties and with a predictable geometry. The packages can be large, most of them larger than normal well spacing. Thus, despite the meagerness of the information available, reservoir geology can be predictable.

Reservoirs often behave predictably too. The knowledge of flow patterns from one field in a specific depositional environment can be applied to reservoirs with a similar depositional environment elsewhere. The understanding of the flow geology is critical. It is not enough to study the geology of the reservoir; the production geologist should also have a shrewd idea as to how the geology influences fluid flow within the field. This involves data integration, graphical overlays of production data onto geological maps, and cross sections. In this way, a large number of observations are built up and finally integrated into a flow geology scheme. The work is hard, painstaking, and takes months to do properly. Nevertheless, it is great fun as the production geologist will be immersed in detective work, collecting clues toward a greater understanding of how the field behaves.

Progress is made when it is established that instead of acting as a big tank of oil, the oil field behaves as several independent drainage cells. Some of these drainage cells may contain large enough volumes of stranded oil to target with infill wells. The process of framing and screening volumes within a reservoir this way is the main method of locating the remaining oil in fields.

SOME PREDICTIONS FOR THE FUTURE OF PRODUCTION GEOLOGY

The challenge for the future is to get more oil out of our fields. This will happen by better locating the remaining oil, in getting more reliable estimates of the target volumes involved, and in harnessing both advanced and inexpensive technologies to recover the oil.

There will be more emphasis on flow geology and data integration techniques for locating bypassed oil in oil fields. At the moment, data integration is mostly a pencil and paper job. The production geologist would surely benefit from computer applications to help them conduct data integration in a more efficient manner. A combination of a data integration application with a 3-D geocellular modeling package would provide the geologist with a very powerful tool for improving oil recovery in oil fields.

The use of 4-D seismic surveys as a means of highlighting bypassed oil volumes will become even more common than it is at the moment. The results of these surveys will then be compared with data integration analysis of the flow geology so as to further validate the location of stranded oil volumes.

The production geologist would also gain from access to a more comprehensive knowledge database on the common patterns of flow behavior in specific depositional environments. Only a limited number of technical papers describe the detailed flow geology of specific oil fields. It is to be hoped that the geological community will make more case histories available in the future.

DOI:10.1306/13161228M913372

Better ways will be found to reduce the uncertainty on the size estimates of remaining oil volumes in order to make it less financially risky to target them. As the production geologist gets more involved in the analysis of the geological influences on the flow performance of the reservoir, there will be closer liaison with the reservoir engineers to help constrain these volumes. More emphasis will be given to the "top down" approach to building reservoir simulation models. The geologist will provide the reservoir engineer with several alternative scenarios to get feedback as to which elements of the flow geology are critical for a successful history match.

There will be an increasing focus on production at smaller scales, that is, at the level of the drainage cell within a reservoir. Small-scale simulation models of drainage cells, allied to an outcrop analog, may give an indication as to whether a drainage cell contains significant target volumes or merely a large number of dispersed uneconomic stranded oil pockets.

New technology will help to improve oil recovery, but perhaps the emphasis will be on the more widespread use of the existing technologies that are known to be effective. For instance, the more extensive use of enhanced oil recovery (EOR) techniques such as miscible floods could go a long way towards improving oil recovery globally, particularly the world's giant oil fields. There will also be an increasing use of low-cost technologies such as through tubing drilling for offshore wells. These will help the industry to access the smaller stranded oil volumes and to make more of these opportunities profitable.

References

Aase, N. E., P. A. Bjorkum, and P. H. Nadeau, 1996, The effect of grain-coating microquartz on preservation of reservoir porosity: AAPG Bulletin, v. 80, no. 10, p. 1654–1673.

Abdallah, W., et al., 2007, Fundamentals of wettability: Oilfield Review, Summer 2007, v. 19, no. 2, p. 44–61.

Abrahamsen, P., O. Lia, and H. Omre, 1992, An integrated approach to prediction of hydrocarbon in place and recoverable reserve with uncertainty measures: Presented at the European Petroleum Computer Conference, Society of Petroleum Engineers, May 24–27, Stavanger, Norway, SPE Paper 24276, 11 p.

Adams, J. E., and M. L. Rhodes, 1960, Dolomitization by seepage refluxion: AAPG Bulletin, v. 44, p. 1912–1920.

Adams, J. T., and C. Dart, 1998, The appearance of potential sealing faults on borehole images, *in* G. Jones, Q. J. Fisher, and R. J. Knipe, eds., Faulting, fault sealing and fluid flow in hydrocarbon reservoirs: Geological Society (London) Special Publication 147, p. 71–86.

Aguilera, R., 1980, Naturally fractured reservoirs: Tulsa, PennWell Books, 521 p.

Ahlbrandt, T. S., and S. G. Fryberger, 1981, Introduction to eolian deposits, *in* P. A. Scholle and D. R. Spearing, eds., Sandstone depositional environments, AAPG Memoir 31, p. 11–47.

Ainsworth, R. B., M. Sanlung, and S. T. C. Duivenvoorden, 1999, Correlation techniques, perforation strategies, and recovery factors: An integrated 3-D reservoir modeling study, Sirikit field, Thailand: AAPG Bulletin, v. 83, no. 10, p. 1535–1551.

Alberty, M., 1992, Standard interpretation, *in* D. Morton-Thompson and A. M. Woods, eds., Development geology reference manual: AAPG Methods in Exploration Series 10, p. 180–185.

Alexander, L. L., and J. W. Handschy, 1998, Fluid flow in a faulted reservoir system: Fault trap analysis for the Block 330 field in Eugene Island, South Addition, offshore Louisiana: AAPG Bulletin, v. 82, no. 3, p. 387–411.

Al-Khalifa, M. A., 2004, Advances in generating and ranking integrated geological models for fluvial reservoir: Presented at the Society of Petroleum Engineers Asia Pacific Conference on Integrated Modeling for Asset Management, March 29–30, Kuala Lumpur, Malaysia, SPE Paper 86999, 5 p.

Ali-Ali, Z. A., and B. A. Stenger, 2001, A case history on integrated fracture modeling in a giant field: Presented at the Society of Petroleum Engineers Annual Technical Conference and Exhibition, September 30–October 3, New Orleans, SPE Paper 71340, 13 p.

Allan, U. S., 1989, Model for hydrocarbon migration and entrapment within faulted structures: AAPG Bulletin, v. 73, no. 7, p. 803–811.

Allen, G. P., and J. L. C. Chambers, 1998, Sedimentation in the modern and Miocene Mahakam delta: Indonesian Petroleum Association, 236 p.

Alpay, O. A., 1972, A practical approach to defining reservoir heterogeneity: Journal of Petroleum Technology, SPE Paper 3608, v. 24, no. 7, p. 841–848.

Alpay, O. A., 1973, Application of aerial photographic interpretation to the study of reservoir natural fracture systems: Journal of Petroleum Technology, SPE Paper 2567, v. 25, no. 1, p. 37–45.

Alsharhan, A. S., 2006, Sedimentological character and hydrocarbon parameters of the middle Permian to Early Triassic Khuff Formation, United Arab Emirates: GeoArabia, v. 11, p. 121–158.

Ambrose, W. A., N. Tyler, and M. J. Parsley, 1991, Facies heterogeneity, pay continuity, and infill potential in barrier-island, fluvial, and submarine-fan reservoirs: Examples from the Texas Gulf Coast and Midland Basin, *in* A. D. Miall and N. Tyler, eds., The three-dimensional facies architecture of terrigenous clastic sediments and its implication for hydrocarbon discovery and recovery: SEPM Concepts in Sedimentology and Paleontology 3, p. 13–21.

Ambrose, W. A., F. P. Wang, M. S. Akhter, and J. Skolnakorn, 1997, Geologic controls on remaining oil in Miocene transgressive-barrier, coastal-plain, and mixed-load fluvial systems in the Miocene Norte area, Lake Maracaibo, Venezuela: Presented at the Society of Petroleum Engineers Annual Technical Conference and Exhibition, October 5–8, San Antonio, Texas, SPE Paper 38662, 10 p.

Anderson, W. G., 1986, Wettability literature survey: 1. Rock/oil/brine interactions and the effects of core handling on wettability: Journal of Petroleum Technology, SPE 13932, v. 38, no. 10, p. 1125–1144.

Anderson, W. G., 1987, Wettability literature survey: 6. The effects of wettability on waterflooding: Journal of Petroleum Technology, SPE 16471, v. 39, no. 12, p. 1605–1622.

Anderton, R., 1995, Sequences, cycles and other nonsense: Are submarine fan models any use in reservoir geology?, *in* A. J. Hartley and D. J. Prosser, eds., Characterization of deep marine clastic systems: Geological Society (London) Special Publication 94, p. 5–11.

Antonellini, M., and A. Aydin, 1994, Effect of faulting on fluid flow in porous sandstones: Petrophysical properties: AAPG Bulletin, v. 78, no. 3, p. 355–377.

Antonellini, M., and A. Aydin, 1995, Effects of faulting on fluid flow in porous sandstones: Geometry and spatial distribution: AAPG Bulletin, v. 79, no. 5, p. 642–670.

Aquino, J. A. L., J. M. Ruiz, M. A. F. Flores, and J. H. García, 2003, The Sihil field: Another giant below Canterell, offshore Campeche, Mexico, *in* M. T. Halbouty, ed., Giant oil and gas fields of the decade 1990–1999: AAPG Memoir 78, p. 141–150.

Archer, J. S., and C. G. Wall, 1986, Petroleum engineering principles and practice: London, Graham & Trotman, 350 p.

Arps, J. J., 1964, Engineering concepts useful in oil finding: AAPG Bulletin, v. 48, no. 2, p. 157–165.

Atchley, S. C., L. W. West, and J. R. Sluggett, 2006, Reserves growth in a mature oil field: The Devonian Leduc Formation at Innisfail field, south-central Alberta, Canada: AAPG Bulletin, v. 90, no. 8, p. 1153–1169.

Atkinson, C. D., J. H. McGowen, S. Bloch, L. L. Lundell, and P. N. Trumbly, 1990, Braidplain and deltaic reservoir, Prudhoe Bay field, Alaska, *in* J. H. Barwis, J. G. McPherson, and R. J. Studlick, eds., Sandstone petroleum reservoirs: Berlin, Springer-Verlag, p. 7–29.

Attanasi, E. D., and D. H. Root, 1994, The enigma of oil and gas field growth: AAPG Bulletin, v. 78, no. 3, p. 321–332.

Aydin, A., and A. M. Johnson, 1978, Development of faults as zones of deformation bands and as slip surfaces in sandstones: Pure and Applied Geophysics, v. 116, p. 931–942.

Ayyalasomayajula, P., N. Silpngarmlers, and J. Kamath, 2005, Well deliverability predictions for a low-permeability gas/condensate reservoir: Presented at the Society of Petroleum Engineers Annual Technical Conference and Exhibition, October 9–12, Dallas, Texas, SPE Paper 95529, 10 p.

Bahar, A., H. Ates, M. H. Al-Deeb, S. E. Salem, H. Badaam, and M. Kelkar, 2003, Practical approach in modeling naturally fractured reservoir: A field case study: Presented at the Society of Petroleum Engineers Annual Technical Conference and Exhibition, October 5–8, Denver, Colorado, SPE Paper 84078, 10 p.

Bahorich, M., and S. Farmer, 1995, 3-D seismic discontinuity for faults and stratigraphic features: The coherency cube: The Leading Edge, v. 14, p. 1053–1058.

Bailey, W. R., T. Manzocchi, J. J. Walsh, K. Keogh, D. Hodgetts, J. Rippon, P. A. R. Nell, S. Flint, and J. A. Strand, 2002, The effect of faults on the 3-D connectivity of reservoir bodies: A case study from the East Pennine coalfield, UK: Petroleum Geoscience, v. 8, no. 3, p. 263–277.

Bajsarowicz, C. J., 1992, Core alteration and preservation, *in* D. Morton-Thompson and A. M. Woods, eds., Development geology reference manual: AAPG Methods in Exploration Series 10, p. 127–130.

Barnett, J. A. M., J. Mortimer, J. H. Rippon, J. J. Walsh, and J. Watterson, 1987, Displacement geometry in the volume containing a single normal fault: AAPG Bulletin, v. 71, no. 8, p. 925–937.

Bater, L., 2003, The North Cormorant field, Block 211/21a, UK North Sea, *in* J. G. Gluyas and H. M. Hichens, eds., United Kingdom oil and gas fields, commemorative millennium volume: Geological Society (London) Memoir 20, p. 315–325.

Beaubouef, R. T., 2004, Deep-water leveed channel complexes of the Cerro Toro formation, Upper Cretaceous, southern Chile: AAPG Bulletin, v. 11, p. 1471–1500.

Beaubouef, R. T., C. Rossen, F. B. Zelt, M. D. Sullivan, D. C. Mohrig, and G. D. C. Jennette, 1999, Field guide for AAPG Hedberg field Research Conference, April 15–20, 1999, Deep-water sandstones, Brushy Canyon Formation, west Texas: AAPG Continuing Education Course Note Series 40, 48 p.

Beliveau, D., 1995, Heterogeneity, geostatistics, horizontal wells, and blackjack poker: Journal of Petroleum Technology, v. 47, no. 4, SPE Paper 30745, p. 1068–1074.

Bell, J. S., 1990, Investigating stress regimes in sedimentary basins using information from oil industry wire-line logs and drilling records, *in* A. C. Morton, A. Hurst, and M. A. Lovell, eds., Geological applications of wireline logs: Geological Society (London) Special Publication 48, p. 305–325.

Bell, J. S., and E. A. Babcock, 1986, The stress regime of the western Canadian basin and implications for hydrocarbon production: Bulletin of Canadian Petroleum Geologists, v. 34, p. 364–378.

Bell, J. S., and D. I. Gough, 1979, Northeast–southwest compressive stress in Alberta— Evidence from oil wells: Earth and Planetary Science Letters, v. 45, p. 475–482.

Bengtson, C. A., 1981, Statistical curvature analysis techniques for structural interpretation of dipmeter data: AAPG Bulletin, v. 65, no. 2, p. 312–332.

Bengtson, C. A., 1982, Structural and stratigraphic uses of dip profiles in petroleum exploration, *in* M. T. Halbouty, ed., The deliberate search for the subtle trap: AAPG Memoir 32, 351 p.

Berg, R. R., 1975, Capillary pressures in stratigraphic traps: AAPG Bulletin, v. 59, no. 6, p. 939–956.

Bernabé, Y., D. T. Fryer, and R. M. Shively, 1994, Experimental observations of the elastic and inelastic behaviour of porous sandstones: Geophysical Journal International, v. 199, p. 403–418.

Bertotti, G., N. Hardebol, J. K. Taal-van Koppen, and S. M. Luthi, 2007, Toward a quantitative definition of mechanical units: New techniques and results from an outcropping deep-water turbidite succession (Tanqua-Karoo Basin, South Africa): AAPG Bulletin, v. 91, no. 8, p. 1085–1098.

Bhatt, A., and H. B. Helle, 2002, Determination of facies from well logs using modular neural networks: Petroleum Geoscience, v. 8, no. 3, p. 217–228.

Bhattacharya, J. P., and R. G. Walker, 1992, Deltas, *in* R. G. Walker and N. P. James, eds., Facies models response to sea level change: Geological Association of Canada, 2d edition, p. 157–177.

Bjornstad, T., K. Garder, I. Hundere, and O. B. Michelsen, 1990, Tracer tests in oil appraisal and reservoir evaluation: State of the art, *in* J. Kleppe, E. W. Berg, A. T. Buller, O. Hjelmeland, and O. Torsaeter, eds., North Sea oil and gas reservoirs II, 1989: London, Graham & Trotman, p. 261–270.

Black, R. C., H. J. Poelen, M. J. Roberts, and S. E. Roddy, 1999, Tern field development: A marriage of new technologies for business benefit, *in* A. J. Fleet and S. A. R. Boldy, eds., Petroleum geology of northwest Europe: Proceedings of the 5th Conference, Geological Society (London), p. 1063–1073.

Blackbourn, G. A., 1990, Cores and core logging for geologists: Caithness, Whittles Publishing, 120 p.

Bleakly, D. C., 1992, Core orientation, *in* D. Morton-Thompson and A. M. Woods, eds., Development geology reference manual: AAPG Methods in Exploration Series 10, p. 122–124.

Bloch, G., M. El Deeb, H. Badaam, F. Cailly, G. Lecante, and A. Meunier, 2003, Seismic facies analysis for fracture detection: A powerful technique: Presented at the Middle East Oil Show, Society of Petroleum Engineers, June 9–12, 2003, Bahrain, SPE Paper 81526, 5 p.

Bloch, S., R. H. Lander, and L. Bonnell, 2002, Anomalously high porosity and permeability in deeply buried sandstone reservoirs: Origin and predictability: AAPG Bulletin, v. 86, no. 2, p. 301–328.

Blunt, M. J., K. Liu, and M. R. Thiele, 1996, A generalized streamline method to predict reservoir flow: Petroleum Geoscience, v. 2, p. 259–269.

Boak, J. M., 1992a, Conversion of well log data to subsurface stratigraphic and structural information, *in* D. Morton-Thompson and A. M. Woods, eds., Development geology reference manual: AAPG Methods in Exploration Series 10, p. 289–293.

Boak, J. M., 1992b, Geological cross sections, *in* D. Morton-Thompson and A. M. Woods, eds., Development geology reference manual: AAPG Methods in Exploration Series 10, p. 300–304.

Borah, I., 1992, Drill stem testing, *in* D. Morton-Thompson and A. M. Woods, eds., Development geology reference manual: AAPG Methods in Exploration Series 10, p. 131–139.

Bourke, L. T., 1992, Sedimentological borehole image analysis in clastic rocks: A systematic approach to interpretation, *in* A. Hurst, C. M. Griffiths, and P. F. Worthington, eds., Geological application of wireline logs II: Geological Society Special Publication 65, p. 31–42.

Bouvier, J. D., C. H. Kaars-Sijpesteijn, D. F. Kluesner, C. C. Onyejekwe, and R. C. Van der Pal, 1989, Three-dimensional seismic interpretation and fault sealing investigations, Nun River field, Nigeria: AAPG Bulletin, v. 73, p. 1397–1414.

Boyd, A., H. Darling, J. Tabanou, R. Davis, C. Flaum, J. Klein, R. M. Sneider, A. Sibbit, and J. Singer, 1995, The lowdown on low-resistivity pay: Oilfield Review, Autumn 1995, p. 4–18.

Braithwaite, C. I. M., J. D. Marshall, and T. C. Holland, 1989, Improving recovery from the Dunlin field, U.K. northern North Sea: Presented at the 54th Annual Technical Conference of the Society of Petroleum Engineers, San Antonio, Texas, SPE Paper 19878, 18 p.

Bramwell, N. P., G. Caillet, L. Meciani, N. Judge, M. Green, and P. Adam, 1999, Chalk exploration, the search for a subtle trap, *in* A. J. Fleet and S. A. R. Boldy, eds., Petroleum geology of northwest Europe: Proceedings of the 5th Conference, Geological Society (London), p. 911–937.

Brasher, J. E. and K. R. Vagle, 1996, Influence of lithofacies and diagenesis on Norwegian North Sea chalk reservoirs: AAPG Bulletin, v. 80, no. 5, p. 746–768.

Bridge, J. S., 2003, Rivers and flood plains: Forms, processes and sedimentary record: Oxford, Blackwell, 491 p.

Bridge, J. S., and S. D. Mackey, 1993, A theoretical study of fluvial sandstone body dimensions, *in* S. S. Flint and I. D. Bryant, eds., Geological modeling of hydrocarbon reservoirs: International Association of Sedimentologists, Special Publication 15, p. 213–236.

Briggs, R. J., R. P. Baron, R. J. Fulleylove, and M. S. Wright, 1987, Development of heavy oil reservoirs: Journal of Petroleum Technology, SPE Paper 15748, v. 40, no. 2, p. 206–214.

Bromley, R. G., 1996, Trace fossils, biology, taphonomy and applications, 2d ed.: London, Chapman & Hall, 361 p.

Brown, A., 2003, Capillary effects on fault-fill sealing: AAPG Bulletin, v. 87, no. 3, p. 381–395.

Brown, A. R., 2004, Interpretation of three-dimensional seismic data, 6th ed.: AAPG Memoir 42, SEG (Society of Exploration Geophysicists) Investigations in Geophysics 9, 560 p.

Brown, A. R., C. G. Dahm, and R. J. Graebner, 1981, A stratigraphic case history using three-dimensional seismic data in the Gulf of Thailand: Geophysical Prospecting, v. 29, no. 3, p. 327–349.

Brown, L. F., and W. L. Fisher, 1977, Seismic stratigraphic interpretation of depositional systems, examples from Brazil rift and pull-apart basins, *in* C. E. Paynton, ed., Seismic stratigraphy—Applications to hydrocarbon exploration: AAPG Memoir 26, p. 213–248.

Bryant, I. D., and S. S. Flint, 1993, Quantitative clastic reservoir geological modelling: Problems and perspectives, *in* S. S. Flint and I. D. Bryant, eds., The geological modeling of hydrocarbon reservoirs and outcrop analogs: International Association of Sedimentologists Special Publication 15, p. 3–20.

Bryant, I. D., and S. E. Livera, 1991, Identification of unswept oil volumes in a mature field by using integrated data analysis: Ness Formation, Brent field, UK North Sea, *in* A. M. Spencer, ed., Generation, accumulation and production of Europe's hydrocarbons: European Association of Petroleum Geoscientists Special Publication 1, p. 75–88.

Buckley, S. E., and M. C. Leverett, 1942, Mechanism of fluid displacement in sands: Petroleum Transactions of the American Institute of Mining Metallurgical and Petroleum Engineers, v. 146, p. 107–116.

Budding, M. C., A. H. M. Paardekam, and S. J. van Rossem, 1992, 3-D connectivity and architecture in sandstone reservoirs: Presented at the International Meeting on Petroleum Engineering, Society of Petroleum Engineers, March 24–27, Beijing, China, SPE Paper 22342, 9 p.

Burnett, M., 1996, 3-D seismic expression of a shallow fluvial system in west central Texas, *in* P. Weimer and T. L. Davis, eds.: AAPG Studies in Geology 42 and SEG (Society of Exploration Geophysicists) Geophysical Developments Series 5, p. 45–56.

Bush, K. R., J. Wakefield, M. Shepherd, and K. Beattie, 2001, Fife field UK—Rejuvenation of a mature asset: Prepared for presentation at the Offshore Europe Conference, Aberdeen, Society of Petroleum Engineers, September 4–7, 2001, SPE Paper 71846, 22 p.

Cacas, M. C., J. M. Daniel, and J. Letouzey, 2001, Nested geological modelling of naturally fractured reservoirs: Petroleum Geoscience, v. 7, p. S43–S52.

Cameron, G. I. F., 1992, Analysis of dipmeter data for sedimentary orientation, *in* A. Hurst, C. M. Griffiths, and P. F. Worthington, eds., Geological application of wireline logs II: Geological Society Special Publication 65, p. 141–154.

Campbell, K. J., 1999, Deepwater geohazards: How significant are they?: The Leading Edge, v. 18, p. 514–519.

Cant, D. J., 1982, Fluvial facies models, *in* P. A. Scholle and D. R. Spearing, eds., Sandstone depositional environments: AAPG Memoir 31, p. 115–137.

Cantrell, D. L., and R. M. Hagerty, 1999, Microporosity in Arab Formation carbonates, Saudi Arabia: GeoArabia, v. 4, no. 2, p. 129–154.

Capen, E. C., 1976, The difficulty of assessing uncertainty: Journal of Petroleum Technology, SPE Paper 5579, v. 28, no. 8, p. 843–850.

Capen, E. C., 1992, Dealing with exploration uncertainties, *in* R. Steinmetz, ed., The business of petroleum exploration: AAPG Treatise of Petroleum Geology, Handbook of Petroleum Geology, p. 29–61.

Caplan, M. L., and T. F. Moslow, 1999, Depositional origin and facies variability of a Middle Triassic barrier island complex, Peejay field, northeastern British Columbia: AAPG Bulletin, v. 83, no. 1, p. 128–154.

Carter, D. C., 2003, 3-D seismic geomorphology: Insights into fluvial reservoir deposition and performance, Widuri field, Java Sea: AAPG Bulletin, v. 87, no. 6, p. 909–934.

Carter, D. C., W. Kortlang, M. Smelcer, and J. C. Troncoso, 1998, An integrated approach to horizontal well design and planning in Widuri field, offshore southeast Sumatra, Indonesia: Proceedings of the Indonesian Petroleum Association, 26th Annual Convention, May 1998, v. 2, p. 135–162.

Cazier, E. C., A. B. Hayward, G. Espinosa, J. Velandia, J.-F. Mugniot, and W. G. Leel Jr., 1995, Petroleum geology of the Cusiana field, Llanos Basin foothills, Colombia: AAPG Bulletin, v. 79, no. 10, p. 1444–1462.

Chambers, R. L., J. M. Yarus, and K. B. Hird, 2000, Petroleum geostatistics for nongeostatisticians: Part 1: Leading Edge, v. 19, p. 474–479.

Chandler, M. A., G. Kocurek, D. J. Goggin, and L. W. Lake, 1989, Effects of stratigraphic heterogeneity on permeability in eolian sandstone sequence, Page Sandstone, northern Arizona: AAPG Bulletin, v. 73, no. 5, p. 658–668.

Chapin, M. A., and D. F. Mayer, 1991, Constructing a three-dimensional rock property model of fluvial sandstones in the Peoria field, Colorado, *in* A. D. Miall and N. Tyler, eds., The three dimensional facies architecture of terrigenous clastic sediments and its implication for hydrocarbon discovery and recovery: SEPM Concepts in Sedimentology and Paleontology 3, p. 160–171.

Chapin, M. A., P. Davies, J. L. Gibson, and H. S. Pettingill, 1994, Reservoir architecture of turbidite sheet sandstones in laterally extensive outcrops, Ross Formation, western Ireland, *in* P. Weimer, A. H. Bouma, and B. F. Perkins, eds., Submarine fans and turbidite systems: Gulf Coast Section SEPM Foundation 15th Annual Research Conference, v. 15, p. 53–68.

Chatzis, I., N. R. Morrow, and H. T. Lim, 1983, Magnitude and detailed structure of residual oil saturation: Society of Petroleum Engineers Journal, SPE Paper 10681, v. 36, no. 2, p. 311–326.

Cheatham, C., 1992, Wellbore trajectory, *in* D. Morton-Thompson and A. M. Woods, eds., Development geology reference manual: AAPG Methods in Exploration Series 10, p. 71–75.

Cheung, P. S.-Y., and D. Heliot, 1990, Workstation-based fracture evaluation using borehole imaging and wire-line logs: Presented at the Society of Petroleum Engineers Annual Technical Conference and Exhibition, September 23–26, New Orleans, Louisiana, SPE Paper 20573, 10 p.

Childs, C., J. J. Walsh, and J. Watterson, 1997, Complexity in fault zones and its implications for fault seal prediction, *in* P. Moller-Pederson and A. G. Koestler, eds., Hydrocarbon seals: Importance for exploration and production: Norwegian Petroleum Society Special Publication 7, p. 61–72.

Chilingar, G. V., and T. F. Yen, 1983, Some notes on wettability and relative permeabilities of carbonate reservoir rocks: II: Energy Sources, v. 7(1), no. 7, p. 67–75.

Choquette, P. W., and L. C. Pray, 1970, Geologic nomenclature and classification of porosity in sedimentary carbonates: AAPG Bulletin, v. 54, no. 2, p. 207–250.

Christensen, J. R., E. H. Stenby, and A. Skauge, 1988, Review of WAG field experience: Presented at the International Petroleum Conference and Exhibition of Mexico, Society of Petroleum Engineers, March 3–5, Villahermosa, Mexico, SPE Paper 39883, 14 p.

Ciftci, B. N., A. A. Aviantara, N. F. Hurley, and D. R. Kerr, 2004, Outcrop-based three-dimensional modeling of the Tensleep Sandstone at Alkali Creek, Bighorn Basin, Wyoming, *in* G. M. Grammer, P. M. Harris, and G. P. Eberli, eds., Integration of outcrop and modern analogs in reservoir modeling: AAPG Memoir 80, p. 235–259.

Clark, I., 1979, Practical geostatistics: London, Applied Science Publishers, 125 p.

Clark, M. S., J. D. Melvin, R. K. Prather, and A. W. Marino, 1997, Characterization and exploitation of the distal margin of a layered, low-permeability turbidite reservoir, Yowlumne field, San Joaquin Basin, California:

Houston Geological Society Bulletin, v. 40, no. 4, p. 12–15.

Clark, N. J., 1969, Elements of petroleum reservoirs: Dallas, American Institute of Mining, Metallurgical and Petroleum Engineers, 250 p.

Clark, N. J., H. M. Shearin, W. P. Scultz, K. Garms, and J. L. Moore, 1958, Miscible drive—Its theory and application: Journal of Petroleum Technology, v. 10, SPE Paper 1036-G, p. 11–20.

Clemetsen, R., A. R. Hurst, R. Knarud, and H. Omre, 1990, A computer program for evaluation of fluvial reservoirs, *in* J. Kleppe, E. W. Berg, A. T. Buller, O. Hjelmeland, and O. Torsaeter, eds., North Sea oil and gas reservoirs II: London, Graham & Trotman, p. 373–385.

Coleman, J. M., and L. D. Wright, 1975, Modern river deltas: Variability of processes and sand bodies, *in* M. L. Broussard, ed., Deltas, models for exploration: Houston Geological Society, p. 99–149.

Collinson, J. D., 1969, The sedimentology of the Grindslow shales and the Kinderscout Grit: A deltaic complex in the Namurian of northern England: Journal of Sedimentary Petrology, v. 39, p. 194–221.

Collinson, J. D., 1996, Alluvial sediments, *in* H. G. Reading, ed., Sedimentary environments: Processes, facies and stratigraphy, 3d ed.: Oxford, Blackwell Science, p. 37–82.

Cone, M. P., and D. G. Kersey, 1992, Porosity, *in* D. Morton-Thompson and A. M. Woods, eds., Development geology reference manual: AAPG Methods in Exploration Series 10, p. 204–209.

Conn, I., 2006, Energy trends and technologies for the coming decades: Address to the Harvard University Center for the Environment, March 8, 2006, 40 p.: www.bp.com (accessed October 2008).

Conway, R., 2007, Well design and intervention, *in* J. Wils and E. C. Neilson, eds., The technical and legal guide to the UK oil and gas industry: Aberdeen, Aberlour Press, 960 p.

Coogan, A. H., D. G. Bebout, and C. Maggio, 1972, Depositional environments and geologic history of Golden Lane and Poza Rica trend, Mexico, an alternative view: AAPG Bulletin, v. 56, no. 8, p. 1419–1447.

Cook, C. C., and K. Brekke, 2004, Productivity preservation through hydraulic propped fractures in the Eldfisk North Sea Chalk field: SPE Reservoir Evaluation and Engineering, v. 7, no. 2, SPE Paper 88031-PA, p. 105–114.

Cook, G., A. Chawathé, D. Larue, H. Legarre, and E. Ajayi, 1999, Incorporating sequence stratigraphy in reservoir simulation: An integrated study of the Meren E-01/MR-05 sands in the Niger Delta: Presented at the Society of Petroleum Engineers Reservoir Simulation Symposium, February 14–17, Houston, Texas, SPE Paper 51892, 12 p.

Corbett, K., M. Friedman, and J. Spang, 1987, Fracture development and mechanical stratigraphy of Austin Chalk, Texas: AAPG Bulletin, v. 71, p. 17–28.

Corbett, P. W. M., P. S. Ringrose, J. L. Jensen, and K. S. Sorbie, 1992, Laminated clastic reservoirs: The interplay of capillary pressure and sedimentary architecture: Presented at the Society of Petroleum Engineers Annual Technical Conference and Exhibition, October 4–7, Washington, D.C., SPE Paper 24699, 12 p.

Cosentino, L., 2001, Integrated reservoir studies: Paris, Editions Technip, 310 p.

Cosentino, L., V. Coury, J.-M. Daniel, E. Manceau, C. Ravenne, I. F. P. Van Lingen, J. Cole, and M. Sengul, 2001, Integrated study of a fractured Middle East reservoir with stratiform super-K intervals: 2. Upscaling and dual media simulation: Presented at the 2001 Society of Petroleum Engineers Middle East Oil Show, Bahrain, SPE Paper 68184, 13 p.

Crabaugh, M., and G. Kocurek, 1993, Entrada Sandstone: An example of a wet eolian system, *in* K. Pye, ed., Dynamics and environmental context of eolian sedimentary systems: Geological Society (London) Special Publication 72, p. 103–126.

Craig, D. H., 1988, Caves and other features of Permian karst in San Andres dolomite, Yates field reservoir, west Texas, *in* N. P. James and P. W. Choquette, eds., Paleokarst: Berlin, Springer-Verlag, p. 342–363.

Craig, F. F., 1971, The reservoir engineering aspects of waterflooding, monograph series: Richardson, Texas, Society of Petroleum Engineers, 144 p.

Cronin, B. T., A. Hurst, H. Celik, and I. Turkmen, 2000, Superb exposures of a channel, levee and overbank complex in an ancient, deep-water slope environment: Sedimentary Geology, v. 132, p. 205–216.

Dake, L. P., 1978, Fundamentals of reservoir engineering: Developments in Petroleum Science, 8: Amsterdam, Elsevier, 444 p.

Dake, L. P., 1994, The practice of reservoir engineering: Developments in Petroleum Science, 36: Amsterdam, Elsevier, 556 p.

Dalley, R. M., E. E. A. Gevers, G. M. Stampli, D. J. Davies, C. N. Gastaldi, P. R. Ruijetnberg, and G. J. D. Vermeer, 1989, Dip and azimuth displays for 3-D seismic interpretation: First Break, v. 7, p. 86–95.

Damsleth, E., C. B. Tjolsen, H. Omre, and H. H. Haldorsen, 1992, A two-stage stochastic model applied to a North Sea reservoir: Journal of Petroleum Technology, v. 44, no. 4, SPE Paper 20605, p. 402–408.

Damsleth, E., V. Sangolt, and G. Aamodt, 1998, Sub-seismic faults can seriously affect fluid flow in the Njord field off western Norway—A stochastic fault modeling case study: Presented at the 1998 Society of Petroleum Engineers Annual Technical Conference and Exhibition, September 27–30, 1998, New Orleans, SPE Paper 49024, 49 p.

Datta-Gupta, A., 2000, Streamline simulation: A technology update: Journal of Petroleum Technology, SPE Paper 65604, v. 52, no. 12, p. 68–73.

Davies, D. K., B. P. J. Williams, and R. K. Vessell, 1993, Dimensions and quality of reservoirs originating in low and high sinuosity channel systems, Lower Cretaceous Travis Peak Formation, east Texas, U.S.A., *in* C. P. North and D. J. Prosser, eds., 1993, Characterization of fluvial and eolian reservoirs: Geological Society Special Publication 73, p. 95–121.

Davis, E. F., and J. C. Shepler, 1969, Reservoir pressure data used to justify infill drilling in a low permeability reservoir: Journal of Petroleum Technology, SPE 2260, v. 21, no. 3, p. 267–273.

Davis, J. C., 1986, Statistics and data analysis in geology: New York, John Wiley & Sons, 656 p.

Davison, I., and R. S. Haszeldine, 1984, Orienting conventional cores for geological purposes: A review of methods, Journal of Petroleum Geology, v. 7, no. 4, p. 461–466.

de la Grandville, B. F., 1982, Appraisal and development of a structural and stratigraphic trap oil field with reservoirs in glacial to periglacial clastics, *in* M. T. Halbouty, ed., The deliberate search for the subtle trap: AAPG Memoir 32, p. 267–286.

de Raaf, J. F. M., H. G. Reading, and R. G. Walker, 1965, Cyclic sedimentation in the lower Westphalian of north Devon, England: Sedimentology, v. 4, p. 1–52.

de Rojas, I., 1987, Geological evaluation of San Diego Norte Pilot Project, Zuata area, Orinoco oil belt, Venezuela: AAPG Bulletin, v. 71, no. 10, p. 1294–1303.

de Waal, H., and R. Calvert, 2003, Overview of global 4-D seismic implementation strategy: Petroleum Geoscience, v. 9, p. 1–6.

Demirmen, F., 2005, Reliability and uncertainty in reserves growth: How the industry fails, and a vision for improvement: Presented at the Society of Petroleum Engineers Hydrocarbon Economics and Evaluation Symposium, April 3–5, 2005, Dallas, Texas, SPE Paper 94680, 9 p.

Dennis, H., J. Baillie, T. Holt, and D. Wessel-Berg, 2000, Hydrodynamic activity and tilted oil-water contacts in the North Sea, *in* K. Ofstad, E. J. Kittilsen, and P. Alexander-Marrack, eds., Improving the exploration process by learning from the past: Norwegian Petroleum Society Special Publication 9, p. 171–185.

Dennis, H., P. Bergmo, and T. Holt, 2005, Tilted oil-water contacts: Modelling the effects of aquifer heterogeneity, *in* A. G. Doré and B. A. Vining, eds., Petroleum geology: Northwest Europe and global perspectives: Proceedings of the 6th Petroleum Geology Conference, Geological Society (London), v. 1, p. 145–158.

Dershowitz, W. S., and H. H. Herda, 1992, Interpretation of fracture spacing and intensity, *in* J. R. Tillerson and W. R. Wawersik, eds., Rock mechanics: Rotterdam, Balkema, p. 757–766.

Deutsch, C. V., 2002, Geostatistical reservoir modeling: Oxford, Oxford University Press, 384 p.

Deutsch, C. V., and A. G. Journel, 1992, GSLIB, Geostatistical software library and user's guide: Oxford, Oxford University Press, 384 p.

D'Heur, M., 1986, The Norwegian chalk fields, *in* A. M. Spencer, ed., Habitat of hydrocarbons on the Norwegian Continental Shelf: London, Graham & Trotman, p. 77–89.

D'Heur, M., 1991, West Ekofisk field—Norway central graben, North Sea: AAPG Bulletin, v. 75, no. 5, p. 946–968.

Dincau, A. R. 1998, Prediction and timing of production induced fault seal breakdown in the South Marsh Island 66 gas field: Gulf Coast Association of Geological Societies Transactions, v. 48, p. 21–32.

Doligez, B., H. Beucher, F. Geffroy, and R. Eschard, 1999, Integrated reservoir characterization: Improvement in heterogeneous stochastic modeling by integration of additional external constraints, *in* R. Schatzinger and J. Jordan, eds., Reservoir characterization—Recent advances: AAPG Memoir 71, p. 333–342.

Domínguez, R., 2007, Structural evolution of the Penguins cluster, UK northern North Sea, *in* S. J. Jolley, D. Barr, J. J. Walsh, and R. J. Knipe, eds., Structurally complex reservoirs: Geological Society (London) Special Publication 292, p. 25–48.

Donofrio, R. R., 1998, North American impact structures hold giant field potential: Oil & Gas Journal, May 11, v. 96, p. 69–83.

Donselaar, M. E., and I. Overeem, 2008, Connectivity of fluvial point bar deposits: An example from the Miocene Huesca fluvial fan, Ebro basin, Spain: AAPG Bulletin, v. 92, no. 9, p. 1109–1129.

Doré, G., and J. Robbins, 2005, The Buzzard field, *in* A. G. Doré and B. A. Vining, eds., Petroleum geology: Northwest Europe and global perspectives: Proceedings of the 6th Petroleum Geology Conference, Geological Society (London), v. 1, p. 241–252.

Doughty, P. T., 2003, Clay smear seals and fault sealing potential of an exhumed growth fault, Rio Grande rift, New Mexico: AAPG Bulletin, v. 87, no. 3, p. 427–444.

Downton, G., A. Hendricks, T. S. Klausen, and D. Pafitis, 2000, New directions in rotary steerable drilling: Oilfield Review, Spring 2000, v. 12, no. 1, p. 18–29.

Doyen, P. M., 1988, Porosity from seismic data: A geostatistical approach: Geophysics, v. 53, no. 10, p. 1263–1276.

Doyle, J. D., and M. L. Sweet, 1995, Three-dimensional distribution of lithofacies, bounding surfaces, porosity, and permeability in a fluvial sandstone—Gypsy Sandstone of northern Oklahoma: AAPG Bulletin, v. 79, no. 1, p. 70–95.

Dreyer, T., 1993, Geometry and facies of large-scale flow units in fluvial-dominated fan-delta-front sequences, *in* M. Ashton, ed., Advances in reservoir geology: Geological Society Special Publication 69, p. 135–174.

Driscoll, V. J., 1974, Recovery optimization through infill drilling concepts, analysis and field results: Presented at the Fall Meeting of the Society of Petroleum Engineers of the American Institute of Mining Metallurgical and Petroleum Engineers, October 6–9, Houston, Texas, SPE Paper 4977, 16 p.

Dromgoole, P., and R. Speers, 1997, Geoscore: A method for quantifying uncertainty in field reserve estimates: Petroleum Geoscience, v. 3, p. 1–12.

Dubrule, O., 1998, Geostatistics in petroleum geology: AAPG Continuing Education Course Note Series 38, 52 p.

Dubrule, O., 2003, Geostatistics for seismic data integration in Earth models: EAGE (European Association of Geoscientists and Engineers) Distinguished Instructor Series 6, 265 p.

Dubrule, O., and E. Damsleth, 2001, Achievements and challenges in petroleum geostatistics: Petroleum Geoscience, v. 7, p. S1–S7.

Durham, L. S., 2006, Top down or bottoms up? Getting a grip on reservoirs: AAPG Explorer, v. 27, October 2006, p. 12,14,16.

Durrani, A. J., B. M. Escovedo, W. Ordemann, J. A. Bickley,

B. J. Tepper, and J. M. Simon, 1994, Rejuvenation of 30-year-old McAllen Ranch field—An application of cross-functional team management: Journal of Petroleum Technology, SPE Paper 24872, v. 46, no. 12, p. 1065–1072.

Dutton, S. P., W. A. Flanders, and M. D. Barton, 2003, Reservoir characterization of a Permian deep-water sandstone, East Ford field, Delaware basin, Texas: AAPG Bulletin, v. 87, no. 4, p. 609–627.

Dyke, C. G., 1992, Stress insensitive natural fracture permeability within hydrocarbon reservoirs, *in* J. A. Hudson, ed., Eurock '92, International Society of Rock Mechanics Symposium, London: London, Thomas Telford, p. 281–286.

Dyke, C. G., B. Wu, and D. Milton-Taylor, 1995, Advances in characterizing naturally fractured permeability from mud-log data: SPE Formation Evaluation, v. 10, no. 3, SPE Paper 25022, p. 160–166.

Ebanks Jr., W. J., M. H. Scheihing, and C. D. Atkinson, 1992, Flow units for reservoir characterization, *in* D. Morton-Thompson and A. M. Woods, eds., Development geology reference manual: AAPG Methods in Exploration Series 10, p. 282–285.

Eberli, G. P., J. L. Masaferro, and J. F. Sarg, 2004, Seismic imaging of carbonate reservoirs and systems, *in* G. P. Eberli, J. L. Masaferro, and J. F. Sarg, eds., Seismic imaging of carbonate reservoirs and systems: AAPG Memoir 81, p. 1–9.

Edie, R. W., and J. M. Andrichuk, 2003, Meander belt entrapment of hydrocarbons at Saddle Lake, Alberta and an untested in situ combustion scheme for recovery of heavy oil: Bulletin of Canadian Petroleum Geology, v. 51, no. 3, p. 253–274.

Edwards, M. B., 1976, Growth faults in upper Triassic deltaic sediments, Svalbard: AAPG Bulletin, v. 60, no. 3, p. 341–355.

Ehrenberg, S. N., 2004, Factors controlling porosity in Upper Carboniferous-Lower Permian carbonate strata of the Barents Sea: AAPG Bulletin, v. 88, no. 12, p. 1653–1676.

Ehrenberg, S. N., G. P. Eberli, M. Keramati, and S. A. Moallemi, 2006, Porosity-permeability relationships in interlayered limestone-dolostone reservoirs: AAPG Bulletin, v. 90, no. 1, p. 91–114.

Ehrenberg, S. N., P. H. Nadeau, and A. A. M. Aqrawi, 2007, A comparison of Khuff and Arab reservoir potential throughout the Middle East: AAPG Bulletin, v. 91, no. 3, p. 275–286.

Eisenberg, L. I., M. V. Langston, and R. E. Fitzmorris, 1994, Reservoir management in a hydrodynamic environment, Iagufu-Hedinia area, Southern Highlands, Papua New Guinea: Presented at the Society of Petroleum Engineers Asia Pacific Oil and Gas Conference, November 7–10, Melbourne, Australia, SPE Paper 28750, 17 p.

Emery, F., and K. J. Myers, 1996, Sequence stratigraphy: Oxford, Blackwell Science, 297 p.

England, W. A., and C. Townsend, 1998, The effects of faulting on production from a shallow marine reservoir: Presented at the 1998 Society of Petroleum Engineers Annual Technical Conference and Exhibition, September 27–30, 1998, New Orleans, SPE Paper 49023, 16 p.

Eschard, R., P. Lemouzy, C. Bacchiana, G. Désaubliaux, J. Parpant, and B. Smart, 1998, Combining sequence stratigraphy, geostatistical simulations, and production data for modeling a fluvial reservoir in the Chaunoy field (Triassic, France): AAPG Bulletin, v. 82, no. 4, p. 545–568.

Eslinger, E., and D. Pevear, 1988, Clay minerals for petroleum geologists and engineers: SEPM Short Course Notes 22, 433 p.

Eubanks, L. G., 1987, North Sabine Lake field: Complex deposition and reservoir morphology of lower Hackberry (Oligocene), southwest Louisiana: AAPG Bulletin, v. 71, no. 10, p. 1162–1170.

Evamy, B. D., J. Haremboure, P. Kamerling, W. A. Knaap, F. A. Molloy, and P. H. Rowlands, 1978, Hydrocarbon habitat of Tertiary Niger Delta: AAPG Bulletin, v. 62, no. 1, p. 1–39.

Færseth, R. B., 2006, Shale smear along large faults: Continuity of smear and fault seal capacity: Journal of the Geological Society (London), v. 163, p. 741–751.

Fanchi, J. R., R. L. Christiansen, and M. J. Heymans, 2002, Estimating oil reserves with oil/water transition zones: SPE Reservoir Evaluation and Engineering, v. 5, no. 4, SPE Paper 79210, p. 311–316.

Fayers, F. J., S. Arbabi, and K. Aziz, 1995, Challenges in reservoir engineering from prospects for horizontal wells: Petroleum Geoscience, v. 1, p. 13–23.

Feldman, H. R., M. R. Gibling, A. W. Archer, W. G. Wightman, and W. P. Lanier, 1995, Stratigraphic architecture of the Tonganoxie Paleovalley fill (lower Virgilian) in northeastern Kansas: AAPG Bulletin, v. 79, no. 7, p. 1019–1043.

Felsenthal, M. and F. J. Gangle, 1975, A case study of thief zones in a California waterflood: Journal of Petroleum Technology, SPE Paper 5363, v. 27, no. 11, p. 1385–1391.

Fielding, C. R., 1984, A coal depositional model for the Durham Coal Measures of NE England: Journal of the Geological Society (London), v. 141, p. 919–931.

Fisher, Q. J. 2005, Recent advances in fault seal analysis as an aid to reservoir characterization and production simulation modelling: Presented at the Society of Petroleum Engineers Europec/EAGE (European Association of Geoscientists and Engineers) Annual Conference, June 13–16, Madrid, Spain, SPE Paper 94460, 8 p.

Fisher, Q. J., and S. J. Jolley, 2007, Treatment of faults in production simulation models, *in* S. J. Jolley, D. Barr, J. J. Walsh, and R. J. Knipe, eds., Structurally complex reservoirs: Geological Society (London) Special Publication 292, p. 219–233.

Fisher, Q. J., and R. J. Knipe, 1988, Fault sealing processes in siliciclastic sediments, *in* H. Jones, Q. J. Fisher, and R. J. Knipe, eds., Faulting, fault sealing and fluid flow in hydrocarbon reservoirs: Geological Society (London) Special Publication 147, p. 117–134.

Fisher, W. R., 1991, Future supply potential of U.S. oil and natural gas: The Leading Edge, v. 11, p. 15–21.

Fisk, H. N., 1961, Bar-finger sands of the Mississippi delta, *in* J. A. Peterson and J. C. Osmond, eds., Geometry of sandstone bodies: AAPG Symposium, SP22, p. 29–52.

Follows, E., 1997, Integration of inclined pilot hole core

with horizontal image logs to appraise an eolian reservoir, Auk field, central North Sea: Petroleum Geoscience, v. 3, p. 43–55.

Fontaine, J. M., R. Cussey, J. Lacaze, R. Lanaud, and L. Yapaudjian, 1987, Seismic interpretation of carbonate depositional environments: AAPG Bulletin, v. 71, no. 3, p. 281–297.

Forsman, N. F., T. R. Gerlach, and N. L. Anderson, 1996, Impact origin of the Newporte Structure, Williston Basin, North Dakota: AAPG Bulletin, v. 80, no. 5, p. 721–730.

Fossen, H., and A. Bale, 2007, Deformation bands and their influence on fluid flow: AAPG Bulletin, v. 91, no. 12, p. 1685–1700.

Fossen, H., and J. Hesthammer, 1998, Structural geology of the Gullfaks field, northern North Sea, *in* M. P. Coward, H. Johnson, and T. S. Daltaban, eds., Structural geology in reservoir characterization: Geological Society Special Publication 127, p. 231–261.

Fossen, H., T. S. Johansen, J. Hesthammer, and A. Rotevatn, 2005, Fault interaction in porous sandstone and implications for reservoir management, examples from southern Utah: AAPG Bulletin, v. 89, no. 12, p. 1593–1606.

Fossen, H., R. A. Schultz, Z. K. Shipton, and K. Mair, 2007, Deformation bands in sandstone: A review: Journal of the Geological Society (London), v. 164, p. 755–769.

Fowler, J. N., E. Guritno, P. Sherwood, M. J. Smith, S. Algar, I. Busono, G. Goffey, and A. Strong, 2004, Depositional architecture of Recent deep-water deposits in the Kutei Basin, East Kalimantan, *in* R. J. Davies, J. A. Cartwright, S. A. Stewart, S. A. Lappin, and J. R. Underhill, eds., 3-D seismic technology, applications to the exploration of sedimentary basins: Geological Society (London) Memoir 29, p. 25–33.

Foxford, K. A., J. J. Walsh, J. Watterson, I. R. Garden, S. C. Guscott, and S. D. Burley, 1998, Structure and content of the Moab fault zone, Utah, U.S.A., and its implications for fault seal prediction, *in* H. Jones, Q. J. Fisher, and R. J. Knipe, eds., Faulting, fault sealing and fluid flow in hydrocarbon reservoirs: Geological Society (London) Special Publication 147, p. 87–103.

Francis, A., 2006a, Understanding stochastic inversion: 1: First Break, v. 24, p. 69–77.

Francis, A., 2006b, Understanding stochastic inversion: 2: First Break, v. 24, p. 79–84.

Fretwell, P. N., W. Gordon Canning, J. Hegre, R. Labourdette, and M. Sweatman, 2007, A new approach to 3-D geological modeling of complex sand injectite reservoirs: The Alba field, United Kingdom central North Sea, *in* A. Hurst and J. Cartwright, eds., Sand injectites: Implications for hydrocarbon exploration and production: AAPG Memoir 87, p. 119–127.

Frey, R. W., and S. G. Pemberton, 1984, Trace fossil facies models, *in* R. G. Walker, ed., Facies models: Geological Association of Canada, p. 47–72.

Friend, P. F., M. J. Slater, and R. C. Williams, 1979, Vertical and lateral building of river sandstone bodies, Ebro Basin, Spain: Journal of the Geological Society of London, v. 136, p. 39–46.

Fryberger, S. G., L. F. Krystinik, and C. J. Schenk, 1990, Modern and ancient eolian deposits: Petroleum exploration and production: Rocky Mountain section, SEPM, Denver, Colorado, 233 p.

Fu, Z., Q. Du, P. Zhang, B. Ji, and W. Lu, 1998, Improving recovery efficiency of water-drive channel sandstone reservoir by drilling horizontal wells laterally: Presented at the Society of Petroleum Engineers International Conference on Horizontal Well Technology, November 1–4, Calgary, Alberta, Canada, SPE Paper 50392, 3 p.

Fulljames, J. R., L. J. J. Zijerveld, and R. C. M. W. Fransen, 1997, Fault seal processes: Systematic analysis of fault seals over geological and production time scales, *in* P. Moller-Petersen and A. G. Koestler, eds., Hydrocarbon seals, importance for exploration and production: Norwegian Petroleum Society Special Publication 7, p. 51–79.

Galloway, W. E., 1975, Process framework for describing the morphologic and stratigraphic evolution of deltaic depositional systems, *in* M. L. Broussard, ed., Deltas, models for exploration: Houston Geological Society, p. 87–98.

Galloway, W. E., 1986, Reservoir facies architecture of microtidal barrier systems: AAPG Bulletin, v. 70, no. 7, p. 787–808.

Galloway, W. E., 1989, Genetic stratigraphic sequences in basin analysis: I. Architecture and genesis of flooding-surface bounded depositional units: AAPG Bulletin, v. 73, no. 2, p. 125–142.

Galloway, W. E., and D. K. Hobday, 1996, Terrigenous clastic depositional systems: Applications to petroleum, coal, and uranium exploration: New York, Springer-Verlag, 489 p.

Garland, C. R., P. Haughton, R. F. King, and T. P. Moulds, 1999, Capturing reservoir heterogeneity in a sand-rich submarine fan, Miller field, *in* A. J. Fleet and S. A. R. Boldy, eds., Petroleum geology of northwest Europe: Proceedings of the 5th Conference, Geological Society (London), v. 2, p. 1199–1208.

Gaucher, D. H., and D. C. Lindley, 1960, Waterflood performance in a stratified, five-spot reservoir—A scaled-model study: Petroleum Transactions of the American Institute of Mining Metallurgical and Petroleum Engineers, v. 219, p. 208–215.

Gauthier, B. D. M., and S. D. Lake, 1993, Probabilistic modeling of faults below the limit of seismic resolution in Pelican field, North Sea, offshore United Kingdom: AAPG Bulletin, v. 77, no. 5, p. 761–777.

Gaynor, G. C., and R. M. Sneider, 1992, Effective pay determination, *in* D. Morton-Thompson and A. M. Woods, eds., Development geology reference manual, AAPG Methods in Exploration Series 10, p. 286–288.

Geehan, G. W., and A. J. Pearce, 1994, Geological reservoir heterogeneity databases and their application to integrated reservoir description, *in* J. O. Aasen, E. W. Berg, A. T. Buller, O. Hjelmeland, R. M. Holt, J. Kleppe, and O. Torsaeter, eds., North Sea oil and gas reservoirs III, 1994: Dordrecht, Kluwer Academic Publishers, p. 131–140.

Geehan, G. W., T. F. Lawton, S. Sakurai, H. Klob, T. R. Clifton, K. F. Inman, and K. E. Nitzberg, 1986, Geologic prediction of shale continuity, Prudhoe Bay field, *in*

L. W. Lake and H. B. Carroll, eds., Reservoir characterization: Orlando, Florida, Academic Press, p. 63–82.

Gerhardt, J. H., and H. H. Haldorsen, 1989, On the value of information: Presented at Offshore Europe, Society of Petroleum Engineers, September 5–8, Aberdeen, United Kingdom, SPE Paper 19291, 11 p.

Gibling, M. R. 2006, Width and thickness of fluvial channel bodies and valley fills in the geological record: A literature compilation and classification: Journal of Sedimentary Research, v. 76, p. 731–770.

Gibson, R. G., 1994, Fault-zone seals in siliciclastic strata of the Columbus basin, offshore Trinidad: AAPG Bulletin, v. 78, no. 9, p. 1372–1385.

Gidley, J. L., 1992, Stimulation, *in* D. Morton-Thompson and A. M. Woods, eds., Development geology reference manual, AAPG Methods in Exploration Series 10, p. 469–473.

Ginsburg, R. N., and N. P. James, 1974, Holocene carbonate sediments of continental shelves, *in* C. A Burk and C. L. Drake, eds., The geology of continental margins, New York, Springer-Verlag, p.137–155.

Gluyas, J. G., and S. W. Garrett, 2005, Better recovery through better reservoir characterization: Overview, *in* A. G. Doré and B. A. Vining, eds., Petroleum geology: Northwest Europe and global perspectives: Proceedings of the 6th Petroleum Geology Conference, Geological Society (London), v. 1, p. 361–365.

Gluyas, J. G., A. G. Robinson, D. Emery, S. M. Grant, and N. H. Oxtoby, 1993, The link between petroleum emplacement and sandstone cementation, *in* J. R. Parker, ed., Petroleum geology of northwest Europe: Proceedings of the 4th Conference, Geological Society (London), p. 1395–1402.

Gluyas, J. G., and R. Swarbrick, 2004, Petroleum geoscience: Malden, Blackwell Publishing, 376 p.

Gluyas, J. G., and J. R. Underhill, 2003, The Staffa field, Block 3/8b, UK North Sea, *in* J. G. Gluyas and H. M. Hichens, eds., United Kingdom oil and gas fields, commemorative millennium volume: Geological Society (London) Memoir 20, p. 327–333.

Goetz, J. F., 1992, Dipmeters, *in* D. Morton-Thompson and A. M. Woods, eds., Development geology reference manual: AAPG Methods in Exploration Series 10, p. 158–162.

Goulty, N. R., 2003, Reservoir stress path during depletion of Norwegian chalk oil fields: Petroleum Geoscience, v. 9, p. 233–241.

Grajales-Nishimura, J. M., E. Cedillo-Pardo, C. Rosales-Dominguez, D. J. Morán-Zenteno, W. Alvarez, P. Claeys, J. Ruíz-Moralles, J. Garcia-Hernández, P. Padilla-Avila, and A. Sánchez-Ríos, 2000, Chicxulub impact: The origin of reservoir and seal facies in the southeastern Mexico oil fields: Geology, v. 28, p. 307–310.

Grajales-Nishimura, J. M., E. Cedillo-Pardo, R. Martinez-Ibarra, J. Garcia-Hernández, and S. Hernandez-Garcia, 2002, Thick carbonate breccia deposition in the western margin of the Yucatan Platform, offshore Campeche (Cantarell field) during the K/T boundary Chicxulub event and its oil exploration importance: Conference Abstracts, Geological Society of America Annual Meeting, Denver 2002, v. 34, no. 6, p. 545.

Grammer, G. M., P. M. Harris, and G. P. Eberli, 2004, Integration of outcrop and modern analogs in reservoir modeling: Overview with examples from the Bahamas, *in* G. M. Grammer, P. M. Harris, and G. P. Eberli, eds., Integration of outcrop and modern analogs in reservoir modeling: AAPG Memoir 80, p. 1–22.

Grötsch, J., and C. Mercadier, 1999, Integrated 3-D reservoir modeling based on 3-D seismic: The Tertiary Malampaya and Camago build-ups, offshore Palawan, Philippines: AAPG Bulletin, v. 83, no. 11, p. 1703–1728.

Guangming, Z., and Z. Quanheng, 1982, Buried-hill oil and gas pools in the north China Basin, *in* M. T. Halbouty, ed., The deliberate search for the subtle trap: AAPG Memoir 32, p. 317–335.

Gupta, R., and H. D. Johnson, 2002, High-resolution facies architecture of heterolithic tidal deposits: An integrated outcrop and electrofacies analysis of a complex reservoir, *in* M. Lovell and N. Parkinson, eds., Geological applications of well logs: AAPG Methods in Exploration 13, p. 161–184.

Guscott, S., K. Russell, A. Thickpenny, and R. Poddubiuk, 2003, The Scott field, Blocks 15/21a, 15/22, UK North Sea, *in* J. G. Gluyas and H. M. Hichens, eds., United Kingdom oil and gas fields, commemorative millennium volume: Geological Society (London) Memoir 20, p. 467–481.

Haas, A., and O. Dubrule, 1994, Geostatistical inversion— A sequential method of stochastic reservoir modelling constrained by seismic data: First Break, v. 12, p. 561–569.

Halbouty, M. T., 1991, East Texas field— U.S.A. East Texas Basin, Texas, *in* E. A. Beaumont and N. H. Foster, eds., Stratigraphic traps: II. Treatise of petroleum geology of oil and gas fields: Tulsa, AAPG, p. 189–206.

Haldorsen, H. H., 1986, Simulator parameter assignment and the problem of scale in reservoir engineering, *in* L. W. Lake and H. B. Carroll, eds., Reservoir characterization: Orlando, Florida, Academic Press, p. 293–340.

Haldorsen, H. H., D. M. Chang, and S. H. Begg, 1987, Discontinuous vertical permeability barriers: A challenge to engineers and geologists, *in* J. Kleppe, E. W. Berg, A. T. Buller, O. Hjelmeland, and O. Torsaeter, eds., North Sea oil and gas reservoirs I: London, Graham & Trotman, p. 127–151.

Haldorsen, H. H., and E. Damsleth, 1990, Stochastic modeling: Journal of Petroleum Technology, v. 42, no. 4, SPE Paper 2031, p. 404–412.

Haldorsen, H. H., and E. Damsleth, 1993, Challenges in reservoir characterization: AAPG Bulletin, v. 77, no. 4, p. 541–551.

Haldorsen, H. H., and L. W. Lake, 1984, A new approach to shale management in field-scale models: Presented at the 1982 Society of Petroleum Engineers Annual Technical Conference and Exhibition, New Orleans, SPE Paper 10976, 16 p.

Haller, D., and F. Porturas, 1998, How to characterize fractures in reservoirs using borehole and core images: Case studies, *in* P. K. Harvey and M. A. Lovell, eds., Core-log integration: Geological Society (London) Special Publication 136, p. 249–259.

Halley, R. B., P. M. Harris and A. C. Hine, 1983, Bank

Margin, *in* P. A. Scholle, D. G. Bebout, and C. H. Moore, eds., Carbonate depositional environments: AAPG Memoir 33, p. 463–506.

Hamilton, D. E., 1992, Using and improving surface models built by computer, *in* D. Morton-Thompson and A. M. Woods, eds., Development geology reference manual: AAPG Methods in Exploration Series 10, p. 431–440.

Hamilton, D. S., and N. Z. Tadros, 1994, Utility of coal seams as genetic stratigraphic sequence boundaries in nonmarine basins: An example from the Gunnedah Basin, Australia: AAPG Bulletin, v. 78, no. 2, p. 267–286.

Hamilton, D. S., M. H. Holtz, P. Ryles, T. Lonergan, and M. Hillyer, 1998, Approaches to identifying reservoir heterogeneity and reserve growth opportunities in a continental-scale bed-load fluvial system: Hutton sandstone, Jackson field, Australia: AAPG Bulletin, v. 82, no. 12, p. 2192–2219.

Hamilton, D. S., et al., 2002, Reactivation of mature oil fields through advanced reservoir characterization: A case history of the Budare field, Venezuela: AAPG Bulletin, v. 86, no. 7, p. 1237–1262.

Hamilton, D. S., R. Barba, M. H. Holtz, J. Yeh, M. Rodriguez, M. Sánchez, P. Calderon, and J. Castillo, 2003, Horizontal-well drilling in the heavy-oil belt, eastern Venezuela Basin: A postmortem of drilling experiences, *in* T. R. Carr, P. Mason, and C. T. Feazel, eds., Horizontal wells: Focus on the reservoir: AAPG Methods in Exploration 14, p. 127–141.

Hancock, N. J., 1992, Quick-look lithology from logs, *in* D. Morton-Thompson and A. M. Woods, eds., Development geology reference manual: AAPG Methods in Exploration Series 10, p. 174–179.

Hardage, B. A., 1992, Checkshots and vertical seismic profiles, *in* D. Morton-Thompson and A. M. Woods, eds., Development geology reference manual: AAPG Methods in Exploration Series 10, p. 401–403.

Harker, S. D., 1998, The palingenesy of the Piper oil field, UK North Sea: Petroleum Geoscience, v. 4, p. 271–286.

Harker, S. D., S. C. H. Green, and R. S. Romani, 1991, The Claymore field, Block 14/19, UK North Sea, *in* I. L. Abbots, ed., United Kingdom oil and gas fields, 25 years commemorative volume: Geological Society (London) Memoir 14, p. 269–278.

Harker, S. D., G. Richardson, L. E. Sides, and R. Cooper, 2003, Alwyn north Triassic main gas-condensate: Drilling deeper promotes production: Petroleum Geoscience, v. 9, no. 2, p. 133–143.

Harris, P. M., and W. S. Kowalik, eds., 1994, Satellite images of carbonate depositional environments: Examples of reservoir and exploration scale geologic facies variation: AAPG Methods in Exploration Series 11, 147 p.

Hart, B. S., D. M. Sibley, and P. B. Flemings, 1997, Seismic stratigraphy, facies architecture and reservoir character of a Pleistocene shelf-margin delta complex, Eugene Island Block 330 field, offshore Louisiana: AAPG Bulletin, v. 81, no. 3, p. 380–397.

Hartman, J. A., and D. D. Paynter, 1979, Drainage anomalies in Gulf Coast Tertiary sandstones: Journal of Petroleum Technology, SPE Paper 7532, v. 31, no. 10, p. 1313–1322.

Hartmann, D. J., and E. A. Beaumont, 1999, Predicting reservoir system quality and performance, *in* E. A. Beaumont and N. H. Foster, eds., Treatise of petroleum geology—Handbook of petroleum geology: Exploring for oil and gas traps: Tulsa, AAPG, p. 9-1–9-154.

Harvey, P. J., 1993, Porosity identification using amplitude variations with offset in Jurassic carbonate, offshore Nova Scotia: The Leading Edge, v. 12, p. 180–184.

Haskell, N., et al., 1999, Delineation of geologic drilling hazards using 3-D seismic attributes: The Leading Edge, v. 18, p. 373–383.

Hawkins, J. T., 1992, Relative permeability, *in* D. Morton-Thompson and A. M. Woods, eds., Development geology reference manual: AAPG Methods in Exploration Series 10, p. 226–228.

Heald, M. T., and R. E. Larese, 1974, Influence of coatings on quartz cementation: Journal of Sedimentary Research, v. 44, p. 1269–1274.

Hearn, C. L., W. J. Ebanks Jr., R. S. Tye, and V. Ranganathan, 1984, Geological factors influencing reservoir performance of the Hartzog Draw field: Journal of Petroleum Technology, v. 36, p. 1335–1344.

Heffer, K. J., and A. B. Dowokpor, 1990, Relationship between azimuths of flood anisotropy and local earth stresses in oil reservoirs, *in* J. Kleppe, E. W. Berg, A. T. Buller, O. Hjelmeland, and O. Torsaeter, eds., North sea oil and gas reservoirs: II, 1989: London, Graham & Trotman, p. 251–260.

Heffer, K. J., P. R. King, and A. D. W. Jones, 1999, Fracture modelling as part of integrated reservoir characterization: Presented at the Middle East Oil Show and Conference, Society of Petroleum Engineers, February 20–23, Bahrain, SPE Paper 53347, 11 p.

Heidelberg Academy of Sciences and Humanity, 2004, The world stress map project—A service for earth system management, 102 p: http://www.world-stress-map.org/ (accessed October 15, 2008).

Hempton, M., J. Marshall, S. Sadler, N. Hogg, R. Charles, and C. Harvey, 2005, Turbidite reservoirs of the Sele Formation, central North Sea: Geological challenges for improving production, *in* A. G. Doré and B. A. Vining, eds., Petroleum geology: Northwest Europe and global perspectives: Proceedings of the 6th Petroleum Geology Conference, Geological Society (London), v. 1, p. 449–459.

Hennings, P. H., J. E. Olson, and L. B. Thompson, 2000, Combining outcrop data and three-dimensional structural models to characterize fractured reservoirs: An example from Wyoming: AAPG Bulletin, v. 84, no. 6, p. 830–849.

Henriquez, A., and C. Jourdan, 1995, Challenges in modelling reservoirs in the North Sea and on the Norwegian Shelf: Petroleum Geoscience, v. 1, no. 4, p. 327–336.

Herries, R. D., 1993, Contrasting styles of fluvial-eolian interaction at a downwind erg margin: Jurassic Kayenta-Navajo transition, northeastern Arizona, U.S.A., *in* C. P. North and D. J. Prosser, eds., Characterization of fluvial and eolian reservoirs: Geological Society Special Publication 73, p. 199–218.

Hesthammer, J., and H. Fossen, 1997, Seismic attribute analysis in structural interpretation of the Gullfaks field, northern North Sea: Petroleum Geoscience, v. 3, no. 1, p. 13–26.

Hesthammer, J., and H. Fossen, 2000, Uncertainties associated with fault sealing analysis: Petroleum Geoscience, v. 6, p. 37–45.

Hesthammer, J., P. A. Bjørkum, and L. Watts, 2002, The effect of temperature on sealing capacity of faults in sandstone reservoirs: Examples from the Gullfaks and Gullfaks Sør fields, North Sea: AAPG Bulletin, v. 86, no. 10, p. 1733–1751.

Heum, O. R., 1996, A fluid dynamic classification of hydrocarbon entrapment: Petroleum Geoscience, v. 2, no. 2, p. 145–158.

Hillier, A. P., 2003, The Leman field, Blocks 49/26, 49/27, 49/28, 53/1, 53/2, UK North Sea, *in* J. G. Gluyas and H. M. Hichens, eds., United Kingdom oil and gas fields, commemorative millennium volume: Geological Society (London) Memoir 20, p. 761–770.

Hillis, R. R., 1998, The influence of fracture stiffness and the in situ stress field on the closure of natural fractures: Petroleum Geoscience, v. 4, p. 57–65.

Hillis, R. R., 2001, Coupled changes in pore pressure and stress in oilfields and sedimentary basins: Petroleum Geoscience, v. 7, no. 4, p. 419–425.

Hillis, R. R., and E. J. Nelson, 2005, In situ stresses in the North Sea and their applications: Petroleum geomechanics from exploration to development, *in* A. G. Doré and B. A. Vining, eds., Petroleum geology: Northwest Europe and global perspectives: Proceedings of the 6th Petroleum Geology Conference, Geological Society (London), v. 1, p. 551–564.

Hirsche, K., J. Porter-Hirsche, L. Mewhort, and R. Davis, 1997, The use and abuse of geostatistics: The Leading Edge, v. 16, p. 253–260.

Hirsche, K., S. Boerner, C. Kalkomey, and C. Gastaldi, 1998, Avoiding pitfalls in geostatistical reservoir characterization: A survival guide: The Leading Edge, v. 17, p. 493–504.

Hodgetts, D., J. Imber, C. Childs, S. Flint, J. Howell, J. Kavanagh, P. Nell, and J. Walsh, 2001, Sequence-stratigraphic responses to shoreline-perpendicular growth faulting in shallow marine reservoirs of the Champion field, offshore Brunei Darussalam, South China Sea: AAPG Bulletin, v. 85, no. 3, p. 433–457.

Holditch, S. A., 1992, Well completions, *in* D. Morton-Thompson and A. M. Woods, eds., Development geology reference manual: AAPG Methods in Exploration Series 10, p. 463–468.

Hollund, K., P. Mostad, B. F. Nielsen, L. Holden, J. Gjerde, M. G. Contursi, A. J. McCann, C. Townsend, and E. Sverdrup, 2002, Havana—A fault modelling tool, *in* A. G. Koestler and R. Hunsdale, eds., Hydrocarbon seal quantification: Norwegian Petroleum Society Special Publication 11, p. 157–171.

Holtz, M., S. C. Ruppel, and C. R. Hocott, 1992, Integrated geologic and engineering determination of oil-reserve growth potential in carbonate reservoirs: Journal of Petroleum Technology, v. 44, no. 11, SPE Paper 22900, p. 1250–1257.

Holtz, M. H., and D. S. Hamilton, 1998, Reservoir characterization methodology to identify reserve growth potential: Presented at the International Petroleum Conference and Exhibition of Mexico, Society of Petroleum Engineers, March 3–5, Villahermosa, Mexico, SPE Paper 39867, 13 p.

Holtz, M. H., J. Garcia, S. Sakurai, T. F. Wawrzyniec, D. C. Jennette, and J. J. Bolanas, 2002, Production and pressure characteristics of gas reservoirs, Veracruz basin, Mexico: Gulf Coast Association of Geological Societies Transactions, v. 52, p. 407–419.

Hoover, A. R., T. Burkhart, and P. B. Flemings, 1999, Reservoir and production analysis of the K40 sand, South Timbalier 295, offshore Louisiana, with comparison to time-lapse (4-D) seismic results: AAPG Bulletin, v. 83, no. 10, p. 1624–1641.

Hornby, B. E., 1995, Use of full-waveform sonic data to image near-borehole structural features: Petroleum Geoscience, v. 1, p. 109–114.

Hsu, K., A. Brie, and R. A. Plumb, 1987, A new method for fracture identification using array sonic tools: Journal of Petroleum Technology, SPE Paper 14397, v. 39, no. 6, p. 677–683.

Huang, V., P. S. Ringrose, and K. S. Sorbie, 1995, Capillary trapping mechanisms in water-wet laminated rocks: SPE Reservoir Engineering, v. 10, SPE Paper 28942, p. 287–292.

Hubbert, M. K., 1953, Entrapment of petroleum under hydrodynamic conditions: AAPG Bulletin, v. 37, p. 1954–2026.

Hunter, R. E., 1977, Basic types of stratification in small eolian dunes: Sedimentology, v. 24, no. 3, p. 361–387.

Hurst, A., 1993, Sedimentary flow units in hydrocarbon reservoirs: Some shortcomings and a case for high-resolution permeability data, *in* S. S. Flint and I. D. Bryant, eds., The geological modeling of hydrocarbon reservoirs and outcrop analogs: International Association of Sedimentologists Special Publication 15, p. 191–204.

Hurst, A., and J. Cartwright, 2007, Relevance of sand injectites to hydrocarbon exploration and production, *in* A. Hurst and J. Cartwright, eds., Sand injectites: Implications for hydrocarbon exploration and production: AAPG Memoir 87, p. 1–19.

Hurst, A., and D. Goggin, 1995, Probe permeametry: An overview and bibliography: AAPG Bulletin, v. 79, no. 3, p. 463–473.

Hurst, A., and A. Milodowski, 1996, Thorium distribution in some North Sea sandstones: Implications for petrophysical evaluation: Petroleum Geoscience, v. 2, no. 1, p. 69–68.

Hurst, A., and P. Nadeau, 1995, Clay microporosity in reservoir sandstones: An application of quantitative electron microscopy in petrophysical evaluation, AAPG Bulletin, v. 79, no. 4, p. 563–573.

Ikoku, C. U., 1984, Natural gas reservoir engineering: New York, John Wiley & Sons, 526 p.

Illing, V. C., 1939, Some factors in oil accumulation: Journal of the Institute of Petroleum, v. 25, p. 201–225.

Inglis, T. A., 1987, Directional drilling: London, Graham & Trotman, 260 p.

Isaaks, E. H., and R. M. Srivastava, 1989, An introduction to applied geostatistics: Oxford, Oxford University Press, 561 p.

Iverson, W. P., 1992, Fracture identification from well logs: Presented at the Society of Petroleum Engineers Rocky Mountain Regional Meeting, May 18–21, Casper, Wyoming, SPE Paper 24351, 9 p.

Jackson, M. L. W., and Ambrose, W. A., 1989, Influence of reservoir heterogeneity on gas-resource potential for geologically based infill drilling, Brooks and I-92 reservoirs, Frio Formation, South Texas: Transactions of the Gulf Coast Association of Geological Societies, v. 39, p. 127–139.

Jackson, R. G., 1975, Hierarchical attributes and a unifying model of bed forms composed of cohesionless material and produced by shearing flow: Geological Society of America Bulletin, v. 86, p. 1523–1533.

Jahn, F., M. Cook, and M. Graham, 1998, Hydrocarbon exploration and production: Developments in petroleum science 46: Amsterdam, Elsevier, 384 p.

James, D. M. D., C. Childs, J. Watterson, and J. J. Walsh, 1997, Discussion on a model for the structure and development of fault zones: Reply: Journal of the Geological Society (London), v. 154, no. 2, p. 366–368.

James, W. R., L. H. Fairchild, G. P. Nakayama, S. J. Hippler, and P. J. Vrolijk, 2004, Fault-seal analysis using a stochastic multi-fault approach: AAPG Bulletin, v. 88, no. 7, p. 885–904.

Jamison, W. R., 1997, Quantitative evaluation of fractures on Monkshood Anticline, a detachment fold in the foothills of Western Canada: AAPG Bulletin, v. 81, no. 7, p. 1110–1132.

Jamison, W. R., and D. W. Stearns, 1982, Tectonic deformation of Wingate Sandstone, Colorado National Monument: AAPG Bulletin, v. 66, no. 12, p. 2584–2608.

Jardine, D., and J. W. Wilshart, 1982, Carbonate reservoir description: Presented at the International Petroleum Exhibition and Technical Symposium of the Society of Petroleum Engineers, March 18–26, 1982, Beijing, China, SPE Paper 10010, 35 p.

Jennings, J., 1992, Surface production equipment, *in* D. Morton-Thompson and A. M. Woods, eds., Development geology reference manual: AAPG Methods in Exploration Series 10, p. 485–487.

Jennings, J. B., 1987, Capillary pressure techniques: Application to exploration and development geology: AAPG Bulletin, v. 71, no. 10, p. 1196–1209.

Jensen, J. L., D. V. Hinkley, and L. W. Lake, 1987, A statistical study of reservoir permeability: Distributions, correlations, and averages: Society of Petroleum Engineers Formation Evaluation, v. 2, SPE Paper 14270, p. 461–468.

Jensen, J. L., L. W. Lake, P. W. M. Corbett, and D. J. Goggin, 1997, Statistics for petroleum engineers and geoscientists: New Jersey, Prentice Hall, 416 p.

Jordan, D. W., and W. A. Pryor, 1992, Hierarchical levels of heterogeneity in a Mississippi river meander belt and application to reservoir systems: AAPG Bulletin, v. 76, no. 10, p. 1601–1624.

Journel, A. G., and J. J. Gomez-Hernandez, 1989, Stochastic imaging of the Wilmington Clastic Sequence: SPE Formation Evaluation, SPE Paper 19857, v. 8, no. 1, p. 33–40.

Kamal, M. M., 1983, Interference and pulse testing—A review: Journal of Petroleum Technology, SPE Paper 10042, v. 35, no. 12, p. 2257–2270.

Kamath, J., R. F. Meyer, and F. M. Nakagawa, 2001, Understanding waterflood residual oil saturation of four carbonate rock types: Presented at the 2001 Society of Petroleum Engineers Annual Technical Conference and Exhibition, September 30–October 3, 2001, New Orleans, Louisiana, SPE Paper 71505, 10 p.

Karasek, R. M., R. L. Vaughan, and T. T. Masuda, 2003, The Beryl field, Block 9/13, UK North Sea, *in* J. G. Gluyas and H. M. Hichens, eds., United Kingdom oil and gas fields, commemorative millennium volume: Geological Society (London) Memoir 20, p. 153–166.

Kennedy, W. J., 1987, Sedimentology of Late Cretaceous–Paleocene Chalk reservoirs, North Sea central graben, *in* J. Brooks and K. Glennie, eds., Petroleum geology of northwest Europe 1987: London, Graham & Trotman, p. 469–481.

Kerans, C., F. J. Lucia, and R. K. Senger, 1994, Integrated characterization of carbonate ramp reservoirs using Permian San Andres Formation outcrop analogs: AAPG Bulletin, v. 78, no. 2, p. 181–216.

Kheshgi, H., F. Cappelen, A. Lee, S. Crookshank, A. Heilbrunn, T. Mikus, W. Robson, B. Senior, T. Stileman, and L. Warren, 2006, Carbon dioxide capture and geological storage: Contributing to climate change solutions: Presented at the Society of Petroleum Engineers Conference on Health, Safety and Environment in Oil and Gas Exploration and Production, April 2–4, 2006, Abu Dhabi, SPE Paper 98583, 5 p.

Kiessling, W., E. Flügel, and J. Golonka, 1999, Paleoreef maps: Evaluation of a comprehensive database on Phanerozoic reefs: AAPG Bulletin, v. 83, no. 10, p. 1552–1587.

King, P. R., 1990, The connectivity and conductivity of overlapping sand bodies, *in* J. Kleppe, E. W. Berg, A. T. Buller, O. Hjelmeland, and O. Torsaeter, eds., North Sea oil and gas reservoirs II, 1989: London, Graham & Trotman, p. 353–362.

Kneller, B., M. Dykstra, and P. Thompson, 2007, Collapse of submarine channel levees, examples from outcrop and subsurface, and reservoir implications: Abstracts of the 2007 AAPG Annual Convention and Exhibition, p. 77.

Knipe, R. J., 1997, Juxtaposition and seal diagrams to help analyze fault seals in hydrocarbon reservoirs: AAPG Bulletin, v. 81, no. 2, p. 187–195.

Knott, S. D., 1994, Fault zone thickness versus displacement in the Permo-Triassic sandstones of NW England: Journal of the Geological Society, v. 151, p. 17–25.

Knott, S. D., A. Beach, P. J. Brockbank, J. L. Brown, J. E. McCallum, and A. I. Welbon, 1996, Spatial and mechanical controls on normal fault populations: Journal of Structural Geology, v. 18, no. 2/3, p. 359–372.

Kocurek, G., 1981, Erg reconstruction: The Entrada Sandstone (Jurassic) of northern Utah and Colorado: Palaeogeography, Palaeoclimatology, Palaeoecology, v. 36, no. 1–2, p. 125–153.

Kocurek, G., 1996, Desert eolian systems, *in* H. G. Reading, ed., Sedimentary environments: Processes, facies and stratigraphy, 3d ed.: Oxford, Blackwell Science, p. 125–153.

Kocurek, G., and J. Nielson, 1986, Conditions favourable for the formation of warm-climate eolian sand sheets: Sedimentology, v. 33, no. 6, p. 795–816.

Kopaska-Merkel, D. C., S. D. Mann, and J. W. Schmoker, 1994, Controls on reservoir development in a shelf carbonate: Upper Jurassic Smackover Formation of Alabama: AAPG Bulletin, v. 78, no. 6, p. 938–959.

Kortekaas, T. F. M., 1985, Water/oil displacement characteristics in cross-bedded reservoir zones: Society of Petroleum Engineers Journal, SPE Paper 12112, v. 25, no. 6, p. 917–926.

Kosters, M., and M. E. Donselaar, 2003, Syn-sedimentary faulting and the formation of localized reservoir sands: Carboniferous examples from the Campine Basin, Belgium and the Dutch offshore: Petroleum Geoscience, v. 9, no. 4, p. 309–319.

Krause, F. F., H. N. Collins, D. A. Nelson, S. D. Machemer, and P. R. French, 1987, Multiscale anatomy of a reservoir: Geological characterization of Pembina-Cardium pool, west-central Alberta, Canada: AAPG Bulletin, v. 71, no. 10, p. 1233–1260.

Kromah, M. J., J. Liou, and D. G. MacDonald, 2005, Step change in reservoir simulation breathes life into a mature oil field: Prepared for the 2005 Society of Petroleum Engineers Latin American and Caribbean Petroleum Engineering Conference, Rio de Janeiro, SPE Paper 94940, 8 p.

Krueger, R. F., 1986, An overview of formation damage and well productivity in oilfield operations: Journal of Petroleum Technology, SPE Paper 10029, v. 38, no. 2, p. 131–152.

Krumbein, W. C., and G. D. Monk, 1942, Permeability as a function of the size parameters of unconsolidated sand: Petroleum Transactions of the American Institute of Mining Metallurgical and Petroleum Engineers, v. 151, p. 153–163.

Krystinik, L. F., 1990, Development geology in eolian reservoirs, *in* S. G. Fryberger, L. F. Krystinik, and C. J. Schenk, eds., Modern and ancient eolian deposits: Petroleum Exploration and Production, Rocky Mountain section, SEPM, Denver, Colorado, p. 13-1–13-12.

Kuhn, O., S. W. Smith, K. Van Noort, and B. Loiseau, 2003, The Fulmar field, Blocks 30/16, 30/11b, UK North Sea, *in* J. G. Gluyas and H. M. Hichens, eds., United Kingdom oil and gas fields, commemorative millennium volume: Geological Society (London) Memoir 20, p. 563–585.

Kulander, B. R., S. L. Dean, and B. J. Ward Jr., 1990, Fractured core analysis: Interpretation, logging and use of natural induced fractures in core: AAPG Methods in Exploration Series 8, 88 p.

Kunka, J. M., G. Williams, B. Cullen, J. Boyd-Gorst, G. R. Dyer, J. A. Garnham, A. Warnock, A. Davis, and P. Lynes, 2003, The Nelson field, Blocks 22/11, 22/6a, 22/7, 22/12a, UK North Sea, *in* J. G. Gluyas and M. H. Hichens, eds., United Kingdom oil and gas fields, commemorative millennium volume: Geological Society (London) Memoir 20, p. 617–646.

Lake, L. W., and H. B. Carroll, eds., 1986, Reservoir characterization: Orlando, Florida, Academic Press, 659 p.

Lambert, B., B. C. Duval, Y. Grosjean, I. M. Umar, and P. Zaugg, 2003, The Peciko case history: Impact of an evolving geologic model on the dramatic increase in reserves in the Mahakam delta, *in* M. T. Halbouty, ed., Giant oil and gas fields of the decade 1990–1999: AAPG Memoir 78, p. 297–320.

Landes, K. K., J. J. Amoruso, L. J. Charlesworth Jr., F. Heany, and P. J. Lesperance, 1960, Petroleum reservoirs in basement rocks: AAPG Bulletin, v. 44, no. 10, p. 1682–1691.

Larue, D. K., 2004, Outcrop and waterflood simulation modeling of the 100-foot channel complex, Texas, and the Ainsa II channel complex, Spain: Analogs to multistory and multilateral channelized slope reservoirs, *in* G. M. Grammer, P. M. Harris, and G. P. Eberli, eds., Integration of outcrop and modern analogs in reservoir modeling, AAPG Memoir 80, p. 337–364.

Larue, D. K., and F. Friedmann, 2005, The controversy concerning stratigraphic architecture of channelized reservoirs and recovery by waterflooding: Petroleum Geoscience, v. 11, p. 131–146.

Larue, D. K., and J. Hovadik, 2006, Connectivity of channelized reservoirs: A modelling approach: Petroleum Geoscience, v. 12, no. 4, p. 291–308.

Larue, D. K., and H. Legarre, 2004, Flow units, connectivity, and reservoir characterization in a wave-dominated reservoir: Meren reservoir, Nigeria: AAPG Bulletin, v. 88, no. 3, p. 303–324.

Laudeman, S. K., 1992, Conducting a reservoir simulation study: An overview, *in* D. Morton-Thompson and A. M. Woods, eds., Development Geology Reference Manual: AAPG Methods in Exploration Series 10, p. 536–540.

Lawton, D. E., and P. P. Roberson, 2003, The Johnston gas field, Blocks 43/26a, 43/27a, UK Southern North Sea, *in* J. G. Gluyas and H. M. Hichens, eds., United Kingdom oil and gas fields, commemorative millennium volume: Geological Society (London) Memoir 20, p. 749–759.

Leach, H. M., N. Herbert, A. Los, and R. L. Smith, 1999, The Schiehallion development *in* A. J. Fleet and S. A. R. Boldy, eds., Petroleum geology of northwest Europe: Proceedings of the 5th Conference, Geological Society (London), p. 683–692.

Le Blanc Sr., R. J., 1977, Distribution and continuity of sandstone reservoirs: 1: Journal of Petroleum Technology, SPE Paper 6137, v. 29, no. 7, p. 776–792.

Lee, W. J., 1992, Pressure transient testing, *in* D. Morton-Thompson and A. M. Woods, eds., Development geology reference manual: AAPG Methods in Exploration Series 10, p. 477–481.

Lehner, F. K., and F. K. Pilaar, 1997, The emplacement of clay smears in synsedimentary normal faults: Inferences and field observations near Frechen Germany, *in* P. Moller-Pederson and A. G. Koestler, eds., Hydrocarbon seals: Importance for exploration and production: Norwegian Petroleum Society Special Publication 7, p. 15–38.

Lennon, R. B., 1990, Midway-Sunset field—U.S.A. San Joaquin Basin, California *in* E. A. Beaumont and N. H. Foster, eds., Structural traps: III. Tectonic fold and fault traps: Treatise of Petroleum Geology of Oil and Gas Fields: Tulsa, AAPG, p. 221–241.

Leveille, G. P., R. Knipe, C. More, D. Ellis, G. Dudley, G. Jones, Q. J. Fisher, and G. Allinson, 1997, Compartmentalization of Rotliegends gas reservoirs by sealing faults, Jupiter fields area, southern North Sea, *in* K. Ziegler, P. Turner, and S. R. Daines, eds., Petroleum geology of the southern North Sea: Future potential: Geological Society Special Publication 123, p. 87–104.

Levell, B. K., J. H. Braakman, and K. W. Rutten, 1988, Oil-bearing sediments of Gondwana glaciation in Oman: AAPG Bulletin, v. 72, no. 7, p. 775–796.

Levorsen, A. I., 1967, Geology of petroleum: San Francisco, Freeman and Company, 724 p.

Lindquist, S. J., 1983, Nugget Formation reservoir characteristics affecting production in the overthrust belt of southwestern Wyoming: Journal of Petroleum Technology, v. 35, no. 7, p. 1355–1365.

Lindsay, N. G., F. C. Murphy, J. J. Walsh, and J. Watterson, 1993, Outcrop studies of shale smear on fault surfaces, *in* S. S. Flint and I. D. Bryant, eds., The geological modelling of hydrocarbon reservoirs and outcrop analogs: International Association of Sedimentologists Special Publication 15, p. 113–123.

Lisle, R. J., 1994, Detection of zones of abnormal strains in structures using Gaussian curvature analysis: AAPG Bulletin, v. 78, no. 12, p. 1811–1819.

Liu, Y., A. Harding, W. Abriel, and S. Strebelle, 2004, Multiple-point simulation integrating wells, three dimensional seismic data, and geology: AAPG Bulletin, v. 88, no. 7, p. 905–921.

Lomando, A. J., 1992, The influence of solid reservoir bitumen on reservoir quality: AAPG Bulletin, v. 76, no. 8, p. 1137–1152.

Lønøy, A., 2006, Making sense of carbonate pore systems: AAPG Bulletin, v. 90, no. 9, p. 1381–1405.

Lopez, J. A., G. Partyka, N. L. Haskell, and S. E. Nissen, 1997, Identification of deltaic facies with 3-D seismic coherency and the spectral decomposition cube: A study from South Marsh Island Area, Gulf of Mexico: Gulf Coast Association of Geological Societies Transactions, v. 47, p. 305–309.

Lorenz, J. C., 1999, Stress-sensitive reservoirs: Journal of Petroleum Technology, SPE Paper 50977, v. 51, no. 1, p. 61–63.

Loth, W. D., 1998, HPHT wells: Perspective on drilling and completion from the field: UK Government Health and Safety Executive, offshore technology report: Sudbury, HSE Books, 47 p.

Loucks, R. G., 1999, Paleocave carbonate reservoirs: Origins, burial-depth modifications, spatial complexity and reservoir implications: AAPG Bulletin, v. 83, no. 11, p. 1795–1834.

Loucks, R. G., and J. H. Anderson, 1980, Depositional facies and porosity development in Lower Ordovician Ellenburger dolomite, Puckett field, Pecos County Texas, *in* R. B. Halley and R. G. Loucks, eds., Carbonate reservoir rocks: SEPM Core Workshop 1, p. 1–31.

Lowry, P., C. D. Jenkins, and D. J. Phelps, 1993, Reservoir scale sand body architecture of Pliocene turbidite sequences, Long Beach Unit, Wilmington oil field, California: Presented at the Society of Petroleum Engineers Annual Technical Conference and Exhibition, October 3–6, Houston, Texas, SPE Paper 26440, 10 p.

Lucia, F. J., 1995, Rock-fabric/petrophysical classification of carbonate pore space for reservoir characterization: AAPG Bulletin, v. 79, no. 9, p. 1275–1300.

Lucia, F. J., 1999, Carbonate reservoir characterization: Berlin, Springer-Verlag, 226 p.

Luthi, S. M., 1992, Borehole imaging devices *in* D. Morton-Thompson and A. M. Woods, eds., Development geology reference manual: AAPG Methods in Exploration Series 10, p. 163–166.

Luthi, S. M., 2001, Geological well logs: Their use in reservoir modelling: Berlin, Springer-Verlag, 373 p.

Luthi, S. M., and P. Souhaité, 1990, Fracture apertures from electrical borehole scans: Geophysics, v. 55, no. 7, p. 821–833.

Lyon, T. A., J. Fuller, B. Granier, M. Hozayen, A. Al-Riyami, A. Khalaf, and B. Thiebot, 1998, Integrated study of a faulted and fractured reservoir: Presented at the 8th Abu Dhabi International Petroleum Exhibition and Conference, Society of Petroleum Engineers, November 1998, SPE Paper 49453, 23 p.

Ma, S., J. Zhang, N. Jin, and Z. Wang, 1999, The 3-D architecture of point bar and the forming and distribution of remaining oil: Presented at the Society of Petroleum Engineers Asia Pacific Improved Oil Recovery Conference, October 25–26, Kuala Lumpur, Malaysia, SPE Paper 57308, 7 p.

MacGregor, A. G., and P. Trussell, 2003, The Magnus field—Extending field life through enhanced oil recovery: Abstracts of the AFES/PESGB (Aberdeen Formation Evaluation Society/Petroleum Exploration Society of Great Britain) Extending Field Life Conference, Aberdeen 2003.

Machel, H. G., 2004, Concepts and models of dolomitization: A critical reappraisal, *in* C. J. R. Braithwaite, G. Rizzi, and G. Darke, eds., The geometry and petrogenesis of dolomite reservoirs: Geological Society (London) Special Publication 235, p. 7–63.

Maerten, L., P. Gillespie, and J.-M. Daniel, 2006, Three-dimensional geomechanical modeling for constraint of subseismic fault simulation: AAPG Bulletin, v. 90, no. 9, p. 1337–1358.

Major, R. P., and M. H. Holtz, 1997, Identifying fracture orientation in a mature carbonate platform reservoir: AAPG Bulletin, v. 81, no. 7, p. 1063–1069.

Major, R. P., D. G. Bebout, and P. M. Harris, 1996, Facies heterogeneity in a modern ooid sand shoal—An analog for hydrocarbon reservoirs: Bureau of Economic Geology, Geological Circular 96-1, University of Texas at Austin, 30 p.

Mäkel, G. H., 2007, The modelling of fractured reservoirs: Constraints and potential for fracture network geometry and hydraulics analysis, *in* S. J. Jolley, D. Barr, J. J. Walsh, and R. J. Knipe, eds., Structurally complex reservoirs: Geological Society (London) Special Publication 292, p. 375–403.

Maliva, R. G., and J. A. D. Dickson, 1992, Microfacies and diagenetic controls of porosity in Cretaceous/Tertiary chalks, Eldfisk field, Norwegian North Sea: AAPG Bulletin, v. 76, no. 11, p. 1825–1838.

Mansoori, J., 1994, A review of basic upscaling procedures:

advantages and disadvantages, *in* J. M. Yarus and R. L. Chambers, eds., Stochastic modeling and geostatistics: AAPG Computer Applications in Geology 3, p. 65–74.

Manzocchi, T., J. J. Walsh, P. Nell, and G. Yielding, 1999, Fault transmissibility multipliers for flow simulation models: Petroleum Geoscience, v. 5, no. 1, p. 53–63.

Manzocchi, T., A. E. Heath, J. J. Walsh, and C. Childs, 2002, The representation of two phase fault-rock properties in flow simulation models: Petroleum Geoscience, v. 8, no. 2, p. 119–132.

Marfurt, K. J., R. L. Kirlin, S. L. Farmer, and M. S. Bahorich, 1998, 3-D seismic attributes using a semblance-based coherency algorithm: Geophysics, v. 63, no. 4, p. 1150–1165.

Markello, J. R., R. Koepnick, and L. E. Waite, 2006, The carbonate analogs through time (CATT) hypothesis—A systematic and predictive look at Phanerozoic carbonate reservoirs: AAPG Search and Discovery Article 40185, 5 p.

Márquez, L. J., M. González, S. Gamble, E. Gomez, M. A. Vivas, H. M. Bressler, L. S. Jones, S. M. Ali, and G. S. Forrest, 2001, Improved reservoir characterization of a mature field through an integrated multi-disciplinary approach, LL-04 reservoir, Tia Juana field, Venezuela: Presented at the Society of Petroleum Engineers Annual Technical Conference and Exhibition, September 30–October 3, New Orleans, Louisiana, SPE Paper 71355, 10 p.

Martin, J. H., 1993, A review of braided fluvial hydrocarbon reservoirs: The petroleum engineer's perspective, *in* J. L. Best and C. S. Bristow, eds., Braided rivers: Geological Society Special Publication 75, p. 333–367.

Masaferro, J. L., R. Bourne, and J.-C. Jauffred, 2003, 3D visualization of carbonate reservoirs: The Leading Edge, v. 19, no. 1, p. 18–25.

Masalmeh, S. K., I. Abu Shiekah, and X. D. Jing, 2005, Improved characterization and modeling of capillary transition zones in carbonate reservoirs: Presented at the International Petroleum Technology Conference, Society of Petroleum Engineers, November 21–23, Doha, Qatar, SPE Paper 10238, 16 p.

Matthews, C. S., 1983, Steamflooding: Journal of Petroleum Technology, SPE Paper 9993, v. 35, no. 3, p. 465–471.

McCaleb, J. A., 1979, The role of the geologist in exploitation: Vail Colorado: AAPG Reservoir Fundamentals School.

McCubbin, D. G., 1992, Barrier Islands, strand-plains, *in* P. A. Scholle and D. R. Spearing, eds., Sandstone depositional environments: AAPG Memoir 31, p. 247–279.

McInally, A. T., T. Redondo-Lopez, J. Garnham, J. Kunka, A. D. Brooks, L. S. Hansen, F. Barclay, and D. Daves, 2003, Optimizing 4-D fluid imaging: Petroleum Geoscience, v. 9, no. 1, p. 91–101.

Mckie, T., and P. Audretsch, 2005, Depositional and structural controls on Triassic reservoir performance in the Heron cluster, ETAP, central North Sea, *in* A. G. Doré and B. A. Vining, eds., Petroleum geology: Northwest Europe and global perspectives: Proceedings of the 6th Petroleum Geology Conference, Geological Society (London), p. 285–297.

McPherson, J. G., G. Shanmugan, and R. J. Moiola, 1988, Fan deltas and braid deltas: Conceptual problems, *in* W. Nemec and R. J. Steel, eds., Fan deltas: Sedimentology and tectonic settings: Glasgow, Blackie and Son, p. 14–22.

McRae, L. E., and M. H. Holtz, 1995, Strategies for optimizing incremental recovery from mature reservoirs in Oligocene Frio fluvial-deltaic sandstones, Rincon field, South Texas: Transactions of the Gulf Coast Association of Geological Societies, v. 45, p. 423–431.

Medeiros, M. F., 1992, Measurement while drilling, *in* D. Morton-Thompson and A. M. Woods, eds., Development geology reference manual: AAPG Methods in Exploration Series 10, p. 88–90.

Meehan, D. N., 1994, Geological steering of horizontal wells: Journal of Petroleum Technology, SPE 29242, v. 46, no. 10, p. 848–852.

Megson, J., and R. Hardman, 2001, Exploration for and development of hydrocarbons in the Chalk of the North Sea: A low permeability system: Petroleum Geoscience, v. 7, no. 1, p. 3–12.

Megson, J., and T. Tygesen, 2005, The North Sea Chalk: An underexplored and underdeveloped play, *in* A. G. Doré and B. A. Vining, eds., Petroleum geology: Northwest Europe and global perspectives: Proceedings of the 6th Petroleum Geology Conference, Geological Society (London), v. 1, p. 159–168.

Meyer, F. O., and R. C. Price, 1992, A new Arab-D depositional model, Ghawar field, Saudi Arabia: Presented at the Society of Petroleum Engineers 8th Middle East Oil Show, SPE Paper 25576, 10 p.

Meyer, F. O., R. C. Price, and S. M. Al-Raimi, 2000, Stratigraphic and petrophysical characteristics of cored Arab-D super-k intervals, Hawiyah Area, Ghawar field, Saudi Arabia: GeoArabia, v. 5, no. 3, p. 355–384.

Miall, A. D., 1988, Reservoir heterogeneities in fluvial sandstones: Lessons from outcrop studies: AAPG Bulletin, v. 72, no. 6, p. 682–697.

Miall, A. D., 1996, The geology of fluvial deposits: Berlin, Springer-Verlag, 582 p.

Miall, A. D., 2006, Reconstructing the architecture and sequence stratigraphy of the preserved fluvial record as a tool for reservoir development: A reality check: AAPG Bulletin, v. 90, no. 7, p. 989–1002.

Miskimins, J. L., 2003, Analysis of hydrocarbons production in a critically-stressed reservoir: Presented at the Society of Petroleum Engineers Annual Technical Conference and Exhibition, October 5–8, Denver, Colorado, SPE Paper 84457, 8 p.

Mitchum, R. M., 1977, Seismic stratigraphy and global changes of sea level, part 11: Glossary of terms used in seismic stratigraphy, *in* C. E. Payton, ed., Seismic Stratigraphy— Applications to Hydrocarbon Exploration: AAPG Memoir 26, p. 205–212.

Mitra, S., 1988, Effects of deformation mechanisms on reservoir potential in central Appalachian overthrust belt: AAPG Bulletin, v. 72, no. 5, p. 536–554.

Mjøs, R., O. Walderhaug, and E. Prestholm, 1993, Crevasse splay sandstone geometries in the Middle Jurassic

Ravenscar Group of Yorkshire, UK, *in* M. Marzo and C. Puigdefabregas, eds., Alluvial sedimentation: Special Publication of the International Association of Sedimentologists 17, p. 167–184.

Monicard, R. P., 1980, Properties of reservoir rocks: Core analysis: Houston, Institute Francais du Petrole Publication, Gulf Publishing Company, 184 p.

Montgomery, S. L., 1998, Thirtyone Formation, Permian Basin, Texas: Structural and lithologic heterogeneity in a Lower Devonian chert reservoir: AAPG Bulletin, v. 82, no. 1, p. 1–24.

Montgomery, S. L., 1998, Permian Clear Fork Group, North Robertson unit: Integrated reservoir management and characterization for infill drilling: 1. Geologic analysis: AAPG Bulletin, v. 82, no. 10, p. 1797–1814.

Moore, D., ed., 1993, Productive low resistivity well logs of the offshore Gulf of Mexico: Houston, Texas, New Orleans and Houston Geological Society, 260 p.

Morris, P. H., S. N. J. Payne, and D. P. J. Richards, 1999, Micropalaeontological biostratigraphy of the Magnus Sandstone Member (Kimmeridgian to early Volgian), Magnus field, UK North Sea, *in* R. W. Jones and M. D. Simmons, eds., Biostratigraphy in production and development geology: Geological Society (London) Special Publication 152, p. 55–73.

Morrow, N. R., 1987, A review of the effects of initial saturation, pore structure and wettability on oil recovery by waterflooding, *in* J. Kleppe, E. W. Berg, A. T. Buller, O. Hjelmeland, and O. Torsaeter, eds., North Sea oil and gas reservoirs I: London, Graham & Trotman, p. 179–181.

Morton, A. and A. Hurst, 1995, Correlation of sandstones using heavy minerals: An example from the Statfjord Formation of the Snorre field, northern North Sea, *in* R. E. Dunay and E. A. Hailwood, eds., Non-biostratigraphical methods of dating and correlation: Geological Society Special Publication 89, p. 3–22.

Morton, A., W. O'B. Knox, and C. Hallsworth, 2002, Correlation of reservoir sandstones using quantitative heavy mineral analysis: Petroleum Geoscience, v. 8, no. 3, p. 251–262.

Moscariello, A., 2003, The Schooner field, Blocks 44/26a, 43/30a, UK North Sea, *in* J. G. Gluyas and H. M. Hichens, eds., United Kingdom oil and gas fields, commemorative millennium volume: Geological Society (London) Memoir 20, p. 811–824.

Moses, V., and D. G. Springham, 1982, Bacteria and the enhancement of oil recovery: London, Applied Science Publishers, 178 p.

Mulvany, P. S., 1992, A model for classifying and interpreting logs of boreholes that intersect faults in stratified rocks: AAPG Bulletin, v. 76, no. 6, p. 895–903.

Munthe, K. L., H. Omre, L. Holden, E. Damsleth, K. Heffer, T. S. Olsen, and J. Watterson, 1993, Subseismic faults in reservoir description and simulation, Presented at the Society of Petroleum Engineers Annual Technical Conference and Exhibition, October 3–6, Houston, Texas, SPE Paper 26500, 8 p.

Muskat, M., and R. D. Wyckoff, 1934, An approximate theory of water-coning in oil production: Petroleum Transactions of the American Institute of Mining Metallurgical and Petroleum Engineers, v. 114, p. 144–163.

Myers, K. J. and N. J. Milton, 1996, Concepts and principles of sequence stratigraphy, *in* D. Emery and K. Myers, eds., Sequence stratigraphy: Oxford, Blackwell Science, 297 p.

Nagel, W. A., and K. A. Byerley, 1992, Core-log transformations and porosity-permeability relationships, *in* D. Morton-Thompson and A. M. Woods, eds., Development geology reference manual: AAPG Methods in Exploration Series 10, p. 214–217.

Nagtegaal, P. J. C., 1979, Relationship of facies and reservoir quality in Rotliegends desert sandstones, southern North Sea region: Journal of Petroleum Geology, v. 2, p. 145–158.

Narr, W., 1996, Estimating average fracture spacing in subsurface rock: AAPG Bulletin, v. 80, no. 10, p. 1565–1586.

Narr, W. and I. Lerche, 1984, A method for estimating subsurface fracture density in core: AAPG Bulletin, v. 68, no. 5, p. 637–648.

Narr, W., and J. Suppe, 1991, Joint spacing in sedimentary rocks: Journal of Structural Geology, v. 13, p. 1037–1048.

Narr, W., D. S. Schechter, and L. B. Thompson, 2006, Naturally fractured reservoir characterization: Society of Petroleum Engineers, 115 p.

Neate, C. J., 1996, Integrating new technology—The development of the Barque Extension structure: Presented at the European Petroleum Conference, Society of Petroleum Engineers, October 22–24, Milan, Italy, SPE Paper 36871, 6 p.

Nedkvitne, T., D. A. Karlsen, K. Bjørlykke, and S. R. Larter, 1993, Relationship between reservoir diagenetic evolution and petroleum emplacement in the Ula field, North Sea: Marine and Petroleum Geology, v. 10, no. 3, p. 255–270.

Needham, D. T., G. Yielding, and B. Freeman, 1996, Analysis of fault geometry and displacement patterns, *in* P. G. Buchanan and D. A. Nieuwland, eds., 1996, Modern developments in structural interpretation, validation and modelling: Geological Society Special Publication 99, p. 189–199.

Neilson, J. E., N. H. Oxtoby, M. D. Simmons, I. R. Simpson, and N. K. Fortunatova, 1998, The relationship between petroleum emplacement and carbonate reservoir quality: Examples from Abu Dhabi and the Amu Darya Basin: Marine and Petroleum Geology, v. 15, p. 57–72.

Nelson, R. A., 1985, Geologic analysis of naturally fractured reservoirs: Houston, Gulf Publishing, 256 p.

Neumann, A. C., and L. S. Land, 1975, Lime mud deposition and calcareous algae in the Bight of Abaco, Bahamas: A budget: Journal of Sedimentary Petrology, v. 45, no. 4, p. 763–786.

Newendorp, P. D., 1996, Decision analysis for petroleum exploration: Aurora, Colorado, Planning Press, 668 p.

Nguyen, D. H., and V. L. Hung, 2003, Hydrocarbon geology of the Cuu Long Basin—Offshore Vietnam: Conference Abstracts, AAPG International Conference and Exhibition, Barcelona, p. A43.

Nilsen, T. H., 1982, Alluvial fan deposits, *in* P. A. Scholle and D. R. Spearing, eds., Sandstone depositional environments: AAPG Memoir 31, p. 49–86.

Nind, T. E. W., 1989, Hydrocarbon reservoir and well performance: London, Chapman & Hall, 368 p.

Nishikiori, N., and Y. Hayashida, 1999, Investigation and modeling of complex water influx into the sandstone reservoir, Khafji oil field, Arabian Gulf: Presented at the Society of Petroleum Engineers Annual Technical Conference and Exhibition, October 3–6, Houston, Texas, SPE Paper 56428, 16 p.

Noah, J. T., G. S. Hofland, and K. Lemke, 1992, Seismic interpretation of meander channel point-bar deposits using realistic seismic modeling techniques: The Leading Edge, v. 11, p. 13–18.

Nordgård Bolås, H. M., and C. Hermanrud, 2002, Rock stress in sedimentary basins—Implications for trap integrity, *in* A. G. Koestler and R. Hunsdale, eds., Hydrocarbon seal quantification: Norwegian Petroleum Society Special Publication 11, p. 17–35.

North, C. P., 1996, The prediction and modelling of subsurface fluvial stratigraphy, *in* P. A. Carling and M. R. Dawson, eds., Advances in fluvial dynamics and stratigraphy: Chichester, John Wiley & Sons, p. 395–508.

North, C. P., and D. J. Prosser, 1993, Characterization of fluvial and eolian reservoirs: Problems and approaches, *in* C. P. North and D. J. Prosser, eds., Characterization of fluvial and eolian reservoirs: Geological Society Special Publication 73, p. 1–6.

North, C. P., and K. S. Taylor, 1996, Ephemeral-fluvial deposits: integrated outcrop and simulation studies reveal complexity: AAPG Bulletin, v. 80, no. 6, p. 811–830.

North, F. K., 1985, Petroleum geology: Boston, Allen and Unwin, 619 p.

Nunns, A. G., 1991, Structural restoration of seismic and geologic sections in extensional regimes: AAPG Bulletin, v. 75, no. 2, p. 278–297.

Nurmi, R., 1996, Horizontal highlights: Middle East Well Evaluation Review, no. 16, p. 8–25.

Nygaard, E., K. Lieberkind, and P. Frykman, 1983, Sedimentology and reservoir parameters of the Chalk Group in the Danish central graben: Geologie en Mijnbouw, v. 62, no. 1, p. 177–190.

Ohen, H. A., and D. G. Kersey, 1992, Permeability, *in* D. Morton-Thompson and A. M. Woods, eds., Development geology reference manual: AAPG Methods in Exploration Series 10, p. 210–213.

Olea, R. A., 1994, Fundamentals of semivariogram estimation, modeling and usage, *in* J. M. Yarus and R. L. Chambers, eds., Stochastic modeling and geostatistics: AAPG Computer Applications in Geology 3, p. 27–35.

O'Mara, P. T., M. Merryweather, and D. S. Cooper, 2003, The Tyne gas fields, Block 44/18a, UK North Sea, *in* J. G. Gluyas and H. M. Hichens, eds., United Kingdom oil and gas fields, commemorative millennium volume: Geological Society (London) Memoir 20, p. 851–860.

Osleger, D. A., R. Barnaby, and C. Kerans, 2004, A laterally accreting grainstone margin from the Albian of northern Mexico: Outcrop model for Cretaceous carbonate reservoirs, *in* G. M. Grammer, P. M. Harris, and G. P. Eberli, eds., Integration of outcrop and modern analogs in reservoir modeling: AAPG Memoir 80, p. 93–107.

Ottesen, S., C. Townsend, and K. M. Øverland, 2005, Investigating the effect of varying fault geometry and transmissibility on recovery: Using a new workflow for structural uncertainty modeling in a clastic reservoir, *in* P. Boult and J. Kaldi, eds., Evaluating fault and cap rock seals: AAPG Hedberg Series 2, p. 125–136.

Ozgen, C., T. Firincioglu, and A. Albertoni, 2003, Fracture characterization for integrated studies: A new approach and its applications: Presented at the Society of Petroleum Engineers Annual Technical Conference and Exhibition, October 5–8, 2003, Denver, SPE Paper 84413, 15 p.

Partyka, G., J. Gridley, and J. Lopez, 1999, Interpretational applications of spectral decomposition in reservoir characterization: The Leading Edge, v. 18, no. 3, p. 353–360.

Parry, C. C., P. K. J. Whitley, and R. D. H. Simpson, 1981, Integration of palynological and sedimentological methods in facies analysis of the Brent Formation, *in* L. V. Illing and G. D. Hobson, eds., Petroleum geology of the continental shelf of northwest Europe: London, Heyden, p. 205–215.

Peacock, D. C. P., and D. J. Sanderson, 1994, Geometry and displacement of relay ramps on normal fault systems: AAPG Bulletin, v. 78, no. 2, p. 147–165.

Peacock, D. C. P., Q. J. Fisher, E. J. M. Willemse, and A. Aydin, 1998, The relationship between faults and pressure solution seams in carbonate rocks and the implications for fluid flow, *in* G. Jones, Q. J. Fisher, and R. J. Knipe, eds., Faulting and fluid flow in hydrocarbon reservoirs: Geological Society (London) Special Publication 147, p. 105–115.

Pedersen, P., R. Hauge, and E. Berg, 1994, The Veslefrikk field, *in* J. O. Aasen, E. W. Berg, A. T. Buller, O. Hjelmeland, R. M. Holt, J. Kleppe, and O. Torsaeter, eds., North Sea oil and gas reservoirs III: Dordrecht, Kluwer Academic Publishers, p. 51–73.

Petford, N., and K. J. W. McCaffrey, eds., 2003, Hydrocarbons in crystalline rocks: Geological Society (London) Special Publication 214, 242 p.

Pettingill, H. S., and P. Weimer, 2001, Global deep-water exploration: Past, present and future frontiers, *in* R. H. Fillon, N. C. Rosen, P. Weimer, A. Lowrie, H. W. Pettingill, R. L. Phair, H. H. Roberts, and B. Van Hoorn, eds., Petroleum systems of deep-water basins: Global and Gulf of Mexico experience: GCS-SEPM Foundation, p. 1–22.

Pickering, G., D. C. P. Peacock, D. J. Sanderson, and J. M. Bull, 1997, Modeling tip zones to predict the throw and length characteristics of faults: AAPG Bulletin, v. 81, no. 1, p. 82–89.

Pieters, J., and G. J. A. Por, 1995, Total system modeling—A tool for effective reservoir management of multiple fields with shared facilities: Presented at the Offshore Europe Conference, Society of Petroleum Engineers, September 5–8, 1995, Aberdeen, SPE Paper 30442, 10 p.

Pittman, E. D., 1971, Microporosity in carbonate rocks: AAPG Bulletin, v. 55, no. 10, p. 1873–1878.

Pittman, E. D., and D. N. Lumsden, 1968, Relationship between chlorite coatings on quartz grains and porosity, Spiro sand, Oklahoma: Journal of Sedimentary Research, v. 38, no. 2, p. 668–670.

Posamentier, H. W., 2004, Seismic geomorphology: Imaging elements of depositional systems from shelf to deep basin using 3-D seismic data: Implications for exploration and development, *in* R. J. Davies, J. A. Cartwright, S. A. Stewart, S. A. Lappin, and J. R. Underhill, eds., 3-D seismic technology, applications to the exploration of sedimentary basins: Geological Society (London) Memoir 29, p. 11–24.

Pranter, M. J., N. F. Hurley, T. L. Davis, M. A. Raines, and S. C. Wehner, 2004, Dual-lateral horizontal wells target bypassed pay in the San Andres Formation, Vacuum field, New Mexico: AAPG Bulletin, v. 88, no. 1, p. 99–113.

Pranter, M. J., A. I. Ellison, R. D. Cole, and P. E. Patterson, 2007, Analysis and modeling of intermediate-scale reservoir heterogeneity based on a fluvial point-bar outcrop analog, Williams Fork Formation, Piceance Basin, Colorado: AAPG Bulletin, v. 91, no. 7, p. 1025–1051.

Preston, J., A. Hartley, M. Hole, S. Buck, J. Bond, M. Mange, and J. Still, 1998, Integrated whole-rock trace element geochemistry and heavy mineral chemistry studies: Aids to the correlation of continental red-bed reservoirs in the Beryl field, UK North Sea: Petroleum Geoscience, v. 4, no. 1, p. 7–16.

Primmer, T., 2005, The Schiehallion field—6 years on from first oil: Petroleum Exploration Society of Great Britain Newsletter, February 2005.

Prosser, D. J., and R. Maskall, 1993, Permeability variation within eolian sandstone: A case study using core cut sub-parallel to slipface bedding, the Auk field, central North Sea, UK: Characterization of fluvial and eolian reservoirs: Geological Society Special Publication 73, p. 377–397.

Pucknell, J. K., and W. H. Broman, 1994, An evaluation of Prudhoe Bay horizontal and high-angle wells after 5 years of production: Journal of Petroleum Technology, v. 46, no. 2, SPE Paper 26596, p. 150–156.

Qi, L., T. R. Carr, and R. H. Goldstein, 2007, Geostatistical three-dimensional modeling of oolite shoals, St. Louis Limestone, southwest Kansas: AAPG Bulletin, v. 91, no. 1, p. 69–96.

Rabia, H., 1985, Oilwell drilling engineering, principles and practice: London, Graham & Trotman, 400 p.

Ratcliffe, K. T., J. Martin, T. J. Pearce, A. D. Hughes, D. E. Lawton, D. S. Wray, and F. Bessa, 2006, A regional chemostratigraphically defined correlation framework for the late Triassic TAG-I Formation in Blocks 402 and 405a, Algeria: Petroleum Geoscience, v. 12, no. 1, p. 3–12.

Rawnsley, K., and L. Wei, 2001, Evaluation of a new method to build geological models of fractured reservoirs calibrated to production data: Petroleum Geoscience, v. 7, no. 1, p. 23–33.

Reading, H. G., 1996, Sedimentary environments: Processes, facies and stratigraphy, 3d ed.: Oxford, Blackwell Science, 704 p.

Reed, B., 1992, Offshore rigs, *in* D. Morton-Thompson and A. M. Woods, eds., Development geology reference manual: AAPG Methods in Exploration Series 10, p. 65–66.

Reistroffer, J. R., and N. Tyler, 1991, Depositional environments and reservoir compartmentalization within the Frio zone 21-B reservoir, Tijerina-Canales-Blucher field, South Texas (abs.): Transactions of the Gulf Coast Association of Geological Societies, v. 41, p. 559–560.

Reynolds, A. D., 1999, Dimensions of paralic sandstone bodies: AAPG Bulletin, v. 83, no. 2, p. 211–229.

Reynolds, A. D., et al., 1998, Implications of outcrop geology for reservoirs in the Neogene productive series: Apsheron Peninsula, Azerbaijan: AAPG Bulletin, v. 82, no. 1, p. 25–29.

Reynolds, H., and G. Watson, 2003, String design and application in through-tubing rotary drilling (TTRD): Presented at the Society of Petroleum Engineers Latin American and Carribean Petroleum Engineering Conference in Port-of-Spain, April 27–30, 2003, Trinidad, SPE Paper 81096, 14 p.

Richardson, J. G., D. G. Harris, R. H. Rossen, and G. Van Hee, 1978, The effect of small, discontinuous shales on oil recovery: Journal of Petroleum Technology, v. 30, no. 11, SPE Paper 6700, p. 1531–1537.

Richardson, J. G., J. B. Sangree, and R. M. Sneider, 1987, Meandering stream reservoirs: Journal of Petroleum Technology, v. 39, no. 12, p. 1501–1502.

Richardson, J. G., J. B. Sangree, and R. M. Sneider, 1988a, Eolian dunes: Journal of Petroleum Technology, v. 40, no. 1, p. 11–12.

Richardson, J. G., J. B. Sangree, and R. M. Sneider, 1988b, Coastal barrier reservoirs: Journal of Petroleum Technology, v. 40, no. 9, p. 1127–1128.

Richardson, J. G., J. B. Sangree, and R. M. Sneider, 1989, Sand-rich deltas: Journal of Petroleum Technology, v. 41, no. 2, p. 157–158.

Rider, M. H., 1996, The geological interpretation of well logs: Caithness, Whittles Publishing, 300 p.

Rijks, E. J. K., and J. C. E. M. Jauffred, 1991, Attribute extraction: An important application in any detailed 3D interpretation study: Leading Edge, v. 10, no. 9, p. 11–19.

Robinson, J. W., and P. J. McCabe, 1997, Sandstone-body and shale-body dimensions in a braided fluvial system: Salt Wash Sandstone Member (Morrison Formation), Garfield County, Utah: AAPG Bulletin, v. 81, no. 8, p. 1267–1291.

Rogers, S. J., J. H. Fang, C. L. Karr, and D. A. Stanley, 1992, Determination of lithology from well logs using a neural network: AAPG Bulletin, v. 76, no. 5, p. 731–739.

Rose, P. R., 1992, Uncertainties impacting reserves, revenue and costs, *in* D. Morton-Thompson and A. M. Woods, eds., Development geology reference manual: AAPG Methods in Exploration Series 10, p. 25–29.

Rotevatn, A., H. Fossen, J. Hesthammer, T. E. Aas, and J. A. Howell, 2007, Are relay ramps conduits for fluid flow? Structural analysis of a relay ramp in Arches National Park, Utah, *in* L. Lonergan, R. Jolly, K. Rawnsley, and

D. J. Sanderson, eds., Fractured reservoirs: Geological Society (London) Special Publication 270, p. 55–71.

Rowbotham, P. S., D. Marion, P. Lamy, E. Insalaco, P. A. Swaby, and Y. Boisseau, 2003, Multidisciplinary stochastic impedance inversion: Integrating understanding and capturing reservoir uncertainty: Petroleum Geoscience, v. 9, no. 4, p. 287–294.

Rowntree, D., 2003, Statistics without tears: London, Penguin Books, 205 p.

Russell, B., 2002, An AVO primer: AAPG Search and Discovery Article 40051, 7 p.

Ryseth, A., H. Fjellbirkeland, I. K. Osmundsen, Å. Skålnes, and E. Zachariassen, 1998, High-resolution stratigraphy and seismic attribute mapping of a fluvial reservoir: Middle Jurassic Ness Formation, Oseberg field: AAPG Bulletin, v. 82, no. 9, p. 1627–1651.

Sahni, A., 2003, Case studies of uncertainty analysis in the seismic to reservoir simulation workflow: Presented at the Society of Petroleum Engineers Annual Technical Conference and Exhibition, October 5–8, Denver, Colorado, SPE Paper 84188, 6 p.

Salem, A. M., S. Morad, L. F. Mato, and I. S. Al-Asam, 2000, Diagenesis and reservoir-quality evolution of fluvial sandstones during progressive burial and uplift: Evidence from the Upper Jurassic Boipeba Member, Reconcavo Basin, Northeastern Brazil: AAPG Bulletin, v. 84, no. 7, p. 1015–1040.

Saller, A. H., J. T. Noah, A. P. Ruzuar, and R. Schneider, 2004, Linked lowstand delta to basin-floor fan deposition, offshore Indonesia: An analog for deep-water reservoir systems: AAPG Bulletin, v. 88, no. 1, p. 21–46.

Salvador, A., and H. J. Leon, 1992, Quiriquire field—Venezuela, Eastern Venezuela (Maturin) Basin, *in* N. H. Foster and E. A. Beaumont, eds., Stratigraphic traps III: AAPG Treatise of Petroleum Geology, p. 313–332.

Sarem, A. M. S., 1992, Waterflooding, *in* D. Morton-Thompson, and A. M. Woods, eds., Development geology reference manual: AAPG Methods in Exploration Series 10, p. 523–526.

Satter, A., J. E. Varnon, and M. T. Hoang, 1994, Integrated reservoir management: Journal of Petroleum Technology, v. 46, no. 12, SPE Paper 22350, p. 1057–1064.

Sayers, C. M., and P. M. T. M. Schutjens, 2007, An introduction to reservoir geomechanics: The Leading Edge, v. 26, no. 5, p. 597–601.

Schlumberger, 1981, Dipmeter interpretation, volume 1—Fundamentals: New York, Schlumberger, 61 p.

Schmechel, D. J., and M. J. McGuire, 1986, Rediscovering an abandoned reservoir: The Lakota Formation at Lost Soldier field, Sweetwater County, Wyoming: Presented at the Society of Petroleum Engineers Rocky Mountain Regional Meeting, May 19–21, Billings, Montana, SPE Paper 15180, 5 p.

Schmoker, J. W., and C. J. Schenk, 1994, Regional porosity trends of the Upper Jurassic Norphlet Formation in southwestern Alabama and vicinity, with comparisons to formations of other basins: AAPG Bulletin, v. 78, no. 2, p. 166–180.

Scholle, P. A., 1977, Chalk diagenesis and its relation to petroleum exploration: Oil from chalks, a modern miracle?: AAPG Bulletin, v. 61, no. 7, p. 982–1009.

Schowalter, T. T., 1979, Mechanics of secondary hydrocarbon migration and entrapment: AAPG Bulletin, v. 63, no. 5, p. 723–760.

Schutjens, P. M. T. M., T. L. Blanton, J. W. Martin, B. C. Lehr, and M. N. Baaijens, 1998, Depletion-induced compaction of an overpressured reservoir sandstone: Experimental approach: Society of Petroleum Engineers Paper 47277, 10 p.

Schutjens, P. M. T. M., T. H. Hanssen, M. H. H. Hettema, J. Merour, P. de Bree, J. W. A. Coremans, and G. Helliesen, 2004, Compaction-induced porosity/permeability reduction in sandstone reservoirs: Data and model for elasticity-dominated deformation: Presented at the 2001 Society of Petroleum Engineers Annual Technical Conference and Exhibition, September 30–October 3, New Orleans, SPE Reservoir Evaluation and Engineering, SPE Paper 88441, v. 7, no. 3, p. 202–216.

Scorer, J., N. Fuller, J. Malcolm, and J. Bruner, 2005, The MacCulloch field: Improving reservoir characterization through time-lapse seismic, *in* A. G. Doré and B. A. Vining, eds., Petroleum geology: Northwest Europe and global perspectives: Proceedings of the 6th Petroleum Geology Conference, Geological Society (London), v. 2, p. 1517–1531.

Segall, P., 1989, Earthquakes triggered by fluid extraction: Geology, v. 17, p. 942–946.

Serra, O., 1984, Fundamentals of well-log interpretation: 1. The acquisition of logging data: Amsterdam, Elsevier, 423 p.

Sexton, W. J., and M. O. Hayes, 1996, Holocene deposits of reservoir-quality sand along the central South Carolina coastline: AAPG Bulletin, v. 80, no. 6, p. 831–854.

Shanley, K. W., 2004, Fluvial reservoir description for a giant, low-permeability gas field: Jonah field, Green River Basin, Wyoming, U.S.A., *in* J. W. Robinson and K. W. Shanley, eds., Jonah field: Case study of a giant tight-gas fluvial reservoir: AAPG Studies in Geology 52, p. 159–182.

Shaw, A., A. Reynolds, and E. Warren, 1996, Integrated description of a complex low net/gross sandstone reservoir: Upper subzones of the Endicott field, north slope of Alaska: Presented at the European 3-D Reservoir Modeling Conference, Society of Petroleum Engineers, April 16–17, Stavanger, Norway, SPE Paper 35496, 13 p.

Shebi, M. A., 1995, The impact of reservoir heterogeneity on fluid flow in the Tensleep Sandstone of the Bighorn Basin: Resources of southwestern Wyoming, field conference guidebook: Wyoming Geological Association, p. 343–359.

Shelton, J. W., 1971, Correlation sections and log maps in determination of sandstone trends: AAPG Bulletin, v. 56, no. 8, p. 1541–1544.

Shepherd, M., 2007, A method for validating the volume of drainage cells in oil fields: Abstracts of the 2007 DEVEX Conference, Aberdeen May 15–16, 2007, 45 p.

Shepherd, M., C. J. Kearney, and J. H. Milne, 1990, Magnus field, *in* E. A. Beaumont and N. H. Foster, eds., Structural traps: II. Traps associated with tectonic faulting: Treatise of petroleum geology of oil and gas fields: Tulsa, AAPG, p. 95–125.

Shepherd, M., K. Wesnaes, and N. Neill, 2003, The Angus field redevelopment— Abandoned but not forgotten: Petroleum Exploration Society of Great Britain Newsletter, December 2003.

Shew, R. D., G. M. Tiller, C. J. Hackbarth, D. R. Rollins, and C. D. White, 1995, Characterization and modeling of channel and thin-bedded turbidite prospects in the Gulf of Mexico: Integration of outcrops, modern analogs, and subsurface data: Presented at the Society of Petroleum Engineers Annual Technical Conference and Exhibition, October 22–25, 1995, Dallas, Texas, SPE Paper 30535, 7 p.

Shinn, E. A., 1983, Tidal flat environment, *in* P. A. Scholle, D. G. Bebout, and C. H. Moore, eds., Carbonate depositional environments: AAPG Memoir 33, p. 171–210.

Shipton, Z. K., and P. A. Cowie, 2001, Damage zone development over micron to kilometer scales in high-porosity Navajo sandstone, Utah: Journal of Structural Geology, v. 23, p. 1825–1844.

Shipton, Z. K., J. P. Evans, K. R. Robeson, C. B. Forster, and S. Snelgrove, 2002, Structural heterogeneity and permeability in faulted eolian sandstone: Implications for subsurface modeling of faults: AAPG Bulletin, v. 86, no. 5, p. 863–883.

Shirzadi, S. G., and A. S. Lawai, 1993, Multidisciplinary approach for targeting new wells in the Prudhoe Bay field: Journal of Petroleum Technology, v. 45, no. 10, SPE Paper 26093, p. 988–993.

Sibley, D. M., and S. S. Mastoris, 1994, Rethinking conventional field development: 3-D shows direct detection of bypassed oil reserves: Presented at the International Petroleum Conference and Exhibition of Mexico, Society of Petroleum Engineers, October 10–13, 1994, Veracruz, Mexico, SPE Paper 28719, 11 p.

Sills, S. R., 1992, Drive mechanisms and recovery, *in* D. Morton-Thompson and A. M. Woods, eds., Development geology reference manual: AAPG Methods in Exploration Series 10, p. 518–522.

Sinclair, I. K., J. E. Evans, E. A. Albrechtsons, and L. J. Sydora, 1999, The Hibernia oilfield— Effects of episodic tectonism on structural character and reservoir compartmentalisation, *in* A. J. Fleet and S. A. R. Boldy, eds., Petroleum geology of northwest Europe: Proceedings of the 5th Conference, Geological Society (London), p. 517–528.

Slatt, R. M., and W. E. Galloway, 1992, Geological heterogeneities, *in* D. Morton-Thompson and A. M. Woods, eds., Development geology reference manual: AAPG Methods in Exploration Series 10, p. 278–281.

Slentz, L. W., 1981, Geochemistry of reservoir fluids as a unique approach to optimum reservoir management: Presented at the Middle East Technical Conference and Exhibition, Society of Petroleum Engineers, March 9–12, 1981, Bahrain, SPE Paper 9582, 14 p.

Smalley, P. C., and N. A. Hale, 1996, Early identification of reservoir compartmentalization by combining a range of conventional and novel data types, Society of Petroleum Engineers Formation Evaluation, v. 11, no. 1, SPE Paper 30533, p. 163–170.

Smallwood, D. D., 1992, Artificial lift, *in* D. Morton-Thompson and A. M. Woods, eds., Development geology reference manual: AAPG Methods in Exploration Series 10, p. 458–497.

Smith, S., M. Bentley, D. Southwood, A. Spence, and T. Wynn, 2005, Why reservoir models so often disappoint: Some lessons learned: Abstracts of the Future of Geological Modelling in Hydrocarbon Development Conference 2005, Geological Society (London), 40 p.

Smolen, J. J., 1992a, Wire-line formation testers, *in* D. Morton-Thompson and A. M. Woods, eds., Development geology reference manual: AAPG Methods in Exploration Series 10, p. 154–157.

Smolen, J. J., 1992b, Production logging, *in* D. Morton-Thompson and A. M. Woods, eds., Development geology reference manual: AAPG Methods in Exploration Series 10, p. 488–491.

Sneider, R. M., 2003, Worldwide examples of low resistivity pay: Houston Geological Society Bulletin, v. 45, no. 6, p. 47–59.

Sneider, R. M., and J. S. Sneider, 2001, New oil in old places: The value of mature-field development, *in* M. W. Downey, J. C. Threet, and W. A. Morgan, eds., Petroleum provinces of the twenty-first century: AAPG Memoir 74, p. 63–84.

Sneider, R. M., F. H. Richardson, D. D. Paynter, R. E. Eddy, and I. A. Wyant, 1977, Predicting reservoir rock geometry and continuity in Pennsylvanian reservoirs, Elk City, Oklahoma: Journal of Petroleum Technology, v. 29, no. 7, SPE Paper 6138, p. 851–866.

Sneider, R. M., C. N. Tinker, and L. D. Meckel, 1978, Deltaic environment reservoir types and their characteristics: Journal of Petroleum Technology, v. 30, no. 11, p. 1538–1546.

Sorensen, S., M. Jones, R. F. P. Hradman, W. K. Leutz, and P. H. Schwarz, 1986, Reservoir characteristics of high and low-productivity chalks from the central North Sea, *in* A. M. Spencer, ed., Habitat of hydrocarbons on the Norwegian Continental Shelf: London, Graham & Trotman, p. 91–110.

Spaak, P., J. Almond, S. Salahudin, Z. Mohd Salleh, and O. Tosun, 1999, Fulmar: A mature field revisited, *in* A. J. Fleet and S. A. R. Boldy, eds., Petroleum geology of northwest Europe: Proceedings of the 5th Conference, Geological Society (London), p. 1089–1100.

Srivastava, R. M., 1994, An overview of stochastic methods for reservoir characterization, *in* J. M. Yarus and R. L. Chambers, eds., Stochastic modeling and geostatistics, AAPG Computer Applications in Geology 3, p. 3–16.

Stancliffe, R. P. W., and M. W. A. van der Kooij, 2001, The use of satellite-based radar interferometry to monitor production activity at the Cold Lake heavy oil field, Alberta, Canada: AAPG Bulletin, v. 85, no. 5, p. 781–794.

Staples, R., P. Hague, G. Cooke, P. Ashton, J. Stammeijer, S. Jolley, T. Stevens, and J. Marshall, 2002, Integrating 4-D seismic to optimize production: Presented at the European Petroleum Conference, Society of Petroleum Engineers, October 29–31, Aberdeen, United Kingdom, SPE Paper 78346-MS, 9 p.

Stearns, D. W., and M. Friedman, 1972, Reservoirs in fractured rock: Geologic exploration methods, *in* R. E. King, ed., Stratigraphic oil and gas fields, AAPG Memoir 16, p. 82–106.

Steffens, G. S., R. C. Shipp, B. E. Prather, A. Nott, J. L. Gibson, and C. D. Winker, 2004, The use of near-sea floor 3D seismic data in deep-water exploration and production, *in* R. J. Davies, J. A. Cartwright, S. A. Stewart, S. A. Lappin, and J. R. Underhill, eds., 3D seismic technology, applications to the exploration of sedimentary basins: Geological Society (London) Memoir 29, p. 35–43.

Stenger, B., T. Pham, N. Al-Afaleg, and P. Lawrence, 2003, Tilted original oil/water contact in the Arab-D reservoir, Ghawar field, Saudi Arabia: GeoArabia, v. 8, no. 1, p. 9–42.

Sternlof, K. R., J. R. Chapin, D. D. Pollard, and L. J. Durlofsky, 2004, Permeability effects of deformation band arrays in sandstone: AAPG Bulletin, v. 88, no. 9, p. 1315–1329.

Stewart, S. A., and S. J. Holt, 2004, Improved drilling performance through integration of seismic, geological and drilling data, *in* R. J. Davies, J. A. Cartwright, S. A. Stewart, S. A. Lappin, and J. R. Underhill, eds., 3D seismic technology, applications to the exploration of sedimentary basins: Geological Society (London) Memoir 29, p. 303–310.

Stewart, S. A., and R. Podolski, 1998, Curvature analysis of gridded geological surfaces, *in* M. P. Coward, T. S. Daltaban, and H. Johnson, eds., Structural geology in reservoir characterization: Geological Society (London) Special Publication 209, p. 133–147.

Stockbridge, C. P., and D. I. Gray, 1991, The Fulmar field, Blocks 30/16 and 30/11b, UK North Sea, *in* I. L. Abbots, ed., United Kingdom oil and gas fields, 25 years commemorative volume: Geological Society (London) Memoir 14, p. 309–316.

Strebelle, S., and K. Payrazyan, 2002, Modeling of a deepwater turbidite reservoir conditional to seismic data using multiple-point geostatistics: Presented at the Society of Petroleum Engineers Annual Technical Conference, September 29–October 2, 2002, San Antonio, Texas, SPE Paper 77425, 10 p.

Streit, J. E., 2002, Estimating fluid pressures that can induce reservoir failure during hydrocarbon depletion: Presented at the Society of Petroleum Engineers/ISRM (International Society for Rock Mechanics) Rock Mechanics Conference, October 20–23, 2001, Irving, Texas, SPE Paper 78226, 7 p.

Stuart, I. A., 2002, The Armada development, UK central North Sea: The Fleming, Drake and Hawkins gas-condensate fields, *in* J. G. Gluyas and H. M. Hichens, eds., United Kingdom oil and gas fields, commemorative millennium volume: Geological Society (London) Memoir 20, p. 139–151.

Sturrock, S. J., 1996, Biostratigraphy, *in* D. Emery and K. Myers, eds., Sequence stratigraphy: Oxford, Blackwell Science, p. 89–107.

Sun, S. Q., and R. Sloan, 2003, Quantification of uncertainty in recovery efficiency predictions: Lessons learned from 250 mature carbonate fields: Presented at the Society of Petroleum Engineers Annual Technical Conference and Exhibition, October 5–8, 2003, Denver, SPE Paper 84459, 15 p.

Sun, T., P. Meakin, T. Jossang, and K. Schwarz, 1996, A simulation model of meandering rivers: Water Resources Research, v. 32, pt. 9, p. 2937–2954.

Surlyk, F., T. Dons, C. K. Clausen, and J. Higham, 2003, Upper Cretaceous, *in* D. Evans, C. Graham, A. Armour, and P. Bathurst, eds., Millennium atlas: Petroleum geology of the central and northern North Sea: Geological Society (London), p. 213–233.

Swanson, D. C., 1979, Deltaic deposits in the Pennsylvanian upper Morrow Formation of the Anadarko Basin: Tulsa Geological Society Special Publication 1, p. 115–168.

Tankersley, T. H., and M. W. Waite, 2002, Reservoir modeling for horizontal exploitation of a giant heavy oil field—Challenges and lessons learned: Presented at the Society of Petroleum Engineers International Thermal Operations and Heavy Oil Symposium and International Horizontal Well Technology Conference, November 4–7, 2002, Calgary, Canada, SPE Paper 78957, 6 p.

Taylor, K. G., R. L. Gawthorpe, and J. C. Van Wagoner, 1995, Stratigraphic control on laterally persistent cementation, Book Cliffs, Utah: Journal of the Geological Society, v. 152, p. 225–228.

Taylor, S. R., 1996, 3-D modeling to optimize production at the successive stages of field life: Society of Petroleum Engineers Formation Evaluation, v. 11, no. 6, SPE Paper 35501, p. 205–210.

Taylor, S. R., and J. F. Lapre, 1987, North Sea chalk diagenesis: Its effect on reservoir location and properties, *in* J. Brooks And K. Glennie, eds., Petroleum geology of northwest Europe: London, Graham & Trotman, p. 483–495.

Tearpock, D. J., and R. E. Bischke, 2003, Applied subsurface geological mapping with structural methods, 2d ed.: New Jersey, Prentice Hall, 822 p.

Tennyson, M. E., 2005, Growth history of oil reserves in major California oil fields during the twentieth century, *in* T. S. Dyman, J. W. Schmoker, and M. Verma, eds., Geologic engineering and assessment studies of reserves growth: U.S. Geological Survey Bulletin 2172-H, 26 p.

Teufel, L. W., D. W. Rhett, and H. E. Farrell, 1991, Effect of reservoir depletion and pore pressure drawdown on in situ stress and deformation in the Ekofisk field, North Sea, *in* J. C. Rogiers, ed., Rock mechanics as an multidisciplinary science: Rotterdam, Balkema, p. 63–72.

Thomas, J. M. D., and R. Bibby, 1991, The depletion of the Rannoch-Etive sand unit in Brent sands reservoirs in the North Sea, *in* W. Linville, ed., Reservoir characterization III: Tulsa, PennWell Books, 1008 p.

Thore, P., A. Shtuka, M. Lecour, T. Ait-Ettajer, and R. Cognot, 2002, Structural uncertainties: Determination, management, and applications: Geophysics, v. 67, no. 3, p. 840–852.

Tiab, D., and E. C. Donaldson, 2004, Petrophysics, 2d ed.: Amsterdam, Elsevier, 889 p.

Tingay, M., B. Müller, and J. Renecker, 2005, Understanding tectonic stress in the oil patch: The World Stress Map Project: Leading Edge, v. 24, no. 12, p. 1276–1282.

Tinker, S. W., 1996, Building the 3-D jigsaw puzzle, applications of sequence stratigraphy to 3-D reservoir characterization, Permian Basin: AAPG Bulletin, v. 80, no. 4, p. 460–484.

Tinkler, J. C., 1992, Oil and gas contracts, *in* D. Morton-Thompson and A. M. Woods, eds., Development geology reference manual: AAPG Methods in Exploration Series 10, p. 16–18.

Todnem, A. C., L. Arnesen, and R. Gaasø, 2005, 4D seismic and through tubing drilling and completion wells extend life on the Gullfaks field: Presented at the Society of Petroleum Engineers/IADC (International Association of Drilling Contractors) Drilling Conference, February 23–25, Amsterdam, Netherlands, SPE Paper 92551, 10 p.

Toublanc, A., S. Renaud, J. E. Sylte, C. K. Clausen, T. Eiben, and G. Nådland, 2005, Ekofisk field: Fracture permeability evaluation and implementation in the flow model: Petroleum Geoscience, v. 11, p. 321–330.

Treiber, L. E., D. L. Archer, and W. W. Owens, 1972, A laboratory evaluation of the wettability of fifty oil-producing reservoirs: Presented at the Society of Petroleum Engineers 46th Annual Fall Meeting, October 3–6, New Orleans, SPE Journal, SPE Paper 3526, v. 12, no. 6, p. 531–540.

Trocchio, J. T., 1990, Investigation of Fateh Mishrif fluid-conductive faults: Journal of Petroleum Technology, v. 42, no. 8, SPE Paper 17992, p. 1038–1045.

Tucker, M. E., and V. P. Wright, 1990, Carbonate sedimentology: London, Blackwell, 496 p.

Tuttle, M. L. W., R. R. Charpentier, and M. E. Brownfield, 1999, The Niger Delta petroleum system: Niger Delta Province, Nigeria, Cameroon, and Equatorial Guinea, Africa: U.S. Geological Survey Open File Report 99-501, 210 p: http://greenwood.cr.usgs.gov/energy/WorldEnergy/OF99-50H/ChapterA.html (accessed September 1, 2008).

Twiss, R. J., and E. M. Moores, 1992, Structural geology: New York, W. H. Freeman and Company, 532 p.

Tye, R. S., 2004, Geomorphology: An approach to determining subsurface reservoir dimensions: AAPG Bulletin, v. 88, no. 8, p. 1123–1147.

Tye, R. S., J. P. Bhattacharya, J. A. Lorsong, S. T. Sindelar, D. G. Knock, D. D. Puls, and R. A. Levinson, 1999, Geology and stratigraphy of fluvio-deltaic deposits in the Ivishak Formation: Applications for development of Prudhoe Bay field, Alaska: AAPG Bulletin, v. 83, no. 10, p. 1588–1623.

Tye, R. S., and J. J. Hickey, 2001, Permeability characterization of distributary mouth bar sandstones in Prudhoe Bay field, Alaska: How horizontal cores reduce risk in developing deltaic reservoirs: AAPG Bulletin, v. 85, no. 3, p. 459–475.

Tyler, N., and W. A. Ambrose, 1986, Facies architecture and production characteristics of strand-plain reservoirs in North Markham-North Bay City field, Frio Formation, Texas: AAPG Bulletin, v. 70, no. 7, p. 809–829.

Tyler, N., and R. J. Finley, 1991, Architectural controls on the recovery of hydrocarbons from sandstone reservoirs, *in* A. D. Miall and N. Tyler, eds., The three dimensional facies architecture of terrigenous clastic sediments and its implication for hydrocarbon discovery and recovery: SEPM Concepts in Sedimentology and Paleontology, v. 3, p. 1–5.

Tyler, N., J. Crispin Gholston, and W. A. Ambrose, 1987, Oil recovery in a low-permeability, wave-dominated, Cretaceous, deltaic reservoir, Big Wells (San Miguel field, South Texas): AAPG Bulletin, v. 71, no. 10, p. 1171–1195.

Underhill, J. R., and M. A. Partington, 1993, Use of genetic sequence stratigraphy in defining and determining a regional tectonic control on the "Mid-Cimmerian Unconformity"—Implications for North Sea basin development and the global sea level chart, *in* P. Weimer and H. W. Posamentier, eds., Siliciclastic sequence stratigraphy: Recent developments and applications: AAPG Memoir 58, p. 449–484.

Underwood, C. A., M. L. Cooke, J. A. Simo, and M. A. Muldoon, 2003, Stratigraphic controls on vertical fracture patterns in Silurian dolomite, northeastern Wisconsin: AAPG Bulletin, v. 87, p. 121–142.

Vail, P. R., R. M. Mitchum Jr., R. G. Todd, J. M. Widmier, S. Thompson III, J. B. Sangree, J. N. Bubb, and W. G. Hatelid, 1977, Seismic stratigraphy and global changes of sea level, *in* C. E. Clayton, ed., Seismic stratigraphy and applications to hydrocarbon exploration: AAPG Memoir 26, p. 49–212.

Van Heerden, I. L., and H. H. Roberts, 1988, Facies development of Atchafalaya delta, Louisiana: A modern bayhead delta: AAPG Bulletin, v. 72, no. 4, p. 439–453.

Van Lingen, P. P., and S. Knight, 1997, Evaluation of capillary entrapment within reservoir flow units: Presented at the Society of Petroleum Engineers Annual Technical Conference and Exhibition, October 5–8, 1997, San Antonio, Texas, SPE Paper 38934, 16 p.

Van Wagoner, J. C., H. W. Posamentier, R. M. Mitchum, P. R. Vail, J. F. Sarg, T. S. Loutit, and J. Hardenbol, 1988, An overview of sequence stratigraphy and key definitions, *in* C. K. Wilgus, B. J. Hastings, H. Posamentier, J. C. Van Wagoner, C. A. Ross, and C. G. St. C. Kendall, eds., Sea level changes: An integrated approach: SEPM Special Publication 42, p. 39–45.

Van Wagoner, J. C., R. M. Mitchum, K. M. Campion, and V. D. Rahmanian, 1990, Siliciclastic sequence stratigraphy in well logs, cores and outcrops, concepts for high-resolution correlation of time and facies: AAPG Methods in Exploration Series 7, 55 p.

Varnai, P., 1998, Three-dimensional seismic stratigraphic expression of Pliocene–Pleistocene turbidite systems, northern Green Canyon (offshore Louisiana), northern Gulf of Mexico: AAPG Bulletin, v. 82, no. 5B, p. 986–1012.

Vaughan, O., R. Jones, and S. Plahn, 2007, Reservoir management aspects of the rejuvenation of the Forties field, UKCS: Presented at Offshore Europe, Society of Petroleum Engineers, September 4–7, 2007, Aberdeen, SPE Paper 109012, 20 p.

Vavra, C. L., J. G. Kaldi, and R. M. Sneider, 1992, Geological applications of capillary pressure: A review: AAPG Bulletin, v. 76, no. 6, p. 840–850.

Verma, M. H., 2000, The significance of field growth and

the role of enhanced oil recovery: U.S. Geological Survey Fact Sheet FS-115-00, 4 p.

Verma, M. H., G. F. Ulmishek, and A. P. Gilbershtein, 2000, Oil and gas reserve growth—A model for the Volga-Ural Province, Russia: Presented at the Society of Petroleum Engineers/AAPG Western Regional Meeting, June 19–22, 2000, Long Beach, California, SPE Paper 62616, 10 p.

Viniegra-O, F., and C. Castillo-Tejero, 1970, Golden Lane fields, Veracuz, Mexico, *in* M. T. Halbouty, ed., Geology of giant petroleum fields: AAPG Memoir 14, p. 309–325.

Vining, M. R., 1997, Reserves growth in a mixed sequence of deltaic and barrier-island Frio sandstones: Umbrella Point field, Chambers County, Texas: Gulf Coast Association of Geological Societies Transactions, v. 47, p. 611–619.

Wach, G. D., C. S. Lolley, D. S. Mims, and C. A. Sellers, 2004, Well placement, cost reduction, and increased production using reservoir models based on outcrop, core, well-log, seismic data, and modern analogs: Onshore and offshore western Trinidad, *in* G. M. Grammer, P. M. Harris, and G. P. Eberli, eds., Integration of outcrop and modern analogs in reservoir modeling: AAPG Memoir 80, p. 279–307.

Walker, R. G., ed., 1984, Facies models: Toronto, Canada, Geological Association of Canada, 317 p.

Walker, R. G., and D. J. Cant, 1984, Sandy fluvial systems, *in* R. G. Walker, ed., Facies models: Toronto, Canada, Geological Association of Canada, p. 71–89.

Walsh, J. J., J. Watterson, A. E. Heath, and C. Childs, 1998, Representation and scaling of faults in fluid flow models: Petroleum Geoscience, v. 4, no. 3, p. 241–251.

Walsh, J. J., C. Childs, and T. Manzocchi, 2002, The incorporation of fault properties in flow models for hydrocarbon migration and reservoir production: Lecture notes, European Association of Geologists and Engineers 2002 Workshop, Florence.

Wang, Y., and A. R. Kovscek, 2003, Integrating production history into reservoir models using streamline-based time-of-flight ranking: Petroleum Geoscience, v. 9, no. 2, p. 163–174.

Warren, J. E., and H. S. Price, 1961, Flow in heterogeneous porous media: Society of Petroleum Engineers Journal, v. 1, no. 3, SPE Paper 1579, p. 153–169.

Watterson, J., J. J. Walsh, P. A. Gillespie, and S. Easton, 1996, Scaling systematics of fault sizes on a large-scale range fault map: Journal of Structural Geology, v. 18, no. 2/3, p. 199–214.

Watts, N. L., 1987, Theoretical aspects of cap-rock and fault seals for single and two phase columns: Marine and Petroleum Geology, v. 4, no. 4, p. 274–307.

Webb, J. E., 1974, Relation of oil migration to secondary clay cementation, Cretaceous sandstones, Wyoming: AAPG Bulletin, v. 58, no. 11, p. 2245–2249.

Weber, K. J., 1982, Influence of common sedimentary structures on fluid flow in reservoir models: Journal of Petroleum Technology, v. 34, no. 3, SPE Paper 9247, p. 655–672.

Weber, K. J., 1986, How heterogeneity affects oil recovery, *in* L. W. Lake and Carroll Jr., eds., Reservoir characterisation: Orlando, Florida, Academia Press, p. 487–554.

Weber, K. J., 1987, Computation of initial well productivities in eolian sandstone on the basis of a geological model, Leman gas field, UK, *in* R. W. Tillman and K. J. Weber, eds., Reservoir sedimentology: Society of Economic Paleontologists and Mineralogists Special Publication 40, p. 333–354.

Weber, K. J., 1993, The use of 3-D seismic in reservoir geological modelling, *in* S. S. Flint and I. D. Bryant, eds., The geological modelling of hydrocarbon reservoirs and outcrop analogs: International Association of Sedimentologists Special Publication 15, p. 181–188.

Weber, K. J., 1995, Perched hydrocarbon-water contacts—A common but poorly understood phenomenon: Abstracts of the 57th EAGE Conference, Glasgow, F035.

Weber, K. J., 1999, The prize—What's possible?: Petroleum Geoscience, v. 5, no. 2, p. 135–144.

Weber, K. J., and L. C. van Geuns, 1990, Framework for constructing clastic reservoir simulation models: Journal of Petroleum Technology, v. 42, no. 10, p. 1248–1253, 1296–1297.

Weber, K. J., L. J. Urai, W. F. Pilaar, F. Lehner, and R. G. Precious, 1978, The role of faults in hydrocarbon migration and trapping in Nigerian growth fault structures: 10th Annual Offshore Technology Conference Proceedings, v. 4, p. 2643–2653.

Weimer, P., and R. M. Slatt, 2004, Petroleum systems of deep-water settings: SEG/EAGE (Society of Exploration Geophysicists/European Association of Geoscientists and Engineers) Distinguished Instructor Series 7, 465 p.

Wennberg, O. P., M. Azizzadeh, A. A. M. Aqrawi, E. Blanc, P. Brockbank, K. B. Lyslo, N. Pickard, L. D. Salem, and T. Svånå, 2007, The Khaviz anticline: An outcrop analog to giant fractured Asmari Formation reservoirs in SW Iran, *in* L. Lonergan, R. Jolly, K. Rawnsley, and D. J. Sanderson, eds., Fractured reservoirs: Geological Society (London) Special Publication 270, p. 23–42.

Werren, E. G., R. D. Shew, E. R. Adams, and R. J. Stancliffe, 1990, Meander-belt reservoir geology, mid-dip Tuscaloosa, Little Creek field, Mississippi, *in* J. H. Barwis, J. G. McPherson, and R. J. Studlick, eds., Sandstone petroleum reservoirs: Berlin, Springer, p. 85–107.

Westphal, H., G. P. Eberli, L. B. Smith, G. M. Grammer, and J. Kislak, 2004, Reservoir characterization of the Mississippian Madison Formation, Wind River basin, Wyoming: AAPG Bulletin, v. 88, no. 4, p. 405–432.

Wetzelaer, A., A. Hildebrandt, S. D. Coutts, and S. W. Veeman, 1996, Locating the remaining oil in the Brent field prior to depressurization: Presented at the European Petroleum Conference, Society of Petroleum Engineers, October 22–24, 1996, Milan, Italy, SPE Paper 36891, 7 p.

Whitebay, L., 1992, Conventional coring, *in* D. Morton-Thompson and A. M. Woods, eds., Development geology reference manual: AAPG Methods in Exploration Series 10, p. 115–118.

Whittaker, A., 1992, Mudlogging: The mudlog, *in* D. Morton-Thompson and A. M. Woods, eds., Development geology reference manual: AAPG Methods in Exploration Series 10, p. 101–103.

Willemse, E. J. M., D. D. Pollard, and A. Aydin, 1996, Three-dimensional analyses of slip distributions on normal fault arrays with consequences for fault scaling: Journal of Structural Geology, v. 18, no. 2–3, p. 295–309.

Williams, G. D., S. J. Kane, T. S. Buddin, and A. J. Richards, 1997, Restoration and balance of complex folded and faulted rock volumes: Tectonophysics, v. 273, no. 3–4, p. 203–218.

Williams, R. R., and A. D. Milne, 1991, The Thistle field, Blocks 211/18a and 211/19, UK North Sea, *in* I. L. Abbots, ed., United Kingdom oil and gas fields, 25 years commemorative volume: Geological Society (London) Memoir 14, p. 199–207.

Willis, B. J., and C. D. White, 2000, Quantitative outcrop data for flow simulation: Journal of Sedimentary Research, v. 70, no. 4, p. 788–802.

Willis, B. J., J. P. Bhattacharya, S. L. Gabel, and C. D. White, 1999, Architecture of a tide-influenced delta in the Frontier Formation of central Wyoming: Sedimentology, v. 46, no. 4, p. 667–688.

Wilson, H. H., 1977, "Frozen in" hydrocarbon accumulations or diagenetic traps—Exploration targets: AAPG Bulletin, v. 61, no. 4, p. 483–491.

Wilson, J. L., 1980, A review of carbonate reservoirs, *in* A. D. Miall, ed., Facts and principles of world petroleum occurrence: Canadian Society of Petroleum Geologists Memoir 6, p. 95–115.

Wong, T.-F., and P. Baud, 1999, Mechanical compaction of porous sandstone: Oil and Gas Science and Technology Review, v. 54, p. 785–797.

Wood, L. J., 2004, Predicting tidal sand reservoir architecture using data from modern and ancient depositional systems, *in* G. M. Grammer, P. M. Harris, and G. P. Eberli, eds., Integration of outcrop and modern analogs in reservoir modeling: AAPG Memoir 80, p. 45–66.

Woods, A. M., 1992, Fishing, *in* D. Morton-Thompson and A. M. Woods, eds., Development geology reference manual: AAPG Methods in Exploration Series 10, p. 83–86.

Worden, R. H., and S. Morad, 2000, Quartz cementation in oil field sandstones: A review of the key controversies, *in* R. H. Worden and S. Morad, eds., Quartz cementation in sandstones: International Association of Sedimentologists Special Publication 29, p. 1–20.

Worden, R. H., N. H. Oxtoby, and P. C. Smalley, 1998, Can oil emplacement prevent quartz cementation in sandstones?: Petroleum Geoscience, v. 4, no. 2, p. 129–137.

Worthington, P. E., 2000, Recognition and evaluation of low-resistivity pay: Petroleum Geoscience, v. 6, no. 1, p. 77–92.

Worthington, P. E., 2002, Applications of saturation-height functions in integrated reservoir description, *in* M. A. Lovell and N. Parkinson, eds., Geological application of well logs: Tulsa, AAPG, p. 75–89.

Worthington, P. F., and L. Cosentino, 2005, The role of cutoffs in integrated reservoir studies: Society of Petroleum Engineers, Reservoir Evaluation and Engineering, v. 8, no. 4, SPE Paper 84387, p. 276–290.

Xiao Guang, T., and H. Zuan, 1991, Buried hill discoveries of the Damintun depression in North China: AAPG Bulletin, v. 75, no. 4, p. 780–794.

Xu, W., T. T. Tran, R. M. Srivastava, and A. G. Journel, 1992, Integrating seismic data in reservoir modeling: The collocated cokriging alternative: Presented at the Society of Petroleum Engineers Annual Technical Conference and Exhibition, October 4–7, 1992, Washington, D.C., SPE Paper 24742, 10 p.

Xue, P., 1986, A point bar facies reservoir model—Semi-communicated sand body: Presented at the International Meeting on Petroleum Engineering, Society of Petroleum Engineers, March 17–20, Beijing, China, SPE Paper 14837, 13 p.

Yarus, J. M., and R. L. Chambers, 2006, Practical geostatistics—An armchair overview for petroleum reservoir engineers: Journal of Petroleum Technology, v. 58, no. 11, p. 78–88.

Yassir, N. A., and A. Zerwer, 1997, Stress regimes in the Gulf Coast, offshore Louisiana: Data from well-bore breakout analysis: AAPG Bulletin, v. 81, no. 2, p. 293–307.

Yielding, G., B. Freeman, and D. T. Needham, 1997, Quantitative fault seal prediction: AAPG Bulletin, v. 81, no. 6, p. 897–917.

Yielding, G., J. A. Øverland, and G. Byberg, 1999, Characterization of fault zones for reservoir modeling: An example from the Gullfaks field, northern North Sea: AAPG Bulletin, v. 83, no. 6, p. 925–951.

Zamora Valcare, G., T. Zapata, A. Ansa, and G. Selva, 2006, Three-dimensional modeling and its application for development of the El Portón field, Argentina: AAPG Bulletin, v. 90, no. 3, p. 307–319.

Zeito, G. A., 1965, Interbedding of shale breaks and reservoir heterogeneities: Journal of Petroleum Technology, v. 17, no. 10, p. 1223–1228.

Zeng, H., M. M. Backus, K. T. Barrow, and N. Tyler, 1996, Facies mapping from three-dimensional seismic data: Potential and guidelines from a Tertiary sandstone-shale sequence model, Powderhorn field, Calhoun County, Texas: AAPG Bulletin, v. 80, no. 1, p. 16–45.

Zheng, S.-Y., P. W. M. Corbett, A. Ryseth, and G. Stewart, 2000, Uncertainty in well test and core permeability analysis: A case study in fluvial channel reservoirs, northern North Sea, Norway: AAPG Bulletin, v. 84, no. 12, p. 1929–1954.

Zoback, M. D., and J. C. Zinke, 2002, Production-induced normal faulting in the Valhall and Ekofisk oil fields: Pure and Applied Geophysics, v. 159, no. 1–3, p. 403–420.

Index

1

3

4

A

B

D

E

F

N

O

P

R

S

T

U

V

W

Y

Z